Lecture Notes in Computer Science 4593

Commenced Publication in 1973
Founding and Former Series Editors:
Gerhard Goos, Juris Hartmanis, and Jan van Leeuwen

T0241024

Alex Biryukov (Ed.)

Fast
Software Encryption

14th International Workshop, FSE 2007
Luxembourg, Luxembourg, March 26-28, 2007
Revised Selected Papers

 Springer

Volume Editor

Alex Biryukov
FSTC, University of Luxembourg
6, rue Richard Coudenhove-Kalergi, 1359 Luxembourg-Kirchberg, Luxembourg
E-mail: alex.biryukov@uni.lu

Library of Congress Control Number: 2007933305

CR Subject Classification (1998): E.3, F.2.1, E.4, G.2, G.4

LNCS Sublibrary: SL 4 – Security and Cryptology

ISSN	0302-9743
ISBN-10	3-540-74617-X Springer Berlin Heidelberg New York
ISBN-13	978-3-540-74617-1 Springer Berlin Heidelberg New York

Springer is a part of Springer Science+Business Media

springer.com

© International Association for Cryptologic Research 2007
Printed in Germany

Typesetting: Camera-ready by author, data conversion by Scientific Publishing Services, Chennai, India
Printed on acid-free paper   SPIN: 12115600   06/3180   5 4 3 2 1 0

Preface

Fast Software Encryption 2007 was the 14th annual workshop in the series, which was sponsored by the International Association for Cryptologic Research (IACR) for the sixth time. FSE has become a brand which attracts top research papers on symmetric cryptology. This includes papers on fast and secure primitives for symmetric cryptography, such as the design and analysis of block ciphers, stream ciphers, encryption schemes, hash functions, and message authentication codes (MACs), and on tools for analysis and evaluation. Previous editions of FSE took place in Cambridge, Leuven, Haifa, Rome, New York, Yokohama, Lund, Delhi, Paris, and Graz.

The Fast Software Encryption 2007 workshop was held March 26-28, 2007 in Luxembourg. It was organized by the General Chair Jean-Claude Asselborn (University of Luxembourg) in cooperation with the research lab LACS (Laboratory of Algorithms, Cryptography and Security) of the Computer Science and Communications research unit of the University of Luxembourg. The conference was attended by 160 registered participants from 36 different countries.

There were 104 papers submitted to FSE 2007, from which 28 were selected for presentation. The selection of papers was a challenging task, each submission had at least four reviewers, papers from Program Committee members having at least five. About 450 reviews were written by the committee and the external reviewers. The discussion phase was very fruitful, leading to more than 400 discussion comments in total, with several discussions going beyond 20 comments. I would like to thank the Program Committee and the external reviewers, who did an excellent job. It was a real pleasure to work with this team.

The conference program also featured an invited talk by Jean-Charles Faugére on the topic "Groebner Bases. Applications in Cryptology." The traditional rump session with short informal presentations of recent results was chaired by Joan Daemen.

We would also like to thank the following people: Thomas Baignères and Matthieu Finiasz as the authors of the iChair review software; Dmitry Khovratovich for his help with the conference Web site and compilation of the proceedings; Volker Müller, Michel Carpentier, Christian Hutter, and SIU for videotaping the talks and providing a wireless LAN for the participants. We would like to thank the students of the Lycée Technique "Ecole de Commerce et de Gestion" and our secretaries Elisa Ferreira, Ragga Eyjolfsdottir, and Mireille Kies for their help in the organization of the workshop. We would also like to thank IACR and in particular Helena Handschuh, Shai Halevi, and Bart Preneel for constant support. Thanks to Britta Schlüter for the public relations work. Finally we are grateful to our sponsors FNR — Luxembourg National Research Fund — and the University of Luxembourg as well as the Centre de Culture et de Rencontre Neumünster, Ministry of Culture, Research and Universities.

March 2007   Alex Biryukov

FSE 2007

March 26–28, 2007, Luxembourg City, Luxembourg

Sponsored by
the International Association for Cryptologic Research (IACR)

General Chair

Jean-Claude Asselborn, University of Luxembourg, Luxembourg

Program Chair

Alex Biryukov, University of Luxembourg, Luxembourg

Program Committee

Frederik Armknecht	NEC, Germany
Steve Babbage	Vodafone, UK
Alex Biryukov (chair)	University of Luxembourg, Luxembourg
Claude Carlet	University of Paris 8 and INRIA, France
Nicolas Courtois	University College London, UK
Joan Daemen	STMicroelectronics, Belgium
Orr Dunkelman	K.U.Leuven, Belgium
Henri Gilbert	France Telecom, France
Louis Granboulan	EADS, France
Helena Handschuh	Spansion, France
Jin Hong	Seoul National University, Korea
Seokhie Hong	CIST, Korea
Tetsu Iwata	Nagoya University, Japan
Thomas Johansson	Lund University, Sweden
Antoine Joux	DGA and University of Versailles, France
Pascal Junod	Nagravision, Switzerland
Charanjit Jutla	IBM T.J. Watson Research Center, USA
John Kelsey	NIST, USA
Lars R. Knudsen	Technical University of Denmark, Denmark
Stefan Lucks	University of Mannheim, Germany
Mitsuru Matsui	Mitsubishi Electric, Japan
Willi Meier	FHNW, Switzerland

Kaisa Nyberg   Nokia and Helsinki University of Technology,
                             Finland
Elisabeth Oswald   University of Bristol, UK
Josef Pieprzyk   Macquarie University, Australia
Bart Preneel   K.U.Leuven, Belgium
Greg Rose   Qualcomm, USA
Palash Sarkar   Indian Statistical Institute, India
Serge Vaudenay   EPFL, Switzerland

Subreviewers

Elena Andreeva   Cameron McDonald
Thomas Baignères   Florian Mendel
Gregory V. Bard   Marine Minier
Côme Berbain   Joydip Mitra
Guido Bertoni   Jean Monnerat
Olivier Billet   Alp Öztarhan
Nick Bone   Sylvain Pasini
Christophe De Cannière   Ludovic Perret
Chris Charnes   Thomas Peyrin
Lily Chen   Gilles Piret
Scott Contini   Thomas Popp
Morris Dworkin   Norbert Pramstaller
Martin Feldhofer   Emmanuel Prouff
Matthieu Finiasz   Christian Rechberger
Benedikt Gierlichs   Matt Robshaw
Sylvain Guilley   Allen Roginsky
Philip Hawkes   Martin Schläffer
Christoph Herbst   Yannick Seurin
Katrin Hoeper   Nicolas Sendrier
Deukjo Hong   Igor Shparlinski
Alexandre Karlov   Soren Steffen Thomsen
Nathan Keller   Dirk Stegemann
Alexander Kholosha   Ron Steinfeld
Dmitry Khovratovich   Jaechul Sung
Jongsung Kim   Daisuke Suzuki
Andrew Klapper   Emin Tatli
Özgül Küçük   Charlotte Vikkelsoe
Ulrich Kühn   Martin Vuagnoux
Changhoon Lee   Ralf-Philipp Weinmann
Svetla Nikova   Christopher Wolf
Stefan Mangard   Hongjun Wu
Stéphane Manuel   Jin Yuan
Krystian Matusiewicz   Erik Zenner
Alexander Maximov

Table of Contents

Fast Talks: Block Cipher Design

Block Cipher Cryptanalysis

Stream Cipher Cryptanalysis (II)

Cryptanalysis of Hash Functions (II)

Theory of Stream Ciphers

Side Channel Attacks

MACs and Small Block Ciphers

Producing Collisions for PANAMA, Instantaneously

Joan Daemen and Gilles Van Assche

STMicroelectronics, Zaventem, Belgium
gro.noekeon@noekeon.org

Abstract. We present a practical attack on the PANAMA hash function that generates a collision in 2^6 evaluations of the state updating function. Our attack improves that of Rijmen and coworkers that had a complexity 2^{82}, too high to produce a collision in practice. This improvement comes mainly from the use of techniques to transfer conditions on the state to message words instead of trying many message pairs and using the ones for which the conditions are satisfied. Our attack works for any arbitrary prefix message, followed by a pair of suffix messages with a given difference. We give an example of a collision and make the collision-generating program available. Our attack does not affect the PANAMA stream cipher, that is still unbroken to the best of our knowledge.

Keywords: symmetric cryptography, hash function, collision.

1 Introduction

A cryptographic hash function maps a message of arbitrary length to a fixed-size output called a digest. One of the requirements for a cryptographic hash function is collision-resistance: it should be infeasible to to find two different messages that give the same digest.

PANAMA can be used both as a hash function and as a stream cipher. In the scope of this paper, we will consider only on the hash part. Our attack does not have impact on the security of the PANAMA stream cipher. Internally, PANAMA has a state and a buffer, which evolve using a state updating function. For every block of message, the state updating function transforms the state and the buffer. We describe PANAMA in Sec. 2.

In this article, we describe a method to produce collisions for the PANAMA hash function that refines the method of Rijmen and coworkers [2] and reduces the workload from 2^{82} to 2^6 applications of the state updating function. We can therefore generate collisions quasi instantaneously. Furthermore, there are many degrees of freedom in the produced messages. The attack works for any initial value of the state. This means that one can find a collision with a pair of messages $(M_1|M, M_1|M^*)$ with an arbitrary prefix M_1. Here, the message parts M and M^* have a fixed difference $M' = M + M^*$. Furthermore, the attacker

A. Biryukov (Ed.): FSE 2007, LNCS 4593, pp. 1–18, 2007.

can append an arbitrary suffix M_2 to both collision messages, independently of M_1, M and M^*. We discuss the structure of the attack in Sec. 3.

Like in [2], we use a differential trail (called differential path in [2]) that leads to a zero difference in state and buffer. A differential trail specifies both the message differences and the differences in the state and in the buffer. For a pair of messages to follow the right differences in the state, a subset of the state bits must satisfy specific conditions. In the attack in [2], part of these conditions were transferred to equations on message words while the remaining ones were satisfied by trying many different message pairs and picking out those for which these conditions happened to be satisfied. In our attack, we transfer *all* conditions to equations on message bits using some simple new techniques explained in Sec. 4. The transfer of equations has negligible workload.

Although very similar, our trail is different from that of [2]. We chose a trail such that the conditions on the state are more easily transferable to equations on the message bits. We describe it in Sec. 5 and all its conditions and their transfer in Sec. 6.

2   Description of PANAMA

The internal memory of PANAMA is composed of 273 32-bit words (hereby denoted words) and is organized in two parts: [1]

- the *state*, with 17 words denoted a_0 through a_{16}, and
- the *buffer*, which is an array of 32×8 words, denoted $b_{i,j}$ with $0 \leq i \leq 31$ and $0 \leq j \leq 7$. (Note that b_i indicates a block of 8 words $b_{i,0} \ldots b_{i,7}$.)

The + sign applied on bits denotes the exclusive *or* (xor) operation and on words the bitwise xor. In subscripts of the state a, it denotes modulo-17 addition.

The message to hash is padded and divided into blocks of 8 words (i.e., 256 bits) each. It is processed as follows. First, both the state and the buffer are initialized to 0. Then, for each message block $p = (p_0, p_1, \ldots, p_7)$ (i.e., for each round), the following operations are applied:

- the state undergoes a non-linear transformation $\theta \circ \pi \circ \gamma$, with

$$\gamma : a_i \leftarrow a_i + (a_{i+1} + \overline{0})a_{i+2} + \overline{0},$$
$$\pi : a_i \leftarrow a_{7i \bmod 17} \ggg i(i+1)/2,$$
$$\theta : a_i \leftarrow a_i + a_{i+1} + a_{i+4},$$

where the invisible multiplication indicates the bitwise *and*, $\overline{0}$ denotes the word with 32 bits 1, and $\ggg$ cyclic right shift of the bits within a word;
- the least significant bit of a_0 is flipped: $a_0 \leftarrow a_0 + 1$;
- the message block is xored into the state:

$$a \leftarrow a + f_{i \rightarrow s}(p) \Leftrightarrow a_{i+1} \leftarrow a_{i+1} + p_i, \ 0 \leq i \leq 7;$$

– eight words of the buffer are xored into the state:

$$a \leftarrow a + f_{\text{b}\to\text{s}}(b_{16}) \quad \Leftrightarrow \quad a_{i+9} \leftarrow a_{i+9} + b_{16,i}, \ 0 \le i \le 7;$$

– the buffer undergoes a linear feedback shift register (LFSR) step:

$$b_i \leftarrow b_{i-1 \bmod 32} \quad (i \neq 25),$$
$$b_{25} \leftarrow b_{24} + r(b_{31}),$$

where the function r is defined as $Y = r(X) \Leftrightarrow Y_j = X_{j+2 \bmod 8}$;
– the message block is xored into the buffer: $b_{0,i} \leftarrow b_{0,i} + p_i, \ 0 \le i \le 7$.

After all the message blocks are processed, 33 extra rounds are performed, called blank rounds. These rounds use the state updating function, with the difference that a part of the state (instead of a message block) is input into the buffer: $b_{0,i} \leftarrow b_{0,i} + a_{i+1}, \ 0 \le i \le 7$.

Finally, the digest is extracted from the state after the blank rounds.

3 Structure of the Attack

The first thing to note is that the presence of the blank rounds makes it hard to produce a collision in the digest if there is a difference in either the state or the buffer after all the message blocks are input. Due to the invertibility of the state updating function such a difference will not cancel out. Moreover, the lack of external input and the propagation properties of the state updating function give the attacker almost no control over the final difference. Therefore, our goal is to produce a collision in both the state and the buffer before the blank rounds.

We produce a collision by following a trail. Two instances of PANAMA process two different messages (p and $p + dp$), which have a given difference (dp). The trail also specifies the differences in the state (da) and in the buffer (db) between the two instances of PANAMA, at each round. So, not only the two messages must have the given difference, they must also produce the right difference in the state and in the buffer.

We shall now describe the general structure of the trail used in the scope of this article. We will first talk about the sequence of message differences, then about the differences in the state.

In the sequel, the round numbers are specified between brackets in superscript: $.^{(i)}$. The convention is that $p^{(i)}$ is the message block processed during round i, and $a^{(i)}$ is the value of the state after round i.

3.1 Collision in the Buffer

The buffer evolves independently from the state and is linear. As noticed in [2], the following message difference sequence gives a collision in the buffer for any x:

$$dp^{(1)} = x, \ dp^{(8)} = r(x), \ dp^{(33)} = x, \text{ all other differences } 0. \tag{1}$$

After 32 rounds, we have $db_{24} = r(x)$ and $db_{31} = x$. After the 33rd round, we get:

$$db_{25} \leftarrow db_{24} + r(db_{31}) = r(x) + r(x) = 0,$$
$$db_0 \leftarrow db_{31} + dp^{(33)} = x + x = 0.$$

Thanks to the linearity of the buffer, any combination of shifted instances of the sequence (1) results in a collision in the buffer. In [2], two such sequences are used, one distant of two rounds from the other. In this paper, we instead use three such sequences at three consecutive rounds. More precisely, the message sequence is as follows (only non-zero differences are indicated):

$$\begin{aligned}
(dp^{(1)}, dp^{(2)}, dp^{(3)}) &= (d^{(1)}, d^{(2)}, d^{(3)}), \\
(dp^{(8)}, dp^{(9)}, dp^{(10)}) &= (r(d^{(1)}), r(d^{(2)}), r(d^{(3)})), \\
(dp^{(33)}, dp^{(34)}, dp^{(35)}) &= (d^{(1)}, d^{(2)}, d^{(3)}).
\end{aligned}$$

3.2   Collision in the State

The state is influenced both by the message blocks and by the buffer words in b_{16}. Let us summarize the sequence of differences that are xored into the state, both from the message block and from b_{16}:

I Rounds $r = i + 0$: State gets difference $dp^{(r)}$ $\phantom{= dp^{(r-17)}} = d^{(i)}$
II Rounds $r = i + 7$: State gets difference $dp^{(r)}$ $\phantom{= dp^{(r-17)}} = r(d^{(i)})$
III Rounds $r = i + 17$: State gets difference $db_{16}^{(r)} = dp^{(r-17)} = d^{(i)}$
IV Rounds $r = i + 24$: State gets difference $db_{16}^{(r)} = dp^{(r-17)} = r(d^{(i)})$
V Rounds $r = i + 32$: State gets difference $dp^{(r)}$ $\phantom{= dp^{(r-17)}} = d^{(i)}$

with $1 \le i \le 3$.

After the three rounds in each of the five sequences described above, we will make sure that we have a collision in the state. These are called *subcollisions*. After the last subcollision, we have both a collision in the state and in the buffer, and we are thus guaranteed to obtain the same digest after the blank rounds.

Before we explain how to obtain a subcollision, we need to detail the properties of the difference propagation in γ, the only non-linear operation of the state updating function.

3.3   Difference Propagation Through γ

Since γ is composed only of bitwise operations, we will only talk about γ as if it operates on 17 bits in this current subsection. The actual γ on words can be seen as 32 such operations in parallel.

Assume that the input of one instance of γ is a, while the input of the other instance is $a + da$. For a given input difference da, not all output differences are possible. The output difference $dc = (dc_0, \ldots, dc_{16})$ is determined by the following equation:

$$dc_i = \gamma_i(da) + da_{i+1}a_{i+2} + da_{i+2}a_{i+1} + 1,$$

where $\gamma_i(a) = a_i + (a_{i+1} + 1)a_{i+2} + 1$ denotes a particular output bit of γ.

Hence, we can obtain an output difference dc from a given input difference da only if a satisfies some conditions. These are as follows:

$$\text{If } da_{i+1} = 1 \text{ and } da_{i+2} = 0, \text{ then } a_{i+2} = dc_i + \gamma_i(da) + 1; \qquad (2)$$

$$\text{If } da_{i+1} = 0 \text{ and } da_{i+2} = 1, \text{ then } a_{i+1} = dc_i + \gamma_i(da) + 1; \qquad (3)$$

$$\text{If } da_{i+1} = 1 \text{ and } da_{i+2} = 1, \text{ then } a_{i+1} + a_{i+2} = dc_i + \gamma_i(da) + 1. \qquad (4)$$

We call conditions of type (2) and (3) *simple* conditions and conditions of type (4) *two-bit parity* conditions. We call a differential (da, dc) for which the set of conditions has a solution a *possible differential*.

Note that the input difference da fully determines the positions of the state bits a_i that are subject to conditions. Assume that we have n consecutive 1s in the pattern da, i.e., we have $da_i = da_{i+n+1} = 0$ and in between $da_{i+l} = 1$ ($1 \leq l \leq n$). Then there are simple conditions on a_i and on a_{i+n+1}, and $n - 1$ two-bit parity conditions on $a_{i+l} + a_{i+l+1}$ ($1 \leq l < n$). This can be applied to all such patterns in da.

From this follows that the number of conditions is equal to the Hamming weight of da plus the number of 001 patterns in da. (For the particular case of $da = 11111111111111111$, there are 16 independent two-bit parity conditions.) We denote by $w(da)$ the number of conditions due to da.

3.4   Specifying the Trail

For our attack to work, we wish to determine equations on the message bits that imply the five subcollisions. In the previous subsection we have shown that given a possible differential (da, dc) over γ, we obtain conditions on input bits of γ.

Consider now subcollision I. Before the first round, there is no difference in the state, hence $da^{(0)} = 0$. At the input of the second round, the message difference appears in the state: $da^{(1)} = f_{i \rightarrow s}(d^{(1)})$. This determines the input difference of γ in round 2. We now need to specify the output of γ in the second round, but we can equivalently specify $da^{(2)}$, as the other operations are linear. After the third round, the fact that we have a collision in the state imposes that $da^{(3)} = 0$, yielding at the output of the third round a difference equal to $f_{i \rightarrow s}(d^{(3)})$. Hence a value for $da^{(2)}$ must be chosen such that differentials $(f_{i \rightarrow s}(d^{(1)}), \pi^{-1} \circ \theta^{-1}(da^{(2)} + f_{i \rightarrow s}(d^{(2)})))$ and $(da^{(2)}, \pi^{-1} \circ \theta^{-1}(f_{i \rightarrow s}(d^{(3)})))$ over γ are possible. For a given message difference sequence $d^{(1)}, d^{(2)}, d^{(3)}$ there may be several, one or none such values of $da^{(2)}$. Note that the first differential imposes conditions on $a^{(1)}$ and the second one on $a^{(2)}$.

As θ and π are linear, it follows that a possible differential over the state-updating function imposes conditions on bits of the state $a^{(i)}$. Doing this for differentials over more rounds is more difficult and we avoid it in out attack. Therefore, for each round in which there is non-zero input difference in the state, we need to know the output difference.

For subcollisions II to V, applying the same reasoning leads to following round differentials, which we write as differentials over $\theta \circ \pi \circ \gamma$ for compactness:

$$\mathbf{I}\left(f_{i \to s}(d^{(1)}), \quad da^{(2)} + f_{i \to s}(d^{(2)})\right) \quad \left(da^{(2)}, \quad f_{i \to s}(d^{(3)})\right)$$

$$\mathbf{II}\left(f_{i \to s}(r(d^{(1)})), \quad da^{(9)} + f_{i \to s}(r(d^{(2)}))\right) \quad \left(da^{(9)}, \quad f_{i \to s}(r(d^{(3)}))\right)$$

$$\mathbf{III}\left(f_{b \to s}(d^{(1)}), \quad da^{(19)} + f_{b \to s}(d^{(2)})\right) \quad \left(da^{(19)}, \quad f_{b \to s}(d^{(3)})\right)$$

$$\mathbf{IV}\left(f_{b \to s}(r(d^{(1)})), \quad da^{(26)} + f_{b \to s}(r(d^{(2)}))\right) \quad \left(da^{(26)}, \quad f_{b \to s}(r(d^{(3)}))\right)$$

$$\mathbf{V}\left(f_{i \to s}(d^{(1)}), \quad da^{(34)} + f_{i \to s}(d^{(2)})\right) \quad \left(da^{(34)}, \quad f_{i \to s}(d^{(3)})\right)$$

Hence, the trail is fully determined by the sequence $(d^{(1)}, d^{(2)}, d^{(2)})$, and the 5 state differences $da^{(2)}, da^{(9)}, da^{(19)}, da^{(26)}$ and $da^{(34)}$. Because the structure of the subcollisions I and V are equal, we can fix $da^{(34)} = da^{(2)}$.

3.5   Symmetric Patterns

Like in [2], we use differences with words that are either 0 or $\bar{0}$. This causes the intra-word rotations in π to have no influence on the difference pattern, as all other operations in the state updating function work in a bitwise fashion.

Let us translate this in the case of the word-oriented γ. We can view Equations (2)–(4) as 32 parallel conditions on the bits of the state words. Thanks to the fact that all the difference words da_i are either 0 or $\bar{0}$, the words a_j on which these conditions apply are the same for the 32 bits; either all or none of the 32 bits of a word a_j are affected by a condition. Hence, the equations can be written word-wise. Note however that this does not restrict the value of the state or of the message words to be either 0 or $\bar{0}$, only the differences.

4   Techniques for Equation Transfer

For a given trail, we have seen in Sec. 3.3 how to express conditions on the state a to get the right output differences. In this section, we explain how to transfer these equations to the message words that the attacker can choose.

We will see that the equations are never transferred to more than two rounds before the start of the subcollision, so there is no overlap between the equations derived from different subcollisions and hence we can satisfy them sequentially.

As the discussion below is generic for all five subcollisions, let us use a common convention. For $j \in \{1, 8, 18, 25, 33\}$, we denote the various stages of the state transformation with the following symbols:

$$\xrightarrow{+p^{(j-2)}} \mathrm{N} \xrightarrow{\gamma} \mathrm{O} \xrightarrow{\theta \circ \pi} \mathrm{P} \xrightarrow{+p^{(j-1)}} \mathrm{Q}$$

$$\mathrm{Q} \xrightarrow{\gamma} \mathrm{R} \xrightarrow{\theta \circ \pi} \mathrm{S} \xrightarrow{+p^{(j)}} \mathrm{T}$$

$$\mathrm{T} \xrightarrow{\gamma} \mathrm{U} \xrightarrow{\theta \circ \pi} \mathrm{V} \xrightarrow{+p^{(j+1)}} \mathrm{W}$$

$$\mathrm{W} \xrightarrow{\gamma} \mathrm{X} \xrightarrow{\theta \circ \pi} \mathrm{Y} \xrightarrow{+p^{(j+2)}} \mathrm{Z}.$$

At a given time, the corresponding italic letter denotes the state value.

Although we will follow a reasoning going backwards from Z down to T, Q or N, the attack works in practice in the forward direction. As the conditions are being satisfied, the state is updated with the known values.

For each subcollision, we wish to have a collision in the state at time Z, hence to have $dZ = 0$. This determines $dY = f_{i \rightarrow s}(dp^{(j+2)})$ or $dY = f_{b \rightarrow s}(db_{16}^{(j+1)})$, together with the difference pattern dX via $\pi^{-1} \circ \theta^{-1}$. The trail also specifies the patterns dW and dT (and indirectly dU). The differential (dT, dU) over γ implies conditions on T and (dW, dX) on W. The attacker must satisfy them by choosing appropriate values for $p^{(j-2)}$, $p^{(j-1)}$, $p^{(j)}$ and $p^{(j+1)}$.

The conditions are either simple, i.e., of type $W_i = target$ or two-bit parity conditions $W_i + W_{i+1} = target$ where "$target$" is a known value.

In the sequel, we often speak about the left and right hand sides of an equation. As a convention, the left hand side contains one isolated variable to be solved, while the right hand side contains other variables that are determined from other equations or set to arbitrary values.

4.1   Immediate Satisfaction in W

Simple conditions on words W_1 through W_8 can be satisfied by setting the value of $p^{(j+1)}$ accordingly. We call this *immediate satisfaction*:

$$W_i = target \rightarrow p_{i-1}^{(j+1)} = V_i + target \text{ (if } 1 \le i \le 8). \tag{5}$$

The value of V_i is determined by the value of T.

A two-bit parity condition $W_i + W_{i+1} = target$ can be satisfied whenever (at least) one of the two words can be modified through p, that is, when $0 \le i \le 8$. For $i = 0$ or $i = 8$, we have:

$$W_0 + W_1 = target \rightarrow p_0^{(j+1)} = V_0 + V_1 + target, \text{ and}$$
$$W_8 + W_9 = target \rightarrow p_7^{(j+1)} = V_8 + V_9 + target.$$

If $1 \le i \le 7$, however, the value of another message word must be taken into account. This other message word must be treated as known and its value may either be fixed by other conditions or set to an arbitrary value. We have either

$$W_i + W_{i+1} = target \rightarrow p_{i-1}^{(j+1)} = V_i + V_{i+1} + p_i^{(j+1)} + target, \text{ or}$$
$$W_i + W_{i+1} = target \rightarrow p_i^{(j+1)} = V_i + V_{i+1} + p_{i-1}^{(j+1)} + target.$$

4.2   Bridge from W to T

The conditions on W that cannot be satisfied immediately can be transferred to equations at time U via $\pi^{-1} \circ \theta^{-1}$.

A condition on some W_i can be converted into an equation in three words of U. For instance, assume we have to satisfy $W_{10} = 0$. We know that

$$W_{10} = (U_2 \lll 3) + (U_9 \lll 45) + (U_{13} \lll 91).$$

In this case, U_2 will be influenced directly by the message words $p^{(j)}$, and this makes it an ideal candidate for immediate satisfaction in T. So, let us isolate this variable and write:

$$U_2 = ((U_9 \lll 45) + (U_{13} \lll 91)) \ggg 3.$$

Remember that the terms U_9 and U_{13} at the right hand side are treated as known values.

In more general terms, a simple condition on W_i is converted into an equation on $U_{7i}^\pi + U_{7(i+1)}^\pi + U_{7(i+4)}^\pi$, with $U_i^\pi = U_i \lll i(i+1)/2$. One can choose to isolate one of the three variables U_{7i}, $U_{7(i+1)}$ or $U_{7(i+4)}$. Then, the cyclic rotation on the left hand side can be replaced by its inverse on the right hand side. For instance, the isolation of U_{7i} gives the following:

$$W_i = target \rightarrow U_{7i} = (U_{7(i+1)}^\pi + U_{7(i+4)}^\pi + target) \ggg 7i(7i+1)/2, \ i \neq 0, \quad (6)$$

while the constant 1 must be taken care of in the case of the word $i = 0$, for instance we can isolate U_0 as $U_0 = U_7^\pi + U_{11}^\pi + target + 1$.

Similarly, a two-bit parity condition on $W_i + W_{i+1}$ is converted into an equation on $U_{7i}^\pi + U_{7(i+2)}^\pi + U_{7(i+4)}^\pi + U_{7(i+5)}^\pi$. Again, one can choose to isolate either of the four variables.

An equation of the type $U_i = target$ can be written as $T_i + (\overline{0} + T_{i+1})T_{i+2} = target + \overline{0}$. One can transfer the equation on T_i by treating T_{i+1} and T_{i+2} as known values:

$$U_i = target \rightarrow T_i = (\overline{0} + T_{i+1})T_{i+2} + target + \overline{0}. \quad (7)$$

Combining the substitutions (6) and (7) is called a *bridge*. Together with the immediate satisfaction, this is the technique we used most often in our collision-generating algorithm.

Of course, all the conditions on T_1 through T_8 are immediately satisfiable by setting the appropriate value in $p^{(j)}$ as in Equation (5) with W and V replaced by T and S, respectively.

4.3   Side Bridge

An interesting special case is an equation on $U_0 = target$. Since an equation on T_0 cannot be immediately satisfied via p_j, we can instead create two equations on T_1 and T_2. We can choose from three ways of creating two equations:

$$U_0 = target \rightarrow T_1 = T_0 + target \ \text{and} \ T_2 = \overline{0}$$
$$U_0 = target \rightarrow T_1 = 0 \qquad\qquad \text{and} \ T_2 = T_0 + target + \overline{0}$$
$$U_0 = target \rightarrow T_1 = T_0 + target \ \text{and} \ T_2 = T_0 + target + \overline{0}$$

In the sequel, this technique is called a *side bridge*. Note that this technique can also be used on other word positions.

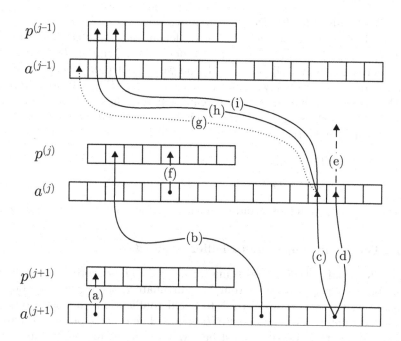

Fig. 1. Schematic illustration of some of the equation transfer techniques. (a) A condition in one of the words $a_{1...8}^{(j+1)}$ can be immediately satisfied via $p^{(j+1)}$. (b) Otherwise, this condition has to be bridged to the previous round. Preferably, it is bridged to one of the words $a_{1...8}^{(j)}$ so that it can be satisfied via $p^{(j)}$. (c)-(d) Sometimes, a bridge (c) must be accompanied by an additional equation (d) to remove circular dependencies. (e) Of course, this additional equation must be also be bridged or satisfied somewhere (not shown explicitly). (f) Conditions can also appear in round j; here is an example of such a condition that can be immediately satisfied via $p^{(j)}$. (g)-(h)-(i) Bridging a condition to the word 0 (g) does not allow immediate satisfaction. Instead, it is possible to side-bridge it to two equations, one on word 1 (h) and one on word 2 (i).

4.4   Dependency Removal

As said in the beginning of Sec. 4, when solving an equation, all terms on the right hand side must be known. This imposes constraints on the order in which the equations are solved. In some cases, it may be necessary to remove circular dependencies on the way from U to T. For instance, assume that we have three equations on U:

$$U_4 = target, \tag{8}$$
$$U_6 = target, \tag{9}$$
$$U_8 = U_4 + target \text{ (up to rotation).} \tag{10}$$

When transferring from U to T, (8) requires T_5 and T_6 to be known, hence that (9) is already solved for T_6. In turn, (9) requires T_7 and T_8 to be known, hence that (10) is solved for T_8. But (10) requires that T_4 is known so that (8) must be solved. To remove this circular set of dependencies, we can force $T_7 = \bar{0}$ so that $U_6 = T_6 + (T_7 + \bar{0})T_8 + \bar{0} = T_6 + \bar{0}$ does not depend on T_8 any more. The dependency of (9) on (10) is removed, and we can solve (9), then (8), then (10).

In more general terms, we can set $T_{i+1} = \bar{0}$ (resp. $T_{i+2} = 0$) for the equation on $U_i = target$, so that the converted equation on T_i has no dependency on T_{i+2} (resp. T_{i+1}). In the sequel, this technique is called *dependency removal*. To sum up, one can apply either:

$$U_i = target \rightarrow T_i = target + \bar{0} \text{ and } T_{i+1} = \bar{0}, \text{ or}$$

$$U_i = target \rightarrow T_i = target + \bar{0} \text{ and } T_{i+2} = 0.$$

The dependency removal and the other techniques are illustrated in Fig. 1.

4.5   The Conditions Due to Differential (dT, dU)

The differential (dT, dU) fixes some words of T. These words are therefore not available for bridges coming from W. Fortunately, there are some degrees of freedom on the way from W to T to avoid conflicts. For instance, one should choose adequately the component of U that will be isolated on the left hand side. For the attack to work, there must be a way to satisfy and bridge the equations in a non-conflicting way.

The equations on T in turn can be bridged to equations on Q. The conditions on Q_1 through Q_8 can then be satisfied via $p^{(j-1)}$.

However, one must pay careful attention to equation dependencies. When for example choosing $p^{(j)}$ to satisfy a condition in W, it has an impact on the state at time T and subsequent rounds. Indeed, changing the value of T_i influences the values of U_{i-2}, U_{i-1} and U_i. Hence care must be taken, when solving equations that are the result of bridges, that equations solved earlier are not affected. In general, dependency problems become more difficult to manage as the number of bridges grows.

4.6   Solving the Equations by Correction

When all bridges are determined along with their dependencies, solving the equations is fairly simple and does not involve much more than an evaluation of the state updating function. This is thanks to the fact that the equations are made linear once the right hand side is determined. Let us illustrate this with an example.

Assume that we need to satisfy $W_9 = 0$. Since immediate satisfaction is not possible, we decide to bridge this condition to $T_6 = target$ (via an equation on U_6). At this stage, we do not calculate the actual value of *target*, but we influence T_6 through $p_5^{(j)}$. So, we evaluate the state updating function for round j with $p_5^{(j)} = 0$ and then the state updating function for round $j + 1$. After this, we obtain W_9^*, which may not be zero as we wished. However the linearity of the equations implies that we can reuse the value W_9^* (up to a rotation) to determine $p_5^{(j)}$. Since U_6 is rotated by $6(6+1)/2 = 21$ positions, we can simply set $p_5^{(j)} = W_9^* \ggg 21$ and we automatically get $W_9 = 0$.

In short, the linearity allows us to satisfy a condition by correcting the corresponding message word. The right value of the message word is determined (up to a rotation) by the correction to bring to the state word under condition.

The reasoning for satisfying and bridging equations is done backwards in time. In contrast, solving them by correction works forwards in time, and the rounds are evaluated sequentially. The state updating function may be evaluated several times, an extra evaluation being needed for each correction.

The dependencies between equations play an important role in the order in which the corrections are applied. In the same way the right hand side of an equation must be known when solving it, a correction shall not affect equations that have been satisfied earlier.

5   The Chosen Trail

To choose a suitable trail, an important parameter to consider is the number of conditions on T and on W in the five subcollisions. Actually, this number should be split in the number of conditions that can be immediately satisfied and those that need to be bridged.

To make a trail easy to exploit using the techniques described in Sec. 4, the application of immediate satisfaction and the bridges coming from the later rounds should as much as possible fit within the 8 message words each round. One can chain bridges over several rounds, but as more rounds are bridged dependencies become increasingly difficult to manage. Generally speaking, the higher the number of conditions to bridge, the more difficult it will be to fit them all on a small number of rounds.

The number of conditions in the first subcollision is $w(da^{(1)})$ in T and $w(da^{(2)})$ in W. The number of conditions for the other four subcollisions are $(w(da^{(i)}),$ $w(da^{(i+1)}))$ with $i = 8, 18, 25$ and 33. Note that the fifth subcollision is identical to the first one. We split each of these numbers $w(da)$ as $w(da) = w_{is}(da) + w_b(da)$, where $w_{is}(da)$ is the number of conditions that can be immediately satisfied and $w_b(da)$ that must be bridged.

As a heuristic criterion, we minimize the maximum number of conditions to bridge:

$$W_b = \max_i\{w_b(da^{(i)})\}.$$

We have searched exhaustively through all $255^2 256$ patterns $(d^{(1)}, d^{(2)}, d^{(3)})$ and selected the one that has a collision trail with a minimal W_b. This resulted in the trail determined by following values:

- $d^{(1)} = 00000101$, $d^{(2)} = 11010000$, $d^{(3)} = 01111011$,
- $da^{(2)} = 10111110000000101$,
- $da^{(9)} = 00100011101011000$,

- $da^{(19)} = 00010111001100000$ and
- $da^{(26)} = 11111100111010010$,

where each digit represents a word either all-zero (0) or all-one (1); the word positions increase from left to right. The number of conditions that can be immediately satisfied or bridged is given in Table 1 below. As a comparison, the trail used in [2] has $W_b = 7$, while our trail has $W_b = 5$.

Table 1. Number of conditions in the trail used in our attack

	I		II		III		IV	
	$da^{(1)}$	$da^{(2)}$	$da^{(8)}$	$da^{(9)}$	$da^{(18)}$	$da^{(19)}$	$da^{(25)}$	$da^{(26)}$
w_{is}	2	6	3	5	0	5	0	8
w_b	1	3	0	4	3	3	3	5

6   Equation Transfer in the Chosen Trail

In this section, we describe in detail how the equations are transferred to the message words using the techniques in Sec. 4 for the trail specified in Sec. 5.

6.1   Subcollisions I and V

For round 1, we have the following differences:

$db_{16}^{(0)} = 00000000$	$da^{(0)} = 0000000000000000 = dQ$
$dp^{(1)} = 00000101$	**Round 1**
$da^{(1)} = 00000010100000000 = dT$	$dU = 00001001100000000$

From the pattern in dT, we can see that we have three conditions on T_5, T_7 and T_9, namely

$$T_5 = dU_4 + \gamma_4(dT) + \bar{0} = 0, \tag{11}$$

$$T_7 = dU_6 + \gamma_6(dT) + \bar{0} = dU_5 + \gamma_5(dT) + \bar{0} = 0, \tag{12}$$

$$T_9 = dU_8 + \gamma_8(dT) + \bar{0} = \bar{0}. \tag{13}$$

The equations (11) and (12) can be immediately satisfied in round 1 via $p_4^{(1)}$ and $p_6^{(1)}$, whereas (13) must be bridged to Q:

$$T_9 = \bar{0} \rightarrow Q_2 = ((\bar{0} + Q_3)Q_4 + \bar{0}) + (R_6^\pi + R_{12}^\pi + \bar{0}) \ggg 3, \tag{14}$$

which can be immediately satisfied in round 0 via $p_1^{(0)}$. Note that this implies that $Q_3, Q_4, \ldots, Q_8, Q_{12}, Q_{13}$ and Q_{14} are known when solving for Q_2.

Then for round 2, we have the following differences:

$db_{16}^{(1)} = 00000000$	
$dp^{(2)} = 11010000$	**Round 2**
$da^{(2)} = 1011111000000101 = dW$	$dX = 01110010000001111$

From the pattern in dW, this imposes nine conditions on $W_0 + W_{16}$, W_1, $W_2 + W_3$, $W_3 + W_4$, $W_4 + W_5$, $W_5 + W_6$, W_7, W_{13} and W_{15}. We will not detail the right hand sides of the corresponding equations, as they can easily be found as explained in Sec. 3.3. The equations on the words W_1 through W_7 can be immediately satisfied via $p^{(2)}$. The other equations are solved as follows:

- The condition on $W_0 + W_{16}$ can be transferred to an equation on $U_4^\pi + U_7^\pi + U_{10}^\pi + U_{11}^\pi$. We isolate U_4 on the left hand side and transfer it to an equation on T_4, which can be immediately satisfied via $p_3^{(1)}$. (Here, T_5, T_6, ..., T_{13} must be known.)
- The condition on W_{15} is transferred to an equation on $U_3^\pi + U_{10}^\pi + U_{14}^\pi$, from which we isolate U_3, transfer it T_3 and satisfy it via $p_2^{(1)}$. Notice that when isolating T_3, this means putting T_4 and T_5 on the right hand side. The value of T_4 must thus be determined before that of T_3, hence the condition on W_{15} may only be solved after the one on $W_0 + W_{16}$. (Here, T_4, T_5, T_{10}, T_{11}, T_{12}, T_{14}, T_{15} and T_{16} must be known.)
- Similarly, the condition on W_{14} becomes an equation on $U_0^\pi + U_6^\pi + U_{13}^\pi$, then on U_6, then on T_6, then on $p_5^{(1)}$. (Here, T_0, T_1, T_2, T_7, T_8, T_{13}, T_{14} and T_{15} must be known.)

Finally in round 3, the differences in the state cancel. We process subcollision V in the same way, using the message blocks 32 rounds later.

6.2 Subcollision II

For round 8, we have the following differences:

$db_{16}^{(7)} = 00000000$	$da^{(7)} = 0000000000000000 = dQ$
$dp^{(8)} = 00010100$	**Round 8**
$da^{(8)} = 0000101000000000 = dT$	$dU = 00100010000000000$

Hence, we have three conditions on T_3, T_5 and T_7, which can be immediately satisfied via $p_2^{(8)}$, $p_4^{(8)}$ and $p_6^{(8)}$. This is summarized in the table below.

Condition on	via	then on	satisfied via
T_3, T_5, T_7			$p_2^{(8)}$, $p_4^{(8)}$, $p_6^{(8)}$

Then for round 9, we have the following differences:

$db_{16}^{(8)} = 00000000$	
$dp^{(9)} = 01000011$	**Round 9**
$da^{(9)} = 00100011101011000 = dW$	$dX = 10101001000011000$

Here we have nine conditions on W_1, W_3, W_5, $W_6 + W_7$, $W_7 + W_8$, W_9, W_{11}, $W_{12} + W_{13}$ and W_{14}. As usual, the conditions on W_1 through W_8 can be immediately satisfied via $p^{(9)}$.

Condition on	via	then on	satisfied via
$W_1, W_3, W_5, W_6 + W_7, W_7 + W_8$			$p_0^{(9)}, p_2^{(9)}, p_{4\ldots6}^{(9)}$
$W_{12} + W_{13}$	$U_0^\pi + U_{10}^\pi + U_{13}^\pi + U_{16}^\pi$	T_1 and T_2, using side bridge	$p_0^{(8)}$ and $p_1^{(8)}$
W_9	$U_2^\pi + U_6^\pi + U_{12}^\pi$	T_6	$p_5^{(8)}$
W_{14}	$U_3^\pi + U_7^\pi + U_{13}^\pi$	T_{13}, with $T_8 = \overline{0}$	$p_7^{(8)}$
T_{13}	$R_0^\pi + R_6^\pi + R_{14}^\pi$	Q_6	$p_5^{(7)}$
W_{11}	$U_3^\pi + U_9^\pi + U_{16}^\pi$	T_9, with $T_{11} = 0$	
T_9	$R_2^\pi + R_6^\pi + R_{12}^\pi$	Q_2, with $Q_1 = \overline{0}$	$p_0^{(7)}$ and $p_1^{(7)}$
T_{11}	$R_3^\pi + R_9^\pi + R_{16}^\pi$	Q_3	$p_2^{(7)}$

For the condition on W_{14}, we transfer it to an equation on $U_3^\pi + U_7^\pi + U_{13}^\pi$. We cannot isolate U_3 or U_7 and transfer it to T_3 or T_7 since we already have an equation on both. Instead, we isolate U_{13} and transfer the condition to T_{13}. We also remove the dependency of U_7 on T_9 by setting $T_8 = \overline{0}$, since T_9 will be needed below.

The condition on W_{11} can be transferred to an equation on $U_3^\pi + U_9^\pi + U_{16}^\pi$. Since T_3 is already busy, we instead isolate U_9 and transfer the condition to T_9. However, U_9 also depends on T_{10}, which is influenced by Q_2; but Q_2 is needed to satisfy the condition on T_9. To remove this circular dependency, we force $T_{11} = 0$ so that $U_9 = T_9 + \overline{0}$ does not depend on T_{10} any more.

The condition on T_9 is bridged to a condition on Q_2. However, Q_2 influences R_0, whose value is needed to solve for Q_6; but Q_6 influences R_6, whose value is needed to solve for Q_2. We force $Q_1 = \overline{0}$ so that $R_0 = Q_0 + \overline{0}$ no longer depends on Q_2.

Finally in round 10, the differences in the state cancel.

6.3   Subcollision III

For round 18, we have the following differences:

$db_{16}^{(17)} = 00000101$	$da^{(17)} = 0000000000000000 = dQ$
$dp^{(18)} = 00000000$	**Round 18**
$da^{(18)} = 0000000000000101 = dT$	$dU = 00000000000010011$

We have three conditions on T_0, T_{13} and T_{16}, which cannot be immediately satisfied in round 18.

Condition on	via	then on	satisfied via
T_0	$R_0^\pi + R_7^\pi + R_{11}^\pi$	Q_7	$p_6^{(17)}$
T_{13}	$R_0^\pi + R_6^\pi + R_{13}^\pi$	Q_6	$p_5^{(17)}$
T_{14}	$R_3^\pi + R_{10}^\pi + R_{14}^\pi$	Q_3	$p_2^{(17)}$

Then for round 19, we have the following differences:

$db_{16}^{(18)} = 11010000$	
$dp^{(19)} = 00000000$	**Round 19**
$da^{(19)} = 0001011100110000 = dW$	$dX = 0111101110010000$

This implies conditions on W_2, W_4, $W_5 + W_6$, $W_6 + W_7$, W_8, W_9, $W_{10} + W_{11}$, W_{12}.

Condition on	via	then on	satisfied via
W_2, W_4, $W_5 + W_6$, $W_6 + W_7$, W_8			$p_1^{(19)}$, $p_{3...5}^{(19)}$, $p_7^{(19)}$
W_{12}	$U_6^\pi + U_{10}^\pi + U_{16}^\pi$	T_6	$p_5^{(18)}$
W_9	$U_2^\pi + U_6^\pi + U_{12}^\pi$	T_2, with $T_4 = 0$	$p_1^{(18)}$ and $p_3^{(18)}$
$W_{10} + W_{11}$	$U_2^\pi + U_3^\pi + U_{13}^\pi + U_{16}^\pi$	T_3	$p_2^{(18)}$

To solve the conditions on W_9 and W_{10}, we need to remove some dependencies. On the one hand, the condition on W_9 is transferred to an equation on $U_2^\pi + U_6^\pi + U_{12}^\pi$, of which we want to isolate U_2 and transfer to T_2; this requires to know T_3 and T_4. On the other hand, the condition on $W_{10} + W_{11}$ becomes an equation on $U_2^\pi + U_3^\pi + U_{13}^\pi + U_{16}^\pi$; to isolate U_3 and transfer the equation to T_3, we need to know U_2. This dependency can be removed by forcing $T_4 = 0$ so that U_2 does not depend on T_3.

Finally in round 20, the differences in the state cancel.

6.4   Subcollision IV

For round 25, we have the following differences:

$db_{16}^{(24)} = 00010100$	$da^{(24)} = 0000000000000000 = dQ$
$dp^{(25)} = 00000000$	**Round 25**
$da^{(25)} = 0000000000010100 = dT$	$dU = 0000000001110100$

We have three conditions on T_{11}, T_{13} and T_{15}, which cannot be immediately satisfied in round 25.

Condition on	via	then on	satisfied via
T_{13}	$R_0^\pi + R_6^\pi + R_{13}^\pi$	Q_6, with $Q_2 = 0$	$p_5^{(24)}$ and $p_1^{(24)}$
T_{15}	$R_3^\pi + R_{10}^\pi + R_{14}^\pi$	Q_3	$p_2^{(24)}$
T_{11}	$R_3^\pi + R_9^\pi + R_{16}^\pi$	side bridge to $Q_0 = 0$ and to Q_1	$p_0^{(24)}$ for Q_1
Q_0	$O_0^\pi + O_7^\pi + O_{11}^\pi$	N_7	$p_6^{(23)}$

The condition on T_{11} cannot be satisfied neither in round 25 nor in round 24. As we transfer it to an equation on $R_3^\pi + R_9^\pi + R_{16}^\pi$, we cannot isolate R_3 as it would conflict with the condition on R_{15}. We instead isolate R_{16} and side-bridge it to $Q_0 = 0$ and to an equation on Q_1. The equation on Q_1 can be satisfied in round 24 via $p_0^{(24)}$. Consequently, $Q_0 = 0$ must be bridged to round 23.

To be able to solve for T_{13}, we must prevent Q_1 from influencing R_0. Hence, we force $Q_2 = 0$.

Then for round 26, we have the following differences:

$db_{16}^{(25)} = 01000011$	
$dp^{(26)} = 00000000$	**Round 26**
$da^{(26)} = 1111110011101001 0 = dW$	$dX = 11000101010010000$

This implies conditions on $W_0 + W_1$, $W_1 + W_2$, $W_2 + W_3$, $W_3 + W_4$, $W_4 + W_5$, W_6, W_7, $W_8 + W_9$, $W_9 + W_{10}$, W_{11}, W_{13}, W_{14} and W_{16}. Among them, the conditions on W_1 through W_8 can be immediately satisfied via $p^{(26)}$.

Condition on	via	then on	satisfied via
$W_0 + W_1, W_1 + W_2, W_2 + W_3, W_3 + W_4, W_4 + W_5, W_6, W_7, W_8 + W_9$			$p^{(26)}$
$W_9 + W_{10}$	$U_6^\pi + U_9^\pi + U_{12}^\pi + U_{13}^\pi$	T_6, with $T_8 = 0$	$p_5^{(25)}$ and $p_7^{(25)}$
W_{13}	$U_0^\pi + U_6^\pi + U_{13}^\pi$	side bridge to T_1 and T_2	$p_0^{(25)}$ and $p_1^{(25)}$
W_{16}	$U_0^\pi + U_4^\pi + U_{10}^\pi$	T_4	$p_3^{(25)}$
W_{11}	$U_3^\pi + U_9^\pi + U_{16}^\pi$	T_3	$p_2^{(25)}$
W_{14}	$U_3^\pi + U_7^\pi + U_{13}^\pi$	T_7	$p_6^{(25)}$

In the condition on $W_9 + W_{10}$, we set $T_8 = 0$ to remove the dependency of U_6 on T_7. This way, the value of T_7 can be determined after that of T_6.

Note that the condition on W_{13} becomes an equation on $U_0^\pi + U_6^\pi + U_{13}^\pi$. Since U_6 is already taken, we isolate U_0 and side-bridge it to T_1 and T_2.

Finally in round 27, the differences in the state cancel.

7 Example of Collision and Workload

We wrote a program that produces collisions based on the trail described in Sec. 5 and using the equation transfer described in Sec. 6 [4]. The workload of the collision-generating function is about 65 applications of the state-updating function: 35 for the 35 message inputs and 30 additional ones for the bridges and some XORs.

An example of pair of collision messages is given in Table 2, which was obtained using our program. Each line represents a message block; for each block, the words in hexadecimal must be read from left to right. The first message is given by the hexadecimal digits in Table 2, while the second message is obtained by xoring with $\bar{0} = \text{ffffffff}$ all the underlined words.

Table 2. Example of pair of messages that produce a collision

$p^{(0)}$	002911b8	f4046c0d	18be4673	67847de2	4ae13b51	3d6c1b7e	2cd6267d	72ae641d
$p^{(1)}$	69522bd8	5f903d84	25558553	c194e805	1f7427d8	37edf3e4	bc922535	01eb3a6b
$p^{(2)}$	0e8257d3	2ea67fd6	0682df75	c21387fe	caa1b829	ccc994ba	9d03bd1c	00992518
$p^{(3)}$	01244898	305e252b	440d462c	491c5b2e	4d061f8b	4db745f9	15473f0e	54de79dc
$p^{(4)}$	39b355bc	2d1261f0	074d4fca	4dc8390e	6443663d	66bb5f6d	428b7e94	26a61a31
$p^{(5)}$	701f5092	5d037474	7a5a4baf	767d758d	450940b5	12383ea4	3b253990	1e1f71d5
$p^{(6)}$	6e5d785e	1ad4176a	63cb2040	6bfc19fc	7f965d80	7ff57876	4e455002	323b054b
$p^{(7)}$	24168c78	d6646fb1	9a2ac8f2	030a45b1	301c3921	e58d996a	56ae7f7d	0732105a
$p^{(8)}$	69bd59fc	6e3b4bdf	1adc0aac	22ee5482	4062e4cf	85f91c0a	45b21fe0	f25f2094
$p^{(9)}$	d7992b2c	1a491c5e	8dc2afaf	3bf6154e	a8ab7031	797d40fa	475d1ef4	e842e121
$p^{(10)}$	4cad0094	314f2b74	5e14301d	4df21075	494469e5	2e405ddc	13667210	1cd05258
$p^{(11)}$	366b5346	66c441da	42305df2	7eb75e5b	60327a81	2c3b3ba0	15a12e7f	54220e5c
$p^{(12)}$	3ef673cb	0822691d	59913a36	409d0de9	12e16f49	798b6174	121f0502	73da3555
$p^{(13)}$	58b077d2	26ca08ac	3699151a	09021b0b	7bb90ef7	57724ba9	139d0f26	70494f23
$p^{(14)}$	692c4a40	4a80585b	187e5da3	16c57533	689955b9	3cd52635	13e96788	40803068
$p^{(15)}$	5db27fad	33ea62e1	23c91a2a	48cc15d5	575331b2	60bf1732	5c674a5d	3cd6190a
$p^{(16)}$	0fbf7ae5	2f14185a	6ad630dc	047e26e9	422d0f77	54dc195d	368e05eb	0d6662b5
$p^{(17)}$	79836169	75ef70c5	2cae43f4	2c49396c	3c613693	e13226d5	5bc5e69d	288f3f57
$p^{(18)}$	3a615feb	58d1d14c	00795183	c49baa76	5e9d7604	79f7f59b	19166db2	617207a2
$p^{(19)}$	6b723ed5	f5fe7f4e	401d5fa4	9acbcbfe	038420bc	e3aac878	202e7da1	5b28d301
$p^{(20)}$	440246c3	18d7068f	6be842d6	5039652a	542c5a21	19530314	6bcb5da9	0fc946d4
$p^{(21)}$	0e127504	5f1e7011	28336601	78742718	249e328d	2b0c3dae	11f406bb	5dd55373
$p^{(22)}$	6ad4001c	5a9f6260	4cd460ca	5fa41c20	205913cf	127e075d	003555d6	07cf042f
$p^{(23)}$	67322c45	6d225953	1af46629	0ecc37d7	46cf7da8	01d35159	b428c608	3a2d3d0d
$p^{(24)}$	4fe67839	304d058a	9ffce0a5	09751255	37e6124e	24851e01	591d2784	252a2fd9
$p^{(25)}$	23c3189f	3362c465	d6437d3f	d4bccbbe	507872ed	f78a65dd	aaa618d1	556224c8
$p^{(26)}$	581ecd2f	305c16ce	83fde1d9	6b9f1da2	7a1f06d4	efbfe9b6	5fdcde8c	018136fc
$p^{(27)}$	0c7b1785	50052d8e	0c153981	380717b0	773b727d	06334fbf	728245ee	251f74b1
$p^{(28)}$	1d1842e4	62705d85	34927fe9	19da7c12	50646f07	4d546b8b	39ce12c6	3bb12fec
$p^{(29)}$	4c852466	513e15e2	6d697ca3	6a155ef3	4ff85bb9	5c4603f4	486a5290	30046806
$p^{(30)}$	17965e32	5e7368b9	470e0ff4	73d95c04	1f163002	182f532d	4d67752c	596871e0
$p^{(31)}$	4ad4041e	2cf76301	3f4a3974	0a4a00f8	5ed01d43	4e573590	4f68140b	587675a2
$p^{(32)}$	66fa4037	aa864c4d	49bb6092	6f117408	74ad1b53	4eae041c	5d2462b5	05881d37
$p^{(33)}$	5579473e	7cfe6737	ebed6b2a	912f3f6a	dd8bfb4b	329eae68	96076905	6f3c52cc
$p^{(34)}$	06e8849c	5f456809	102bfd9d	527ab906	a1d33100	72aa5ea1	8ab21c2b	68f50f55
$p^{(35)}$	45c52997	39607312	345919ca	263d7857	3b971002	40276cb6	138a726c	29593908

hash result:

$h(p)$	45d93522	0168bdcd	e830f65a	6e46f3e9	1bb0bbd6	3d37a576	718f4032	0c65079f

8 Conclusions

In this paper, we have explained how to refine the attack [2] in order to produce collisions in PANAMA using only about 2^6 evaluations of the state updating function. As noted in [2], PANAMA gives too many degrees of freedom per round to the attacker.

One could consider to fix PANAMA to be resistant against this type of attack. We have actually done this in [3] and the result is RADIOGATÚN. Its design was based on the insight obtained from the attack in [2] and the possibility of the attack in this paper. This lead us to reduce the number of message words injected each round from 8 to 3, giving an attacker much less freedom per round and requiring more bridges with accompanying dependency problems for a trail with similar complexity. More importantly, we have added feedback from the state to the buffer, making the buffer evolution during hashing become nonlinear. This makes the split in nicely separated subcollisions no longer possible. For more explanations on the evolution from PANAMA to RADIOGATÚN we refer to Appendix A of [3].

Interestingly, our attack on PANAMA can be seen as an application of *trail backtracking* [3]. In this context, we have defined a metric of a trail called its *backtracking depth*. As we explain in [3], the backtracking depth gives a good idea of the number of rounds that must be bridged at the worst point in the trail. The backtracking depth of the trail we used in this paper turns out to be only 2. This suggests that the conditions can be satisfied at any round in the trail by the message words injected immediately before it and those before the previous round and hence that the number of bridges is rather limited. In the design of RADIOGATÚN one of the main criteria is exactly the non-existence of collision trails with low backtracking depth.

References

1. Daemen, J., Clapp, C.S.K.: Fast hashing and stream encryption with PANAMA. In: Vaudenay, S. (ed.) FSE 1998. LNCS, vol. 1372, pp. 60–74. Springer, Heidelberg (1998)
2. Rijmen, V., Van Rompay, B., Preneel, B., Vandewalle, J.: Producing Collisions for PANAMA. In: Matsui, M. (ed.) FSE 2001. LNCS, vol. 2355, pp. 37–51. Springer, Heidelberg (2002)
3. Bertoni, G., Daemen, J., Peeters, M., Van Assche, G.: "RADIOGATÚN a Belt-and-Mill Hash Function", presented at the NIST Second cryptographic hash workshop (August 2006) available from http://radiogatun.noekeon.org/
4. Program to generate collisions for PANAMA: available from http://radiogatun.noekeon.org/panama

Cryptanalysis of FORK-256

Krystian Matusiewicz[1], Thomas Peyrin[2], Olivier Billet[2], Scott Contini[1], and Josef Pieprzyk[1]

[1] Centre for Advanced Computing, Algorithms and Cryptography,
Department of Computing, Macquarie University
{kmatus,scontini,josef}@ics.mq.edu.au
[2] France Telecom Research and Development
Network and Services Security Lab
{thomas.peyrin,olivier.billet}@orange-ftgroup.com

Abstract. In this paper we present a cryptanalysis of a new 256-bit hash function, FORK-256, proposed by Hong *et al.* at FSE 2006. This cryptanalysis is based on some unexpected differentials existing for the step transformation. We show their possible uses in different attack scenarios by giving a 1-bit (resp. 2-bit) near collision attack against the full compression function of FORK-256 running with complexity of 2^{125} (resp. 2^{120}) and with negligible memory, and by exhibiting a 22-bit near pseudo-collision. We also show that we can find collisions for the full compression function with a small amount of memory with complexity not exceeding $2^{126.6}$ hash evaluations. We further show how to reduce this complexity to $2^{109.6}$ hash computations by using 2^{73} memory words. Finally, we show that this attack can be extended with no additional cost to find collisions for the full hash function, i.e. with the predefined IV.

Keywords: hash functions, cryptanalysis, FORK-256, micro-collisions.

1 Introduction

Most of the dedicated hash functions published in the last 15 years follow more or less closely ideas used by R. Rivest in the design MD4 [13,14] and MD5 [15]. Using terminology from [16], their step transformations are all based on source-heavy Unbalanced Feistel Networks (UFN) and employ bitwise boolean functions. Apart from MD4 and MD5 other examples include RIPEMD [12], HAVAL [21], SHA-1 [10] and also SHA-256 [11]. A very nice feature of all these designs is that they are very fast in software implementations on modern 32-bit processors and only use a small set of basic instructions executed by modern processors in constant-time like additions, rotations, and boolean functions [4].

However, traditional wisdom says that monoculture is dangerous, and this proved to be also true in the world of hash functions. Ground-breaking attacks on MD4, MD5 by X. Wang *et al.* [19,17] were later refined and applied to attack SHA-0 [20] and SHA-1 [18] as well as some other hash functions. Since source-heavy UFNs with Boolean functions seem to be susceptible to attacks similar to Wang's because only one register is changed after each step and the attacker can

A. Biryukov (Ed.): FSE 2007, LNCS 4593, pp. 19–38, 2007.

manipulate it to a certain extent, one could try designing a hash function using the other flavour of UFNs, namely target-heavy UFNs where changes in one register influence many others. This is the case with the hash function Tiger [1] tailored for 64 bit platforms and designed in 1995, and a recently proposed FORK-256 [3] which is the focus of this paper.

The paper is organized as follows. In the next section, we briefly describe FORK-256. Then, in Section 3, we discuss some properties of the step transformation of the compression function. In Section 4 we investigate a special kind of rather pathological differentials in the step transformation. We analyse those differentials in details and derive an efficient necessary and sufficient condition for their existence. Effectiveness of this test allows a fast research of suitable configurations. Section 5 studies simple paths using those differentials and shows how to use them to efficiently find near-collisions for the compression function. In Section 6, we then show how to exploit local differentials studied in Section 4 to construct a high-level differential path for the full function as well as for its various simplified variants. Finally, in Section 7 we present two algorithms for finding collisions against FORK-256's compression function, and show in Section 8 how this method can be extended to find collisions for the full hash function.

Notation. Throughout the paper we use the following notations. Unless stated otherwise, all words are 32-bit words and are sometimes though of as elements of $\mathbb{Z}_{2^{32}}$ or $\mathbb{Z}_2^{32}$.

$x + y$	addition in $\mathbb{Z}$ or $\mathbb{Z}_{2^{32}}$ depending on the context,
$x - y$	subtraction in $\mathbb{Z}$ or $\mathbb{Z}_{2^{32}}$,
$x \oplus y$	bitwise xor of two words,
$x^{\lll a}$	rotation of bits of the word x by a positions to the left,
$R_{j,i}$	value of register $R \in \{A, \ldots, H\}$ in branch $j = 1, \ldots, 4$ at step i,
$h_w(x)$	Hamming weight of word x.

2   Description of FORK-256

FORK-256 is a new dedicated hash function proposed by Hong *et al.* [3,2]. It is based on the classical Merkle-Damgård iterative construction with a compression function that maps 256 bits of state CV_n and 512 bits of message M to 256 bits of a new state CV_{n+1}. For the complete description we refer to [3].

The compression function uses a set $\{\text{BRANCH}_j\}_{j=1,2,3,4}$ of four branches running in parallel, each one of them using a different scheduling of sixteen 32 bit message blocks M_i, $i = 0, \ldots, 15$ by permuting them through σ_j. The same set of chaining variables $\text{CV} = (A_0, B_0, C_0, D_0, E_0, F_0, G_0, H_0)$ is used in the four branches. After computing outputs of parallel branches $h_j = \text{BRANCH}_j(\text{CV}, M)$ the compression function updates the set of chaining variables according to the formula

$$\text{CV} := \text{CV} + [(h_1 + h_2) \oplus (h_3 + h_4)] \ ,$$

where '+' and '$\oplus$' are performed word-wise. This construction can be seen as further extension of the design principle of two parallel lines used in RIPEMD [12].

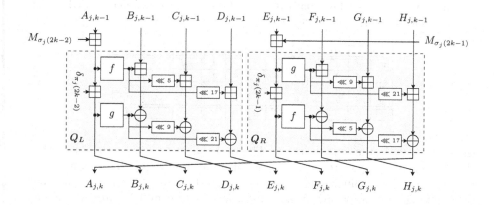

Fig. 1. Step transformation of branch j of FORK-256. Q-structures are marked with frames.

Each branch function BRANCH$_j$, $j = 1, 2, 3, 4$ consists of eight steps. In each step $k = 1, \ldots, 8$ the branch function updates its own copy of eight chaining variables using the step transformation depicted in Fig. 1. $R_{j,i}$ denotes the value of the register $R \in \{A, \ldots, H\}$ in j-th branch after step i and all $A_{j,0}, \ldots, H_{j,0}$ are initialised with corresponding values of eight chaining variables $A_0, \ldots, H_0$.

The functions f and g are defined as $f(x) = x + \left(x^{\lll 7} \oplus x^{\lll 22}\right)$ and $g(x) = x \oplus \left(x^{\lll 13} + x^{\lll 27}\right)$, respectively. Finally, the constants $\delta_0, \ldots, \delta_{15}$ are given in Table 6 and permutations σ_j and π_j are defined in Table 7 of Appendix A.

3   Preliminary Observations on FORK-256

As seen in the previous section, FORK-256 uses four parallel branches operating on the same initial state and using the same blocks of messages but in a different order. This seems to be the strength of FORK-256 since the first reported efforts to break it was limited to two of the four branches [8]. In other words, the main difficulty in cryptanalysing FORK-256 comes from the fact that the same message blocks are input in each of the four branches in a permuted fashion. Thus, while one or maybe two branches may be easily dealt with, the effect of the difference is difficult to cancel in the remaining branches. There are however some specific differential characteristics of interest.

The first one, as noted by [9] and [8], overcomes the issue by applying the same *modular* difference d to every message block. Hence, just after the fourth step has been completed, if the internal state has the same difference d on all of its eight 32-bit words, there is a collision after the eighth step. This behavior, summarized in Table 1, renders the use of four branches with message reordering as a mean to protect against differential analysis ineffective since the same difference is applied to every message block and the same differential pattern is occurring

Table 1. A four steps differential pattern to force an inner collision for FORK-256. The table shows the pattern in one branch and its probability to occur for each step.

step	ΔA	ΔB	ΔC	ΔD	ΔE	ΔF	ΔG	ΔH	ΔM_L	ΔM_R	Prob.
in	d	d	d	d	d	d	d	d	$-d$	$-d$	
1	d	0	d	d	d	0	d	d	$-d$	$-d$	P_d^6
2	d	0	0	d	d	0	0	d	$-d$	$-d$	P_d^4
3	d	0	0	0	d	0	0	0	$-d$	$-d$	P_d^2
out	0	0	0	0	0	0	0	0			1

simultaneously in the four branches. The probability P_d is the probability that the difference d propagates without modification in one step. This comes from the fact that the modular difference has to pass through a '$\oplus$'. Indeed, modular differences do not propagate without modification whenever '$\oplus$' are used in the design of the hash function (just as xor differences do not propagate without modification whenever '$+$' are used). This probability can be computed exactly for any given difference d, and this computation is given in Appendix B. Note that it does propagate without modification when it enters the step in the register A or the register E of the internal state, but with probability P_d otherwise. The overall probability for the differential pattern of Table 1 to occur is thus P_d^{12} for each branch.

Table 2. A seven steps differential pattern to get an inner near-collision for FORK-256

step	ΔA	ΔB	ΔC	ΔD	ΔE	ΔF	ΔG	ΔH	ΔM_L	ΔM_R	Prob.
in	0	d	0	0	0	0	0	0	0	0	
1	0	0	d	0	0	0	0	0	0	0	P_d
2	0	0	0	d	0	0	0	0	0	0	P_d
3	0	0	0	0	d	0	0	0	0	0	P_d
4	0	0	0	0	0	d	0	0	0	0	P'
5	0	0	0	0	0	0	d	0	0	0	P_d
6	0	0	0	0	0	0	0	d	0	0	P_d
out	d	0	0	0	0	0	0	0			P_d

Another way to deal with the four branches simultaneously is to apply a difference on the IV instead of the message M. This type of collisions is called a pseudo-collision. For the compression function h of any iterated hash function, a pseudo-collision can be expressed as $h(IV, M) = h(IV', M')$, where $(IV', M') \neq (IV, M)$. In the case of FORK-256, differences in the words of the internal state register do not diffuse identically, see the description of the states of FORK-256's step function in Fig. 1. More precisely, only the differences in the words A and E will spread to the other registers in the next step. The other differences (in the words B, C, D, F, G, H) only shift one word to the right. Hence, by applying a difference to the second word of IV, the difference propagates without spreading during three steps. Note that it propagates without being modified with probability P_d only, just as for the first differential pattern. During the fourth step however, the difference most likely spreads to the three internal registers F, G, and H in all four branches. However, we show in the next section that there is a way to prevent the spread of the difference from registers A and E.

4   Micro-collisions in Q_L and Q_R

The step transformation described in Section 2 can be logically split into three parts: addition of message words, two parallel mixing structures Q_L and Q_R and a final permutation of registers (see Fig. 1). The key role is played by the two structures Q_L and Q_R as they are the main source of diffusion in the compression function.

In the next paragraphs, we describe a way of finding differentials of the form $(\Delta A, 0, 0, 0) \rightarrow (\Delta A, 0, 0, 0)$ in Q_L and show that it works for Q_R as well. The idea is to look for pairs of inputs to the register A and appropriate input values of registers B, C and D such that the output differences in registers B, C, D are equal to zero in spite of non-zero differences at the outputs of functions f and g. Such situation is possible if we have three simultaneous *micro-collisions* i.e. differences in g cancel out differences from f in all three registers B, C, D.

4.1   Necessary and Sufficient Condition for Micro-collisions

Let us denote $y = f(x)$, $y' = f(x')$ and $z = g(x + \delta)$, $z' = g(x' + \delta)$. We have a micro-collision in the first line if the equation

$$(y + B) \oplus z = (y' + B) \oplus z' \tag{1}$$

is satisfied for given y, y', z, z' and some constant B. Our aim is to find the set of all constants B for which (1) is satisfied. Let us first introduce three different representations of differences between two numbers x and x' of $\mathbb{Z}_{2^{32}}$.

- The first representation is the xor difference. We treat it as a vector of $\mathbb{Z}_2^{32}$ representing bits of $x \oplus x'$ and denote it as $\Delta^{\oplus}(x, x') \in \{0, 1\}^{32}$.
- The second one is the integer difference between the two numbers x and x', which we denote by $\partial x := x - x'$. Note that $-2^{32} < \partial x < 2^{32}$.
- The third one is the signed binary representation which uses digits from the set $\{-1, 0, 1\}$. A pair (x, x') has signed binary representation $\Delta^{\pm}(x, x') = (x_0 - x'_0, x_1 - x'_1, \ldots, x_{31} - x'_{31})$, i.e. the i-th component is the result of the subtraction of corresponding i-th bits of x and x'.

A simple but important observation is that a difference with signed representation $(r_0, r_1, \ldots, r_{31})$ has a xor difference of $(|r_0|, |r_1|, \ldots, |r_{31}|)$, that is the xor difference has ones in those places where the signed difference has a non-zero digit, either -1 or 1. The relationship between integer and signed binary representations is more interesting. An integer difference ∂x corresponds to a signed binary representation $(r_0, \ldots, r_{31})$ if $\partial x = \sum_{i=0}^{31} 2^i \cdot r_i$ where $r_i \in \{-1, 0, 1\}$. Of course this correspondence is one-to-many because of the value-preserving transformations of signed representations $(*, 0, 1, *) \leftrightarrow (*, 1, -1, *)$ and $(*, 0, -1, *) \leftrightarrow$

$(*, -1, 1, *)$ that can stretch or shrink chunks of ones. Consider an example: assume we work with 4-bit words and let $\Delta^{\pm}(11, 2) = (1, 0, 0, 1)$, $\Delta^{\pm}(14, 5) = (1, 0, 1, -1)$, and $\Delta^{\pm}(12, 3) = (1, 1, -1, -1)$. All these binary signed representations correspond to the integer difference $\partial x = 9$. Note that we can go from one pair of values to another by adding an appropriate constant, e.g. $(12, 3) = (11 + 1, 2 + 1)$. This addition preserves the integer difference but can modify the signed binary representation.

We are now equipped with the necessary tools and go back to our initial problem. Rewriting (1) as $(y + B) \oplus (y' + B) = z \oplus z'$, we can easily see that the signed difference $\Delta^{\pm}(y + B, y' + B)$ can have non-zero digits only in those places where the xor difference $\Delta^{\oplus}(z, z')$ has ones. This narrows down the set of possible signed binary representations that can "fit" into the xor difference of a particular form to $2^{h_w(\Delta^{\oplus}(z, z'))}$. But since a single signed binary representation corresponds to a unique integer difference, there are also only $2^{h_w(\Delta^{\oplus}(z, z'))}$ integer differences ∂y that "fit" into the given xor difference $\Delta^{\oplus}(z, z')$ and what is important, integer differences are preserved when adding a constant B.

Thus, to check whether a particular difference $\partial y = y - y'$ may "fit" into xor difference we need to solve the following problem: given $\partial y = y - y'$, $-2^{32} < \partial y < 2^{32}$ and a set of positions $I = \{k_0, k_1, \ldots, k_m\} \subset \{0, \ldots, 31\}$ (that is determined by non-zero bits of $\Delta^{\oplus}(z, z')$), decide whether it is possible to find a binary signed representation $r = (r_0, \ldots, r_{31})$ corresponding to ∂y such that:

$$\partial y = \sum_{i=0}^{m} 2^{k_i} \cdot r_{k_i} \quad \text{where } r_{k_i} \in \{-1, 1\} . \tag{2}$$

Replacing t_i by $(r_{k_i} + 1)/2$, this equation can be rewritten in the equivalent form:

$$\partial y + \sum_{i=0}^{m} 2^{k_i} = 2^{k_0+1} t_0 + 2^{k_1+1} t_1 + \cdots + 2^{k_m+1} t_m , \tag{3}$$

where $t_i \in \{0, 1\}$. Deciding if there are numbers t_i that satisfy (3) is an instance of the knapsack problem and since it is superincreasing—weights are powers of two, we can do this very efficiently. This gives us a computationally efficient necessary condition for a micro-collision in a line: if $\partial y = y - y'$ cannot be represented as (2), no constant B exists and there is no solution to (1). Moreover, we can show that this condition is also sufficient: if we can find a solution of (2), then there exists a constant B that modifies the signed difference so that it "fits" the prescribed xor pattern.

Observe that since the solution of the superincreasing knapsack problem (3) is unique, so is the solution of (2). This means that we know the unique signed representation $\Delta^{\pm}(u, u + \partial y) = (r_0, \ldots, r_{31})$ that is compatible with the xor difference $\Delta^{\oplus}(z, z')$ and yields the integer difference ∂y. However, a unique signed representation corresponds to a number of concrete pairs $(u, u + \partial y)$. If at a particular position $j \in I$ we have $r_j = -1$, we know that in this position the value of j-th bit of u has to change from 1 to 0. Similarly, if we have $r_j = 1$, the j-th bit of u should change from 0 to 1. The rest of the bits of u (corresponding to positions with zeros in $\Delta^{\pm}(u, u + \partial y))$ can have arbitrary values. That way, we can easily determine the set $\mathcal{U}$ of all such values u. It is clear that $\mathcal{U}$ always contains at least one element.

Now, since $u = y + B$ for all $u \in \mathcal{U}$, the set $\mathcal{B}$ of all constants B satisfying (1) is simply $\mathcal{B} = \{u - y : u \in \mathcal{U}\}$. This reasoning also shows that if we can have a micro-collision in a line, there are $|\mathcal{B}| = 2^{32 - h_w(z \oplus z')}$ constants that yield the micro-collision if the most significant bit of $z \oplus z'$ is zero and $|\mathcal{B}| = 2^{32 - h_w(z \oplus z') + 1}$ if the MSB of $z \oplus z'$ is one. The difference is caused by the fact that if $31 \in I$, we do not need to change u_{31} in a particular way (i.e. either $1 \rightarrow 0$ or $0 \rightarrow 1$) as any change is fine because we do not introduce carries.

Finally, since we didn't use any properties of functions f and g, the same argument applies not only to micro-collisions in Q_R but also to the same structure with any functions in place of f and g.

5   A First Attempt with a Simple Differential Path

In Section 3 we have seen that the seven step differential pattern of Table 2 with a modular difference d happens simultaneously in the four branches with probability P_d^{12} as soon as we have a micro-collision in each branch. For this, we need the registers (E, F, G, H) to reach a prescribed value at the fourth step in the four branches, which can be easily computed thanks to the method given in Section 3. Note that P_d is the probability that the modular difference is unchanged when a '$\oplus$' is involved. Also, we do not care about the difference after the fourth step, because it never spreads again. So the exponent 12 comes from the fact that the difference has to go unchanged through the first three steps in the four branches. We show here how to force these registers to take on prescribed values.

5.1   Near-Collision at the Seventh Round

Our main tool here is a good scheduling in the determination of each message block so as to be able to force the four quadruplets of each branch to their required values.

To do this, we study the relationships between message blocks and IV blocks with this last quadruplet. Before getting into details of the attack, let us emphasize the following fact that simplifies the study of these relationships in the branch j. Forcing the value of the quadruplet $(E_{j,3}, F_{j,3}, G_{j,3}, H_{j,3})$ is equivalent to setting the value of the quadruplet $(E_{j,3}, F_{j,3}, F_{j,2}, F_{j,1})$. This fact can be easily checked by going backwards in the threads of the FORK-256's step transformation, which can be translated in the following sequence of equations:

$$
\begin{aligned}
F_{j,2} &= \left(G_{j,3} \oplus f(F_{j,3})\right) - g(F_{j,3} - \delta_{\pi_j(5)}), \\
G_{j,2} &= \left(H_{j,3} \oplus f(F_{j,3})^{\lll 5}\right) - g(F_{j,3} - \delta_{\pi_j(5)})^{\lll 9}, \\
F_{j,1} &= \left(G_{j,2} \oplus f(F_{j,2})\right) - g(F_{j,2} - \delta_{\pi_j(3)}).
\end{aligned}
\tag{4}
$$

Table 3. Relationship between the words of the quadruplets in each branch and the message blocks and IV. The symbols '*' and 'x' denote a degree of freedom in setting the value of a word W by adjusting the corresponding parameter P when all the remaining parameters of the row have already been fixed. The 'x' is used to emphasize that the parameter P can be used *directly* to set word W to its target value.

	IV								message block M_i															
	A	B	C	D	E	F	G	H	0	1	2	3	4	5	6	7	8	9	10	11	12	13	14	15
$E_{4,3}$	*	x		*	*		*	*	*					*		*				x	*			*
$E_{3,3}$	*	x		*	*		*	*						*	*			*		*	x	*	*	
$E_{2,3}$	*	x		*	*		*	*					x					*	*		*		*	*
$E_{1,3}$	*	x		*	*		*	*	*	*	*	*	*			x								
$F_{4,3}$	*		x		*				x	*				*						*				
$F_{3,3}$	*		x		*							x			*	*			*					
$F_{2,3}$	*		x		*														x	*			*	*
$F_{1,3}$	*		x		*				*	*	*	*		x										
$G_{4,3}\leftrightarrow F_{4,2}$	*		x											*			x					x		
$G_{3,3}\leftrightarrow F_{3,2}$	*		x												*				x				x	
$G_{2,3}\leftrightarrow F_{2,2}$	*		x																	x			*	
$G_{1,3}\leftrightarrow F_{1,2}$	*		x								*			x										
$H_{4,3}\leftrightarrow F_{4,1}$			x																			x		
$H_{3,3}\leftrightarrow F_{3,1}$			x													x								
$H_{2,3}\leftrightarrow F_{2,1}$			x																					x
$H_{1,3}\leftrightarrow F_{1,1}$			x									x												

Table 3 summarizes those relationships. In the left column, there are words of the quadruplets that we would like to force to some predetermined values, and each row shows the dependence of one word on the message blocks and the IV registers.

Considering Table 3, we propose the following algorithm to sequentially assign values to the message blocks and IV values so that the four quadruplets in all four branches take on the prescribed values. We can refine our algorithm to help the difference to propagate without a change. For instance, we can force the input of the function g to zero in order to be sure that the difference is unchanged.

1. *Initialize.* Choose A_0 randomly.

2. *Adjust.* Do the following four assignments: $M_0 := -(A_0 + \delta_{\pi_1(0)})$, $M_{14} := -(A_0 + \delta_{\pi_2(0)})$, $M_7 := -(A_0 + \delta_{\pi_3(0)})$, $M_5 := -(A_0 + \delta_{\pi_4(0)})$.

3. *Force words $G_{j,3}$.* Choose D_0 so that $F_{3,2}$ gets the correct value. Then, choose M_3, M_9, and M_8 in turn so that $F_{1,2}$, $F_{2,2}$ and $F_{4,2}$ get their correct values, respectively.

4. *Force words $H_{j,3}$.* (If this step has been run 2^{32} times, return to step 1.) Randomly choose H_0. Adjust M_{11} and M_1 to prevent the difference from being modified in the second step of FORK-256 in the branches 2 and 4, respectively. Then, set the words E_0, M_{15}, M_6, and M_{12} so that $F_{1,1}$, $F_{2,1}$, $F_{3,1}$, and $F_{4,1}$ get their correct values, respectively.

5. *Force words $F_{j,3}$.* Set C_0 so that $F_{4,3}$ gets its correct value. Then, set M_{10} and M_2 in turn so that $F_{2,3}$ and $F_{3,3}$ get their correct values, respectively.

Now, $F_{1,3}$ is assigned a random value. If this value is the correct one, continue to the next step, otherwise, return to the Step 4.

6. *Force words $E_{1,3}$, $E_{2,3}$, and $E_{4,3}$.* (If this step has been run 2^{32} times, return to Step 1.) Choose a random value for G_0. Fix B_0 so that $E_{4,3}$ takes the correct value. Fix M_4 so that $E_{2,3}$ takes the correct value. Check the random value taken by $E_{1,3}$: if this is the expected value, go to Step 7, else go back to Step 6.

7. *Force word $E_{3,3}$.* (If this step has been run 2^{32} times, go to Step 1.) Choose M_{13} at random. Check the random value taken by $E_{3,3}$: if this is the expected value, output all messages M_i and all IV blocks. Otherwise, go back to step 7.

Notice that the algorithm makes a few *independent* exhaustive searches in spaces of size 2^{32}. Almost no memory is required, and the average time complexity is 2^{32} applications of one fourth of FORK-256, that is about 2^{30} computations of the hash function. Now, for the attack to succeed, the difference d has to propagate unmodified up to step 3. Since the probability to propagate in one branch is P_d, and taking into account the fact that we took care of it in the first step (Step 2 of the algorithm) and in two of the four branches of the second step (Step 4 of the algorithm), the overall probability is P_d^6.

We eventually remark that the word F_0 of the IV does not modify the four targeted quadruplets. Hence, in the output of our algorithm, we can make F_0 to take any of the 2^{32} possible values and the result remains valid. That is, our algorithm outputs 2^{32} pairs $\{(M, \mathrm{IV}), (M, \mathrm{IV}')\}$ such that after the seventh step differences only appear in registers $A_{j,7}$ in all four branches.

Finally, our algorithm outputs 2^{32} solutions with a complexity equivalent to $2^{30} \cdot P_d^{-6}$ hash computations, and the average cost of computing a solution pair is thus about $1/4 \cdot P_d^{-6}$.

5.2   Choosing the Difference

In the two previous paragraphs, we saw that a useful difference has to fulfill two constraints. The first one is that micro-collisions must happen in all four branches of FORK-256 in the fourth step. The second one is that the probability P_d of propagating without modification should be as high as possible. Since differences with small Hamming weight yield bigger probabilities P_d, we checked all differences with Hamming weights one and two, and we finally chose the difference $d = \mathrm{0x00000404}$. For this choice of difference we have P_d of about 2^{-3}, and a possible set of target values for each branch is:

$$E_{1,3} = \mathrm{0x030e9c3f}, \quad E_{2,3} = \mathrm{0x7e24de5c}, \quad E_{3,3} = \mathrm{0x00fa4d1e}, \quad E_{4,3} = \mathrm{0x20b7363f},$$
$$F_{1,3} = \mathrm{0xa4115fb0}, \quad F_{2,3} = \mathrm{0x10276030}, \quad F_{3,3} = \mathrm{0x35edee6e}, \quad F_{4,3} = \mathrm{0xefc6172f},$$
$$G_{1,3} = \mathrm{0x22c18168}, \quad G_{2,3} = \mathrm{0x4db27e00}, \quad G_{3,3} = \mathrm{0xd81cdc6c}, \quad G_{4,3} = \mathrm{0x8c2c7c00},$$
$$H_{1,3} = \mathrm{0x1816822c}, \quad H_{2,3} = \mathrm{0x27e004db}, \quad H_{3,3} = \mathrm{0xcdc6bd82}, \quad H_{4,3} = \mathrm{0xc7bff8c3}.$$

5.3 Near-Collisions for FORK-256's Compression Function

Seven step reduced version. We focus on a seven step reduced version of FORK-256: the two additions, the xor, and the feed-forward are kept but the eighth step is removed except for the final permutation of registers. It may appear that we can find a collision against this seven steps reduced version of FORK-256, but this is not true. Indeed, we have seen that a difference remains in the internal registers $A_{j,7}$. Those differences have their lowest bit set to 1 exactly at the same position as the lowest bit of the difference d initially introduced, in our case the third lowest significant bit. These bits are shifted to the left by the addition in the first two branches and the last two branches, and the xor cancels them. However, a differential bit reappears at the previous position due to the feed-forward, and we can not get rid of it.

We thus seek 1-bit near collisions, the probability of which has been estimated as follows. We chose a random internal state before the seventh step (i.e. the values $A_{j,6}$, $B_{j,6}$, $C_{j,6}$, and $D_{j,6}$ in each of the four branches), and ran the seventh step transformation, plus the recombination mechanism. After 2^{32} experiments, there was, on the average, 8.96 non zero bits. The probability of a 1-bit near-collision has been evaluated to 2^{-15} (127665 outputs out of the 2^{32} experiments were 1-bit near-collisions). Since the algorithm given in the previous section outputs 2^{32} correct values in $2^{30}/P_d^6 = 2^{49}$ hash computations, the complexity to find a set of 2^{17} distinct 1-bit near-collisions is about 2^{49} hash computations.

Near-Collision for the full compression function. The algorithm studied in the previous paragraphs outputs 2^{32} pairs for which FORK-256's outputs collide on four of the eight 32-bit words with a complexity of 2^{49} hash computations. It remains one bit of difference (at a fixed position) in the second word and three 32-bit words to cancel. The probability of a 1-bit near collision on the second word was experimentally found to be 2^{-15}, and cancelling any of the three remaining 32-bit words was experimentally found to require an average of 2^{31} trials for $d = \text{0x0000404}$. Hence the overall complexity to find a 1-bit near collision is about $2^{49+93+15-32} = 2^{125}$. Similarly, the probability to find a 2-bit near collision was experimentally found to be less than 2^{-10} so that the overall complexity to find such a collision is about $2^{49+93+10-32} = 2^{120}$.

Experimental results. We exhibit a 22-bit near collision on FORK-256's compression function that was obtained by running our algorithm given in Section 5.1 together with the difference and the set of targets specified in Sect. 5.2. (Note that this also leads to a 2-bit near collision for the seven steps reduced version.)

CV_n:0x8406e290 0x5988c6af 0x76a1d478 0x0eb60cea 0xf5c5d865 0x458b2dd1 0x528590bf 0xc3bf98a1

CV_n':0x8406e290 0x5988cab3 0x76a1d478 0x0eb60cea 0xf5c5d865 0x458b2dd1 0x528590bf 0xc3bf98a1

M:0x396eedd8 0x0e8c2a93 0xb961f8a4 0xf0a06fc6 0x9935952b 0xe01d16c9 0xddc60aa4 0x0ac1d8df

　　0xc6fef1d8 0x4c472ca6 0x58d9322d 0x2d087b65 0x7c8e1a26 0x71ba5da1 0xba5d2bfc 0x1988f929

CV_{n+1}:0x9897c70a 0x4e18862d 0xb4725ac1 0xcfc9f92c 0x9aa0637d 0xae772570 0x74dd4af1 0xcd444dd7

CV_{n+1}':0x9897c70a 0x4e1880f9 0x1e677302 0x4c650966 0xf4792bf4 0xae772570 0x74dd4af1 0xcd444dd7

6   Finding High-Level Differential Paths in FORK-256

In this section we return to the question of finding differential paths in four branches of FORK-256 to present a general solution to that problem. If we can avoid mixing introduced by the structures Q_L and Q_R (i.e. we know how to find micro-collisions) and we can assume that differences in registers B, C, D and F, G, H remain unchanged ($P_d = 1$), the only places where differences can change are registers A and E, after the addition of a message word difference. Thus, the values of registers in steps can be simply seen as linear functions of registers of the initial vector $(A_0, \ldots, H_0)$ and message words $M_0, \ldots, M_{15}$.

If we consider the most general case and assume (very optimistically) that any two differences can cancel each other, we are in fact working over $\mathbb{F}_2$ and differences in all registers are $\mathbb{F}_2$-linear combinations of differences $\Delta A_0, \ldots, \Delta H_0$ and $\Delta M_0, \ldots, \Delta M_{15}$ (which are now seen as elements of $\mathbb{F}_2$). Now, output differences $(\Delta A, \ldots, \Delta H)$ of the whole compression function (with feed-forward) are also linear combinations of differences from $S = (\Delta A_0 \ldots, \Delta H_0, \Delta M_0, \ldots, \Delta M_{15})$ and this can be represented by an $\mathbb{F}_2$-linear mapping $(\Delta A, \ldots, \Delta H) = L_{out}(S)$. This means we can find the set S_c of all vectors S of input differences that yield zero output differences at the end of the function simply as the kernel of this map, $S_c = \ker(L_{out})$.

To minimise the complexity of the attack, we want to find high-level paths as short as possible. Since each register difference in each step is a linear function of differences $\Delta A_0 \ldots, \Delta H_0, \Delta M_0, \ldots, \Delta M_{15}$ and there are only 2^{24} of them, the straightforward approach is to enumerate them all and for any desirable subset of registers (e.g. for collisions in two or three branches) count the number of registers containing non-zero differences and pick those input differences S that give the smallest one. Using simple algebra and coding theory techniques we can make this process very efficient. Details can be found in [7].

Differences in registers other than A and E do not contribute to the complexity of the attack that much because they do not require finding micro-collisions. The measure based on the number of differences in registers A and E only corresponds to the number of "difficult" differentials we need to handle that require finding micro-collisions. Experiments show that there is a close correlation between the number of required micro-collisions and the overall length (number of all registers containing differences) of the differential path so it seems sufficient to use the measure based on differences in A and E only. Results of a search for such paths are presented in more details in [7], here we want to discuss an extension of this method.

6.1   More General Variant of Path Finding

We can generalize this approach further. Depending on whether we force a micro-collision to happen in a particular line or not, we have eight different models for each Q-structure. Using the linear model that assumes that all differences cancel each other, we can express output differences of each Q_L-structure as

$$\Delta A_{i+1} = \Delta A_i \ , \qquad\qquad \Delta C_{i+1} = \Delta C_i + q_C \cdot \Delta A_i \ ,$$
$$\Delta B_{i+1} = \Delta B_i + q_B \cdot \Delta A_i \ , \qquad\qquad \Delta D_{i+1} = \Delta D_i + q_D \cdot \Delta A_i \ .$$

where $q_B, q_C, q_D \in \mathbb{F}_2$ are fixed coefficients characterizing the Q_L-structure. The same is true for Q_R-structures. This means that we have 8^{64} possible linear models of FORK-256 when we allow such varied micro-collisions to happen. Allowing for micro-collisions in only selected lines decreases the number of active Q-structures, however, at the expense of additional conditions required to cancel differences coming from different parts of the structure.

Results of our search for such paths are summarized in Table 4. They show that by introducing such an extended model of Q-structures we can significantly decrease the number of necessary micro-collisions compared to the case when we require micro-collisions in all three lines simultaneously. Of special interest is the result showing that under favourable conditions, collisions can be achieved by using a single difference in M_{12} with six micro-collisions in the path. We show how to use this scenario to generate near-collisions but also collisions for the full compression function in Section 7.

Table 4. Minimal numbers m of Q-structures with micro-collisions for different scenarios of finding generalized high-level differential paths. Q-structures are numbered from 1 to 64 where 1 corresponds to Q_L in the first step of branch 1 and 64 to Q_R in the last step of branch 4. Notation N:110 means that in Q-structure number N input difference to A (resp. E) propagates to the second and third register but not to fourth (e.g. to B, C or F, G resp.) For example, differential path from Fig. 2 is encoded as 13:110, 31:111, 40:000, 47:111, 50:000, 57:000.

Scenario	Branches	m	Differences in	active Q-structures
Pseudo-collisions	1,2,3,4	5	H_0, M_2, M_{11}	12:000, 25:000, 35:001, 41:001, 51:010
Collisions	1,2,3,4	6	M_{12}	13:000, 31:001, 40:000, 47:100, 50:000, 57:000
Pseudo-collisions	1,2,3	2	B_0, M_{12}	8:100, 24:0
	1,2,4	3	H_0, M_{11}	3:000, 51:010, 60:000
	1,3,4	3	H_0, M_2	35:001, 44:000, 51:000
	2,3,4	3	D_0, M_9	36:010, 43:000, 52:000
Collisions	1,2,3	3	M_0, M_3, M_9	1:001, 20:010, 39:100
	1,2,4	4	M_1, M_2	2:001, 9:000, 25:100, 51:000
	1,3,4	5	M_9	10:000, 39:001, 42:001 43:010, 59:000
	2,3,4	5	M_3, M_9	20:010, 27:000, 39:000 57:000, 59:010

7   Collisions for the Full Compression Function

In this section we show how to use a high-level path with differences in M_{12} only presented in Section 6 in order to find very low weight output differences of the FORK-256's compression function. We then show two different strategies to find full collisions faster than the bound given by the birthday paradox.

The key observation is that if we introduce a difference in M_{12} only and are able to find micro-collisions in the first and fifth step of the fourth branch as well as in the fourth step of the third branch, and prevent the propagation of the difference from $A_{1,6}$ to $E_{1,7}$ in the first branch, then the output difference is confined to registers B, C, D, and E of the output, i.e. to at most 128 bits. This behavior is illustrated in Fig. 2. The number of affected bits can be further decreased by a careful selection of the modular difference i.e. differences that are set on few most significant bits guarantee that the difference in output register B is confined to those most significant bits as well.

In the next paragraph, we develop our first strategy which does not require large memory. We show that pairs of messages satisfying the aforementioned constraints can be efficiently found and thus, assuming that the output differences closely follow the uniform distribution, we can expect to find very low weight differences and ultimately a collision. Finally, in the second paragraph, we use another strategy relying on precomputed tables to speed up the process of finding collisions.

7.1 Finding Collisions with Low Memory Requirements

The attack consists of two phases. During the first one, we find simultaneous micro-collisions at the first and fifth steps of the fourth branch as well as at the fourth step of the third branch for a modular difference injected in M_{12}. In the second phase we use free message words M_4 and M_9 that do not interfere with already fixed messages and micro-collisions found in the third and fourth branches in order to find messages yielding no difference in the register $E_{1,7}$. This is a reduced micro-collision in the single thread $D_{1,6} \rightarrow E_{1,7}$ during the seventh step of the first branch. The description below is brief – for more details, see our implementation from [6].

Finding micro-collisions in third and fourth branches. Here we assume that a suitable modular difference d has already been chosen. We proceed as follows:

1. *Fourth branch, first step.* Pick x_1 s.t. the pair $(x_1, x_1 + d)$ gives simultaneous micro-collisions in Q_R at the first step of the branch four, set $M_{12} := x_1 - E_0$ and assign the correct values to F_0, G_0, and H_0 for this micro-collision to happen.

2. *Fourth branch, fifth step.* Assign random values to M_5, M_1, M_8, M_{15}, M_0, M_{13}, and M_{11}. Then compute the first half of the branch, up to the fifth step and find a pair of values $(x_2, x_2 + d^*)$ (where d^* is the modular difference in register $A_{4,4}$) yielding simultaneous micro-collisions in Q_L. Compute the corresponding constants ρ_1, ρ_2, and ρ_3. If no solution exists, repeat this step, otherwise]

   - Set $M_3 := x_2 - A_{4,4}$.
   - Fix $M_{13} := \rho_1 - A_{4,3} - \delta_8$ so that $B_{4,4}$ gets its correct value ρ_1.
   - Fix $M_{15} := [\rho_2 \oplus g(B_{4,4})] - f(B_{4,4} - \delta_8) - A_{4,2} - \delta_{10}$ so that $C_{4,4} = \rho_2$.
   - Similarly, fix M_1 so that $D_{4,4} = \rho_3$.

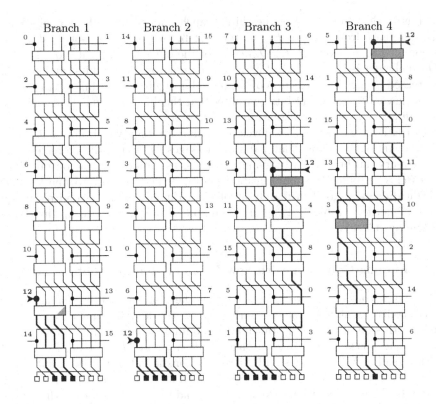

Fig. 2. High level path used to find collisions for FORK-256. Thick lines show the propagation of differences. Q-structures requiring micro-collisions are greyed out. Numbers indicate message ordering.

Adjustments need to be made to M_0 and M_{11} to compensate for the changes in M_1 and M_{15}.

3. *Third branch.* Find a pair of values $(x_3, x_3 + d)$ that causes simultaneous micro-collisions in Q_R in the fourth step and find the corresponding constants λ_1, λ_2, and λ_3. Similar to above, fix M_2 so that $F_{3,3} = \lambda_1$, M_{14} so that $G_{3,3} = \lambda_2$, and M_6 so that $H_{3,3} = \lambda_3$. An adjustment needs to be made to M_{10} to compensate for the change in M_6. Similarly, we have to compensate for the change in M_{14} but we cannot change M_{13} since it is set in the branch four. Instead, we change B_0 so that $E_{3,3}$ gets the correct value x_3.

4. *Fourth branch.* The only modification made in the third branch that can spoil the arrangement of the fourth branch is B_0. It can only change the value $A_{4,4}$. Thus, adjusting M_{11} (again) is enough to put everything back into order.

At the end of this procedure, we have obtained a differential path in the third and fourth branches presented in Fig. 2. For the remaining part of the attack, the key fact is that the values of message words M_4 and M_9 do not alter this path. These 64 bits of freedom are used to perform the second part of the attack.

Single micro-collision in the first branch. We are left with taking care of the first and second branches. Fortunately, the second branch actually does not require any attention: M_{12} appears in the very last step and so only induces differences in the output registers B, C, D, and E. The first branch however requires a single micro-collision in the thread $D_{1,6} \rightarrow E_{1,7}$ during the seventh step, and it seems to us that there is no better way of finding messages causing that micro-collision than by randomly testing message words M_4 and M_9. The success probability of this search heavily depends on the modular difference being used. The two best modular differences we found are displayed in Table 5.

Table 5. Best modular differences d we could find and their probabilities of inducing a single micro-collision in thread $D_{1,6} \rightarrow E_{1,7}$ during the seventh step in the first branch. The number of input values $A_{1,6}$ that may result in the micro-collision is denoted by η.

difference d	η	observed probability
0xdd080000	$2^{21.7}$	$2^{-24.6}$
0x22f80000	$2^{21.7}$	$2^{-24.6}$

Let us analyse the computational complexity of finding this single micro-collision in terms of numbers of full FORK-256 evaluations. Let η denote the number of allowable values for the chosen modular difference in use. By allowable value we mean an input x for which there exist three constants that cause a micro-collision for the pair $(x, x + d)$. For the modular difference $d = $22f80000 we have $\eta = 2^{21.7}$ allowable input values.

Our algorithm to arrange the first branch correctly runs as follows:

1. *Initialize.* Fix M_4 to zero.

2. *Pre-compute table.* Compute all the internal registers up to the seventh step. Then, for each allowable value x, set $A_{1,6} = x$ and go one step backwards to get the corresponding $H_{1,5}$, and store the result into a hash table T.

3. *Search for M_9.* For every possible value of M_9 compute the corresponding value of $H_{1,5}$ and look for a match in T. If there is a match, go to Step 4. When all M_9 are exhausted, increment M_4 and go back to Step 2.

4. *Check.* If current value of M_9 leads to a single micro-collision in thread $D_{1,6} \rightarrow E_{1,7}$ then output the pair (M_4, M_9). Continue Step 3.

Step 2 requires $1/64$ of a full FORK-256 computation for each of the η allowable values. The complexity of this step is thus $\eta/64 = 2^{15.7}$ FORK-256 evaluations. Step 3 requires $1/64$ of full FORK-256 computation for each of the 2^{32} values for M_9. The complexity of this step is thus 2^{26} FORK-256 evaluations. Since Step 4 succeeds with probability $2^{-24.6}$ (see Table 5), we get $2^{7.4}$ solutions for a work effort of 2^{26}. Hence the cost of finding a single solution with our algorithm is about $2^{18.6}$ FORK-256 evaluations.

Experiments. Our C implementation of the algorithm is available for download from [6]. We conducted experiments and verified that for the difference $d =$ 0xdd080000, the distribution of output differences on 108 affected bits (there are 109 bits that may contain differences, but we know that the differences in bit 19 of register B will always cancel out) is very close to uniform [7]. Moreover, after a few days of computations on a Pentium 4 running at 2.8 GHz, we were able to find an output difference of weight 28 [6].

Complexity of the attack. Since at most 108 bits are affected and the distribution of differences is close to uniform, we expect to find a collision after generating 2^{108} pairs. With a work factor of $2^{18.6}$ FORK-256 computations per pair, the total complexity required to find a collision is thus $2^{108} \cdot 2^{18.6} = 2^{126.6}$, which is better than the bound given by the birthday paradox. Additionally, this attack only requires about $2 \cdot 2^{22}$ 32-bit words of memory for storing precomputed inputs for micro-collisions and a hash table of similar size. It also parallelizes perfectly on many computers, each one performing independent computations starting with different seed.

The above complexity estimate is rather conservative, because if we multiply empirical probabilities of single bit differences being zero we get the value of $2^{106.4}$ rather than 2^{108} and thus also a lower complexity of the attack of 2^{125} but one has to be cautious as there is no guarantee that the bits are uncorrelated enough to make this figure accurate. We refer to [7] for details.

7.2   Finding Collisions Faster with Precomputed Tables

In this paragraph we show how to speed up the collision search with the use of precomputed tables. To this end, let us study the spreading process when no difference is involved. During the step transformation, the eight registers are split into two subsets, namely (A, B, C, D) and (E, F, G, H). If we restrict our attention to one of them, let us say (E, F, G, H), we immediately see that the message block M acting on the input register E allows to set the output register F to any value. But what about the action of this message block on one of the three other registers, say, the output register H? The answer is that, on the average, for any input register G, there exists a value of the message block such that the output register H takes any prescribed value. This comes from the observation that for a fixed value of δ and G, the function $\psi_G : y \mapsto \left(g(y)^{\lll 9} + G\right) \oplus f(y + \delta)^{\lll 5}$ is very often a bijection. Hence, for any fixed value of the output register H a table T_H can be built that stores values (G, y) such that $\psi_G(y) = H$. This table can be built during a pre-computation step in time 2^{32} with 2^{32} memory. By building 2^{32} such tables (one for every possible value of H), it is then possible, for any given pair (G, H), to find a message so that G is indeed transformed into H during one half of the step transformation. The cost of the pre-computation is now 2^{64} both in time and memory, but access time is comparable to a single operation. Obviously, as already seen in the previous paragraph, such a table and the freedom given by the incoming message block can be used to fix the value of one of the thread $F \rightarrow G$, $G \rightarrow H$, and $H \rightarrow A$ only.

In the following attack, we use a number of such tables. The first one, T_{10}, is used to control the thread $C_{3,1} \rightarrow D_{3,2}$ through M_{10}, that is $M_{10} = T_{10}(C_{3,1}, D_{3,2})$. Another family of tables, $T_{9,a}$, is used to determine what value of M_9 produces the expected transition $E_{1,4} \rightarrow A_{1,6}$ given a fixed M_{11}, that is $M_9 = T_{9,a}(E_{1,4}, M_{11})$ so that $A_{1,6} = a$, where a is some fixed value. (There are 36 such values for which the probability of a single micro-collision is 2^{-8}, 1236 values with probability 2^{-9} and many more with smaller probabilities.)

As in the previous attack, our goal is to use the high level path of Figure 2 by injecting a modular difference in M_{12} only, and to cause micro-collisions in grayed areas of this figure. To this end, we construct a sequencing allowing to set the message blocks fitting these constraints, but contrary to what is done in the previous attack, we choose the three micro-collisions of the branch three and four in advance. But now, we must ensure that the modular difference in the register $A_{4,4}$ is the same as the one injected in M_{12}. Additionally, we note that for the difference $d = \texttt{0xdd080000}$ we are going to use, we consider around 2^9 values of a for which the difference d does not spread from $A_{1,6}$ to $E_{1,7}$ with highest probability, i.e. a single micro-collision is most likely to happen.

1. *Initialize.* Set M_{12}, F_0, G_0, and H_0 in order to get a micro-collision in the first step of the fourth branch.

2. *Fourth branch.* Set M_1 to fix $B_{4,2}$ to its correct value. Choose a random M_5. Adjust M_8 so that difference d propagates unchanged. Set M_{15} to fix $B_{4,3}$ to its correct value. Adjust M_0 so that difference d propagates unchanged. Set M_{13} to fix $B_{4,4}$ to its correct value. Adjust M_{11} so that difference d propagates unchanged, and set M_3 to fix $A_{4,4}$ to its correct value.

3. *Third branch.* Set M_6 to fix $F_{3,1}$ to its correct value. Choose M_7 randomly. Set M_{14} to fix $F_{3,2}$ to its correct value. Use the hash table T_{10} to set M_{10} so that $E_{3,3}$ gets its correct value. (This is possible because M_5, M_7, M_{13}, and M_{14} are already fixed.) Set M_2 to fix $F_{3,3}$ to its correct value.

4. *First branch.* Choose M_4 randomly. Using the hash table $T_{9,a}$ for some value of a, decide which value M_9 will lead to the value of $A_{1,6}$ equal to a. This value prevents the difference of M_{12} from spreading into $E_{1,7}$ with probability at least 2^{-9}. If the difference spreads into $E_{1,7}$, restart Step 4 with another value of a. After testing around 2^9 such values, difference in the first branch does not spread to $E_{1,7}$ with a high probability.

The complexity of this algorithm is close to $2^{1.6}$ FORK-256 evaluations if we assume access to tables in a single processor operation, with a pre-computation step of complexity about 2^{64} in time and 2^{73} words of memory. Since at most 108 bits of the output differ for the modular difference $\texttt{0xdd080000}$, the algorithm finds a collision in about $2^{109.6}$ FORK-256 computations.

8 Compression Function's Collisions Turned into Hash Ones

Here we show that the last algorithm can be turned into collision finding algorithm for the full hash function, i.e. with a given IV. Our algorithm indeed relies on the fact that three values of IV—namely F_0, G_0, H_0—have specific values. By prepending a well chosen 512-bit message block to the colliding inputs for the compression function we get the expected result for the whole hash function.

Now since the targeted values are three 32-bit words, the probability to reach these value by prepending a random 512-bit message block is 2^{-96}, so we need around 2^{96} FORK-256 computations. This can be done after the execution of our algorithm and thus the overall complexity is dominated by $2^{109.6}$ of the FORK-256 evaluations.

9 Conclusion

In this paper we exposed a number of weaknesses of the compression function of FORK-256. We studied in detail the properties of Q-structures and described very efficient algorithms to finding micro-collisions for them. We further showed how this can be exploited to mount various attacks against FORK-256's compression function. Finally, we showed that the chosen-IV collision-finding attack for the compression function can be extended to find collisions for the full hash function, i.e. with a given IV. We expect that more computational power would allow to investigate slight variations of the attacks we presented, and might improve them significantly.

Although we are intrigued by the design of FORK-256, we think it should not be used in applications that require a high level of security against collision attacks.

Acknowledgements

The second and third authors would like to thank Sebastien Kunz-Jacques for his kind lending of cpu time as well as Gilles Macario-Rat for his help with precomputed tables in paragraph 7.2.

The project was supported by ARC grant DP0663452.

References

1. Anderson, R., Biham, E.: Tiger: A fast new hash function. In: Gollmann, D. (ed.) FSE'96. LNCS, vol. 1039, pp. 121–144. Springer, Heidelberg (1996)
2. Hong, D., Chang, D., Sung, J., Lee, S., Hong, S., Lee, J., Moon, D., Chee, S.: A New Dedicated 256-bit Hash Function: FORK-256. In: Robshaw, M. (ed.) FSE 2006. LNCS, vol. 4047, pp. 195–209. Springer, Heidelberg (2006)
3. Hong, D., Sung, J., Hong, S., Lee, S., Moon, D.: A new dedicated 256-bit hash function: FORK-256. In: First NIST Workshop on Hash Functions (2005)

4. Intel Corporation. Intel 64 and IA-32 architectures optimization reference manual (2006) Appendix C, Instruction latency and throughput. Available from http://developer.intel.com/design/processor/manuals/248966.pdf

5. Lipmaa, H., Walln, J., Dumas, P.: On the additive differential probability of exclusive-or. In: Fast Software Encryption – FSE '04. LNCS, vol. 3017, pp. 317–331. Springer, Heidelberg (2004)

6. Matusiewicz, K., Contini, S., Pieprzyk, J.: Cryptanalysis of FORK-256. Web page, http://www.ics.mq.edu.au/~kmatus/FORK/

7. Matusiewicz, K., Contini, S., Pieprzyk, J.: Weaknesses of the compression function of FORK-256. IACR e-print Archive, report 2006/317, available from http://eprint.iacr.org/2006/317

8. Mendel, F., Lano, J., Preneel, B.: Cryptanalysis of reduced variants of the FORK-256 hash function. In: Abe, M. (ed.) CT-RSA 2007. LNCS, vol. 4377, pp. 85–100. Springer, Heidelberg (2006)

9. Muller, F.: Personal communication (2006)

10. National Institute of Standards and Technology. Secure hash standard (SHS). FIPS 180-1 (April 1995) Replaced by [11].

11. National Institute of Standards and Technology. Secure hash standard (SHS). FIPS 180-2 (August 2002)

12. Preneel, B., Bosselaers, A., Dobbertin, H.: RIPEMD-160: A strenghtened Version of RIPEMD. In: Gollmann, D. (ed.) FSE'96. LNCS, vol. 1039, pp. 71–82. Springer, Heidelberg (1996)

13. Rivest, R.L.: The MD4 Message Digest Algorithm. In: Menezes, A.J., Vanstone, S.A. (eds.) CRYPTO 1990. LNCS, vol. 537, pp. 303–311. Springer, Heidelberg (1991)

14. Rivest, R.L.: The MD4 Message Digest Algorithm. RFC 1320, IETF (April 1992)

15. Rivest, R.L.: The MD5 Message Digest Algorithm. RFC 1321, IETF (April 1992)

16. Schneier, B., Kesley, J.: Unbalanced Feistel networks and block cipher design. In: Gollmann, D. (ed.) FSE'96. LNCS, vol. 1039, pp. 121–144. Springer, Heidelberg (1996)

17. Wang, X., Lai, X., Feng, D., Chen, H., Yu, X.: Cryptanalysis of the Hash Functions MD4 and RIPEMD. In: Cramer, R.J.F. (ed.) EUROCRYPT 2005. LNCS, vol. 3494, pp. 1–18. Springer, Heidelberg (2005)

18. Wang, X., Yin, Y.L., Yu, H.: Finding collisions in the full SHA-1. In: Shoup, V. (ed.) CRYPTO 2005. LNCS, vol. 3621, pp. 17–36. Springer, Heidelberg (2005)

19. Wang, X., Yu, H.: How to break MD5 and other hash functions. In: Cramer, R.J.F. (ed.) EUROCRYPT 2005. LNCS, vol. 3494, pp. 19–35. Springer, Heidelberg (2005)

20. Wang, X., Yu, H., Yin, Y.L.: Efficient collision search attacks on SHA-0. In: Shoup, V. (ed.) CRYPTO 2005. LNCS, vol. 3621, pp. 1–16. Springer, Heidelberg (2005)

21. Zheng, Y., Pieprzyk, J., Seberry, J.: HAVAL – A One-Way Hashing Algorithm with Variable Length of Output. In: Zheng, Y., Seberry, J. (eds.) AUSCRYPT 1992. LNCS, vol. 718, pp. 83–104. Springer, Heidelberg (1993)

A   Additional Details of the Specification of FORK-256

Table 6. Constants $\delta_0, \ldots, \delta_{15}$ used in FORK-256. They are defined as the first 32 bits of fractional parts of binary expansions of cube roots of the first 16 primes.

δ	0	1	2	3	4	5	6	7
0	428a2f98	71374491	b5c0fbcf	e9b5dba5	3956c25b	59f111f1	923f82a4	ab1c5ed5
8	d807aa98	12835b01	243185be	550c7dc3	72be5d74	80deb1fe	9bdc06a7	c19bf174

Table 7. Message and constant permutations used in four branches of FORK-256

j	message permutation σ_j	permutation of constants, π_j
1	0 1 2 3 4 5 6 7 8 9 10 11 12 13 14 15	0 1 2 3 4 5 6 7 8 9 10 11 12 13 14 15
2	14 15 11 9 8 10 3 4 2 13 0 5 6 7 12 1	15 14 13 12 11 10 9 8 7 6 5 4 3 2 1 0
3	7 6 10 14 13 2 9 12 11 4 15 8 5 0 1 3	1 0 3 2 5 4 7 6 9 8 11 10 13 12 15 14
4	5 12 1 8 15 0 13 11 3 10 9 2 7 14 4 6	14 15 12 13 10 11 8 9 6 7 4 5 2 3 0 1

B   Propagation of Modular Differences Through '$\oplus$'

When studying the internal step transformation of FORK-256, the problem appears of computing the probability that a given modular difference d propagates through a '$\oplus$' without being modified. An even more general version of this problem has already been studied at FSE 2004 by Lipmaa, Wallén, and Dumas [5]. Here we give a much weaker version of their result that fits our needs:

Property 1. Given any 32-bit word d, the probability

$$P_d = \Pr_{x,y}\left[\left((x+d)\oplus y\right) = \left(x\oplus y\right)+d\right]$$

where elements x and y are 32-bit words can be expressed as the following matrix product:

$$P_d = L \times M_{d_{31}} \times M_{d_{30}} \times \cdots \times M_{d_0} \times C,$$

where d_i denotes the i-th bit of d and L, C, M_0, and M_1 are defined as:

$$M_0 = \frac{1}{4}\begin{pmatrix} 4 & 0 & 0 & 1 & 0 & 1 & 1 & 0 \\ 0 & 0 & 0 & 1 & 0 & 1 & 0 & 0 \\ 0 & 0 & 0 & 1 & 0 & 0 & 1 & 0 \\ 0 & 0 & 0 & 1 & 0 & 0 & 0 & 0 \\ 0 & 0 & 0 & 0 & 0 & 1 & 1 & 0 \\ 0 & 0 & 0 & 0 & 0 & 1 & 0 & 0 \\ 0 & 0 & 0 & 0 & 0 & 0 & 1 & 0 \\ 0 & 0 & 0 & 0 & 0 & 0 & 0 & 0 \end{pmatrix}, \quad M_1 = \frac{1}{4}\begin{pmatrix} 1 & 0 & 0 & 0 & 0 & 0 & 0 & 0 \\ 1 & 0 & 0 & 1 & 0 & 0 & 0 & 0 \\ 0 & 0 & 0 & 0 & 0 & 0 & 0 & 0 \\ 0 & 0 & 0 & 1 & 0 & 0 & 0 & 0 \\ 1 & 0 & 0 & 0 & 0 & 0 & 1 & 0 \\ 1 & 0 & 0 & 1 & 0 & 4 & 1 & 0 \\ 0 & 0 & 0 & 0 & 0 & 0 & 1 & 0 \\ 0 & 0 & 0 & 1 & 0 & 0 & 1 & 0 \end{pmatrix},$$

$$L = (1\ 0\ 0\ 0\ 0\ 0\ 0\ 0), \quad {}^{T}C = (1\ 1\ 1\ 1\ 1\ 1\ 1\ 1).$$

The Grindahl Hash Functions

Lars R. Knudsen[1], Christian Rechberger[2], and Søren S. Thomsen[1,*]

[1] Technical University of Denmark, DK-2800 Kgs. Lyngby, Denmark
{lars@ramkilde.com, crypto@znoren.dk}
[2] Graz University of Technology, A-8010 Graz, Austria
christian.rechberger@iaik.tugraz.at

Abstract. In this paper we propose the Grindahl hash functions, which are based on components of the Rijndael algorithm. To make collision search sufficiently difficult, this design has the important feature that no low-weight characteristics form collisions, and at the same time it limits access to the state. We propose two concrete hash functions, Grindahl-256 and Grindahl-512 with claimed security levels with respect to collision, preimage and second preimage attacks of 2^{128} and 2^{256}, respectively. Both proposals have lower memory requirements than other hash functions at comparable speeds and security levels.

Keywords: hash functions, Rijndael, AES, design strategy, proposal.

1 Introduction

As a result of a large number of attacks [5,6,11,19,42,44,45,46] on hash functions such as MD5 [41] and SHA-1 [23] of the so-called MD4 family, and general attacks [30,31,33] on the typical construction method [18,37], there is an increasing need for considering alternative construction methods and principles for future hash functions.

In this paper we develop an alternative design strategy for hash functions, and we propose a collection of hash functions named Grindahl[1]. We also propose the two instances Grindahl-256 and Grindahl-512. The main motivation is as follows. Among the many properties a cryptographic hash function is expected to have, collision resistance seems to be the hardest to achieve. Shortcut collision attacks faster than birthday attacks efficiently identify a subset of the message space with the property that a random pair of messages from the subset collide with probability higher than $2^{-n/2}$, where n is the output size of the hash function. A popular tool is differential cryptanalysis. In addition to restricting the

---

* The first two authors have been supported by the European Commission through the IST Programme under Contract IST-2002-507932 ECRYPT. The third author is supported by the Danish Research Council for Technology and Production Sciences, grant no. 274-05-0151.
[1] To be pronounced 'grijndael'.
grind /graɪnd/: to break or crush sth into very small pieces [...] using a special machine.

A. Biryukov (Ed.): FSE 2007, LNCS 4593, pp. 39–57, 2007.

pairs that are chosen from the subset of the message space to those having a certain difference, one might impose restrictions on the relations between actual values of message bits. Usually, the aim is to prevent differences from spreading uncontrollably. Most members of the MD4 family as well as recent proposals like FORK-256 [29] allow this kind of control.

Our aim is to obtain hash functions which do not allow this type of control by using building blocks of the block cipher Rijndael [16]. These well analysed and well understood building blocks are used to ensure that any difference introduced at the input needs to spread over large parts of the state before a collision is possible. While aiming for a security level up to the birthday bound for collision and (second) preimage attacks, we obtain speeds comparable to members of the SHA-2 family [24], but with a fundamentally different design and lower memory requirements for an implementation. Additionally, the computational overhead for small messages compares favourably to other design strategies.

Our design has a structure similar to some of those of J. Daemen (et al.) such as SubHash and StepRightUp [14], Panama [15], and the recent proposal RadioGatún [4]. The similarity lies in the way small pieces of message via a round function sequentially update a state in an invertible fashion.

Apart from being hash function proposals, we describe how our proposals can be turned into compression functions accepting only a fixed-size input. In order to test and develop cryptanalytic methods, variants with reduced cryptographic strength are helpful. For the MD4 family of hash functions, changing the number of rounds serves this purpose very well. Our design allows for such simple modifications and we encourage the reader to analyse such simpler variants. See Appendix B for suggestions.

2   The Grindahl Design

In this section we present our proposal for a collection of hash functions. We start off with a description of the general design strategy which we call "Concatenate-Permute-Truncate". This design principle was first proposed by R. Merkle and used in his hash function Snefru [38], and it requires the existence of a non-linear permutation. We propose a highly parameterisable permutation hence in effect a collection of non-linear permutations, in Section 2.3. The general design and our proposed permutations together form the Grindahl hash functions. Concrete proposals are presented in Section 3.

2.1   General Strategy

The proposal of this paper was conceived from the following general design strategy for an n-bit hash function. In the following, we denote by m the *state size*, and by b the *message size*. We require $m \geq n$ and $b > 0$. We denote by $\text{trunc}_k(x)$ the least significant k bits of x. Let $P : \{0,1\}^{m+b} \to \{0,1\}^{m+b}$ be a non-linear permutation, and let s_0 be the *initial state* with $|s_0| = m$.

The principle behind our design is "Concatenate-Permute-Truncate"; let d be the message (appropriately padded) to be hashed, and split d into t blocks of b bits, i.e. $d = d_1 \| \cdots \| d_t$, $|d_i| = b$. Then for $0 < i \leq t$ do

$$S_i \leftarrow d_i \| s_{i-1} \qquad \text{(Concatenate)} \qquad (1)$$

$$\hat{S}_i \leftarrow P(S_i) \qquad \text{(Permute)} \qquad (2)$$

$$s_i \leftarrow \text{trunc}_m(\hat{S}_i) \qquad \text{(Truncate)} \qquad (3)$$

Hence, a message block is concatenated with the state to form what we shall call the *extended state*, on which some permutation P is applied. Subsequently, the extended state is truncated down to the new state. The steps (1)–(3) form an *input round*.

We define an output transformation consisting of *blank rounds* and a truncation step at the end. Blank rounds are defined as follows. For $t < i \leq t + \nu_{\text{br}}$, $\nu_{\text{br}} \geq 0$, do

$$\hat{S}_i \leftarrow P(\hat{S}_{i-1}).$$

Hence, the blank rounds work only on the extended state, which means that in the processing of the final message block d_t, (3) above can be omitted. Finally, the output of the hash function is $\text{trunc}_n(\hat{S}_{t+\nu_{\text{br}}})$.

2.2   Invertibility

Assuming that the permutation P is invertible, the hash function is not one-way in the sense that for a given output, some initial state and a message producing that output can be easily found. However, this does not directly give rise to proper (second) preimage attacks. If P has sufficient cryptographic properties then an attacker will have no control over the initial state obtained.

The success probability of meet-in-the-middle attacks is affected in part by the value of m above. If no weaknesses of P are exploited then internal collision attacks (collisions before the blank rounds) and meet-in-the-middle attacks have complexity $2^{m/2}$. If one requires that no (second) preimage attacks better than a brute force search exist, then one has to choose $m \geq 2n$.

2.3   Design Approach for the Permutation

A well-known family of permutations is the block cipher algorithm Rijndael [16], a subset of which was adopted as the Advanced Encryption Standard (AES) by the US government in 2001 [25]. What follows is an approach that uses the design principle of Rijndael to build a permutation to be used in our hash proposal. This bears some resemblance to the leak-extraction method of the stream cipher LEX [8] or the message authentication code framework ALRED [17].

We operate with a more general description of the algorithm than Rijndael itself. As in Rijndael, we view the (in our case extended) state as a matrix of bytes, although here the extended state is a matrix α of ν_{rw} rows and ν_{cl} columns of bytes. The entries of α are denoted by $\alpha_{i,j}$, meaning the entry in row i, column

j (numbering starts from zero, and indices are always to be reduced modulo ν_{rw} and ν_{cl}, respectively). Hence, the extended state is the matrix

$$
\alpha = \begin{bmatrix}
\alpha_{0,0} & \alpha_{0,1} & \cdots & \alpha_{0,\nu_{\mathrm{cl}}-1} \\
\alpha_{1,0} & \alpha_{1,1} & \cdots & \alpha_{1,\nu_{\mathrm{cl}}-1} \\
\vdots & \vdots & \ddots & \vdots \\
\alpha_{\nu_{\mathrm{rw}}-1,0} & \alpha_{\nu_{\mathrm{rw}}-1,1} & \cdots & \alpha_{\nu_{\mathrm{rw}}-1,\nu_{\mathrm{cl}}-1}
\end{bmatrix}.
$$

We assume that b is a multiple of 8, and we define $\nu_{\mathrm{mb}} = b/8$ as the number of bytes in a message block. Hence, according to (3) and (1), in the process of truncation followed by concatenation (with the following message block), ν_{mb} extended state bytes are overwritten. These extended state bytes do not have to be computed in all except the last input round.

The reader is expected to be familiar with the transformations defined in the Rijndael specification [16]. Here we adopt the new names introduced in the actual standard [25], i.e. SubBytes, ShiftRows, MixColumns and AddRoundKey.

We do not use AddRoundKey directly in this design. Instead, we introduce a related transformation, AddConstant. We now comment on the four transformations used in the Grindahl design. See also Fig. 1.

SubBytes. The non-linear substitution function SubBytes is defined exactly as in the Rijndael specification, i.e. we use the same S-box.

ShiftRows. The ShiftRows transformation cyclically shifts bytes a number of positions along each row. We introduce the *rotation constants* as the ν_{rw}-tuple $(\rho_0, \rho_1, \ldots, \rho_{\nu_{\mathrm{rw}}-1})$ of integers, $0 \leq \rho_i < \nu_{\mathrm{cl}}$, with the meaning that in row i bytes should be cyclically shifted ρ_i positions to the right. (Hence, with this definition the rotation constants of Rijndael are $(0, 3, 2, 1)$).

MixColumns. The transformation MixColumns is defined as in the Rijndael specification whenever $\nu_{\mathrm{rw}} = 4$. For other values of ν_{rw}, the transformation must be redefined. The important property of maximal difference propagation should be maintained: when a difference is introduced to $k > 0$ bytes in a column before MixColumns, the effect should be that after MixColumns at least $\nu_{\mathrm{rw}} - k + 1$ bytes have changed.

AddConstant. As mentioned we replace the AddRoundKey transformation known from the Rijndael design by AddConstant, which introduces asymmetry to each round: Let α be the extended state matrix, and let M be some matrix of the same size. Then define AddConstant as

$$
\mathsf{AddConstant}(\alpha) = M \oplus \alpha.
$$

We note that if an extended state consists of ν_{cl} equal columns, then by defining e.g., $M = 0$ the extended state will still consist of ν_{cl} equal columns after one round. By flipping some bits of the extended state we can circumvent this property.

The four transformations operate on a matrix of bytes. However, given an invertible mapping from an extended state to a bit string, we may also apply

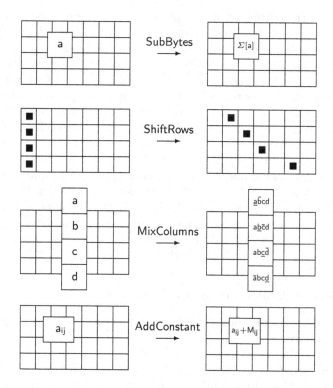

Fig. 1. The four transformations on the extended state. Here an example with an extended state of 4 rows and 7 columns. Σ is the S-box, and example rotation constants are $(1, 2, 3, 5)$.

them on bit strings. The mapping from a bit string to an extended state matrix is done as follows. Let x be an $(8\nu_{\mathrm{rw}}\nu_{\mathrm{cl}})$-bit string. Map this into an extended state α by splitting x into $\nu_{\mathrm{rw}}\nu_{\mathrm{cl}}$ 8-bit chunks $x_0, \ldots, x_{\nu_{\mathrm{rw}}\nu_{\mathrm{cl}}-1}$, and then let $\alpha_{i,j} = x_{i+\nu_{\mathrm{rw}}j}$, $0 \le i < \nu_{\mathrm{rw}}$ and $0 \le j < \nu_{\mathrm{cl}}$. This mapping has a natural inverse.

Given appropriate definitions of the four transformations we define the permutation P as

$$P(\alpha) = \mathsf{MixColumns} \circ \mathsf{ShiftRows} \circ \mathsf{SubBytes} \circ \mathsf{AddConstant}(\alpha).$$

2.4   Birthday Attacks

If P were an ideal permutation, then internal collisions and (second) preimages would have complexity $2^{m/2}$. However, P is obviously not ideal. In fact, it is easy to see that the complexity of e.g. an internal collision attack for this choice of P is at most $2^{(m-b)/2}$: assume that the first column of the matrix is overwritten by the message input. Compute the extended state before the blank rounds of a number of different messages. Now append two constant blocks to all messages. The first constant message block overwrites the first column of the extended state. Then the permutation is applied, where ShiftRows moves ν_{rw} bytes into the first column of the extended state, and subsequently MixColumns mixes the bytes in the column. The second constant message block overwrites this column. This means that if two extended states agree on all bytes except the first column and the bytes that are moved into the first column by the first constant message block, then the two extended states will agree on all bytes after the second constant message block. The expected number of messages needed for this attack to succeed is $2^{(m-b)/2}$. In fact this approach can be generalised to every way of mapping an input message to the extended state.

Since the permutation is invertible, the entire hash function is invertible, and hence meet-in-the-middle attacks can be applied: Compute the intermediate (extended) state for $2^{(m-b)/2}$ different messages all having the same last two blocks. Given a target image of the hash function, compute in the backward direction the intermediate state for the same number of message suffices. With good probability, there is a match between the two sets of intermediate states. This yields a (second) preimage with complexity about $2^{(m-b)/2}$.

2.5   Design Parameters for the Permutation

We now present some considerations with respect to the design parameters introduced above.

Rotation constants. The rotation constants used in ShiftRows should be chosen carefully. Most importantly, the rotation constants should ensure that the entire state depends on the message input as quickly as possible when considering that the first ν_{mb} bytes of the state are overwritten by message input in every round. This means, for instance, that ρ_i, $0 \leq i < \nu_{mb}$, cannot be 0, and that $\rho_i \neq \rho_j$ whenever $i \neq j$.

Several tuples of rotation constants ensure full diffusion after the same (minimum) number of rounds μ. However, of these some are better than others in the sense that a larger part of the state depends on every message byte after $\mu - 1$ rounds, after $\mu - 2$ rounds etc. In Appendix C we suggest a method for choosing rotation constants given a particular geometry of the extended state.

State geometry. Parameters that affect the size m of the state should be chosen with Section 2.4 in mind. Usually designers of hash functions aim for full (second) preimage resistance, meaning these have expected complexity 2^n. We have chosen to accept a lower complexity, since it is already required that n is large enough to resist birthday collision attacks, which have complexity $2^{n/2}$. Other hash functions have been proposed and standardised, for which (second) preimage resistance is lower than for an ideal hash function, e.g., the MDC-2 construction [39] which has preimage complexity at most $2^{3n/4}$ [40]. Appendix A discusses known generic attacks that depend on the state size. By adjusting the state geometry accordingly, resistance against these attacks can be achieved conveniently.

Choices for ν_{rw} and ν_{cl} are a trade-off between two distinct properties. If the two numbers are about the same, diffusion happens faster than with more columns than rows. On the other hand, with a wider state and only a single column being used for message input, the birthday attack as described above has a higher complexity, and hence the extended state need only be slightly larger than the output. For implementation purposes it makes sense to choose ν_{rw} to be a multiple of 4, since then on a 32-bit machine SubBytes and MixColumns can be performed in one by table lookups.

2.6 Design Parameters for the Output Transformation

The number ν_{br} of blank rounds in the output transformation should be chosen such that the last message block affects all output bytes.

It might be desirable that the output transformation have the property that when given only black-box access to it, one is unable to distinguish the output transformation from a pseudo-random function. It is unclear how useful this property is in practice, since the output transformation is invertible (by guessing the discarded state bytes), but it would seem to complicate attempts at finding external collisions.

With ν_{br} blank rounds, the actual number of invocations of P after the last message block is concatenated with the state is $\nu_{br} + 1$. During these rounds, no extended state bytes are overwritten by message input. Hence, if the requirement above is fulfilled after μ rounds, then one should choose $\nu_{br} \geq \mu - 1$. Suggestions for actual values for ν_{br} are given in Sections 3.1 and 3.2.

3 Proposals for Hash Functions

We propose concrete instantiations of the design strategy presented in Section 2. Hash functions with 256-bit and with 512-bit output size are given in Sections 3.1 and 3.2, respectively. For both proposals, the initial state is the all-zero state, and padding is performed as described in Section 3.3. Additionally we give a preliminary security analysis in Section 3.4.

In the following, constant bytes (or elements of $\mathbb{F}_{256}$) are written in hexadecimal using this font, e.g., c5.

3.1 Grindahl-256

Grindahl-256 is defined as follows. Let the parameters have the following values: $\nu_{rw} = 4$, $\nu_{cl} = 13$, $\nu_{mb} = 4$, $\nu_{br} = 8$. Hence, the extended state has 4 rows and 13 columns, and the message input is 32 bits in size. In the truncation/concatenation process these overwrite the contents of the first column of the extended state. The proposal has 8 blank rounds in the output transformation.

The rotation constants used in the ShiftRows transformation are $(1, 2, 4, 10)$, chosen by the method described in Appendix C. For these rotation constants every message byte affects the entire extended state after four rounds. SubBytes and MixColumns are defined as in Rijndael, and AddConstant is defined simply as

$$\alpha_{3,12} \leftarrow \alpha_{3,12} \oplus 01.$$

Security claim. We claim that the effort to find collisions, second preimages and preimages is of the order of 2^{128} iterations.

3.2 Grindahl-512

We also propose the 512-bit hash function Grindahl-512. For this variant, we propose the following parameter values: $\nu_{\text{rw}} = 8$, $\nu_{\text{cl}} = 13$, $\nu_{\text{mb}} = 8$, $\nu_{\text{br}} = 8$. In other words, the state has 8 rows and 13 columns, and the message input is 64 bits in size, corresponding to the first column of the extended state. The output transformation contains 8 blank rounds.

The proposed rotation constants for ShiftRows are $(1, 2, 3, 4, 5, 6, 7, 8)$, which cause full diffusion after three rounds (these were chosen according to Appendix C). MixColumns has to be redefined since the extended state contains 8 rows. We define this transformation as

$$\text{MixColumns}(\alpha) = A \cdot \alpha,$$

with

$$A = \begin{bmatrix} 02 & 0c & 06 & 08 & 01 & 04 & 01 & 01 \\ 01 & 02 & 0c & 06 & 08 & 01 & 04 & 01 \\ 01 & 01 & 02 & 0c & 06 & 08 & 01 & 04 \\ 04 & 01 & 01 & 02 & 0c & 06 & 08 & 01 \\ 01 & 04 & 01 & 01 & 02 & 0c & 06 & 08 \\ 08 & 01 & 04 & 01 & 01 & 02 & 0c & 06 \\ 06 & 08 & 01 & 04 & 01 & 01 & 02 & 0c \\ 0c & 06 & 08 & 01 & 04 & 01 & 01 & 02 \end{bmatrix}.$$

In this product, bytes are to be considered elements of the field $\mathbb{F}_{256}$, which is defined as in Rijndael. This transformation ensures maximal difference propagation, since the error-correcting code over $\mathbb{F}_{256}$ with generator matrix $\begin{bmatrix} I & A^{\mathrm{T}} \end{bmatrix}$ is MDS (see e.g., [35]). SubBytes is defined as in Rijndael, and AddConstant is defined as

$$\alpha_{7,12} \leftarrow \alpha_{7,12} \oplus 01.$$

Security claim. We claim that the effort to find collisions, second preimages and preimages is of the order of 2^{256} iterations.

3.3 Padding Rule

Padding for both proposals is performed as follows. Append a '1'-bit to the message, and then a number of '0'-bits to fill the last message block. Finally, append a 64-bit representation of the number of message blocks in the padded message. This means that both hash functions can digest messages of size at most $2^{64} - 1$ message blocks including the padding itself. (Note the difference between this and most existing padding rules in that the number of message *blocks* is appended rather than the number of message *bits*.)

3.4 Security Analysis

We claim the same security level against collision, preimage and second preimage attacks. We do not know of any (second) preimage attacks with complexity as low as the birthday bound, and we expect the actual security level against these attacks to be higher. However, here we focus on collision resistance.

Collision resistance. Collisions can either be internal or external collisions. In the case of external collisions we believe the number of blank rounds is high enough to rule out any differential with a low number of active bytes. For internal collisions, the complete state which is considerably larger than the output size needs to collide for different input messages. Additionally, we give more evidence that shortcut collision finding methods are unlikely. For Grindahl-256, we give a lower bound for the number of rounds to arrive at an internal collision.

Exhaustive search through all possible input difference patterns and difference propagation patterns shows that Grindahl-256 has the following property.

Property 1. *An internal collision for Grindahl-256 requires at least 6 input rounds. Moreover, any characteristic starting or ending in the extended state with no difference contains at least one round where at least half the extended state bytes (excluding the first column) are active.*

A characteristic leading from a zero-difference state to a collision via different message inputs is given in Appendix D. This characteristic spans 6 rounds, and no shorter characteristic was found in an exhaustive search.

We cannot completely rule out the possibility that high probability characteristics exist, but we believe that a classical differential attack will not be successful.

For Grindahl-512, collisions after 4 input rounds cannot be ruled out with this method. However, due to efficient mixing of the used building blocks, we believe that no low-weight differentials lead to an internal collision.

It remains to be seen if and how methods that improved collision search methods for other hash functions can be adapted to analyse the Grindahl design. Recently developed candidate techniques are message modification as introduced in the cryptanalysis of several members of the MD4 family [44,45,46], neutral bits as used to speed up collision search for SHA-0 [5], internal meet-in-the-middle techniques as used in the analysis of Tiger [32] or the greedy-like approach as pursued in the analysis of SHA-1 in [10]. They all have in common that they exploit the attacker's knowledge of all intermediate states to improve the effectiveness of traditional differential cryptanalysis. We argue that the building blocks used and the limited direct access to the internal state in the Grindahl design make these techniques difficult to apply efficiently.

A potential attack method. An anonymous reviewer proposed such a method tailored to the Grindahl design. In the following, we give a brief sketch of the proposed method. First, a chain of differentials in which in every round the number of active bytes is low must be found. Given such a differential chain, the attack is launched as follows. In the attack, we do not care about actual differences; we only care about whether there is a difference or not.

For a column containing active bytes to satisfy the differential through the MixColumns transformation, some linear constraints must be satisfied. If the actual differences do not matter, these are the only constraints for a full round. Additionally, the fact that new input bytes do not affect some parts of the state for a limited number of rounds can be exploited. This means that often a (small) number of active bytes can be arbitrary (we call these "neutral bytes"), and the remaining ones are fixed ("control bytes"). A neutral byte may later be used as a control byte to ensure that a characteristic is followed. Given that every round introduces up to four neutral bytes, this attack may or may not be applicable. It seems hard to make use of a neutral byte that has been through several S-boxes, since (1) the S-box is non-linear, and so one cannot deterministically select a given output difference by a given input difference, and (2) after a number of S-boxes a message byte affects a large part of the state, and hence it becomes less useful as a control byte. We leave it as an open problem to investigate the feasibility of this attack.

4   Designing Secure Compression Functions

Any proposal following the Grindahl design strategy of hash functions accepting variable length inputs can easily be turned into a compression function for fixed length inputs. Let H be an instance of the Grindahl hash functions with initial state s_0, where padding is omitted. Hence, H accepts only messages of size an integer multiple of b in bits. Let the output size of H be n. Based on H, define a compression function $h : \{0,1\}^{tb} \to \{0,1\}^n$, where t is an integer greater than n/b, as $h(x) = H(x)$.

Note that here, the compression function is defined as taking only one input, whereas usually one thinks of a compression function as taking two inputs, a chaining value and a message block. We think it makes good sense to treat the two inputs as one: in most modes of operation of the compression function an attacker is expected to have the same control over the chaining value as over the message. In practice we would suggest that the two inputs are simply concatenated, and hence if we instead describe the compression function as $h : \{0,1\}^n \times \{0,1\}^{tb-n} \to \{0,1\}^n$, then we would define h as $h(c,x) = H(c\|x)$.

The choice of t implies a certain trade-off between speed and security. By decreasing t security is increased, but speed is reduced since in effect the rate of blank rounds to input rounds increases.

If an additional input (e.g., a salt, a key or a counter) to the compression function is required, then we suggest that this is prepended to the chaining/message input. This way, many of the newly proposed modes of operation which turn a secure compression function into a secure hash function are directly applicable to our proposal. We now describe instances of our design strategy acting as a compression function to be used with modes like Merkle-Damgård [18,37], EMD [2], randomized hashing [28], HAIFA [7] etc.

Compression function mode for concrete proposals. We suggest compression function modes for both Grindahl-256 and Grindahl-512 by letting t above be equal to $40+s$. Here s denotes the number of input blocks used for an additional input, with the possibility of $s = 0$. Hence, the compression functions both take $(40 + s)\nu_{mb}$-byte inputs. This corresponds to $1280+32s$ bits and $2560+64s$ bits for Grindahl-256 and Grindahl-512, respectively. Of these, respectively 1024 and 2048 bits form the message input, and the rest is reserved for chaining input and, if applicable, a salt/key/counter.

5 Implementation

Implementations of Grindahl can directly inherit most of the extensive research done to optimise implementations of the AES block cipher on different platforms. Also side-channel attacks might be an issue if a hash function is used to process secret key material, be it as a key-derivation-function (KDF) or as a hash-based message authentication code (MAC). Here we refer to extensive work done to protect implementations of the AES against these kinds of attacks, e.g. a very fast bit-sliced AES implementation immune to timing attacks [36].

5.1 Software Performance

One usually defines the rate of a hash function based on a block cipher as the number of blocks that are processed for each block encryption. We may do the same here, except that we have to take into account the extended state size, and the fact that ν_{mb} bytes are processed per round, whereas in AES-128, 16 bytes are processed for every 10 rounds. Hence, the rate of a Grindahl instance is $\frac{10\nu_{mb}}{\nu_{rw}\nu_{cl}-\nu_{mb}}$. Here we also take into account that the ν_{mb} bytes that are overwritten by the following message block do not have to be computed. As an example, the rate of both Grindahl-256 and Grindahl-512 is 5/6.

In an optimised software implementation of Grindahl-n on a 32-bit platform, $n/64$ tables of 256 32-bit words are needed. In the discussion (Section 5.4) on memory requirements this is not taken into account as the need for these tables is a consequence of the optimisation only.

Implementation report. Grindahl-256 has been implemented in C on a 32-bit Pentium 4 processor. It runs at about 32 cycles/byte. This might be compared with the Rijndael-128 implementation from the Crypto++ [13] package, which, according to the website, performs at about 33 cycles/byte on a Pentium 4. As expected, performance is similar. For additional comparison, on the same platform SHA-256 is benchmarked at about 45 cycles/byte [13]. As with Rijndael, we expect that hand-optimised implementations will improve performance by a

Table 1. Needed working memory in bits for different hash functions

128-bit security		256-bit security	
name	memory	name	memory
Grindahl-256	416	Grindahl-512	832
SHA-256	1024	SHA-512	2048
RadioGatún-14	812	RadioGatún-27	1566
FORK-256	1280	Whirlpool	1536
LASH-256	1536	LASH-512	3072

factor up to about 2. Grindahl-512 is more suited for 64-bit architectures, and here we expect its speed to be similar to Grindahl-256 on a 32-bit architecture.

Program code and test vectors. The interested reader may find C implementations of and test vectors for Grindahl-256 and Grindahl-512 at [27].

5.2 On Hardware Implementations

For passively powered designs, the number of active registers/logic per clock cycle should be small. In contrast to the MD4 family, the Grindahl hash function design allows for hardware designs with a small data path width without penalties. Also, compared to the MD4 family and other proposals a smaller number of registers is needed in the Grindahl design, which allows for low-cost and low-power implementations. In addition, various trade-offs towards high speed are possible, utilising the many implementation options already pioneered for the AES and benefiting from the smaller number of needed registers.

Regarding low-cost hardware implementations, based on [21] we estimate area requirements to about 5-6,000 gate equivalents for Grindahl-256. This compares favourably with the smallest known SHA-256 implementation [20], which requires more than 10,000 gate equivalents.

5.3 On Hashing Small Messages

The blank rounds add fixed costs even for very small messages. However, in absolute terms, for both Grindahl-256 and Grindahl-512 it is equivalent to only 32 bytes of additional padding. Note that members of the SHA-2 family operate on input blocks of size 64 or 128 bytes. Hence, given the much smaller input block size of 4 (or 8) bytes, small messages are still handled more efficiently.

5.4 Memory Requirements

Due to the small message blocks, the needed working memory for Grindahl-256 and Grindahl-512 is small. This certainly suits implementations in constrained environments. It is interesting to note that implementations of the Grindahl strategy need only $b + m$ bits of working memory. Implementations of the MD4 family, on the other hand, require $b + 2m$ bits of memory, where usually $b = 512$ and m is the same as the output size. The reason for the difference is the feed-forward of the initial state which counters meet-in-the-middle attacks. However, even if this feed-forward operation would be omitted, the general picture would not change.

We see the low memory requirements as an important feature of our proposal, and in Table 1 we give a comparison with other hash functions [1,3,4,24,29] claiming a comparable security level against collision search attacks. Note that we do not consider memory needed to store temporary variables or constants.

6 Conclusion

We proposed the Grindahl hash functions which overcome several identified weaknesses of the commonly used MD4 family of hash functions. The identified weaknesses are addressed on two layers. By using the well analysed building blocks of Rijndael we obtain a design for which we claim that no efficient differential collision structures exist. In addition, we limit access to the internal state with the idea to thwart advanced techniques that improve effectiveness for collision search methods.

We proposed two instantiations of the Grindahl hash collection, Grindahl-256 and Grindahl-512. The claimed security level with respect to collision, preimage and second preimage attacks is 2^{128} and 2^{256}, respectively. Grindahl-256 performs at about the same rate as AES-128 without the key schedule, and on a 64-bit platform we expect that implementations of Grindahl-512 can achieve similar speeds.

Other intriguing implementation aspects of the proposals are very low working memory requirements which aid implementations in constrained environments, as well as the efficient handling of small messages.

Acknowledgments. The authors would like to thank Charanjit Jutla and the anonymous reviewers for many helpful comments.

References

1. Barreto, P.S.L.M., Rijmen, V.: The Whirlpool hashing function (May 2003), available at https://www.cosic.esat.kuleuven.be/nessie/tweaks.html
2. Bellare, M., Ristenpart, T.: Multi-Property-Preserving Hash Domain Extension and the EMD Transform. In: Lai, X., Chen, K. (eds.) ASIACRYPT 2006. LNCS, vol. 4284, pp. 299–314. Springer, Heidelberg (2006)
3. Bentahar, K., Page, D., Saarinen, M.-J.O., Silverman, J.H., Smart, N.: LASH. Presented at Second Cryptographic Hash Workshop, Santa Barbara (August 24-25, 2006)
4. Bertoni, G., Daemen, J., Peeters, M., Van Assche, G.: RADIOGATÚN, a belt-and-mill hash function. Presented at Second Cryptographic Hash Workshop, Santa Barbara (August 24-25, 2006), See http://radiogatun.noekeon.org/

5. Biham, E., Chen, R.: Near-Collisions of SHA-0. In: Franklin, M. (ed.) CRYPTO 2004. LNCS, vol. 3152. Springer, Heidelberg, pp. 290–305 (2004)
6. Biham, E., Chen, R., Joux, A., Carribault, P., Lemuet, C., Jalby, W.: Collisions of SHA-0 and Reduced SHA-1. In: Cramer, R.J.F. (ed.) EUROCRYPT 2005. LNCS, vol. 3494, pp. 36–57. Springer, Heidelberg (2005)
7. Biham, E., Dunkelman, O.: A Framework for Iterative Hash Functions – HAIFA. Presented at Second Cryptographic Hash Workshop, Santa Barbara (August 24-25 2006)
8. Biryukov, A.: The Design of a Stream Cipher LEX. In: Selected Areas in Cryptography, 2006, LNCS. Springer, Heidelberg (to appear)
9. Brassard, G. (ed.): CRYPTO 1989. LNCS, vol. 435, pp. 20–24. Springer, Heidelberg (1990)
10. Cannière, C.D., Rechberger, C.: Finding SHA-1 Characteristics: General Results and Applications. In: Lai, X., Chen, K. (eds.) ASIACRYPT 2006. LNCS, vol. 4284, pp. 1–20. Springer, Heidelberg (2006)
11. Chabaud, F., Joux, A.: Differential Collisions in SHA-0. In: Krawczyk, H. (ed.) CRYPTO 1998. LNCS, vol. 1462, pp. 56–71. Springer, Heidelberg (1998)
12. Cramer, R.J.F. (ed.): EUROCRYPT 2005. LNCS, vol. 3494. Springer, Heidelberg (2005)
13. The Crypto++ website (2007), http://www.cryptopp.com/
14. Daemen, J.: Cipher and hash function design strategies based on linear and differential cryptanalysis. PhD thesis, Katholieke Universiteit Leuven (March 1995)
15. Daemen, J., Clapp, C.S.K.: Fast Hashing and Stream Encryption with PANAMA. In: Vaudenay, S. (ed.) FSE 1998. LNCS, vol. 1372, pp. 60–74. Springer, Heidelberg (1998)
16. Daemen, J., Rijmen, V.: The Block Cipher Rijndael. In: Schneier, B., Quisquater, J.-J. (eds.) CARDIS 1998. LNCS, vol. 1820, pp. 277–284. Springer, Heidelberg (2000)
17. Daemen, J., Rijmen, V.: A New MAC Construction ALRED and a Specific Instance ALPHA-MAC. In: Gilbert, H., Handschuh, H. (eds.) FSE 2005. LNCS, vol. 3557, pp. 1–17. Springer, Heidelberg (2005)
18. Damgård, I.: A Design Principle for Hash Functions. In: Brassard, G. (ed.) CRYPTO 1989. LNCS, vol. 435, pp. 416–427. Springer, Heidelberg (1990)
19. Dobbertin, H.: Cryptanalysis of MD4. Journal of Cryptology 11(4), 253–271 (1998)
20. Feldhofer, M., Rechberger, C.: A Case Against Currently Used Hash Functions in RFID Protocols. Presented at the Workshop on RFID Security (2006)
21. Feldhofer, M., Wolkerstorfer, J., Rijmen, V.: AES Implementation on a Grain of Sand. IEE Proceedings on Information Security 152(1), 13–20 (2005)
22. Ferguson, N., Schneier, B.: Practical Cryptography. Wiley Publishing, Chichester (2003)
23. FIPS 180-1, Secure Hash Standard: Federal Information Processing Standards Publication 180-1, U.S. Department of Commerce/NIST, National Technical Information Service, Springfield, Virginia, Supersedes FIPS 180 (April 1995)
24. FIPS 180-2, Secure Hash Standard. Federal Information Processing Standards Publication 180-2, U.S. Department of Commerce/NIST, National Technical Information Service, Springfield, Virginia, Supersedes FIPS 180 and FIPS 180-1 (August 2002)
25. FIPS 197, Advanced Encryption Standard (AES): Federal Information Processing Standards Publication 197, U.S. Department of Commerce/NIST, National Technical Information Service, Springfield, Virginia (November 2001)

26. Franklin, M. (ed.): CRYPTO 2004. LNCS, vol. 3152. Springer, Heidelberg (2004)
27. The Grindahl web page (2007), http://www.ramkilde.com/grindahl
28. Halevi, S., Krawczyk, H.: Strengthening Digital Signatures via Randomized Hashing. In: Dwork, C. (ed.) CRYPTO 2006. LNCS, vol. 4117, pp. 41–59. Springer, Heidelberg (2006)
29. Hong, D., Chang, D., Sung, J., Lee, S., Hong, S., Lee, J., Moon, D., Chee, S.: A New Dedicated 256-Bit Hash Function: FORK-256. In: Robshaw, M. (ed.) FSE 2006. LNCS, vol. 4047, pp. 195–209. Springer, Heidelberg (2006)
30. Joux, A.: Multicollisions in Iterated Hash Functions. In: Franklin, M. (ed.) CRYPTO 2004. LNCS, vol. 3152, pp. 306–316. Springer, Heidelberg (2004)
31. Kelsey, J., Kohno, T.: Herding Hash Functions and the Nostradamus Attack. In: Vaudenay, S. (ed.) EUROCRYPT 2006. LNCS, vol. 4004, pp. 183–200. Springer, Heidelberg (2006)
32. Kelsey, J., Lucks, S.: Collisions and Near-Collisions for Reduced-Round Tiger. In: Robshaw, M. (ed.) FSE 2006. LNCS, vol. 4047, pp. 111–125. Springer, Heidelberg (2006)
33. Kelsey, J., Schneier, B.: Second Preimages on n-bit Hash Functions for Much Less than 2^n Work. In: Cramer, R.J.F. (ed.) EUROCRYPT 2005. LNCS, vol. 3494, pp. 474–490. Springer, Heidelberg (2005)
34. Lai, X., Chen, K. (eds.): ASIACRYPT 2006. LNCS, vol. 4284. Springer, Heidelberg (2006)
35. MacWilliams, F.J., Sloane, N.J.A.: The Theory of Error-Correcting Codes. North-Holland, Amsterdam (1977)
36. Matsui, M.: How Far Can We Go on the x64 Processors? In: Robshaw, M. (ed.) FSE 2006. LNCS, vol. 4047, pp. 341–358. Springer, Heidelberg (2006)
37. Merkle, R.C.: One Way Hash Functions and DES. In: Brassard, G. (ed.) CRYPTO 1989. LNCS, vol. 435, pp. 428–446. Springer, Heidelberg (1990)
38. Merkle, R.C.: A Fast Software One-Way Hash Function. Journal of Cryptology 3(1), 43–58 (1990)
39. Meyer, C.H., Schilling, M.: Secure program load with Manipulation Detection Code. In: Proceedings of the 6th Worldwide Congress on Computer and Communications Security and Protection (SECURICOM'88), pp. 111–130 (1988)
40. Preneel, B.: Analysis and Design of Cryptographic Hash Functions. PhD thesis, Katholieke Universiteit Leuven (January 1993)
41. RFC 1321: The MD5 Message-Digest Algorithm. Internet Request for Comments 1321, R. Rivest (April 1992)
42. Rijmen, V., Oswald, E.: Update on SHA-1. In: Menezes, A.J. (ed.) CT-RSA 2005. LNCS, vol. 3376, pp. 58–71. Springer, Heidelberg (2005)
43. Robshaw, M. (ed.): FSE 2006 (Revised Selected Papers). LNCS, vol. 4047. Springer, Heidelberg (2006)
44. Wang, X., Lai, X., Feng, D., Chen, H., Yu, X.: Cryptanalysis of the Hash Functions MD4 and RIPEMD. In: Cramer, R.J.F. (ed.) EUROCRYPT 2005. LNCS, vol. 3494, pp. 1–18. Springer, Heidelberg (2005)
45. Wang, X., Yin, Y.L., Yu, H.: Finding Collisions in the Full SHA-1. In: Shoup, V. (ed.) CRYPTO 2005. LNCS, vol. 3621, pp. 17–36. Springer, Heidelberg (2005)
46. Wang, X., Yu, H.: How to Break MD5 and Other Hash Functions. In: Cramer, R.J.F. (ed.) EUROCRYPT 2005. LNCS, vol. 3494, pp. 19–35. Springer, Heidelberg (2005)

A   Resistance of Concatenate-Permute-Truncate to Known Attacks

The well-known Merkle-Damgård construction [18,37] is a construction method for hash functions based on an underlying compression function $f : \{0,1\}^n \times \{0,1\}^b \rightarrow \{0,1\}^n$. The output size of the hash function is n. The main differences in the "Concatenate-Permute-Truncate" design are

- that we allow $m > n$, and hence attacking the internals of the hash function may be different (and hopefully harder) than attacking the hash function itself, and
- that we allow an output transformation, the blank rounds.

If $m = n$ and $\nu_{\mathrm{br}} = 0$, then the design can be seen as a special case of the Merkle-Damgård construction: we iterate "as usual" over a compression function of a certain kind. However, the possibility of varying m and ν_{br} makes it possible to obtain different properties for our design than the properties of hash functions based on the Merkle-Damgård construction.

In the following we describe some known attacks on the Merkle-Damgård construction, and we argue whether or not our design protects against these attacks, and, if applicable, how resistance to these attacks depends on the parameters n, m and ν_{br}. The analysis will assume that P is a black-box permutation, i.e. that the internals of P are unknown.

A.1   The Length-Extension Attack

Let H be a hash function based on the Merkle-Damgård construction. Then H is susceptible to the so-called length-extension attack [22]. Given a collision, i.e. two messages d and d', with $|d| = |d'|$, such that $H(d) = H(d')$, then for any suffix x it holds also that $H(d\|x) = H(d'\|x)$.

This attack can be applied to the design described above only if the collision occurs before the blank rounds. If $m > n$ and we only consider the birthday attack, then an internal collision is harder to find than a collision for the full hash function.

A related property of the Merkle-Damgård construction is that if the length of an unknown message d, and hence the padding p of d, is known, and also $H(d)$ is known, then $H(d\|p\|x)$ can be computed for any suffix x. This attack is particularly a threat in some schemes that use a hash function for message authentication.

The attack can be mounted on a hash function following our design only if the attacker correctly guesses the $b + m - n$ bits that are truncated away. If he does so, he can go backwards through the blank rounds and obtain the extended state after the processing of the last message block.

A.2   Multi-collisions

An efficient method for constructing multi-collisions was described by A. Joux in 2004 [30]. A multi-collision is a set of (at least two) messages that all have the same hash. The attack of Joux has complexity $t2^{n/2}$ to find a 2^t-way multi-collision for an n-bit hash function in the Merkle-Damgård construction. In our design, the complexity would be $t2^{m/2}$. This is to be compared with a brute-force search for multi-collisions, which for even a modest t gets very close to 2^n. Hence, if one wants full resistance against the Joux multi-collision attack, one should choose $m \geq 2n$.

A.3   The Herding Attack

The herding attack [31] by J. Kelsey and T. Kohno shows another slight weakness of the Merkle-Damgård construction. Here, a binary tree of collisions is used to form a hash result h, which an attacker publishes. Subsequently he chooses a message, finds a message linking to one of the leaves of the binary tree, and then he has a complete, partially chosen message d with the hash h.

The complexity of the simplest version of this attack in the Merkle-Damgård construction is about $2^{(2n-5)/3}$. In our design, n can be replaced by m, and so with $m \geq (3n+5)/2$, the complexity of the attack is the same as for a preimage, which is what one would expect from a random hash function.

There is another important version of the attack for which the complexity decreases with increased size of the message d. If the size of d is about 2^r, then the complexity is about $2^{(2n-5)/3-r}$. Hence, for a given upper limit on the size of messages, m must be chosen accordingly if one wants to completely protect against this kind of attack.

A.4   Second Preimage Attack

There is a second preimage attack [33] by J. Kelsey and B. Schneier on the Merkle-Damgård construction. This attack requires finding an expandable message, meaning a set of messages of varying sizes such that all these messages collide internally in the hash function, given some fixed initial value. To find this expandable message one either makes use of fixed points or the ability to find (internal) collisions between a one-block message and a t-block message for varying values of t. Finally, the complexity of the attack depends on the complexity of finding the expandable message, and on the length of the target message, i.e. the message for which one tries to find a second message with the same hash. This complexity amounts to roughly $2^{n/2}+2^{n-k}$, where k is the number of blocks in the target message. In our design, n can be replaced by m, and hence with no upper limit on the message size, m should be at least $2n$ for this attack to have the same complexity as a brute-force search. If the target message can have length at most 2^r blocks, then $m \geq \min(2n, n+r)$ is required.

B   Reduced Variants

We now suggest some methods of reducing the cryptographic strength of the two proposals Grindahl-256 and Grindahl-512, without reducing the output size. The methods can be combined, but some reductions rule out others, or modify the way that others can be applied.

- Increase the size of each message block to more than one column.
- Reduce the number of columns in the extended state. The number of columns should be at least 8 plus twice the number of columns that are overwritten by the message block. Otherwise, as described in Section 2.4, birthday attacks are simplified.
- Reduce the number of blank rounds.

C   A Method for Choosing Rotation Constants

We suggest the following method for choosing rotation constants in the Grindahl hash collection, given a particular geometry of the extended state. Let R_1 be the set of ν_{rw}-tuples of rotation constants that ensure optimal diffusion, i.e. all state bytes depend on all message bytes as quickly as possible. Let $R_2 \subseteq R_1$ be the subset of R_1 of rotation constants that ensure optimal diffusion in the blank rounds, i.e. when no extended state bytes are overwritten by message bytes.

Now let $f_j(d_i)$ be the number of state bytes that message byte d_i affects after j rounds. Let μ be the number of rounds needed for every state byte to be affected by every message byte. Now let $R_3 \subseteq R_2$ be the subset of R_2 of rotation constants for which the sum

$$\sum_{j=1}^{\mu-1} \sum_{i=0}^{\nu_{mb}-1} f_j(d_i)$$

is maximal. Sort R_3 lexicographically, i.e. such that (r_1, r_2, r_3, r_4) comes before (s_1, s_2, s_3, s_4) if and only if $r_1 < s_1$, or $r_1 = s_1$ and $r_2 < s_2$, or $(r_1, r_2) = (s_1, s_2)$ and $r_3 < s_3$, or $(r_1, r_2, r_3) = (s_1, s_2, s_3)$ and $r_4 < s_4$. Choose as rotation constants the first tuple in the sorted R_3.

D   A Characteristic Leading to a Collision in Grindahl-256

Below is given a characteristic that leads from a zero-difference state to a zero-difference state (a collision) via message inputs containing differences. Exact differences are not given, instead a single bit for each byte is given stating whether or not there is a difference on that particular byte.

Evidently, after 6 rounds the state contains no difference (the only difference in the extended state is in the first column). Hence, this characteristic spans 6 rounds, but an additional message block with no difference (e.g., a padding block) is needed before the blank rounds in order for the difference to disappear entirely.

Round no.	Initial state	Message	Mesg. input →	ShiftRows →	MixColumns →
1	0000000000000	1111	1000000000000	0100000000000	0110100000100
	0000000000000		1000000000000	0010000000000	0110100000100
	0000000000000		1000000000000	0000100000000	0110100000100
	0000000000000		1000000000000	0000000000100	0110100000100
2	0110100000100	0111	0110100000100	0011010000010	0111101110111
	0110100000100		1110100000100	0011101000001	0111011110111
	0110100000100		1110100000100	0100111010000	0101111110111
	0110100000100		1110100000100	0100000100111	0111111110111
3	0111101110111	1001	1111101110111	1111110111011	1100010010010
	0111011110111		0111011110111	1101110111101	1000100100101
	0101111110111		0101111110111	0111010111111	1010010010110
	0111111110111		1111111110111	1111110111111	1011110001001
4	1100010010010	1100	1100010010010	0110001001001	1010000000000
	1000100100101		1000100100101	0110001001001	1100000000000
	1010010010110		0010010010110	0110001001001	1000000001001
	1011110001001		0011110001001	1110001001001	1000001000000
5	1010000000000	0000	0010000000000	0001000000000	1000000000000
	1100000000000		0100000000000	0001000000000	1000000000000
	1000000001001		0000000001001	1001000000000	1000000000000
	1000001000000		0000001000000	0001000000000	1001000000000
6	1000000000000	0000	0000000000000	0000000000000	1000000000000
	1000000000000		0000000000000	0000000000000	1000000000000
	1000000000000		0000000000000	0000000000000	1000000000000
	1001000000000		0001000000000	1000000000000	1000000000000
7	1000000000000	0000	0000000000000	0000000000000	0000000000000
	1000000000000		0000000000000	0000000000000	0000000000000
	1000000000000		0000000000000	0000000000000	0000000000000
	1000000000000		0000000000000	0000000000000	0000000000000

Overtaking VEST

Antoine Joux[1,2] and Jean-René Reinhard[3]

[1] DGA
[2] Université de Versailles St-Quentin-en-Yvelines, PRISM
45, avenue des États-Unis, 78035 Versailles Cedex, France
antoine.joux@m4x.org
[3] DCSSI Crypto Lab
51, Boulevard de La Tour-Maubourg
75700 Paris 07 SP, France
jean-rene.reinhard@m4x.org

Abstract. VEST is a set of four stream cipher families submitted by
S. O'Neil, B. Gittins and H. Landman to the eSTREAM call for stream
cipher proposals of the European project ECRYPT. The state of any
family member is made of three components: a counter, a counter diffusor
and a core accumulator. We show that collisions can be found in the
counter during the IV Setup. Moreover they can be combined with a
collision in the linear counter diffusor to form collisions on the whole
cipher. As a consequence, it is possible to retrieve 53 bits of the keyed
state of the stream cipher by performing a chosen IV attack. For the
default member of a VEST family, we present a "long" IV attack which
requires $2^{22.24}$ IV setups, and a "short" IV attack which requires $2^{28.73}$
IV setups on average. The 53 bits retrieved can be used to reduce the
complexity of the exhaustive key search. The chosen IV attack can be
turned into a chosen message attack on a MAC based on VEST.

Keywords: Stream cipher, inner collision, chosen IV attack.

1 Introduction

VEST [8] is a set of four stream cipher families proposed to the eSTREAM
project [6] by S. O'Neil, B. Gittins and H. Landman . VEST-v, with $v \in
\{4, 8, 16, 32\}$, is a family of stream ciphers with expected security respectively
2^{80}, 2^{128}, 2^{160} and 2^{256}, and output rate v bits by clock cycle. All families share
the same design. Only the sizes of the components change to meet the target
security. There also is a selection algorithm which given a parameter called the
family key outputs a specific member of a VEST family.

Recently, VEST specifications have been updated [9]. Compared to the earlier specification [8], changes include some modifications in the parameters used
and also the definition of additional modes, that turn VEST in a MAC or an
authenticated encryption scheme.

In this paper, we point out basic weaknesses of VEST components. The weaknesses can be used to create inner collisions in the algorithm, in a way similar

A. Biryukov (Ed.): FSE 2007, LNCS 4593, pp. 58–72, 2007.

to local collisions in hash function of the SHA family [4]. As a consequence we are able to mount a chosen IV attack against VEST stream cipher that recovers 53 bits of the keyed state. This information enables us to reduce by 53 bits the complexity of an exhaustive key search. The chosen IV attack on VEST stream cipher can also be used as an existential forgery attack against the VEST MAC.

In section 2, we give a description of VEST components and of their internal modes of operation. Then we describe in section 3 the basic weaknesses we found on the counter and linear counter diffusor components. In section 4, we describe two efficient chosen IVs attacks that recover 53 bits of the keyed state of the cipher. In section 5, we use these 53 bits to greatly speed-up exhaustive key search. Finally, in section 6, we describe briefly how to turn the attacks of section 4 into existential forgery attacks against VEST in MAC mode.

2   Description of VEST

Members of the VEST stream cipher families are made of four components:

- a set of 16 non linear feedback shift registers, called *counter*,
- a linear counter diffusor,
- an accumulator,
- a memory-less linear output filter.

There are three main internal modes of operation:

- The Key Setup mode which introduces the key into the cipher state. The state of the cipher is first set to all 0's, then the key is introduced and produces a keyed state. In this mode, the cipher does not output any data.
- The IV Setup mode which introduces an IV into a keyed state. This mode also has no output data.
- The Keystream generation mode which produces a stream of pseudo random bits.

We review in the following the four components of a VEST stream cipher and the Key Setup and IV Setup mechanisms.

2.1   Counter

The counter, a set of 16 registers c_i, $(0 \leq i < 16)$, is the autonomous part of the stream cipher in Keystream generation mode. Each register is a non linear feedback shift register (NLFSR) of size $w = 10$ or 11 bits. As in [8,9], we note $c_{i\,j}^r$ the value of bit j of register i at step r. Their update function is described in Fig. 1.

Each register is updated independently. There are two register modes of operation. While in register counter mode, the registers are autonomously updated. In register keying mode, the update function of a register is disturbed by one bit at each clock cycle. The functions g_i are non-linear and chosen such that the graphs of the registers update functions are made of two cycles of approximately

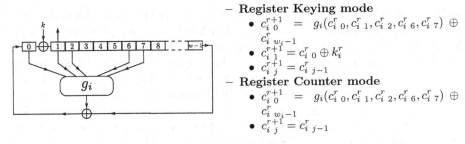

– Register Keying mode

- $c_{i\ 0}^{r+1} = g_i(c_{i\ 0}^r, c_{i\ 1}^r, c_{i\ 2}^r, c_{i\ 6}^r, c_{i\ 7}^r) \oplus c_{i\ w_i-1}^r$
- $c_{i\ 1}^{r+1} = c_{i\ 0}^r \oplus k_i^r$
- $c_{i\ j}^{r+1} = c_{i\ j-1}^r$

– Register Counter mode

- $c_{i\ 0}^{r+1} = g_i(c_{i\ 0}^r, c_{i\ 1}^r, c_{i\ 2}^r, c_{i\ 6}^r, c_{i\ 7}^r) \oplus c_{i\ w_i-1}^r$
- $c_{i\ j}^{r+1} = c_{i\ j-1}^r$

Fig. 1. Register update

same length, the length of all cycles being pairwise relatively prime. Thus, a minimum period of evolution of the set of registers can be guaranteed. 32 non linear functions meeting this requirement are specified in [9, Appendix F], 16 for registers of length 11 and 16 for registers of length 10. For a given member of a family, every register is associated to a non linear function specified for its length. Every cycle, bit 1 is extracted from each of the 16 registers.

2.2 Linear Counter Diffusor

The linear counter diffusor is a 10–bit value used to disturb the core accumulator. Every cycle, the linear counter diffusor is updated linearly with the 16–bit output from the counter. As in [8,9], we note d_j^r the value at step r of bit j of the linear counter diffusor. We note

$$
D^{(r)} = \begin{pmatrix} d_0^r \\ d_1^r \\ \vdots \\ d_9^r \end{pmatrix} \quad C^{(r)} = \begin{pmatrix} c_{0\ 1}^r \\ c_{1\ 1}^r \\ \vdots \\ c_{15\ 1}^r \end{pmatrix}.
$$

The linear counter diffusor update function can be written as

$$
D^{(r+1)} = A \cdot D^{(r)} \oplus M \cdot C^{(r)} \oplus B,
$$

with

$$
A = \begin{bmatrix}
0&1&0&0&0&0&0&0&0&0 \\
0&0&1&0&0&0&0&0&0&0 \\
0&0&0&1&0&0&0&0&0&0 \\
0&0&0&0&1&0&0&0&0&0 \\
0&0&0&0&0&1&0&0&0&0 \\
0&0&0&0&0&0&1&0&0&0 \\
0&0&0&0&0&0&0&1&0&0 \\
0&0&0&0&0&0&0&0&1&0 \\
1&0&0&0&0&0&0&0&0&0 \\
0&0&0&0&0&0&0&0&0&1
\end{bmatrix}
\quad
M = \begin{bmatrix}
0&1&0&0&1&1&0&0&0&0&0&1&0&1&0&0 \\
1&0&1&0&0&0&1&0&1&0&0&0&0&0&1&0 \\
0&0&0&1&1&0&0&1&0&0&1&0&0&0&0&1 \\
1&0&0&1&0&1&0&0&0&1&0&0&1&0&0&0 \\
0&1&0&0&1&0&1&0&0&0&0&0&1&0&0&1 \\
1&0&0&0&0&0&0&1&0&1&0&0&0&1&1&0 \\
0&1&0&0&0&0&0&0&1&0&0&1&0&0&1&1 \\
0&0&1&0&0&1&1&0&0&0&1&0&1&0&0&0 \\
1&0&0&1&0&0&0&1&1&1&0&0&0&0&0&0 \\
0&0&0&0&0&0&0&0&1&0&1&0&1&1&0&1
\end{bmatrix}
\quad
B = \begin{pmatrix} 1 \\ 0 \\ 0 \\ 0 \\ 1 \\ 0 \\ 0 \\ 1 \\ 1 \\ 1 \end{pmatrix}.
$$

2.3 Accumulator

The remainder of the state of a VEST stream-cipher is an accumulator. Every cycle the accumulator state goes through a substitution phase and a permutation phase. Then the output of the counter diffusor is XOR-ed to the first 10 bits of the accumulator state. For a full description of the accumulator, we refer to [9].

2.4 Output Combiner

At each clock cycle, depending on its current state the cipher outputs either 0 or n bits. Each output bit is computed by taking the exclusive-or of 6 bits taken from the accumulator state. The number of output bits is at least 18 times smaller than the size of the accumulator. The design of the core of the algorithm, that is to say the accumulator and this output filter follows the same design strategy as the LEX stream cipher, another candidate to the eSTREAM project [1,2]. The idea is to extract a small part of a state which is updated by one round of a block cipher, the round-key being provided by an autonomous component.

2.5 Key Setup Mode

The Key Setup takes as input a F–bit key K, where F is a multiple of 8, and enters it in the cipher state. At the beginning of the Key Setup, all bits of the state are set to 1. At the end of the Key Setup, the value of the state of the cipher is called the keyed state. During Key Setup, no output is produced by the cipher. The linear counter diffusor and the accumulator work as described above. The registers of the counter go through two phases:

- The first phase of the keysetup introduces K in the registers. First the key is prepended with fifteen 0's, and appended with a single 1 followed by fifteen 0's, creating a key K'. During $F + 16$ steps the registers operate in register keying mode. At each step r, the bits $K'_r, \ldots, K'_{r+15}$ are used to disturb the evolution of registers $0, \ldots, 15$ respectively.
- The second phase of the Key Setup introduces a 8–bit constant in the first 8 registers, and mixes the state of the cipher. During the introduction of the constant, the first 8 registers are in keying mode, register i being disturbed by bit i of the constant, and the last 8 registers are in counter mode. Then the state of the cipher is mixed by going through 31 steps during which all the registers are in counter mode.

The keyed state can be stored and reloaded later into the state of the cipher, if one wants for example to speed up IV setups using the same key. The keyed state is equivalent to the key introduced.

2.6 IV Setup Mode

The IV Setup may be applied after the Key Setup. It takes as input an IV of length W bits, with W a multiple of 8. As for the Key Setup, no output is produced during IV Setup and the accumulator and counter diffusor work as described above. The counter goes through 2 phases:

- During the first phase of the IV Setup, the IV is introduced into the counter. This phase lasts $W/8$ steps. Note that the IV is not introduced in all registers as the key. Instead it only affects 8 registers as the constant introduced in the second phase of the keying process. As a consequence each bit of IV affects a single register. At each clock cycle, one byte of the IV is introduced.
- The second phase is identical to the Key Setup second phase with a different value of the constant.

3   Basic Weaknesses of VEST Components

In this section, we identify two weaknesses of VEST stream ciphers. They concern the differential behaviour of the NLFSRs in keying mode and of the counter diffusor.

3.1   Differential Characteristics of the Registers

During the first phase of IV Setup, the update function of register i, $0 \leq i \leq 7$ can be viewed as a function

$$f_i : \{0,1\}^{w_i} \times \{0,1\}^n \to \{0,1\}^{w_i} \times \{0,1\}^n,$$

which modifies a state with an input IV of length n. The output of the function is the modified state and an output value of length n. Studying the differential behaviour of this function with respect to its second input, we find some kind of imbalance.

The differential patterns relevant within the context of the IV Setup are

$$0 \times \Delta \to 0 \times \alpha :$$

starting from a common value (for example the keyed register state) we introduce a pair of IVs with difference Δ and look for a collision on the register state after the first phase of the IV Setup and a fixed difference α on the output. For an IV pair, we call colliding states the states starting from which we get a collision and the expected difference α on the register output after processing each IV of the pair.

Differential behaviour over one step. In order to understand the source of the imbalance, let us consider the effect of a single bit difference in the register state on one step of the IV Setup. We can distinguish three cases:

- The difference is at position $w-1$: the update moves the difference to position 0,
- The difference is at position $j \notin \{0,1,2,6,7\}$: the update shifts the difference to position $j+1$

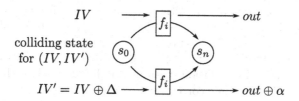

Fig. 2. Differential pattern and colliding states

- The difference is at position $j \in \{0, 1, 2, 6, 7\}$: the update shifts the difference to position $j + 1$ and, depending on the non linear function, may also create a difference at position 0.

We still have to take into account the IV bit introduction at this step. After the computation of the feedback and the shift of the register, but before the output at this step, one bit of IV is XOR-ed at position 1. In our differential setting this bit can be used either to introduce a new difference in position 1 or to correct a difference that was present at position 0 at the end of the previous step.

At the end of each step, the output is the bit in position 1 of the register state.

Local Collisions. We now adopt a strategy very similar to the local collisions used in [4] to attack SHA-0. The key idea is to introduce a single difference in the register state and control its propagation until it vanishes. We first consider the linear part of the update function. We then take into account the non-linear part.

At step 0 we introduce a difference through different IV bits. Thus, after step 0 there is a difference in position 1 of the register. The linear part of the update function consists solely of a rotation of the register bits. After step 1 the difference is in position 2, then after step 2 in position 3 ... and after step $w - 1$ the difference is in position 0. At step w it comes back in position 1, where we can cancel it by using a secondary difference on the IV pair. Thus when the non-linear feedback function is not used, it takes $w + 1$ steps to introduce and then cancel a difference. Looking at the output difference α, taken from bit 1, we see that it consists in a single 1 followed by w 0's. In the sequel, we show how to build collisions following the same pattern, $w + 1$ steps and same value α.

We now add the non linear function (NLF). The collision described above does no longer occur all the time. Indeed, when the NLF is active, i.e. when there is a difference on at least one of its input, there may be a difference on its output. Heuristically, this happens with probability $1/2$. Thus after a step during which the NLF is active, there may be an additional difference in position 0. To prevent the propagation of this difference, we use the IV introduction in position 1 at the next step to cancel it. Combining all the IV bit differences on the $w + 1$ steps we get a corrective pattern. Note that with IV differences on $w + 1$ steps, it is not possible to correct additional differences produced during step w.

Due to the propagation of the initial difference through every position of the register, the number of active steps is at least 5 : steps 1, 2, 6, 7 and w. Heuristically, after each active step there is probability $1/2$ that another difference appear in position 0, thus that the next step is also active. We can expect that some corrective patterns lead to collisions with probability 2^{-5}. Thus, there should be 2^{w-5} colliding states for some corrective patterns.

In practice, $w = 10$ or $w = 11$. So we are able to compute all the differential characteristic of length $w + 1$ of a counter. In fact, for a given IV and starting state, there is at most one IV difference that creates a collision after $w + 1$ steps and has the correct output difference α. We computed all these values and stored for each IV pair, the initial states for which a collision occurs. We performed this exhaustive search for each of the possible non linear functions. This computation takes a few minutes an Intel Celeron 1.4 Ghz.

We observe that for some good pairs of IVs with appropriate difference patterns, the number of colliding states is higher than 2^{w-5}, sometimes exceeding 2^{w-5} by a factor of 2 or more. We give in Table 1 the maximum number of colliding states N_i for the best IV pair for each proposed non linear function in [9, Appendix F]. For the first (resp. last) 16 functions $w = 11$ (resp. 10) and the expected number of colliding states is 64 (resp. 32).

Table 1. Size of the largest colliding states for register update functions

i	N_i	i	N_i	i	N_i	i	N_i	i	N_i	i	N_i	i	N_i	i	N_i
0	127	4	106	8	122	12	102	16	70	20	44	24	59	28	52
1	107	5	107	9	95	13	96	17	67	21	60	25	76	29	64
2	117	6	96	10	90	14	104	18	74	22	62	26	65	30	54
3	128	7	150	11	156	15	136	19	52	23	77	27	54	31	77

3.2   Collision in the Counter Diffusor

In Section 2.2, a description of the linear counter diffusor was given. Matrix M having more columns than rows has a non zero kernel, generated by vectors:

$$(1,0,0,0,1,1,1,1,0,0,0,0,0,0,0,0)^{\mathbf{T}},$$
$$(1,1,1,1,0,1,1,0,1,1,1,0,0,0,0,0)^{T},$$
$$(0,1,1,0,0,0,1,0,1,0,0,1,0,0,0,0)^{T},$$
$$(0,1,0,1,1,0,1,0,1,0,0,0,1,0,0,0)^{T},$$
$$(1,1,0,1,1,0,0,0,0,0,0,0,0,0,1,0)^{T},$$
$$(0,1,0,1,0,0,0,0,0,1,0,0,0,1,0,1)^{T}$$

This clearly contradicts the property of collision freeness in elementary VEST components claimed in [9, Section 5.4]. If the output of the counter at step r differs by a linear combination of these vectors, the contribution of the counters to the update of the diffusion bits $\{d_j^r\}_j$ will be the same. The highlighted kernel vector only uses the first 8 registers which can, to some extent, be controlled and is especially useful in our attack.

4   Partial Keyed State Recovery

In this section, we exploit the vulnerabilities described in the previous section to recover the value of a fixed part of the keyed state. All the attacks described are chosen IVs differential attacks. The first attack assumes no constraints on the IV length. The second attack uses an IV as short as possible and can be used when the length of the IV is constrained to some standard value, namely 128 bits. In fact, we show that an attack is possible as soon as the IV length is greater than 96 bits.

4.1   Attack with Long IVs

We describe in this subsection a first attack which assumes that the IV length is greater than $23 \times 8 = 184$ bits. We note that [9] doesn't specify any limit on the IV length.

The goal of the attack is to exploit the above weaknesses to build a pair of colliding IVs, that is to say a pair of IVs such that the state of the cipher is the same after the Key Setup using a unknown key K and the IV Setup using each IV of the pair. We have studied in the previous section the differential behaviour of the NLFSRs in keying mode and of the counter diffusor. Our attack uses the outlined properties to introduce differences into the counter and then cancel them, while introducing no differences in the core accumulator.

Indeed, during the first phase of the IV Setup, all the NLFSRs are not updated in the same way. NLFSRs 0 to 7 work in keying mode and are disturbed by IV bits, each IV bit being used exactly once, while NLFSRs 8 to 15 work in counter mode. As a consequence, an attacker can only influence the first 8 NLFSRs. In order to introduce no difference in the core accumulator, the counter diffusor state must remain the same while the counter state varies. The counter diffusor update depending linearly on the NLFSRs output through matrix M, this is possible if and only if the counter output difference lies in the kernel of M. Furthermore, as the attacker can only introduce differences in the first 8 NLF-SRs, the counter output difference can only be non zero on the corresponding components. Unfortunately for VEST security, one such element exists in the kernel of M (see section 3.2). Thus, if the outputs of the NLFSRs are different for exactly NLFSRs 0, 4, 5, 6 and 7, the counter diffusor is updated in the same way as if there were no difference. As a consequence, if there is no difference in the counter diffusor and in the core accumulator and if the counter output difference is as above, no difference is introduced in the counter diffusor and in the core accumulator.

Furthermore, we described in section 3.1 differential patterns on the NLFSRs. The output difference for these differential patterns is a single 1 followed by w 0's. If we combine such differential patterns, one pattern for each of the five NLFSRs 0, 4, 5, 6 and 7, making them start at the same step, and introduce no difference on the three NLFSRs 1, 2 and 3, and no difference on NLFSRs 0, 4, 5, 6 and 7 after the differential patterns, we get a differential pattern on the whole counter such that:

- there is a collision on the whole counter state if the starting state of the counter is in a good set;
- the output of the NLFSRs doesn't introduce differences in the counter diffusor and in the core accumulator. Indeed the counter output difference is either all 0's, or equal to the highlighted element of the kernel of M given in 3.2.

In order to make the good set explicit, we first remind that during the IV Setup first phase different NLFSRs are updated independently. Furthermore, we can extract from the IV the part that affects a particular register. Having a global collision on the whole counter state is thus equivalent to five partial collisions on the registers 0, 4, 5, 6 and 7. Thus the set of initial states of the counter that leads to collisions is the cartesian product S of the sets S^r of registers initial states that lead to collisions for the IV pairs used. Note that due to the guaranteed output difference of the partial collisions, this leads to a collision on the whole cipher state at the end of the first phase of the IV Setup. As no difference is introduced in the second phase of the IV Setup, there is a collision on the whole cipher state at the end of the IV Setup.

The idea of the basic attack is to choose for each register an IV pair that maximizes the number of colliding states. This also maximizes the number of initial states leading to a collision on the whole counter. If we choose an initial counter state randomly, the probability it yields a collision is the number of colliding states divided by the total number of states. For the default member of a VEST family the value of this probability is:

$$p \approx 2^{-21.24}.$$

We then build IV pairs of length 23 bytes, such that the first 11 bytes are random and identical, and the last 12 bytes is a fixed IV pair, the composition of the best IV pairs for registers 0, 4, 5, 6 and 7 and some fixed values for registers 1, 2 and 3. We note that on average 11 bits are enough to completely randomize the counter state. As a consequence p is the probability that such an IV pair collides on the IV Setup, resulting in a collision on the whole cipher state after the IV Setup as explained above. To test this collision, we compare the first 32 bits of the keystream output after each IV setup. If a collision occurs in the IV Setup the keystream bits are identical. If there is no collision, the probability of having for two different IVs the same first 32 bits of keystream is 2^{-32}. After approximately $1/p$ pairs of IV setups we find with high probability a pair of IVs yielding the same first keystream bits and with high probability it is because of an inner collision on the counter.

Once such a pair is obtained, it is easy to retrieve the 53 bits of the keyed state in registers 0, 4, 5, 6 and 7 ($w = 11$ for registers 0, 4 and 7, $w = 10$ for registers 5 and 6). We proceed register by register. Having a collision after the IV Setup, the state of register i after step 11 is a colliding state for the IV pair extracted from the colliding IVs by taking the IV part relative to register r, restricted to step 12 to 23. We then make a guess on this value, among the set S^r of colliding states. Backtracking the 11 first steps of the IV Setup, we obtain a candidate for the register value after Key Setup. In order to test this value, starting from the colliding IV pair, we build another IV pair which only differs on its contribution to register r. The first 11 bits are chosen to ensure that the new starting value for register r after the randomizing step belongs to $\mathcal{S}^r$ if the guess is correct. The remaining 12 bits are left unchanged in both element of the IV pair. If the IV setup using this new pair does not give rise to a collision, we eliminate the guessed values from the candidates. The probability for a wrong candidate to pass one test successfully is $\#\mathcal{S}^r/2^{w_r}$. Iterating this procedure we get rapidly rid of the wrong candidates. Repeating this procedure on all interesting registers for some examples we were able to retrieve the value of the registers in the keyed state using a little less than 500 additional IV setups. It is possible to improve this basic approach and eliminate several values at once using a single IV setup. This lowers the number of additional IV setups below 200.

Of course, this attack can easily be adapted to contexts were the IV is longer than 23 bytes. It can also be adapted when using shorter IVs, as long as the IV length is greater than 12. In that case, this version of the attack won't work anymore for all the keys, since the randomizing of the register states isn't guaranteed to turn the keyed register states into an element of $\mathcal{S}$. Thus the attack becomes key-dependent.

The limiting part of the attack is the look-up of a collision on the whole counter. It's complexity is approximately $2/p \approx 2^{22.24}$ IV setups. We implemented this attack. On a Pentium Xeon 2.8 GHz it takes a few minutes on the average to find a pair of colliding IVs.

4.2   Attack with Short IVs

In the previous subsection, we described a differential attack that allows to recover part of the keyed state provided the IV length is long enough. We also mentioned that the attack becomes key-dependent when the IV length is reduced, and that the minimal IV length for which an attack may be possible is 12 bytes. In this subsection we describe an improved approach that works for all keys using IVs of minimal length, i.e. 12 bytes.

The former attack generates random states from the keyed state until one random state falls in a particular set $\mathcal{S}$. The idea of the attack described below is to generate a family of sets $\{\mathcal{S}_i\}_i$ which covers the entire state space.

Indeed, for a particular register r, we can associate to a pair of IVs C_i^r for this register the set of the states starting from which we get the expected register output difference (1 followed by 0's) and a collision on the register state after IV setups using each of the IVs of C_i^r. We saw in section 3.1 that for certain IV pairs C^r the cardinality of the associated set $\mathcal{S}^r$, $\#\mathcal{S}^r$, might be particularly high. The register size w being only 10 or 11 bits, we saw it was possible to compute for each IV pairs with IV length $w+1$ the set of colliding states for this pair. By doing this we get a family of sets. We have verified that the union of these sets covers the whole space of register values for any of the proposed non linear function. Thus, it is possible to select $N^{(r)} < 2^{w_r}$ sets, $\mathcal{S}_1^r, \mathcal{S}_2^r, \ldots, \mathcal{S}_{N^{(r)}}^r$, so that the union of these sets covers $\{0,1\}^{w_r}$. By composing 5 families of sets picked for registers 0, 4, 5, 6 and 7 by cartesian product, we get a family of sets whose union covers all the possible values for the 53–bit part of the keyed state made of these registers. The size of this family $\{S_i\}_i$ is the product of the size of the families for each registers: $N = N^{(0)} N^{(4)} N^{(5)} N^{(6)} N^{(7)}$. The elements of the family are sets of states for which a collision occurs after the IV setup using an IV pair obtained as a combination of the IV pairs for each register.

Suppose that we have such a family $\{S_i\}_i$ and the associated family of IV pairs. By doing two IV setups for each pair, we are bound to find a pair of IV for which there is a collision after the IV Setup. The complexity of this attack in number of IV setups is $2N$ in the worst case. Furthermore, if we first test the pairs whose associated set of colliding states is largest, the expected number of IV setups on the average case becomes smaller.

In order to get the best complexity we have to build for each register a family of covering IV pairs $\{C_i^r\}_i$. In order to improve the average complexity, we use a greedy algorithm to pick up these families. We first build all the colliding sets for all the IV pairs. We then sort them by decreasing cardinality and pick up the first pair. We then remove the states in the colliding set of the picked IV pair from all the unpicked sets and sort again the list of sets by decreasing size. We iterate this step until we get a list of pairs so that every state is in the colliding state of a pair in the list. We give in Table 2 the length of the obtained lists for the functions used by registers 0, 4, 5, 6 and 7 in the default member of a VEST family.

Table 2. Size of the covering families for some non linear functions

function number	covering family size
0	59
1	93
19	77
20	86
2	96

Once we have obtained the families for the five interesting registers, we have to combine them to get a family of IV pairs for the whole counter. In order to get the best average complexity for the attack using these 5 families, we have to test the pairs by decreasing order of their colliding sets of states. For VEST-v default member the number of pairs is $\approx 2^{31.70}$. The generic sorting algorithms are difficult to apply because of the amount of memory required. However, we do not really need to store the IV pairs order, as long as we are able to enumerate them rapidly. Furthermore, the only information required to make the comparison is the size of the colliding states sets for each IV pair. Thus we are left with the following problem:

Given two lists of integers $(a_i)_{1 \leq i \leq N_a}, (b_j)_{1 \leq j \leq N_b}$ sorted by decreasing order, enumerate in decreasing order all the products $a_i b_j$.

In [10] and [3], an algorithm is proposed to solve this problem. This algorithm does not store all the products. Assuming $N_a > N_b$, its time complexity is $O(N_a N_b \log N_b)$ and its memory complexity $O(N_b)$.

Implementing this algorithm and using the 5 families obtained we were able to enumerate the $2^{31.70}$ IV pairs of covering family in decreasing order in less than 2 hours on a 1.4 GHz Celeron M processor. Noting $(C_i)_{1 \leq i \leq N}$ this list of pairs, $(\mathcal{S}_i)_{1 \leq i \leq N}$ the associated list of colliding states and $(N_i = \#\mathcal{S}_i)_{1 \leq i \leq N}$ the decreasing list of the cardinality of these sets, the mean complexity of the collision search, evaluated in number of pairs of IV setups is:

$$ \mathcal{C} = \sum_{i=1}^{N} i \cdot \frac{N_i}{2^{53}}. $$

Computing this sum during the enumeration of the pairs in decreasing order, we obtain for the VEST-v default member $\mathcal{C} \approx 2^{27.73}$.

Once a collision is obtained, we have a small list of candidate values for the keyed state value of every interesting registers. By testing these candidates separately for each register, we are able to retrieve 53 bits of the keyed state. The number of additional IV setups for these tests is negligible before the number of IV setups necessary to find an IV collision.

The attack described in this subsection enables an attacker to retrieve 53 bits of the keyed state in $2^{28.73}$ IV setups on average and $2^{32.70}$ in the worst case. This is slightly more than the complexity of the basic attack. However this attack uses IVs of minimal length. Indeed, at least 12 steps are required to create a collision in registers 0, 4 and 7. With longer IVs, of length between 12 and 23 bytes, we can use the beginning of the IVs as a partial randomizer and add an early abort strategy to the attack of this subsection, to improve the overall complexity of the attack.

5   Key Recovery

In this section, we show how the partial keyed state value recovered by the attacks of the previous section can be used to recover the key of the stream cipher faster than exhaustive search. This enables to evaluate the impact on the cipher security of the collisions that were discovered in the IV Setup mechanism.

5.1   Backtracking the Key Setup Second Phase

We begin by backtracking the second phase of the key setup. As all the bits entering the interesting registers (0, 4, 5, 6 and 7) are known, we are able to retrieve the states of these registers after the key bits introduction. We also know that the registers are set to 1 before the key bits introduction. Thus we are able to perform a meet-in-the-middle attack on the key, even though the key bits are introduced in all the registers through a sliding window mechanism. The same kind of attack disqualified double-DES as a successor of DES.

5.2   Meet-in-the-Middle Attack

One can notice that bits 0 to $l-1$ (resp. l to $F-1$) of the key are introduced in register r between step $15-r$ and step $l-1+15-r$ (resp. $l+15-r$ and

$F - 1 + 15 - r$). By guessing the first l bits or the last $F - l$ bits of the key, one can compute the values of register r before step $l + 15 - r$. This enables us to implement a time/memory tradeoff attack against VEST ciphers. We build the table $\mathcal{A}$ of the 53–bit values of the 5 interesting registers, at steps $l + 15 - r$. This requires 2^l memory.

Then, for each 2^{F-l} values j of the end of the key we perform the following:

- backtrack the register values assuming j: we note x the 53–bit value obtained;
- look for i so that $\mathcal{A}[i] = x$. The probability of the match is 2^{-53};
- for each match, check the key $(i||j)$, where $||$ designates the concatenation.

Complexity. This enables to explore all the keys which sets the states of the interesting registers to the recovered values. This attack recovers the key used by the cipher using $2^{\max(F-53, F-l)}$ time and 2^l memory. The average number of keys to test is 2^{F-53}.

5.3   Key Recovery Through Related-Key Attack

It is also possible to mount a very efficient related key attack. Assume we are given two ciphers keyed with key K and key K' which differs of K only on bit $F - l - 1$. Performing our chosen IV attack, we are able to retrieve for both keys the keyed states of the interesting registers. By guessing the last l bits of the key, we can backtrack their introduction from the keyed state up to the step after the key difference introduction for each register. For the correct guess, there should only be a difference at position 1 of the registers after the backtracking. This happens with probability 2^{-53}. Thus we are able to check our guesses if $l < 53$. For the registers to behave in a random way l should also be larger than their length.

Once we have determined the last l bits of the key, we can iterate this process on the unknown part of the key. For a 128–bit key, taking $l = 16$ we are able to recover the whole key with 8 related keys performing $\approx 2^{26}$ IV setups and $\approx 2^{19}$ partial key introduction backtracking.

5.4   Security Discussion

The attacks described above show that the differential attack result can theoretically be exploited to recover the cipher key faster than exhaustive key search. It seems that this is also the case in more practical attacker models. This breaks VEST cipher when it is used with keys of size of the security parameters. In [9, Section 3.3], the authors of VEST recommend to use keys of size at least twice the security parameter. In this case VEST could be considered as resistant to this attack since the complexity of our time memory tradeoff may remain greater than the security parameter. It will nevertheless fall to the same related-key attack. Anyway, it is usually considered as a bad practice to use cryptosystems where part of the key material can easily be recovered.

The current specification of VEST stream cipher does not forbid the common usage consisting of taking a key the length of the security parameter. In the case of use of a key the length of the security parameter, it even recommends the use of long IVs, which makes our attack more efficient. In its latest version, VEST does not meet its claimed security.

6   Existential Forgeries for VEST Hash MAC Mode

VEST can be used as a keyed hash function using the procedure described in [9, Section 3.4]. The VEST cipher is first keyed. Then the data to be MAC-ed is introduced into the cipher as an IV during the IV Setup, using a different constant for the second phase. Finally $2n$ bits are output by the cipher in counter mode. In order to finish the description of this keyed hash function, a padding should be described to hash messages of arbitrary length. Independently of this padding we can transpose the chosen IV attack into an existential forgery attack.

Indeed, IVs in the previous attacks can be replaced by messages in the current setting. Thus, asking an oracle for approximately $2^{22.24}$ chosen message MACs enables an attacker to retrieve 53 bits of the keyed state of a VEST-v default member. The attacker can then create a pair of messages that collide and ask for the MAC of one of the messages. This MAC is the MAC value for the other message of the pair. This provides an easy existential forgery chosen message attack.

7   Conclusion

In this paper, we showed that despite its apparent complexity, the VEST stream cipher has simple properties which allows for the easy creation of internal collision. The overall result gives an efficient partial key recovery attack against VEST.

Our chosen IV attacks are practical. We were able to generate collisions for both attacks. The simple attack requires about 30 minutes on an Intel Xeon 2.8 GHz, the short IV attack required a few hours on the same machine.

Once again this shows that IV Setup is a very crucial part of a stream cipher security [11,7,5,12].

Following this cryptanalysis, the authors of VEST proposed a modified version of their algorithm. The update function of the linear counter diffusor has been changed so that there is no inner collision in the diffusor during the IV setup. This makes the modified version immune to the differential attack presented in this paper. However, its security remains to be analyzed in the light of the weaknesses presented here.

Acknowledgements. We would like to thank Dan Bernstein for his parallelization of the key recovery attack.

References

1. Biryukov, A.: A new 128 bit key stream cipher : LEX. eSTREAM, ECRYPT Stream Cipher Project, Report 2005/013 (2005), http://www.ecrypt.eu.org/stream
2. Biryukov, A.: The Design of a Stream Cipher LEX. In: Biham, E., Youssef, A. (eds.) Selected Areas in Cryptography – SAC 2006, LNCS, vol. 4356, Springer, Heidelberg (to appear, 2007)
3. Boneh, D., Joux, A., Nguyen, P.: Why Textbook ElGamal and RSA Encryption are Insecure. In: Okamoto, T. (ed.) ASIACRYPT 2000. LNCS, vol. 1976, pp. 30–43. Springer, Heidelberg (2000)
4. Chabaud, F., Joux, A.: Differential Collisions in SHA-0. In: Krawczyk, H. (ed.) CRYPTO 1998. LNCS, vol. 1462, pp. 56–71. Springer, Heidelberg (1998)
5. Cid, C., Gilbert, H., Johansson, T.: Cryptanalysis of Pomaranch. eS-TREAM, ECRYPT Stream Cipher Project, Report 2005/060 (2005) http://www.ecrypt.eu.org/stream
6. ECRYPT. eSTREAM: ECRYPT Stream Cipher Project, IST-2002-507932, http://www.ecrypt.eu.org/stream
7. Jaulmes, E., Muller, F.: Cryptanalysis of ECRYPT Candidates F-FCSR-8 and F-FCSR-H. eSTREAM, ECRYPT Stream Cipher Project, Report 2005/046 (2005), http://www.ecrypt.eu.org/stream
8. O'Neil, S., Gittins, B., Landman, H.: VEST – Hardware-Dedicated Stream Ciphers. eSTREAM, ECRYPT Stream Cipher Project, Report 2005/032 (2005), http://www.ecrypt.eu.org/stream
9. O'Neil, S., Gittins, B., Landman, H.: VEST Ciphers. eSTREAM, ECRYPT Stream Cipher Project (2006), http://www.ecrypt.eu.org/stream/p2ciphers/vest/vest_p2.pdf
10. Schroeppel, R., Shamir, A.: A $T = O(2^{n/2})$, $S = O(2^{n/4})$ algorithm for certain NP-complete problems. SIAM Journal on Computing 10(3), 456–464 (1981)
11. Wu, H., Preneel, B.: Chosen IV Attack on Stream Cipher WG. eS-TREAM, ECRYPT Stream Cipher Project, Report 2005/045 (2005), http://www.ecrypt.eu.org/stream
12. Wu, H., Preneel, B.: Key Recovery Attack on Py and Pypy with Chosen IVs. eSTREAM, ECRYPT Stream Cipher Project, Report 2006/052 (2006), http://www.ecrypt.eu.org/stream

Cryptanalysis of Achterbahn-128/80

María Naya-Plasencia[*]

INRIA, projet CODES, Domaine de Voluceau
78153 Le Chesnay Cedex, France
Maria.Naya_Plasencia@inria.fr

Abstract. This paper presents two key-recovery attacks against Achterbahn-128/80, the last version of one of the stream cipher proposals in the eSTREAM project. The attack against the 80-bit variant, Achterbahn-80, has complexity 2^{61}. The attack against Achterbahn-128 requires $2^{80.58}$ operations and 2^{60} keystream bits. These attacks are based on an improvement of the attack due to Hell and Johansson against Achterbahn version 2. They mainly rely on an algorithm that makes profit of the independence of the constituent registers.

Keywords: stream cipher, eSTREAM, Achterbahn, cryptanalysis, correlation attack, linear approximation, parity check, key-recovery attack.

1 Introduction

Achterbahn [4,6] is a stream cipher proposal submitted to the eSTREAM project. After the cryptanalysis of the first two versions [10,9], it has moved on to a new one called Achterbahn-128/80 [5] published in June 2006. Achterbahn-128/80 corresponds to two keystream generators with key sizes of 128 bits and 80 bits, respectively. Their maximal keystream length is limited to 2^{63}.

We present here two attacks against both generators. The attack against the 80-bit variant, Achterbahn-80, has complexity 2^{61}. The attack against Achterbahn-128 requires $2^{80.58}$ operations and 2^{61} keystream bits. These attacks are based on an improvement of the attack against Achterbahn version 2 and also on an algorithm that makes profit of the independence of the constituent registers.

The paper is organized as follows. Section 2 presents the main specifications of Achterbahn-128/80. Section 3 then describes the general principle of the attack proposed by Hell and Johansson [9] against the previous version of the cipher Achterbahn version 2, since our attacks rely on a similar technique. We also exhibit a new attack against Achterbahn version 2 with complexity 2^{53}, while the best previously known attack had complexity 2^{64}. Section 4 then presents two distinguishing attacks against Achterbahn-80 and Achterbahn-128 respectively.

---

[*] This work was supported in part by the European Commission through the IST Programme under Contract IST-2002-507932 ECRYPT. The information in this document reflects only the author's views, is provided as is and no warranty is given that the information is fit for any particular purpose. The user thereof uses the information at its sole risk and liability.

A. Biryukov (Ed.): FSE 2007, LNCS 4593, pp. 73–86, 2007.

Section 5 describes how this previous distinguishing attacks can be transformed into key-recovery attacks.

2 Main Specifications of Achterbahn-128/80

2.1 Main Specifications of Achterbahn-128

Achterbahn-128 is a keystream generator, consisting of 13 binary nonlinear feedback shift registers (NLFSRs) denoted by $R0, R1, \ldots, R12$. The length of register i is $L_i = 21 + i$ for $i = 0, 1, \ldots, 12$. These NLFSRs are primitive in the sense that their periods T_i are equal to $2^{L_i} - 1$. The sequence which is used as an input to the Boolean combining function is not the output sequence of the NLFSR directly, but a shifted version of itself. The shift amount depends on the register number, but it is fixed for each register. In the following, $x_i = (x_i(t))_{t \geq 0}$ for $0 \leq i \leq 12$ denotes the shifted version of the output of the register i at time t. The output of the keystream generator at time t, denoted by $S(t)$, is the one of the Boolean combining function F with the inputs corresponding to the output sequences of the NLFSRs correctly shifted, i.e. $S(t) = F(x_0(t), \ldots, x_{12}(t))$. The algebraic normal form of the 13-variable combining function F is given in [5].

Its main cryptographic properties are: balancedness, algebraic degree 4, correlation immunity order 8, nonlinearity 3584, algebraic immunity 4.

2.2 Main Specifications of Achterbahn-80

Achterbahn-80 consists of 11 registers, which are the same ones as in the above case, except for the first and the last ones. The Boolean combining function, G, is a sub-function of F :

$$G(x_1, \ldots, x_{11}) = F(0, x_1, \ldots, x_{11}, 0).$$

Its main cryptographic properties are: balancedness, algebraic degree 4, correlation immunity order 6, nonlinearity 896, algebraic immunity 4. As we can see, Achterbahn-128 contains Achterbahn-80 as a substructure.

2.3 The Key-Loading Algorithm

The key-loading algorithm uses the key K of 128/80 bits and an initial value IV of 128/80 bits. The method for initializing the registers is the following one: first of all, all registers are filled with the bits of $K || IV$. After that, register i is clocked $a - L_i$ times where a is the number of bits of $K || IV$, and the remaining bits of $K || IV$ are added to the feedback bit. Then, each register outputs one bit. Those bits are taken as input on the Boolean combining function, which outputs a new bit. This bit is now added to the feedbacks for 32 additional clockings. Then we overwrite the last cell of each register with a 1, in order to avoid the all zero state.

This algorithm has been modified in relation to the previous versions. The aim of this modification is to prevent the attacker from recovering the key K from the knowledge of the initial states of some registers.

3 Attack Against Achterbahn Version 2 in 2^{53}

3.1 Principle of Hell and Johansson Attack

Achterbahn version 2 was the previous version of Achterbahn. The main and most important differences to this last one, which are used by the attack are that:

- it had 10 registers, with lengths between 19 and 32 bits,
- the Boolean function, f, had correlation immunity order 5.

This version has been broken by Hell and Johansson [9] using a quadratic approximation. Their attack is a distinguishing attack that relies on a biased parity-check relation between the keystream bits which holds with probability

$$p = \frac{1}{2}(1 + \eta) \text{ with } |\eta| \ll 1,$$

where η is the bias of the relation. The attack then consists of an exhaustive search on 2^k initial states. For each of those states, the parity-check relation is computed for N samples in order to detect the bias. As noticed in [8], the usual estimate [9,10,11] of the number of samples which are required for distinguishing the keystream,

$$N \sim \frac{1}{\eta^2},$$

is a bit underestimated. Actually, this problem can be seen as a decoding problem where the received word corresponds to the sequence formed by the N parity-check evaluations. And this received word can be seen as the result of the transmission of a codeword through a binary symmetric channel with cross-over probability p. Then, the number of samples N required for decoding is

$$N = \frac{k}{C(p)},$$

where $C(p)$ is the capacity of the channel, i.e.,

$$C(p) = 1 + p \log_2(p) + (1 - p) \log_2(1 - p).$$

Moreover, when $p = \frac{1}{2}(1 + \eta)$ with $|\eta| \ll 1$, we have $C(p) \sim \frac{\eta^2}{2\ln(2)}$, leading to

$$N \sim \frac{2k \ln 2}{\eta^2},$$

where 2^k is the number of possible initial states of the guessing registers, as we will see.

The attack proposed by Hell and Johansson exploits a quadratic approximation q of the combining function f:

$$Q(y_1, \ldots, y_n) = \sum_{j=1}^{s} y_{i_j} + \sum_{i=1}^{m} (y_{j_i} y_{k_i})$$

with m quadratic terms and which satisfies

$$\Pr[F(y_1, \ldots, y_n) = Q(y_1, \ldots, y_n)] = \frac{1}{2}(1 + \varepsilon).$$

We build the parity-check equations, as the ones introduced by [10], that make disappear the quadratic terms by summing up:

$$q(t) = \sum_{j=1}^{s} x_{i_j}(t) + \sum_{i=1}^{m} x_{j_i}(t)x_{k_i}(t)$$

at 2^m different epochs $(t+\tau)$, where τ varies in the set of the linear combinations with $0-1$ coefficients of $T_{j_1}T_{k_1}, T_{j_2}T_{k_2}, \ldots, T_{j_m}T_{k_m}$, where T_i denotes the period of Ri. In the following, this set is denoted by $\langle T_{j_1}T_{k_1}, \ldots, T_{j_m}T_{k_m} \rangle$, i.e.,

$$\mathcal{I} = \langle T_{j_1}T_{k_1}, \ldots, T_{j_m}T_{k_m} \rangle = \left\{ \sum_{i=1}^{m} c_i T_{j_i} T_{k_i}, c_1, \ldots, c_m \in \{0,1\} \right\}.$$

This leads to a parity-check sequence pc defined by:

$$pc(t) = \sum_{\tau \in \mathcal{I}} q(t + \tau) = \sum_{\tau \in \mathcal{I}} (x_{i_1}(t + \tau) + \ldots + x_{i_s}(t + \tau)).$$

We then decimate the sequence $(pc(t))_{t \geq 0}$ by the periods of r sequences among $(x_{i_1}(t))_{t \geq 0}, \ldots, (x_{i_s}(t))_{t \geq 0}$. We can suppose here without loss of generality that the periods of the first r sequences have been chosen. Now a new parity-check, pc_r, can be defined by:

$$pc_r(t) = pc(tT_{i_1} \ldots T_{i_r}).$$

This way, the influence of those r registers on the parity-check $pc_r(t)$ corresponds to the addition of a constant for all $t \geq 0$, so it will be 0 or 1 for all the parity-checks.

Now, the attack consists in performing an exhaustive search for the initial states of the $(s-r)$ remaining registers, i.e. those of indices $i_{r+1}, \ldots, i_s$. For each possible values for these initial states, we compute the sequence:

$$\sigma(t) = \sum_{\tau \in \langle T_{j_1}T_{k_1}, \ldots, T_{j_m}T_{k_m} \rangle} \left[S(tT_{i_1} \ldots T_{i_r} + \tau) + \sum_{j=r+1}^{s} x_{i_j}(tT_{i_1} \ldots T_{i_r} + \tau) \right] \quad (1)$$

We have

$$\Pr[\sigma(t) = 0] \geq \frac{1}{2}(1 + \varepsilon^{2^m}).$$

It has been recently observed by Hell and Johansson that the total bias may be much higher than this bound. However, it can be shown that equality holds in some particular cases, as noted in [7]. An interesting case of equality is when f is v-resilient, and we build parity-checks from the terms appearing in a linear approximation of $(v + 1)$ variables (see Appendix). This also provides the bias of the parity-checks obtained in [9,8] from some quadratic approximations, since they can also be derived from such linear approximations.

This result is going to be used all along our attacks, as we will work with linear approximations of $(v + 1)$ variables.

3.2   Complexity

Using the previously computed bias, we can distinguish the keystream $(S(t))_{t\geq0}$ from a random sequence and also recover the initial states of $(s-r)$ constituent registers.

- We will have 2^m terms in each parity-check. That means that we need to compute $\varepsilon^{-2^{m+1}} \times 2 \times \sum_{j=r+1}^{s}(L_{i_j}-1) \times \ln(2) = 2^{n_b 2^{m+1}} \times 2 \times \sum_{j=r+1}^{s}(L_{i_j}-1) \times \ln(2)$ values of $\sigma(t)$ for mounting the distinguishing attack, where $n_b = \log_2\varepsilon^{-1}$. Besides, $\sigma(t)$ is defined by (1), implying that the attack requires

$$2^{n_b 2^{m+1}+\sum_{j=1}^{r}L_{i_j}} \times 2 \times \sum_{j=r+1}^{s}(L_{i_j}-1) \times \ln(2) + \sum_{i=1}^{m}2^{L_{j_i}+L_{k_i}} \text{ keystream bits,}$$

where L_{i_j} are the lengths of the registers associated to the periods by which we have decimated, and the last term corresponds to the maximal distance between the bits involved in each parity-check.
- Time complexity will be

$$2^{m}2^{n_b 2^{m+1}+\sum_{j=r+1}^{s}(L_{i_j}-1)} \times 2 \times \sum_{j=r+1}^{s}(L_{i_j}-1) \times \ln(2)$$

where $i_{r+1},\ldots,i_s$ are the indices of the registers over whom we have made an exhaustive search and whose initial state we are going to find.

3.3   Example with Achterbahn Version 2

Hell and Johansson [9] have used this attack against Achterbahn version 2 with the following quadratic approximation:

$$Q(x_1,\ldots,x_{10}) = x_1 + x_2 + x_3 x_8 + x_4 x_6.$$

Then, they decimate by the period of the second register, whose length is 22. After that, they make an exhaustive search over the first register, of length 19. Time complexity will be 2^{67} and data complexity 2^{64} (the complexity given in [9], equal to $2^{59.02}$, is obtained by using the estimation $N = \varepsilon^{-2}$ instead of the one given in Section 3.1). Using the small lengths of the registers, time complexity can be reduced below data complexity, so the overall complexity of the attack will be 2^{64}.

3.4   Improvement of the Attack Against Achterbahn Version 2

We are going to improve the previously described attack against Achterbahn version 2 and we reduce the complexity to 2^{53}.

For this attack, we use the idea of associating the variables in order to reduce the number of terms that we will have in the parity-checks. The only negative effect that this could have on the final complexity of the attack is to enlarge the number of required keystream bits; but being careful, we make it stay the same while reducing the time complexity.

The chosen approximation. At first, we searched for all the quadratics approximations of f with one and two quadratic terms, as the original attack presented by Hell and Johansson was based on a quadratic approximation. Finally, after looking for a trade-off between the number of terms, the number of variables, the bias, etc., we found that none quadratic approximation was better for this attack than linear ones. It is worth noticing that, since the combining function f is 5-resilient, any approximation of f involves at least 6 input variables. Moreover, the highest bias corresponding to an approximation of f by a 6-variable function is achieved by a function of degree one as proved in [3]. After analyzing all linear approximations of the Boolean combining function, we found that the best one was:

$$g(x_1, \ldots, x_{10}) = x_8 + x_6 + x_4 + x_3 + x_2 + x_1.$$

We have $f(x_1, \ldots, x_{10}) = g(x_1, \ldots, x_{10})$ with a probability of $\frac{1}{2}(1 + 2^{-3})$.

Parity-checks. Let us build a parity-check as follows:

$$ggg(t) = g(t) + g(t + T_1T_8) + g(t + T_2T_6) + g(t + T_1T_8 + T_2T_6),$$

with

$$g(t) = x_8(t) + x_6(t) + x_4(t) + x_3(t) + x_2(t) + x_1(t).$$

The terms x_8, x_6, x_2, x_1 will disappear and, so, $ggg(t)$ is a sequence that depends uniquely on the sequences x_3 and x_4. Adding four times the approximation has the effect of multiplying the bias four times, so the bias of

$$\sigma(t) = S(t) + S(t + T_1T_8) + S(t + T_2T_6) + S(t + T_1T_8 + T_2T_6)$$

is $2^{-3 \times 4} = 2^{-12}$ because 4 is the number of terms in $ggg(t)$. That means that we will need $2^{3 \times 4 \times 2} \times 2 \times (L_4 - 1) \times \ln(2) = 2^{29}$ values of the parity-check for detecting this bias. If we decimate $ggg(t)$ by the period of register 3, we will need

$$2^{29}T_3 + T_1T_8 + T_2T_6 = 2^{29+23} + 2^{29+19} + 2^{27+22} = 2^{52} \text{ bits of keystream,}$$

and time complexity will be $2^{29} \times 2^{L_4 - 1} = 2^{53}$ as we only guess the initial state of register 4. This complexity is 2^{53} while the complexity of the previous attack was equal to 2^{64}.

4   Distinguishing Attacks Against Achterbahn-128/80

4.1   Distinguishing Attack Against Achterbahn-80

This attack is very similar to the improvement of the attack against Achterbahn version 2 which has been described in the previous section.

Our attack exploits the following linear approximation of the combining function G:

$$\ell(x_1, \ldots, x_{11}) = x_1 + x_3 + x_4 + x_5 + x_6 + x_7 + x_{10}.$$

Since G is 6-resilient, ℓ is the best approximation by a 7-variable function.

For $\ell(t) = x_1(t) + x_3(t) + x_4(t) + x_5(t) + x_6(t) + x_7(t) + x_{10}(t)$, the keystream $(S(t))_{t \geq 0}$ satisfies $\Pr[S(t) = \ell(t)] = \frac{1}{2}(1 - 2^{-3})$.

Parity-checks. Let us build a parity-check as follows:

$$\ell\ell(t) = \ell(t) + \ell(t + T_4 T_7) + \ell(t + T_6 T_5) + \ell(t + T_4 T_7 + T_6 T_5).$$

The terms containing the sequences x_4, x_5, x_6, x_7 vanish in $\ell\ell(t)$, so $\ell\ell(t)$ depends exclusively on the sequences x_1, x_3 and x_{10}.

Adding four times the approximation has the effect of multiplying the bias four times, so the bias of

$$\sigma(t) = S(t) + S(t + T_7 T_4) + S(t + T_6 T_5) + S(t + T_7 T_4 + T_6 T_5)$$

where $(S(t))_{t \geq 0}$ is the keystream, is $2^{-4 \times 3}$.

We now decimate $\sigma(t)$ by the period of the R_{10}, which is involved in the parity-check, so we create like this a new parity-check $\sigma'(t) = \sigma(t(2^{31} - 1))$.

Then, the attack performs an exhaustive search for the initial states of registers 1 and 3. Then we need $2^{3 \times 4 \times 2} \times 2 \times (46 - 2) \times \ln(2) = 2^{30}$ parity-checks $\sigma'(t)$ to detect this bias. Its time complexity is $2^{30} \times 2^{L_1 + L_3 - 2} = 2^{74}$.

The number of keystream bits that we need is $2^{30} \times T_{10} + T_4 T_7 + T_6 T_5 = 2^{61}$.

4.2   Distinguishing Attack Against Achterbahn-128

Now, we present a distinguishing attack against the 128-bit version of Achterbahn which also recovers the initial states of two registers.

We consider the following approximation of the combining function F:

$$\ell(x_0, \ldots, x_{12}) = x_0 + x_3 + x_7 + x_4 + x_{10} + x_8 + x_9 + x_1 + x_2.$$

Then, for $\ell(t) = x_0(t) + x_3(t) + x_7(t) + x_4(t) + x_{10}(t) + x_8(t) + x_9(t) + x_1(t) + x_2(t)$, we have $\Pr[S(t) = \ell(t)] = \frac{1}{2}(1 + 2^{-3})$.

Parity-checks. The period of any sequence obtained by combining the registers 0, 3 and 7 is equal to $\mathrm{lcm}(T_0, T_3, T_7)$, i.e. $2^{59.3}$ as T_0 T_3 and T_7 have common divisors. We are going to denote this value by $T_{0,3,7}$.

If we build a parity check as follows:

$$\ell\ell\ell(t) = \sum_{\tau \in \langle T_{0,3,7}, T_{4,10}, T_{8,9} \rangle} \ell(t + \tau),$$

the terms containing the sequences x_0, x_3, x_7, x_4, x_{10}, x_8, x_9 will disappear from $\ell\ell\ell(t)$, so $\ell\ell\ell(t)$ depends exclusively on the sequences x_1 and x_2:

$$\ell\ell\ell(t) = \sum_{\tau \in \langle T_{0,3,7}, T_{4,10}, T_{8,9} \rangle} \ell(t + \tau)$$

$$= \sum_{\tau \in \langle T_{0,3,7}, T_{4,10}, T_{8,9} \rangle} x_1(t + \tau) + x_2(t + \tau)$$

$$= \sigma_1(t) + \sigma_2(t),$$

where $\sigma_1(t)$ and $\sigma_2(t)$ are the parity-checks computed over the sequences generated by NLFSRs 1 and 2.

Adding eight times the approximation has the effect of multiplying the bias eight times, so the bias of $\sigma(t) = \sum_{\tau \in \langle T_{0,3,7}, T_{4,10}, T_{8,9} \rangle} S(t+\tau)$ where $(S(t))_{t \geq 0}$ is the keystream, is $2^{-8 \times 3}$. So:

$$\Pr[\sigma(t) + \sigma_1(t) + \sigma_2(t) = 1] = \frac{1}{2}(1 - \varepsilon^8).$$

This means that we need $2^{3 \times 8 \times 2} \times 2 \times (45-2) \times \ln(2) = 2^{54}$ values of $\sigma(t) + \sigma_1(t) + \sigma_2(t)$ to detect this bias, when we perform an exhaustive search on registers 1 and 2.

We now describe an algorithm for computing the sum $\sigma(t) + \sigma_1(t) + \sigma_2(t)$ over all values of t. This algorithm has a lower complexity than the trivial algorithm which consists on computing the 2^{54} parity-checks for all the initial states of the registers 1 and 2. Here we use $(2^{54} - 2^8)$ values of t since $(2^{54} - 2^8) = T_2 \times (2^{31} + 2^8)$. We can write it down as follows:

$$\sum_{t'=0}^{2^{54}-2^8-1} \sigma(t') \oplus \ell\ell\ell(t') = \sum_{k=0}^{T_2-1} \sum_{t=0}^{2^{31}+2^8-1} \sigma(T_2 t + k) \oplus \ell\ell\ell(T_2 t + k)$$

$$= \sum_{k=0}^{T_2-1} \sum_{t=0}^{2^{31}+2^8-1} \sigma(T_2 t + k) \oplus \sigma_1(T_2 t + k) \oplus \sigma_2(T_2 t + k)$$

$$= \sum_{k=0}^{T_2-1} \left[(\sigma_2(k) \oplus 1) \left(\sum_{t=0}^{2^{31}+2^8-1} \sigma(T_2 t + k) \oplus \sigma_1(T_2 t + k) \right) + \right.$$

$$\left. \sigma_2(k) \left((2^{31} + 2^8) - \sum_{t=0}^{2^{31}+2^8-1} \sigma(T_2 t + k) \oplus \sigma_1(T_2 t + k) \right) \right],$$

since $\sigma_2(T_2 t + k)$ is constant for a fixed value of k.

At this point, we can obtain $\sigma(t)$ from the keystream and we can make an exhaustive search for the initial state of register 1. More precisely:

- We choose an initial state for register 2, e.g. the all one initial state. We compute and save a binary vector V_2 of length T_2:

$$V_2[k] = \sigma_2(k),$$

where the sequence x_2 is generated from the chosen initial state. The complexity of this step is $T_2 \times 2^3$ operations.
- For each possible initial state of register 1:
  - we compute and save a vector V_1 composed of T_2 integers of 32 bits.

$$V_1[k] = \sum_{t=0}^{2^{31}+2^8-1} \sigma(T_2 t + k) \oplus \sigma_1(T_2 t + k).$$

The complexity of this step is $2^{54} \times (2^4 + 2^5) = 2^{59.58}$ for each possible initial state of register 1, where 2^4 corresponds to the number of operations required for computing each $(\sigma(t) + \sigma_1(t))$ and $(2^{31} + 2^8) \times 2^5 = (2^{31} + 2^8) \times 32$ is the cost of summing up $2^{31} + 2^8$ integers of 32 bits.

- For each possible i from 0 to $T_2 - 1$:
  * we define V_2' of length T_2:

$$V_2'[k] = V_2[k + i \mod T_2].$$

  Actually, $(V_2'[k])_{k<T_2}$ corresponds to $(\sigma_2(k))_{k<T_2}$ when the initial state of register 2 corresponds to the internal state after clocking register 2 i times from the all-one initial state.

  * With the two vectors that we have obtained, we compute:

$$\sum_{k=0}^{T_2-1} \left[(V_2'[k] \oplus 1) V_1[k] + V_2'[k] (2^{31} + 2^8 - V_1[k]) \right]. \qquad (2)$$

When we do this with the correct initial states of registers 1 and 2, we will find the expected bias. The major difference with the classical exhaustive search used in [9,8,10] is that the sequence $V_1[k]$ is computed independently of the choice of the initial state of R_2. As a comparison, the classical algorithm has time complexity 2^{102}.

Table 1. Algorithm for finding the initial states of registers 1 and 2

```
for each possible initial state of R1 do
    for k = 0 to T₂ − 1 do
        V₁[k] = Σ_{t=0}^{2³¹+2⁸−1} σ(T₂t + k) ⊕ σ₁(T₂t + k)
    end for
    for each possible initial i state of R2 do
        for k = 0 to T₂ − 1 do
            V₂′[k] = V₂[k + i mod T₂]
        end for
        Σ_{k=0}^{T₂−1} [(V₂′[k] ⊕ 1) V₁[k] + V₂′[k] (2³¹ + 2⁸ − V₁[k])]
        if we find the bias then
            return the initial states of R1 and R2
        end if
    end for
end for
```

The total time complexity of the attack is going to be:

$$2^{L_1-1} \times \left[2^{54} \times (2^4 + 2^5) + T_2 \times 2 \times T_2 \times 2^5 \right] + T_2 \times 2^3 = 2^{80.58},$$

where $2 \times T_2 \times 2^5$ is the time it takes to compute the sum described by (2). Actually, we can speed up the process by rewriting the sum (2) in the following way

$$\sum_{k=0}^{T_2-1}(-1)^{V_2[k+i]}\left(V_1[k]-\frac{2^{31}+2^8}{2}\right)+T_2\frac{2^{31}+2^8}{2}.$$

The issue is now to find the i that maximizes this sum, this is the same as computing the maximum of the crosscorrelation of two sequences of length T_2. We can do that efficiently using a fast Fourier transform as explained in [1, pages 306-312]. The final complexity will be in $O(T_2 \log T_2)$. Anyway, this does not change our total complexity as the higher term is the first one.

The complexity is going to be, finally:

$$2^{L_1-1} \times \left[2^{54} \times \left(2^4 + 2^5\right) + O(T_2 \log T_2)\right] + T_2 \times 2^3 = 2^{80.58}.$$

The length of keystream needed is $T_{0,3,7} + T_{4,10} + T_{8,9} + 2^{54} < 2^{61}$ bits.

We can apply the algorithm to the attack against Achterbahn-80 described in Section 4.1 and its time complexity will be reduced to:

$$2^{L_1-1} \times \left[2^{30} \times \left(2^3 + 2^{2.59}\right) + O(T_3 \log T_3)\right] + T_3 \times 2^2 = 2^{54.8}.$$

4.3 Attack with a New Keystream Limitation

Recently, the authors of Achterbahn have proposed a new limitation of the keystream length [7], which is 2^{52} for Achterbahn-80 and 2^{56} for Achterbahn-128. Those limitations are not restrictive enough to prevent the cipher from being cryptanalysed. In fact, we can mount an attack against the 128-bit version which is very similar to the last one with the same linear approximation, where the sequences considered for building the parity-checks are generated by only two terms (so R_1 and R_{10}, R_2 and R_9, R_3 and R_8). Then we perform an exhaustive search over registers 0, 4 and 7 with the previously described the algorithm, where we consider register 0 and register 4 together. The complexity is, finally:

$$2^{L_0-1} \times 2^{L_4-1} \times \left[2^{54.63} \times \left(2^4 + 2^{4.7}\right) + O(T_7 \log T_7)\right] + T_7 \times 2^3 = 2^{104}.$$

The length of keystream needed is

$$2^{54.63} + T_{1,10} + T_{2,9} + T_{3,8} = 2^{54.63} + 2^{53} + 2^{53} + 2^{53} < 2^{56} \text{ bits}.$$

For Achterbahn-80 there is also a succesfull attack which is only slightly different from the one we have previously described [12].

5 Recovering the Key

As explained by Hell and Johansson in [8], if we recover the initial states of all the registers, we will be able to retrieve the key as all the initialization steps which do not involve the key become invertible. It is easy to show that once we have found the initial states of two registers, the complexity of finding the remaining ones will be lower (for the other registers appearing in the used approximation it is quite obvious: we apply the same method but simplified, as now we know two variables. For the other registers we can use the same method but with other linear approximations making profit of the already-known variables). Once we have found the initial states of all the registers, we can invert all the initializing steps until the end of the second step, which corresponds to the introduction of the key bits. At this point, there are two methods proposed in [8]. The first one is clocking backwards register i ($|k| - L_i$) times for each i. We do this for all the possibles values of the last $|k| - L_m$ key bits, where L_m is 21 for Achterbahn-128 and 22 for Achterbahn-80. When all the registers have the same first L_m bits, we have found the correct $|k| - L_m$ bits of the key. The second method proposed is a meet-in-the-middle attack with time-memory tradeoff as explained in [10]. It leads to a complexity of:

- For Achterbahn-80: 2^{58} in time or 2^{40} in memory and 2^{40} in time.
- For Achterbahn-128: 2^{107} in time or 2^{40} in memory and 2^{88} in time.

We can do better. We are going to explain the technique for Achterbahn-128. The idea is that we do not need to invert all the clocking steps in the meet-in-the-middle attack, if we split the key into 2 parts composed of the first 40 bits and the last 88 bits, we could make an exhaustive search for the first part and store in a table the states of the registers obtained after applying the initialization process for each set of 40 bits. Then, if we make an exhaustive search through the 88 remaining key bits, and we clock backwards the registers from the known states, we will find a match in the table. But we do not need to make this search over all the 88 remaining bits. Instead, we make it through the last 73 bits (that means that 15 rounds are not inverted). At the end of doing this, we need that, for all i, the first $L_i - 15$ bits of the state of register i match with the last $L_i - 15$ bits of the states of the registers saved in the table. For instance, for the register 0 we will have a match on 6 bits, and for register 12 we will have 18. We do not have to worry about matches coming from wrong values of the 73 bits since the number of such false alarms is:

$$2^{88-15} \times \frac{2^{40}}{2^{329-13 \times 15}} = 2^{-21},$$

as $(329 - 13 \times 15)$ is the number of bits we consider for a match, 2^{40} is the size of the table, and 2^{73} is the number of possibilities for the exhaustive search. As we can see, with such a match we have found 113 bits of the key. The other 15 can be found with very low complexity by clocking the registers until finding the desired state. So the final complexity for the step of retrieving the key in Achterbahn-128 once we have the initial states of all the registers is 2^{73} in time and $2^{40} \times (329 - 13 \times 15) \simeq 2^{48}$ in memory. If we do the same thing with Achterbahn-80, we could have a complexity of 2^{40} in time and 2^{41} in memory.

6   Conclusion

We have proposed an attack against Achterbahn-80 in 2^{55} operations, so we can consider as the total complexity the data complexity which is equal to 2^{61}, since it is bigger. An attack against Achterbahn-128 is also proposed in $2^{80.58}$ where fewer than 2^{61} bits of keystream are required. After that we can recover the key of Achterbahn-80 with a complexity of 2^{40} in time and 2^{41} in memory (the time complexity is less than for the distinguishing part of the attack). For Achterbahn-128 we can recover the key with a complexity of 2^{73} in time and 2^{48} in memory. The complexities of the best attacks against all versions of Achterbahn are summarized in the following table:

Table 2. Attacks complexities against all versions of Achterbahn (Each complexity corresponds to the best key-recovery attack)

version	data complexity	time complexity	references
v1 (80-bit)	2^{32}	2^{55}	[10]
v2 (80-bit)	2^{64}	2^{67}	[9]
v2 (80-bit)	2^{52}	2^{53}	
v80 (80-bit)	2^{61}	2^{55}	
v128 (128-bit)	2^{60}	$2^{80.58}$	

Acknowledgments

The author would like to thank Anne Canteaut for her helpful advice, discussions, suggestions and support. Also, many thanks to Yann Laigle-Chapuy, Andrea Röck and Frederic Didier for their useful comments.

References

1. Blahut, R.E.: Fast Algorithms for Digital Signal Processing. Addison-Wesley, Reading (1985)
2. Canteaut, A., Charpin, P.: Decomposing bent functions. IEEE Transactions on Information Theory 49(8), 2004–2019 (2003)
3. Canteaut, A., Trabbia, M.: Improved fast correlation attacks using parity-check equations of weight 4 and 5. In: Preneel, B. (ed.) EUROCRYPT 2000. LNCS, vol. 1807, pp. 573–588. Springer, Heidelberg (2000)
4. Gammel, B.M., Gottfert, R., Kniffler, O.: The Achterbahn stream cipher. eSTREAM, ECRYPT Stream Cipher Project, Report 2005/002 (2005), http://www.ecrypt.eu.org/stream/ciphers/achterbahn/achterbahn.pdf
5. Gammel, B.M., Gottfert, R., Kniffler, O.: Achterbahn-128/80. eSTREAM, ECRYPT Stream Cipher Project, Report 2006/001 (2006), http://www.ecrypt.eu.org/stream/p2ciphers/achterbahn/achterbahn_p2.pdf
6. Gammel, B.M., Gottfert, R., Kniffler, O.: Status of Achterbahn and tweaks. eSTREAM, ECRYPT Stream Cipher Project, Report 2006/027 (2006), http://www.ecrypt.eu.org/stream/papersdir/2006/027.pdf
7. Gammel, B.M., Gottfert, R., Kniffler, O.: Achterbahn-128/80: Design and analysis. In: ECRYPT Network of Excellence - SASC Workshop Record, pp. 152–165 (2007)

8. Hell, M., Johansson, T.: Cryptanalysis of Achterbahn-128/80. eS-TREAM, ECRYPT Stream Cipher Project, Report 2006/054 (2006), http://www.ecrypt.eu.org/stream/papersdir/2006/054.pdf
9. Hell, M., Johansson, T.: Cryptanalysis of Achterbahn-version 2. eS-TREAM, ECRYPT Stream Cipher Project, Report 2006/042 (2006), http://www.ecrypt.eu.org/stream/ciphers/achterbahn/achterbahn.pdf
10. Johansson, T., Meier, W., Muller, F.: Cryptanalysis of Achterbahn. In: Robshaw, M. (ed.) FSE 2006. LNCS, vol. 4047, pp. 1–14. Springer, Heidelberg (2006)
11. Naya-Plasencia, M.: Cryptanalysis of Achterbahn-128/80. eS-TREAM, ECRYPT Stream Cipher Project, Report 2006/055 (2006), http://www.ecrypt.eu.org/stream/papersdir/2006/055.pdf
12. Naya-Plasencia, M.: Cryptanalysis of Achterbahn-128/80 with a new keystream limitation. eSTREAM, ECRYPT Stream Cipher Project, Report 2007/004 (2007), http://www.ecrypt.eu.org/stream/papersdir/2007/004.pdf

A   On the Biases of Parity-Checks Derived from Linear Approximations

Proposition 1. *Given a v-resilient Boolean function f, the bias of a parity-check built from a $(v + 1)$-variable linear approximation of f with bias ε is ε raised to the power of the number of terms in the parity check.*

Proof. Let f be a v-resilient Boolean function of n variables and $\ell = x_{j_0} + \ldots + x_{j_v} = \alpha \cdot x$ be a linear approximation of f with bias ε. We can now build $g(x_1, \ldots, x_n) = f(x_1, \ldots, x_n) + \ell(x_{j_0}, \ldots, x_{j_v})$. We have

$$\Pr[g(x_1, \ldots, x_n) = 0] = \frac{1}{2}(1 + \varepsilon).$$

Let W denote the subspace of $\mathbf{F}_2^n$ spanned by the basis vectors $e_{j_0}, \ldots, e_{j_v}$, and let V be in direct sum with W. Then, for any n-variable function f, and any $a \in \mathbf{F}_2^{v+1}$, $f_{|a+V}$ denotes the restriction of f to $(a + V)$. In other words, $f_{|a+V}$ is the function of $(n - v - 1)$ variables derived from f when $x_{j_0}, \ldots, x_{j_v}$ are fixed and equal to $a_0, \ldots, a_v$. If we build the parity-checks with g considering the sequences defined by the terms of the linear approximation we will have:

$$\Pr\left[\sum_{\tau \in \langle T_{j_0}, \ldots, T_{j_v} \rangle} g(x_1(t + \tau), \ldots, x_n(t + \tau)) = 0\right]$$

$$= \frac{1}{2^{v+1}} \sum_{a \in \mathbf{F}_2^{v+1}} \Pr\left[\sum_{\tau \in \langle T_{j_0}, \ldots, T_{j_v} \rangle} g_{|a+V}(x_1(t + \tau), \ldots, x_n(t + \tau)) = 0\right].$$

It is quite obvious that the variables appearing in the terms of the sum over τ are independent, as the variables that could be repeated are the $(v + 1)$ fixed ones. So, as all the variables appearing are independent, each sum has the effect of multiplying the corresponding bias by itself 2^{v+1} times. Now we want to show that this bias is also equal to ε. This equivalently means that

$$\frac{1}{2^{n-v-1}} \sum_{x \in a+V} (-1)^{g_{|a}+V(x)} = \frac{1}{2^n} \sum_{x \in \mathbf{F}_2^n} (-1)^{g(x)}.$$

Let $\widehat{f}(a)$ denote the Walsh coefficient of f at point $a \in \mathbf{F}_2^n$, i.e:

$$\widehat{f}(a) = \sum_{x \in \mathbf{F}_2^n} (-1)^{f(x)+ax}.$$

Then, from [2, pages 2005-2006] we have

$$2^{v+1} \sum_{x \in a+V} (-1)^{g_{|a}+V(x)} = \sum_{u \in W} \widehat{g}(u)$$
$$= \sum_{u \in W} \widehat{f}(\alpha + u)$$
$$= \widehat{f}(\alpha) = \widehat{g}(0)$$

since f is v-resilient. So:

$$\Pr\left[\sum_{\tau \in \langle T_{j_0}, \dots, T_{j_v} \rangle} g(x_1(t+\tau), \dots, x_n(t+\tau)) = 0 \right] = \frac{1}{2^{vr+1}} \times 2^{vr+1} \times 0.5(1 + \varepsilon^{2^v}).$$

And then, the bias of the parity check will be ε^{2^v}. It is obvious that if, instead of building the parity-checks by considering each term in the linear approximation separately, we do it by associating several terms, the result will not change. The final bias of the parity-check will also be the bias of the linear approximation raised to the power of the number of terms in the parity-check.   ◇

Differential-Linear Attacks
Against the Stream Cipher Phelix*

Hongjun Wu and Bart Preneel

Katholieke Universiteit Leuven, ESAT/SCD-COSIC
Kasteelpark Arenberg 10, B-3001 Leuven-Heverlee, Belgium
{wu.hongjun,bart.preneel}@esat.kuleuven.be

Abstract. The previous key recovery attacks against Helix obtain the key with about 2^{88} operations using chosen nonces (reusing nonce) and about 1000 adaptively chosen plaintext words (or $2^{35.6}$ chosen plaintext words). The stream cipher Phelix is the strengthened version of Helix. In this paper we apply the differential-linear cryptanalysis to recover the key of Phelix. With 2^{34} chosen nonces and 2^{37} chosen plaintext words, the key of Phelix can be recovered with about $2^{41.5}$ operations.

1 Introduction

Phelix [5] is a fast stream cipher with an embedded authentication mechanism. It is one of the focus ciphers (both software and hardware) of the ECRYPT eS-TREAM project. Phelix is a strengthened version of the stream cipher Helix [1].

Muller has applied differential attack to Helix [2]. He showed that the key of Helix can be recovered faster than by brute force if the attacker can force the initialization vectors to be used more than once. The attack requires about 2^{12} adaptively chosen plaintext words and 2^{88} operations. Paul and Preneel reduced the number of adaptively chosen plaintext words by a factor of at least 3 [4]. Later Paul and Preneel showed that $2^{35.6}$ chosen plaintext words can be used instead of adaptively chosen plaintexts [3]. All these key recovery attacks against Helix require about 2^{88} operations.

Phelix was designed and submitted to the ECRYPT eSTREAM project in 2005. The output function of Helix has been changed so that a larger plaintext diffusion can be achieved in Phelix. The Phelix designers claimed that Phelix is able to resist a differential key recovery attack even if the nonce is reused: "We claim, however, that even in such a case (referring to nonce reuse) it remains infeasible to recover the key" [5].

In this paper, we apply differential-linear cryptanalysis to Phelix assuming nonce reuse (this corresponds to a chosen nonce attack). We show that the key of Phelix can be recovered with a low complexity: 2^{37} chosen plaintext words and $2^{41.5}$ operations. Although the Phelix designers did expect that Phelix would

* This work was supported in part by the Concerted Research Action (GOA) Ambiorics 2005/11 of the Flemish Government and in part by the European Commission through the IST Programme under Contract IST-2002-507932 ECRYPT.

A. Biryukov (Ed.): FSE 2007, LNCS 4593, pp. 87–100, 2007.

loose most of its security properties when the nonce is reused, this paper shows that Phelix is completely insecure in such a setting.

This paper is organized as follows. In Sect. 2, we illustrate the operations of Phelix. Section 3 analyzes how the addend bits affect the differential distribution. Section 4 describes a basic differential key recovery attack on Phelix. The improved attack is given in Sect. 5. We discuss how to strengthen Phelix in Sect. 6. Section 7 concludes this paper.

2   The Stream Cipher Phelix

In this section, we only consider the encryption algorithm of Phelix. The full description of Phelix is given in [5]. The key size and nonce size of Phelix are 256 bits and 128 bits, respectively. The designers claim that there is no attack against Phelix with less than 2^{128} operations.

Phelix updates fives 32-bit words: Z_0, Z_1, Z_2, Z_3 and Z_4. At the ith step, two secret 32-bit words $X_{i,0}$, $X_{i,1}$ and one 32-bit plaintext word P_i are applied to update the internal states. One 32-bit keystream word S_i is generated and is used to encrypt the plaintext P_i. Note that the plaintext is used to update the internal state so that the authentication can be performed. The word $X_{i,0}$ is related to the key, and the word $X_{i,1}$ is related to the key and nonce in a very simple way. Recovering any $X_{i,0}$ and $X_{i,1}$ implies recovering part of the key. One step of Phelix is given in Fig. 1.

3   The Differential Propagation of Addition

In this section, we study how the addend bits affect the differential propagation. The importance of this study is that it shows that the values of the addend bits can be determined by observing the differential distribution of the sum.

Theorem 1. Denote ϕ_i as the ith least significant bit of ϕ. Suppose two positive m-bit integers ϕ and ϕ' differ only at the nth least significant bit position ($\phi \oplus \phi' = 2^n$). Let β be an m-bit random integer (m is much larger than n). Let $\psi = \phi + \beta$ and $\psi' = \phi' + \beta$. For $\beta_n = 0$, denote the probability that $\psi_{n+i} = \psi'_{n+i}$ as $p_{n+i,0}$. For $\beta_n = 1$, denote the probability that $\psi_{n+i} = \psi'_{n+i}$ as $p_{n+i,1}$. Then the difference $\Delta p_{n+i} = p_{n+i,0} - p_{n+i,1} = 2^{-n-i+1}$ ($i > 0$).

Theorem 1 can be proved easily if we consider the bias in the carry bits. We omit the proof here. In Theorem 1, the bias of the differential distribution decreases quickly as the value of n increases. We need another differential property that produces difference with a large bias even for large n. Before introducing that property, we give the following lemma from [6].

Lemma 1. Denote u and v as two random and independent n-bit integers. Let $c_n = (u + v) \gg n$, where c_n denotes the carry bit at the nth least significant bit position. Denote the most significant bit of u as u_{n-1}. Then $\Pr(c_n \oplus u_{n-1} = 0) = \frac{3}{4}$.

The large bias of the differential distribution for large n is given below.

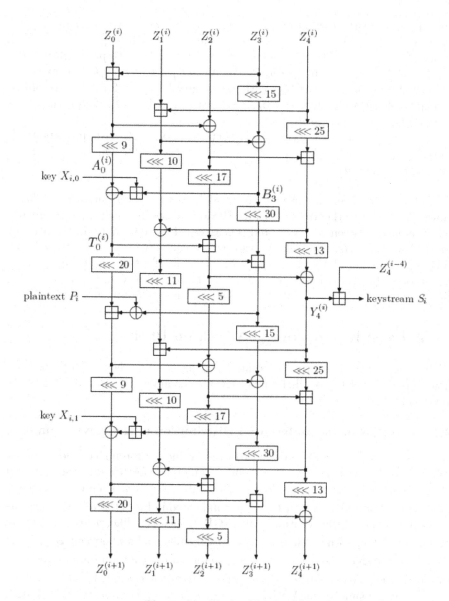

Fig. 1. One block of Phelix [5]

Theorem 2. Denote ϕ_i as the ith least significant bit of ϕ. Suppose two positive m-bit integers ϕ and ϕ' differ only at the nth least significant bit position ($\phi \oplus \phi' = 2^n$). Let β be an m-bit random integer (m is much larger than n). Let $\psi = \phi + \beta$ and $\psi' = \phi' + \beta$. For $\beta_n \oplus \beta_{n-1} = 0$, denote the probability that $\psi_{n+i} = \psi'_{n+i}$ as $\bar{p}_{n+i,0}$. For $\beta_n \oplus \beta_{n-1} = 1$, denote the probability that $\psi_{n+i} = \psi'_{n+i}$ as $\bar{p}_{n+i,1}$. Then the difference $\Delta \bar{p}_{n+i} = \bar{p}_{n+i,0} - \bar{p}_{n+i,1} = 2^{-i} \, (i > 0)$.

Proof. Denote the carry bit at the ith least significant bit position in $\psi = \phi + \beta$ as c_i, and that in $\psi' = \phi' + \beta$ as c'_i. Note that $c'_n = c_n$, thus $c'_n \oplus \beta_n = c_n \oplus \beta_n$. When $c'_n \oplus \beta_n = c_n \oplus \beta_n = 0$, we know that $\psi \oplus \psi' = 2^n$ with probability 1, i.e., $\psi_{n+i} = \psi'_{n+i}$ with probability 1 for $i > 0$. When $c'_n \oplus \beta_n = c_n \oplus \beta_n = 1$, by induction we obtain that $\psi_{n+i} = \psi'_{n+i}$ with probability $1 - 2^{-i+1}$ for $i > 0$. According to Lemma 1, we know that $c_n \oplus \beta_{n-1} = 0$ with probability $\frac{3}{4}$. If $\beta_n \oplus \beta_{n-1} = 0$, then $c_n \oplus \beta_n = 0$ with probability $\frac{3}{4}$, thus $\bar{p}_{n+i,0} = \frac{3}{4} \times 1 + \frac{1}{4} \times (1 - 2^{-i+1}) = 1 - \frac{1}{4} \times 2^{-i+1}$. If $\beta_n \oplus \beta_{n-1} = 1$, then $c_n \oplus \beta_n = 0$ with probability $\frac{1}{4}$, thus $\bar{p}_{n+i,1} = \frac{1}{4} \times 1 + \frac{3}{4} \times (1 - 2^{-i+1}) = 1 - \frac{3}{4} \times 2^{-i+1}$. Then the difference $\Delta\bar{p}_{n+i} = \bar{p}_{n+i,0} - \bar{p}_{n+i,1} = 2^{-i}$ for $i > 0$.

The above two theorems provide the guidelines to recover the key of Phelix. However, these two theorems deal with the ideal cases in which there is only one bit difference between ϕ and ϕ', and β is assumed to be random. In the attacks, we deal with the complicated situation where each bit of $\phi \oplus \phi'$ is biased, and β is a fixed integer. The value of each bit of β will affect the distribution of the higher order bits of $(\phi + \beta) \oplus (\phi' + \beta)$ in a complicated way. In order to simplify the analysis, we will use simulations to obtain these relations in the attacks.

4   A Basic Key Recovery Attack on Phelix

We will first investigate the differential propagation in Phelix. Then we show how to recover the key of Phelix by observing the differential distribution of the keystream.

4.1   The Bias in the Differential Distribution of the Keystream

Assume an attacker can choose an arbitrary value for the nonce, then a nonce can be used more than once. We introduce one-bit difference into the plaintext at the ith step, i.e., $P_i \neq P'_i$, and $P_i \oplus P'_i = 2^n$ ($0 \leq n \leq 31$). Then we analyze the difference between $B_3^{(i+1)}$ and $B_3'^{(i+1)}$ (as indicated in Fig. 1). If all the carry bits are 0 (replacing all the additions with XORs), then the differences only appear at the 9th, 11th, 13th, 15th and 17th least significant bits between $B_3^{(i+1)}$ and $B_3'^{(i+1)}$. Because of the carry bits, the differential distribution becomes complicated. We run the simulation and use the randomly generated $Y_k^{(i)}$ ($4 \geq k \geq 0$), P_i, $X_{i,1}$ in the simulation. With 2^{30} plaintext pairs, we obtain the distribution of $B_3^{(i+1)} \oplus B_3'^{(i+1)}$ in Table 1.

From Table 1, we see that the distribution of $B_3^{(i+1)} \oplus B_3'^{(i+1)}$ is heavily biased. For example, $B_3^{(i+1),8} = B_3'^{(i+1),8}$ with probability close to 1, while $B_3^{(i+1),9} = B_3'^{(i+1),9}$ with probability close to 0. Note that $T_0^{(i+1)} = A_0^{(i+1)} \oplus (B_3^{(i+1)} + X_{i+1,0})$, according to Theorem 2, the distribution of $T_0^{(i+1)} \oplus T_0'^{(i+1)}$ will be affected by the value of $X_{i+1,0}^8 \oplus X_{i+1,0}^9$, thus the distribution of $B_{i+1} \oplus B'_{i+1}$ will be affected by the value of $X_{i,0}^8 \oplus X_{i,0}^9$. By observing the distribution of $S_{i+1} \oplus S'_{i+1}$, it may be possible to determine the value of $X_{i,0}^8 \oplus X_{i,0}^9$. Shifting

Table 1. The probability that $B_3^{(i+1),j} \oplus B_3'^{(i+1),j} = 0$ for $P_i \oplus P_i' = 1$

j	p	j	p	j	p	j	p
0	0.9997	8	1.0000	16	0.5001	24	0.9161
1	0.9998	9	0.0000	17	0.4348	25	0.9470
2	0.9999	10	0.5000	18	0.5000	26	0.9673
3	0.9999	11	0.4375	19	0.5486	27	0.9803
4	1.0000	12	0.5000	20	0.6366	28	0.9883
5	1.0000	13	0.4492	21	0.7283	29	0.9931
6	1.0000	14	0.5000	22	0.8083	30	0.9960
7	1.0000	15	0.4273	23	0.8708	31	0.9977

the one-bit difference between P_i and P_i', we may determine other values of $X_{i,0}^{j+1} \oplus X_{i,0}^j$ for $0 \le j \le 30$, and recover in this way the key $X_{i,0}$. After recovering eight consecutive $X_{i,0}$ values, the 256-bit key can be found immediately.

The above analysis gives a brief idea of the attack. However, the actual attacks are quite complicated due to the interference of many differences. It is very tedious to derive exactly how the distribution of $S_{i+1} \oplus S_{i+1}'$ is affected by the value of $X_{i+1,0}^{j+1} \oplus X_{i+1,0}^j$. On the other hand, it is easy to search for the relation with simulations. In the following, we carried out a simulation to find the relation between the value of $X_{i+1,0}^{j+1} \oplus X_{i+1,0}^j$ and the distribution of $S_{i+1} \oplus S_{i+1}'$.

Let two plaintexts differ only in the ith word, and $P_i \oplus P_i' = 1$. We use the randomly generated $Y_k^{(i)}$ ($4 \ge k \ge 0$), P_i, $X_{i,1}$, $X_{i+1,0}$, $Z_4^{(i-3)}$ in the simulation. Denote with $p_{j,0}^n$ the probability that $S_{i+1}^n \oplus S_{i+1}'^n = 0$ when $X_{i,0}^{j+1} \oplus X_{i,0}^j = 0$. And denote $p_{j,1}^n$ as the probability that $S_{i+1}^n \oplus S_{i+1}'^n = 0$ when $X_{i,0}^{j+1} \oplus X_{i,0}^j = 1$. Let $\Delta \tilde{p}_j^n = (p_{j,0}^n - p_{j,1}^n) \times \frac{N}{\sigma}$, where N denotes the number of plaintext pairs, and $\sigma = \frac{\sqrt{N}}{2}$. Assume that the values of $p_{j,0}^n$ and $p_{j,1}^n$ are close to $\frac{1}{2}$. If $\Delta \tilde{p}_j^n > 4$, the difference between $p_{j,0}^n$ and $p_{j,1}^n$ is larger than 4σ, hence the value of $X_{i,0}^{j+1} \oplus X_{i,0}^j$ can be determined correctly with high probability. For every value of the two bits $X_{i,0}^{j+1}$ and $X_{i,0}^j$, we use 2^{28} pairs to generate $S_{i+1} \oplus S_{i+1}'$, then compute $p_{j,0}^n$ and $p_{j,1}^n$. Thus $N = 2^{29}$, and $\sigma = 2^{13.5}$. We list the large values of $\Delta \tilde{p}_j^n$ below:

For $j = 9$, $\Delta \tilde{p}_9^{13} = 55.7$.

For $j = 10$, $\Delta \tilde{p}_{10}^{13} = 133.9$.

For $j = 14$, $\Delta \tilde{p}_{14}^{17} = 51.5$.

For $j = 15$, $\Delta \tilde{p}_{15}^{19} = -9.1$, $\Delta \tilde{p}_{15}^{22} = 14.9$, $\Delta \tilde{p}_{15}^{23} = -15.7$.

For $j = 16$, $\Delta \tilde{p}_{16}^{19} = -50.8$, $\Delta \tilde{p}_{16}^{21} = 62.0$, $\Delta \tilde{p}_{16}^{22} = 97.7$, $\Delta \tilde{p}_{16}^{23} = -106.6$,
$\Delta \tilde{p}_{16}^{25} = 11.8$, $\Delta \tilde{p}_{16}^{26} = 16.0$, $\Delta \tilde{p}_{16}^{27} = -17.4$.

For $j = 17$, $\Delta \tilde{p}_{17}^{21} = 77.4$, $\Delta \tilde{p}_{17}^{23} = 145.3$, $\Delta \tilde{p}_{17}^{25} = -171.6$, $\Delta \tilde{p}_{17}^{25} = 12.3$,
$\Delta \tilde{p}_{17}^{26} = 28.5$, $\Delta \tilde{p}_{17}^{27} = -30.4$.

For $j = 18$, $\Delta \tilde{p}_{18}^{21} = 80.2$, $\Delta \tilde{p}_{18}^{22} = 179.7$, $\Delta \tilde{p}_{18}^{23} = -241.7$, $\Delta \tilde{p}_{18}^{26} = 32.8$,
$\Delta \tilde{p}_{18}^{27} = -43.7$.

For $j = 19$, $\Delta\tilde{p}_{19}^{22} = 139.6$, $\Delta\tilde{p}_{19}^{23} = -220.6$, $\Delta\tilde{p}_{19}^{26} = 19.0$, $\Delta\tilde{p}_{19}^{27} = -46.5$.
For $j = 20$, $\Delta\tilde{p}_{20}^{23} = -156.7$, $\Delta\tilde{p}_{20}^{25} = -5.7$, $\Delta\tilde{p}_{20}^{26} = 18.3$, $\Delta\tilde{p}_{20}^{27} = -30.6$.
For $j = 21$, $\Delta\tilde{p}_{21}^{25} = -6.8$, $\Delta\tilde{p}_{21}^{26} = 9.5$, $\Delta\tilde{p}_{20}^{27} = -28.5$.

The data given above show that the distribution of $S_{i+1} \oplus S'_{i+1}$ is strongly affected by the value of $X_{i+1,0}^{j+1} \oplus X_{i+1,0}^{j}$.

4.2   Recovering the Key

Note that in the above analysis, when we deal with a particular $X_{i+1,0}^{j+1} \oplus X_{i+1,0}^{j}$, the other bits of $X_{i+1,0}$ are random. In the key recovery attack, the value of $X_{i+1,0}$ is fixed, so we need to consider the interference between the bits $X_{i+1,0}^{j+1} \oplus X_{i+1,0}^{j}$.

We notice that there are many large biases related to $S_{i+1}^{23} \oplus S'^{23}_{i+1}$. However, the values of $X_{i+1,0}^{j+1} \oplus X_{i+1,0}^{j}$ ($15 \le j \le 20$) all have a significant effect on the distribution of $S_{i+1}^{23} \oplus S'^{23}_{i+1}$. It is thus a bit complicated to determine the values of $X_{i+1,0}^{j+1} \oplus X_{i+1,0}^{j}$ ($15 \le j \le 20$).

In the following, we consider the bit $S_{i+1}^{17} \oplus S'^{17}_{i+1}$. Its distribution is dominated by the value of $X_{i+1,0}^{15} \oplus X_{i+1,0}^{14}$. For every value of the two bits $X_{i+1,0}^{15}$ and $X_{i+1,0}^{14}$, we use 2^{30} pairs to generate $S_{i+1} \oplus S'_{i+1}$, then compute $p_{14,0}^{17}$ and $p_{14,1}^{17}$. From the simulation, we found that $p_{14,0}^{17} = 0.50227$ and $p_{14,1}^{17} = 0.50117$. We denote the average of $p_{14,0}^{17}$ and $p_{14,1}^{17}$ as $\bar{p}_{14}^{17}$, i.e., $\bar{p}_{14}^{17} = \frac{p_{14,0}^{17}+p_{14,1}^{17}}{2} = 0.50172$. Running a similar simulation, we found that $p_{13,0}^{17} = 0.50175$ and $p_{13,1}^{17} = 0.50169$. For all the $j \ne 13, j \ne 14$, we found that $p_{j,0}^{17} \approx \bar{p}_{14}^{17}$ and $p_{j,1}^{17} \approx \bar{p}_{14}^{17}$. The value of $X_{i+1,0}^{15} \oplus X_{i+1,0}^{14}$ is recovered as follows: from the keystreams, we compute the fraction for which $S_{i+1}^{17} \oplus S'^{17}_{i+1} = 0$. If it is larger than $\bar{p}_{14}^{17}$, then the value of $X_{i+1,0}^{15} \oplus X_{i+1,0}^{14}$ is considered to be 0; otherwise the value of $X_{i+1,0}^{15} \oplus X_{i+1,0}^{14}$ is considered to be 1.

We now compute the number of plaintext pairs required to determine the value of $X_{i+1,0}^{15} \oplus X_{i+1,0}^{14}$. Suppose that N pairs of plaintexts are used. The standard deviation is $\sigma = \sqrt{N \times \bar{p}_{14}^{17} \times (1 - \bar{p}_{14}^{17})}$. To determine the value of $X_{i+1,0}^{15} \oplus X_{i+1,0}^{14}$ with success rate 0.99, we require that $N \times ((p_{14,0}^{17} - p_{14,1}^{17}) - (p_{13,0}^{17} - p_{13,1}^{17})) > 4.66 \times \sigma$ (The cumulative distribution function of the normal distribution gives value 0.99 at the point 2.33σ). Thus we require that $N > 2^{22.27}$.

We used the Phelix C source code submitted to eSTREAM in the experiments.[1]

Experiment 1. The goal of this experiment is to recover the value of $X_{1,0}^{15} \oplus X_{1,0}^{14}$. Each plaintext has two words P_0 and P_1. For each plaintext pair, the two words differ only in the least significant bit of P_0. N plaintext pairs are used for each key to determine the value of $X_{1,0}^{15} \oplus X_{1,0}^{14}$ as follows: if the fraction of cases for

---

[1] Note that we have found and corrected a small bug in this source code. The output should be computed as $S_i = Y_4^{(i)} + Z_4^{(i-4)}$ in stead of $S_i = Y_4^{(i)} + Z_4^{(i-3)}$.

which $S_1^{17} \oplus S_1'^{17} = 0$ is larger than $\bar{p}_{14}^{17} = 0.50172$, then the value of $X_{1,0}^{15} \oplus X_{1,0}^{14}$ is considered to be 0; otherwise the value of $X_{1,0}^{15} \oplus X_{1,0}^{14}$ is considered to be 1. A random nonce was used for each plaintext pair. We tested 200 keys in the experiment. For $N = 2^{22.3}$, the values of $X_{1,0}^{15} \oplus X_{1,0}^{14}$ of 183 keys are determined correctly. For $N = 2^{25}$, the values of $X_{1,0}^{15} \oplus X_{1,0}^{14}$ of 192 keys are determined correctly.

Experiment 1 shows that the value of $X_{1,0}^{15} \oplus X_{1,0}^{14}$ can be determined successfully by introducing a difference in the least significant bit of P_0, but with a higher error rate. The reason is that other bits of $X_{1,0}$ affects the determination of $X_{1,0}^{15} \oplus X_{1,0}^{14}$ in a subtle way.

We now proceed to recover the other bits of $X_{1,0}$. By rotating the one-bit difference between P_0 and P_0', and using the same threshold value, we can determine the value of $X_{1,0}^{j+1} \oplus X_{1,0}^j$ for $2 \le j \le 3$, $5 \le j \le 10$ and $14 \le j \le 28$.

Thus we are able to recover 23 bits of information on each $X_{i,0}$. Note that the 256-bit key is recovered from eight consecutive $X_{i,0}$. Thus we are able to recover $23 \times 8 = 184$ bits of the key with success rate about $\frac{192}{200} = 0.96$. The number of plaintext pairs required in the attack is about $2^{25} \times 32 \times 8 = 2^{33}$.

We need to improve the above attack in two approaches: recovering more key bits and improving the success rate. The direct approach is to adjust the threshold value for each key bit position. In the following, we illustrate a more advanced approach which recovers the values of $Z_4^{(i)}$ before recovering the key.

5   Improving the Attack on Phelix

In the above attack, we use a random nonce for each plaintext pair, i.e., every nonce is used twice with the same key. When the nonce is used many times with the same key, we can introduce the difference at P_i and recover the value of Z_4^{i-3} by observing the distribution of $S_{i+1} \oplus S_{i+1}'$. Then we proceed to recover $X_{i+1,0}$.

5.1   Recovering $Z_4^{(i)}$

We introduce the difference to the least significant bit of P_i ($P_i \oplus P_i' = 1$). A simulation is carried out to determine the distribution of $Y_4^{(i+1)} \oplus Y_4'^{(i+1)}$. We use the randomly generated $Y_k^{(i)}$ ($4 \ge k \ge 0$), P_i, $X_{i,1}$, $X_{i+1,0}$ in the simulation. Denote $\dot{p}_n$ as the probability that $Y_4^{(i+1),n} \oplus Y_4'^{(i+1),n} = 0$. With 2^{30} pairs, we obtain the values of $\dot{p}_n$ in Table 2.

From Table 2, we notice that $Y_4^{(i+1)} \oplus Y_4'^{(i+1)}$ is heavily biased. For example, $Y_4^{(i+1),2} = Y_4'^{(i+1),2}$ with probability about 0.70291, while $Y_4^{(i+1),3} = Y_4'^{(i+1),3}$ with probability about 0.22246. Note that $S_{i+1} = Y_4^{(i+1)} \oplus Z_4^{(i-3)}$, according to Theorem 2, the distribution of $S_{i+1} \oplus S_{i+1}'$ is affected by the value of $Z_4^{(i-3),3} \oplus Z_4^{(i-3),2}$. Next we carry out simulations to characterize this relation.

We use the randomly generated $Y_k^{(i)}$ ($4 \ge k \ge 0$), P_i, $X_{i,1}$, $X_{i+1,0}$, $Z_4^{(i-3)}$ in the simulation. The one-bit difference is introduced to P_i, i.e., $P_i \oplus P_i' = 2^j$.

Table 2. The probability that $Y_4^{(i+1),j} \oplus Y_4'^{(i+1),j} = 0$ for $P_i \oplus P_i' = 1$

j	$\dot{p}_j - 0.5$	j	$\dot{p}_j - 0.5$	j	$\dot{p}_j - 0.5$	j	$\dot{p}_j - 0.5$
0	0.03326	8	0.00003	16	−0.00003	24	0.00046
1	0.12983	9	0.03517	17	0.00268	25	0.05926
2	0.20291	10	0.00002	18	−0.00001	26	0.15064
3	−0.27754	11	0.00001	19	−0.00266	27	−0.24028
4	−0.00005	12	0.00000	20	−0.00004	28	0.00001
5	0.05663	13	0.02293	21	0.02276	29	0.05770
6	−0.15327	14	−0.00001	22	0.07434	30	0.15508
7	−0.00001	15	−0.00001	23	−0.14414	31	−0.24907

Denote $\ddot{p}_{j,0}^n$ as the probability that $S_{i+1}^n \oplus S_{i+1}''^n = 0$ when $Z_4^{(i-3),j+1} \oplus Z_4^{(i-3),j} = 0$. And denote $\ddot{p}_{j,1}^n$ as the probability that $S_{i+1}^n \oplus S_{i+1}''^n = 0$ when $Z_4^{(i-3),j+1} \oplus Z_4^{(i-3),j} = 1$. For each value of $Z_4^{(i-3),3}$ and $Z_4^{(i-3),2}$, we use 2^{28} plaintext pairs. We find that $\ddot{p}_{2,0}^5 = 0.5461$ and $\ddot{p}_{2,1}^5 = 0.5193$. The large difference between $\ddot{p}_{2,0}^5$ and $\ddot{p}_{2,1}^5$ shows that the value of $Z_4^{(i-3),3} \oplus Z_4^{(i-3),2}$ can be determined with success rate 0.999 with about $2^{13.9}$ plaintext pairs (The cumulative distribution function of the normal distribution gives value 0.999 at the point 3.1σ).

The above approach is able to recover $Z_4^{(0)}$, but the success rate is not that high according to our experiment. In the following, we use a new approach to determine $Z_4^{(0)}$. To reduce the interference between the bits of $Z_4^{(i-3)}$, we recover the least significant bit of $Z_4^{(i-3)}$ first, then proceed to recover the more significant bits bit-by-bit.

We start with determining the value of $Z_4^{(i-3),0}$. Let $P_i \oplus P_i' = 1$. Running the simulation with 2^{28} plaintext pairs, we found that $\ddot{p}_{-1,0}^2 = 0.70296$, and $\ddot{p}_{-1,1}^2 = 0.65422$ (let $Z_4^{(i-3),-1} = 0$). To determine the value of $Z_4^{(i-3),0}$ with success rate 0.999, we need about $2^{12.0}$ plaintext pairs.

Experiment 2. The goal of this experiment is to determine the value of $Z_4^{(0),0}$. Each plaintext has five random words P_i ($0 \leq i \leq 4$). For each plaintext pair, the difference is only in the least significant bit of P_3. N plaintext pairs are used for each key/nonce pair to determine the value of $Z^{(0),0}$ as follows: if the rate that $S_4^2 \oplus S_4'^2 = 0$ is larger than $\frac{\ddot{p}_{-1,0}^2 + \ddot{p}_{-1,1}^2}{2} = \frac{0.70296 + 0.65422}{2} = 0.6786$, then the value of $Z_4^{(0),0}$ is considered to be 0; otherwise the value of $Z_4^{(0),0}$ is considered to be 1. We tested 1000 key/nonce pairs in the experiment. For $N = 2^{12}$, the values of $Z_4^{(0),0}$ of 998 key/nonce pairs are determined correctly. For $N = 2^{13}$, the values of $Z_4^{(0),0}$ of all the key/nonce pairs are determined correctly.

After recovering the value of $Z_4^{(i-3),0}$, we proceed to recover the values of the other bits of $Z_4^{(i-3)}$. Let $Z_4^{(i-3),(n-1\cdots0)}$ denote the n least significant bits of $Z^{(i-3)}$, i.e., $Z_4^{(i-3),(n-1\cdots0)} = Z^{(i-3)} \bmod 2^n$. Let the difference be introduced to the kth least significant bit of P_i, i.e., $P_i \oplus P_i' = 2^k$. Denote $\dot{p}_{Z_4^{(i-3),(n-1\cdots0)}}^{k,j,0}$ as the probability that the value of the jth bit of $(S_{i+1} - Z_4^{(i-3),(n-1\cdots0)}) \oplus (S'_{i+1} - Z_4^{(i-3),(n-1\cdots0)})$ is 0 when $Z_4^{(i-3),n} = 0$. Denote $\dot{p}_{Z_4^{(i-3)},(n-1\cdots0)}^{k,j,1}$ as the probability that the value of the jth bit of $(S_{i+1} - Z_4^{(i-3),(n-1\cdots0)}) \oplus (S'_{i+1} - Z_4^{(i-3),(n-1\cdots0)})$ is 0 when $Z_4^{(i-3),n} = 1$. If the value of $Z_4^{(i-3),(n-1\cdots0)}$ is determined correctly, then $\dot{p}_{Z_4^{(i-3)},(n-1\cdots0)}^{k,j,0} = \dot{p}_0^{k,j,0}$, and $\dot{p}_{Z_4^{(i-3)},(n-1\cdots0)}^{k,j,1} = \dot{p}_0^{k,j,1}$. This property is important for recovering $Z_4^{(i-3)}$.

Let $P_i \oplus P'_i = 2$. We use 2^{28} plaintext pairs in the simulation. We found that $\dot{p}_0^{1,3,0} = 0.66469$ and $\dot{p}_0^{1,3,1} = 0.60220$. It shows that when $Z_4^{(i-3),0} = 0$, if the rate that $S_{i+1}^3 \oplus S'^3_{i+1} = 0$ is larger than $\frac{0.66469+0.60220}{2} = 0.63345$, then the value of $Z_4^{(i-3),1}$ is determined to be 0; otherwise the value of $Z_4^{(i-3),1}$ is determined to be 1. We need about $2^{11.3}$ plaintext pairs to determine the value of $Z_4^{(i-3),1}$ correctly with success rate 0.999. Using the Phelix code in the experiment, we tested 1000 random key/nonce pairs satisfying $Z_4^{(0),0} = 0$, and 2^{12} plaintext pairs are used for each key/nonce pair with the difference $P_3 \oplus P'_3 = 2$. We found that all the 1000 values of $Z_4^{(0),1}$ are determined correctly. If $Z_4^{(i-3),0} = 1$, we observe the third least significant bit of $(S_{i+1} - 1) \oplus (S'_{i+1} - 1)$, and we can determine the value of $Z_4^{(0),1} = 0$ with success rate 0.999 with about $2^{11.3}$ plaintext pairs.

Let $P_i \oplus P'_i = 2^2$, we are able to determine the value of $Z_4^{(i-3),2}$ by observing the fourth least significant bit of $(S_{i+1} - Z_4^{(i-3),(1\cdots0)}) \oplus (S'_{i+1} - Z_4^{(i-3),(1\cdots0)})$. In general, let $P_i \oplus P'_i = 2^j$, then we are able to determine the value of $Z_4^{(i-3),j}$ by observing the $(j+2)$th least significant bit of $(S_{i+1} - Z_4^{(i-3),(j-1\cdots0)}) \oplus (S'_{i+1} - Z_4^{(i-3),(j-1\cdots0)})$ with about 2^{12} plaintext pairs. Thus we are able to recover $Z_4^{(i-3)}$ (except the values of $Z_4^{(i-3),30}$ and $Z_4^{(i-3),31}$) with success rate very close to 1. The number of plaintext pairs required in the above attack is about $2^{12} \times 30 \approx 2^{17}$.

5.2 Recovering $X_{i+1,0}$

After recovering $Z_4^{(i-3)}$ (except $Z_4^{(i-3),31}$ and $Z_4^{(i-3),30}$), we know the value of $(S_{i+1} - Z_4^{(i-3),(29\cdots0)}) \oplus (S'_{i+1} - Z_4^{(i-3),(29\cdots0)})$. Thus we know the value of $Y^{(i+1),j} \oplus Y'^{(i+1),j}$ $(0 \le j \le 30)$. Then we are able to recover $X_{i+1,0}$ more efficiently.

Let two plaintexts differ only in the ith word. And let $P_i \oplus P'_i = 1$. We use the randomly generated $Y_k^{(i)}$ $(4 \ge k \ge 0)$, P_i, $X_{i,1}$, $X_{i+1,0}$, $Z_4^{(i-3)}$ in the simulation. For every value of the two bits $X_{i,0}^{j+1}$ and $X_{i,0}^j$, we use 2^{28} plaintext pairs to generate $Y_4^{(i+1)} \oplus Y_4'^{(i+1)}$, then compute $p_{j,0}^n$ and $p_{j,1}^n$ (suppose that $S_{i+1} = Y_4^{(i+1)}$ since $Z_4^{(i-3)}$ is known). Thus $N = 2^{29}$, and $\sigma = 2^{13.5}$. We list the following two large biases $\Delta\tilde{p}_j^n$:

For $j = 9$, $\Delta\tilde{p}_9^{13} = 144.1$
For $j = 10$, $\Delta\tilde{p}_{10}^{13} = 362.12$

We use $\Delta \tilde{p}_9^{13}$ and $\Delta \tilde{p}_{10}^{13}$ in the attack. Note that the values of $X_{i+1,0}^{10} \oplus X_{i+1,0}^9$ and $X_{i+1,0}^{11} \oplus X_{i+1,0}^{10}$ both affect the distribution of $Y_4^{(i+1),13} \oplus Y_4'^{(i+1),13}$. We carried out a simulation with 2^{30} chosen plaintext pairs to determine how the value of $X_{i+1,0}^9$ affects the value of $Y_4^{(i+1),13}$. If $X_{i+1,0}^9 = 0$, and the values of $X_{i+1,0}^{11} \| X_{i+1,0}^0$ are 00, 11, 01 and 11 (in binary format), we obtain that $Y_4^{(i+1),13} \oplus Y_4'^{(i+1),13} = 0$ with probability 0.53033, 0.52334, 0.51946 and 0.51864, respectively; if $X_{i+1,0}^9 = 1$, we obtain that $Y_4^{(i+1),13} \oplus Y_4'^{(i+1),13} = 0$ with probability 0.52334, 0.53030, 0.51861 and 0.51948, respectively. We thus let $p_{0,0}^{13} = 0.52334$, and $p_{0,1}^{13} = \frac{0.51946+0.51948}{2} = 0.51947$. About $2^{19.3}$ plaintext pairs are required to determine the value of $X_{i+1,0}^{11} \oplus X_{i+1,0}^{10}$ with success rate 0.999.

Experiment 3. Suppose that the value of $Z_4^{(0)}$ is known. The goal of this experiment is to determine the value of $X_{4,0}^{11} \oplus X_{4,0}^{10}$. Each plaintext has five random words P_i ($0 \le i \le 4$). For each plaintext pair, those five words differ only in the least significant bit of P_3. N plaintext pairs are used for each key/nonce pair to determine the value of $X_{4,0}^{11} \oplus X_{4,0}^{10}$ as follows: if the rate that $Y_4^{13} \oplus Y_4'^{13} = 0$ is larger than $\frac{0.52334+0.51947}{2} = 0.52140$, then the value of $X_{4,0}^{11} \oplus X_{4,0}^{10}$ is considered to be 0; otherwise the value of $X_{4,0}^{11} \oplus X_{4,0}^{10}$ is considered to be 1. We tested 1000 key/nonce pairs in the experiment. For $N = 2^{19.3}$, 948 values of 1000 $X_{4,0}^{11} \oplus X_{4,0}^{10}$ are determined correctly. We change the threshold value 0.52140 to 0.52035, then 970 values of 1000 $X_{4,0}^{11} \oplus X_{4,0}^{10}$ are determined correctly for $N = 2^{20}$, 976 values are determined correctly for $N = 2^{21}$, 990 values are determined correctly for $N = 2^{22}$.

Experiment 3 shows that the value of $X_{4,0}^{11} \oplus X_{4,0}^{10}$ can be determined successfully by introducing a difference in the least significant bit of P_3. With 2^{22} chosen pairs, we are able to determine the value of $X_{4,0}^{11} \oplus X_{4,0}^{10}$ with success rate about 0.99.

Then we shift the one-bit difference to recover the values of $X_{1,0}^{j+1} \oplus X_{1,0}^j$ for $2 \le j \le 28$. The threshold value needs to be modified for different values of j. The results are given in Table 3 in Appendix A. Note that according to Experiment 3, the threshold values should be slightly adjusted to achieve high success rate.

The reason that the values of $X_{4,0}^{j+1} \oplus X_{4,0}^j$ cannot be recovered for $j \ge 29$ is that the value of $X_{1,0}^{j+1} \oplus X_{1,0}^j$ cannot affect the distribution of $S_1^{j+3} \oplus S_1'^{j+3}$ since $S_1 \oplus S_1'$ is a 32-bit word. The reason that the number of plaintext required for $j = 9$ is relatively small is that the difference for $j = 13$ is introduced to the most significant bit of the word P_3, thus it causes less difference propagation, and results in a larger bias in the keystream.

Note that the most significant bit of $Y_4^{(i+1)} \oplus Y_4'^{(i+1)}$ is not known since $Z_4^{i-1,31}$ and $Z_4^{i-1,30}$ are not recovered. Thus to determine the value of $X_{1,0}^{29} \oplus X_{1,0}^{28}$, we need to consider the most significant bit of $(S_{i+1} - Z_4^{(i-3),(29\cdots0)}) \oplus (S'_{i+1} - Z_4^{(i-3),(29\cdots0)})$. The threshold value needs to be changed to 0.51128; and the number of plaintext pairs required is $2^{22.1}$.

After recovering the values of $X_{1,0}^{j+1} \oplus X_{1,0}^{j}$ for $2 \le j \le 28$, we proceed to determine the value of $X_{i+1,0}^0$, $X_{i+1,0}^1$ and $X_{i+1,0}^2$.

We start with recovering $X_{i+1,0}^0$. Let $P_i \oplus P'_i = 2^{21}$. Running the simulation with 2^{28} plaintext pairs, we found that $p_{21,0}^2 = 0.51596$, $p_{21,1}^2 = 0.50355$. Thus $2^{15.93}$ plaintext pairs are needed to determine the value of $X_{i+1,0}^0$ with success rate 0.999. Using the Phelix code in the experiment, we introduce the difference $P_3 \oplus P'_3 = 2^{21}$, and set the threshold value as $\frac{0.51596+0.50355}{2} = 0.50975$. We tested 1000 key/nonce pairs in the experiment. With 2^{16} plaintext pairs, all the values of the 1000 $X_{4,0}^0$ are determined correctly.

After determine the value of $X_{i+1,0}^0$, we determine the value of $X_{i+1,0}^1$ as follows. The simulation shows that the value of $X_{i+1,0}^1$ can be determined only when $X_{i+1,0}^0 = 0$. For $X_{i+1,0}^0 = 0$, we set the difference as $P_i \oplus P'_i = 2^{22}$. With 2^{28} chosen plaintext pairs, we found that if $X_{i+1,0}^1 = 0$, then $Y_4^{(i+1),3} = 0$ with rate 0.51528; otherwise $Y_4^{(i+1),3} = 0$ with rate 0.50459. With $2^{16.4}$ plaintext pairs, the value of $X_{i+1,0}^1$ can be determined with success rate 0.999. Using the Phelix code in the experiement, we introduce the difference $P_3 \oplus P'_3 = 2^{22}$, and set the threshold value as $\frac{0.51528+0.50459}{2} = 0.50994$. We tested 1000 key/nonce pairs with $X_{4,0}^0 = 0$ in the experiment. With $2^{16.4}$ plaintext pairs, all the values of the 1000 $X_{4,0}^1$ are determined correctly. It shows that the value of $X_{i+1,0}^1$ can be determined successfully if $X_{4,0}^0 = 0$.

We continue to recover the value of $X_{i+1,0}^2$. We introduce a difference to the 15th least significant bit of P_i, and observe the distribution of $Y_4^{(i-3),4}$. We carry out a simulation with 2^{31} plaintext pairs with $P_3 \oplus P'_3 = 2^{15}$. 2^{31} plaintext pairs are used for each value of $X_{i+1,0}^1 X_{i+1,0}^0$. When $X_{i+1,0}^0 = 0$, if $X_{i+1,0}^1 = 0$, the fraction of values for which $Y_4^{(i-3),4} = 0$ if $X_{i+1,0}^2 = 0$ and $X_{i+1,0}^2 = 1$ are 0.53106 and 0.52613, respectively; if $X_{i+1,0}^1 = 1$, the franction of values for which $Y_4^{(i-3),4} = 0$ if $X_{i+1,0}^2 = 0$ and $X_{i+1,0}^2 = 1$ are 0.52318 and 0.52315, respectively. It shows that the value $X_{i+1,0}^2$ can only be determined if the values of $X_{i+1,0}^1$ and $X_{i+1,0}^1$ are both zero, and $2^{18.6}$ plaintext pairs are required to achieve the success rate 0.999.

In the above attacks, we recovered 28.75 bits of $X_{i+1,0}$: $X_{i+1,0}^{j+1} \oplus X_{i+1,0}^j$ for $2 \le j \le 28$, $X_{i+1,0}^0$, $X_{i+1,0}^1$ (only if $X_{i+1,0}^0 = 0$), and $X_{i+1,0}^1$ (only if $X_{i+1,0}^0 = 0$ and $X_{i+1,0}^1 = 0$). 27 bits of $X_{i+1,0}^{j+1} \oplus X_{i+1,0}^j$ ($2 \le j \le 28$) can be determined according to Table 3. From Experiment 3, we know that if we slightly adjust the threshold value, and use about $2^{2.7}$ times the number of plaintext pairs compared to Table 3, the success rate is about 0.99. The number of plaintext pairs required to determine these 27 bits is thus about $27 \times 2^{22.2} \times 2^{2.7} = 2^{29.7}$. The number of plaintext pairs to determine $X_{i+1,0}^0$, $X_{i+1,0}^1$ and $X_{i+1,0}^2$ is small compared to

$2^{29.7}$. The attack to recover 28.75 bits of $X_{i+1,0}$ requires thus about $2^{32.7}$ chosen plaintext pairs.

After recovering eight consecutive $X_{i+1,0}$, we recovered $28.75 \times 8 = 230$ key bits. To recover the 256-bit key, the amount of operations required is about $2^{256-230+\left(\frac{27 \times 8}{0.01 \times 27 \times 8}\right)} = 2^{41.5}$.

The number of chosen plaintext pairs required in the attack is about $2^{29.7} \times 8 = 2^{32.7}$. The length of each plaintext ranges from 5 to 13 words. Thus the total amount of chosen plaintext required is about $2 \times 2^{32.7} \times \frac{5+13}{2} \approx 2^{37}$ words. (The number of plaintext pairs needed to recover 8 consecutive $Z_4^{(i)}$ is about $2^{17} \times 8 = 2^{25}$. It is small compared to $2^{32.7}$).

6   An Approach to Strengthen Helix and Phelix

In Helix and Phelix, the plaintext is used to affect the internal state of the cipher. In order to achieve a high encryption speed, each plaintext word affects the keystream without passing through sufficient confusion and diffusion layers. This is the intrinsic weakness in the structure of Helix and Phelix. In the following, we provide a method to reduce the effect of such weakness.

The security of the encryption of Helix and Phelix can be improved significantly if a secure one-way function is used to generate the initial state of the cipher from the key and nonce. Then even if the internal state of one particular nonce is recovered, the impact on the security of the encryption is very limited since the key of the cipher is not affected. We believe that such an approach can be applied to improve the security of all the ciphers that use the plaintext to affect the internal state.

However, we must point out that such an approach does not substantially improve the security of the MAC in Helix and Phelix. Once an internal state is recovered, the attacker can forge many messages related to that particular nonce.

7   Conclusion

Phelix is vulnerable to a key recovery attack when chosen nonces and chosen plaintexts are used. The computational complexity of the attack is much less than that of the attack against Helix. Our attack shows that Phelix fails to strengthen Helix in this respect.

We believe that one necessary requirement for a secure general-purpose stream cipher is that the key of the cipher should not be recoverable even if the attacker can control the generation of the nonce. In practice an attacker may gain access to a Phelix encryption device for a while, reuse a nonce and recover the key. We thus consider Phelix as insecure. (When the integrity checking mechanism is not enforced, an attacker can even modify the nonces in ciphertext and obtain repeated nonces.) Muller has pointed out the practical impact of the key recovery attack with reused nonces on the security of Helix in detail [2]. Muller stated clearly the difference between the nonce reusing attack against Helix and that against a synchronous stream cipher since the attack against Helix results in key recovery. The same comments apply to the attacks against Phelix.

Acknowledgements

The authors would like to thank the anonymous reviewers of FSE 2007 for their helpful comments.

References

1. Ferguson, N., Whiting, D., Schneier, B., Kelsey, J., Lucks, S., Kohno, T.: Helix, Fast Encryption and Authentication in a Single Cryptographic Primitive. In: Johansson, T. (ed.) FSE 2003. LNCS, vol. 2887, pp. 330–346. Springer, Heidelberg (2003)
2. Muller, F.: Differential Attacks against the Helix Stream Cipher. In: Roy, B., Meier, W. (eds.) FSE 2004. LNCS, vol. 3017, pp. 94–108. Springer, Heidelberg (2004)
3. Paul, S., Preneel, B.: Solving Systems of Differential Equations of Addition. In: Boyd, C., González Nieto, J.M. (eds.) ACISP 2005. LNCS, vol. 3574, pp. 75–88. Springer, Heidelberg (2005)
4. Paul, S., Preneel, B.: Near Optimal Algorithms for Solving Differential Equations of Addition with Batch Queries. In: Maitra, S., Madhavan, C.E.V., Venkatesan, R. (eds.) INDOCRYPT 2005. LNCS, vol. 3797, pp. 90–103. Springer, Heidelberg (2005)
5. Whiting, D., Schneier, B., Lucks, S., Muller, F.: Phelix: Fast Encryption and Authentication in a Single Cryptographic Primitive. eSTREAM, ECRYPT Stream Cipher Project Report 2005/027 (2005)
6. Wu, H., Preneel, B.: Cryptanalysis of the Stream Cipher ABC v2. In: Selected Areas in Cryptography – SAC 2006. LNCS (to appear)

A   The Complexity to Recover $X_{i+1,0}$ with $Z_4^{(i-3)}$ Known

The number of plaintext pairs and the threshold value required to recover the value of each $X_{i+1,0}^{j+1} \oplus X_{i+1,0}^{j}$ ($2 \leq j \leq 28$) are given in Table 3. Each value n in the second column indicates that the difference is introduced in the nth least significant bit of P_i. Each value n in the third column shows that the nth least significant bit of $Y_4^{(i+1)}$ is used in the attack.

Table 3. The number of plaintext pairs for recovering $X_{i+1,0}^{j+1} \oplus X_{i+1,0}^{j}$

j	Difference position in P_i	Bit position in $Y_4^{(i+1)}$	Threshold value	Plaintext Pairs
2	24	5	0.51101	$2^{24.4}$
3	25	6	0.51110	$2^{23.1}$
4	26	7	0.51120	$2^{22.5}$
5	27	8	0.51125	$2^{22.3}$
6	28	9	0.51091	$2^{22.4}$
7	29	10	0.51116	$2^{23.6}$
8	30	11	0.51562	$2^{20.7}$
9	31	12	0.54353	$2^{18.4}$
10	0	13	0.52141	$2^{19.3}$
11	1	14	0.52099	$2^{19.3}$
12	2	15	0.51850	$2^{19.5}$
13	3	16	0.50998	$2^{21.3}$
14	4	17	0.51107	$2^{21.9}$
15	5	18	0.51128	$2^{22.2}$
16	6	19	0.51129	$2^{22.2}$
17	7	20	0.51131	$2^{22.2}$
18	8	21	0.51128	$2^{22.1}$
19	9	22	0.51117	$2^{21.7}$
20	10	23	0.51149	$2^{22.2}$
21	11	24	0.51172	$2^{22.0}$
22	12	25	0.51187	$2^{22.0}$
23	13	26	0.51191	$2^{22.0}$
24	14	27	0.51185	$2^{22.1}$
25	15	28	0.51129	$2^{22.2}$
26	16	29	0.51129	$2^{22.1}$
27	17	30	0.51131	$2^{22.2}$
28	18	31	0.51130	$2^{22.1}$

How to Enrich the Message Space of a Cipher

Thomas Ristenpart[1] and Phillip Rogaway[2]

[1] Dept. of Computer Science & Engineering, University of California San Diego
9500 Gilman Drive, La Jolla, CA 92093-0404, USA
tristenp@cs.ucsd.edu
http://www-cse.ucsd.edu/users/tristenp
[2] Dept. of Computer Science, University of California Davis
One Shields Avenue, Davis, CA 95616, USA
rogaway@cs.ucdavis.edu
http://www.cs.ucdavis.edu/~rogaway/

Abstract. Given (deterministic) ciphers $\mathcal{E}$ and E that can encipher messages of l and n bits, respectively, we construct a cipher $\mathcal{E}^* = \mathrm{XLS}[\mathcal{E}, E]$ that can encipher messages of $l + s$ bits for any $s < n$. Enciphering such a string will take one call to $\mathcal{E}$ and two calls to E. We prove that $\mathcal{E}^*$ is a strong pseudorandom permutation as long as $\mathcal{E}$ and E are. Our construction works even in the tweakable and VIL (variable-input-length) settings. It makes use of a multipermutation (a pair of orthogonal Latin squares), a combinatorial object not previously used to get a provable-security result.

Keywords: Deterministic encryption, enciphering scheme, symmetric encryption, length-preserving encryption, multipermutation.

1 Introduction

DOMAIN EXTENSION. Consider a cryptographic scheme with a message space $\mathcal{M} = \bigcup_{l \in \mathcal{L}} \{0,1\}^l$ for some set $\mathcal{L}$ of *permissible message lengths*. The scheme can handle any message of $l \in \mathcal{L}$ bits but it can't handle messages of $l^* \notin \mathcal{L}$ bits. Often the set of permissible message lengths $\mathcal{L}$ is what worked out well for the scheme's *designers*—it made the scheme simple, natural, or amenable to analysis—but it might not be ideal for the scheme's *users* who, all other things being equal, might prefer a scheme that works across arbitrary-length messages. To address this issue, one may wish to *extend* the scheme to handle more message lengths. Examples are extending CBC encryption using ciphertext stealing [21] and extending a pseudorandom function F with message space $(\{0,1\}^n)^+$ by appropriately padding the message and calling F.

Our work is about extending the domain of a *cipher*. When we speak of a *cipher* in this paper we mean a *deterministic* map $\mathcal{E} \colon \mathcal{K} \times \mathcal{M} \to \mathcal{M}$ where $\mathcal{M} = \bigcup_{l \in \mathcal{L}} \{0,1\}^l$ and $\mathcal{E}_K(\cdot) = \mathcal{E}(K, \cdot)$ is a *length-preserving permutation*. Such an object is also called an *enciphering scheme*, a *pseudorandom permutation*, an *arbitrary-input-length blockcipher*, or a *deterministic* cipher / encryption scheme. Our goal is to extend a cipher $\mathcal{E} \colon \mathcal{K} \times \mathcal{M} \to \mathcal{M}$ with permissible message

A. Biryukov (Ed.): FSE 2007, LNCS 4593, pp. 101–118, 2007.

lengths $\mathcal{L}$ to a cipher $\mathcal{E}^*\colon \mathcal{K}^* \times \mathcal{M}^* \to \mathcal{M}^*$ with an enlarged set $\mathcal{L}^* \supseteq \mathcal{L}$ of permissible message lengths. Being an *extension* of $\mathcal{E}$, what $\mathcal{E}^*$ does on a string of length $l \in \mathcal{L}$ and key $\langle K, K' \rangle$ must be identical to what $\mathcal{E}$ would do on key K. Note that padding-based methods will not work: even if there *is* a point in the message space of $\mathcal{E}$ that one can pad a plaintext to, padding M to M^* and then applying $\mathcal{E}$ would be length-increasing, and so not a cipher. Unlike signatures, MACs, pseudorandom functions, and semantically secure encryption, there is no obvious way to extend a cipher's domain.

OUR CONTRIBUTION. We show how, with the help of an n-bit blockcipher E, to extend a cipher's set of permissible message lengths from $\mathcal{L} \subseteq [n..\infty)$ to $\mathcal{L}^* = \mathcal{L} + [0..n-1] = \{\ell + i \mid \ell \in \mathcal{L} \text{ and } i \in [0..n-1]\}$. In other words, we enlarge the message space from $\mathcal{M}$ to $\mathcal{M}^* = \mathcal{M} \parallel \{0,1\}^{<n}$ where $\mathcal{M} \subseteq \{0,1\}^{\geq n}$.

We call our construction XLS (eXtension by Latin Squares). Its overhead is two blockcipher calls, eight xor instructions, and two one-bit rotations. This is the work beyond enciphering (or deciphering) a single l-bit string that is needed to encipher (or decipher) an $l + s$ bit one, where $s \in [1..n-1]$. If the message is in the original domain there is no overhead beyond determining this. As an example, if $\mathcal{E} = E$ is an n-bit blockcipher then it will take three blockcipher calls to encipher a $2n - 1$ bit string.

The XLS method is described in Fig. 1. For a message M already in the domain of $\mathcal{E}$, just apply it. Otherwise, suppose that M has length $l + s$ where $l \in \mathcal{L}$ and $s \in [1..n-1]$. To encipher M: apply the blockcipher E to the last full n-bit block of M; mix together the last $2s$-bits; flip the immediately preceding bit; encipher under $\mathcal{E}$ the first l bits; mix together the last $2s$-bits; flip the immediately preceding bit; then apply E to the last full n-bit block. Our recommended instantiation of the mixing step uses three xors and a single one-bit circular rotation.

We prove that XLS works. More specifically, if $\mathcal{E}$ is secure in the sense of a strong pseudorandom permutation (a strong PRP) [17] then $\mathcal{E}^*$ inherits this property. This assumes that the blockcipher E is likewise a strong PRP. The result holds even in the variable-input-length (VIL) setting [3]: if $\mathcal{E}$ is VIL-secure then so is $\mathcal{E}^*$. See Theorem 2. It also holds in the tweakable-enciphering-scheme setting [16]: if $\mathcal{E}$ is tweakable then $\mathcal{E}^*$ inherits this. See Section 7. If one makes the weaker assumption that $\mathcal{E}$ and E are ordinary (not necessarily strong) PRPs, then one can conclude that $\mathcal{E}^*$ is a PRP. See Section 8.

While XLS is relatively simple, it is surprisingly delicate. We show that natural alternative ways of mixing do not work. We show that omitting the bit flip does not work. And attempting to get by without any mixing—say by enciphering the last n bits, the first l bits, then the last n bits—doesn't work even if one demands that the "overlap" in what is enciphered is $n/2$ bits: there is an attack of complexity $2^{n/4}$.

All that said, we develop sufficient conditions on the mixing function that *are* enough to guarantee security, and we provide a mixing function based on multipermutations (also called orthogonal Latin squares [7]). Though conceptually elegant, implementing multipermutations in this setting is slightly complicated, so we provide an alternate mixing function that approximates multipermutations via bit rotations. This comes at the (insignificant) cost of a slightly larger constant in the security reduction. XLS is the first mode of operation to employ multipermutations or approximate multipermutations to yield a provable-security guarantee. Indeed, such mixing functions may prove to be useful in further provable-security contexts.

We comment that we cannot handle messages of length less than n bits (the blocklength of the blockcipher that we use)—for example, we don't know how to encipher a 32-bit string using AES (in an efficient way and with a known and desirable security bound). This is a long open problem [6,12].

RELATED WORK. There are several known methods for turning a blockcipher with message space $\mathcal{M} = \{0,1\}^n$ into a cipher with some message space $\{0,1\}^{\geq n}$. Halevi does this in his EME* and TET constructions [12,13]; Fluhrer and McGrew do it (without a provable-security guarantee) with XCB [20]; Wang, Feng, and Wu do it in HCTR [30]; and Chakraborty and Sarkar do it in HCH [9]. All of these constructions are somewhat complex, and their methods for dealing with "inconvenient-length" strings are non-generic. Constructions of ciphers from n-bit blockciphers that result in a message space like $(\{0,1\}^n)^+$ are offered by Zheng, Matsumoto, and Imai [31], Naor and Reingold [22], Halevi and Rogaway [15,14], Patel, Ramzan, and Sundaram [24], and Chakraborty and Sarkar [8]. One can even view Luby and Rackoff [17] in this light.

Anderson and Biham [2] and Lucks [18,19] make a wide-blocksize cipher out of a stream cipher and a hash function, and Schroeppel provides a cipher [28] that works on an arbitrary message space *de novo*.

When $\mathcal{E} = E$ is an n-bit blockcipher, the XLS construction solves the *elastic blockcipher* problem of Cook, Yung, and Keromytis [11,10], where one wants to extend a blockcipher from n bits to $[n..2n-1]$ bits. The Cook *et. al* solution is heuristic—there is no proof of security—but with XLS we have, for example, an "elastic AES" that provably preserves the security of AES.

When a cipher like CMC or EME [15,14] plays the role of $\mathcal{E}$ in XLS, one gets a cipher with efficiency comparable to that of a mode like EME* [12].

Bellare and Rogaway first defined VIL ciphers [3] and built one (although it is not secure as a *strong* PRP). An and Bellare [1] offer the viewpoint that cryptographic constructions are often aimed at adjusting the domain of a primitive. This viewpoint is implicit in our work.

APPLICATIONS. While primarily interested in the "theoretical" question of how to accomplish domain extension for ciphers, arbitrary-input-length enciphering is a problem with many applications. A well-known application is disk-sector encryption, the problem being addressed by the IEEE Security in Storage Work Group P1619. Another application is saving bandwidth in network protocols: if one has a 53-byte payload to be enciphered, and no IV or sequence number to do it, the best that can be done without increasing the size of the datagram is to encipher this 53-byte string. A related application is the security-retrofitting of legacy communications protocols, where there is a mandated and immutable allocation of bytes in a datagram, this value not necessarily a multiple of, say, 16 bytes. Another application is in a database setting where it should be manifest when two confidential database records are identical, these records having arbitrary length that should not be changed, but nothing else about the records should be leaked. Arbitrary-length enciphering enables bandwidth-efficient use of the encode-then-encipher paradigm of Bellare and Rogaway [4], where one gets authenticity by enciphering strings encoded with redundancy and semantic security by enciphering strings that rarely collide.

2   Preliminaries

BASICS AND NOTATION. For strings $X, Y \in \{0,1\}^*$, we use $X \parallel Y$ or $X Y$ to denote concatenation. We write $X[i]$ for $i \in [1 .. |X|]$ to represent the ith bit of X (thus $X = X[1]X[2]\cdots X[s]$). The complement of a bit b is $\mathtt{flip}(b)$. For a set $\mathcal{C}$ and element X we write $\mathcal{C} \xleftarrow{\cup} X$ for $\mathcal{C} \leftarrow \mathcal{C} \cup \{X\}$. We require that for any set of bit strings $\mathcal{S} \subseteq \{0,1\}^*$, if $X \in \mathcal{S}$ then $\{0,1\}^{|X|} \subseteq \mathcal{S}$.

A *cipher* is a map $\mathcal{E}: \mathcal{K} \times \mathcal{M} \to \mathcal{M}$ where $\mathcal{K}$ is a nonempty set, $\mathcal{M} \subseteq \{0,1\}^*$ is a nonempty set, and $E_K(\cdot) = E(K, \cdot)$ is a length-preserving permutation. The set $\mathcal{K}$ is called the *key space* and the set $\mathcal{M}$ is called the *message space*. We can view the message space as $\cup_{l \in \mathcal{L}}\{0,1\}^l$ where $\mathcal{L} = \{l \mid \exists X \in \mathcal{M} \text{ s.t. } |X| = l\}$. Let $\mathcal{D}$ be the cipher with the same signature as $\mathcal{E}$ and defined by $\mathcal{D}_K(Y) = X$ iff $\mathcal{E}_K(X) = Y$. A *blockcipher* is a cipher with message space $\mathcal{M} = \{0,1\}^n$ for some $n \geq 1$ (the blocksize). For $\mathcal{M} \subseteq \{0,1\}^*$ let $\mathrm{Perm}(\mathcal{M})$ be the set of all length-preserving permutations on $\mathcal{M}$. By selecting $\mathcal{K} = \mathrm{Perm}(\mathcal{M})$ we have a cipher for which a uniformly chosen permutation on $\{0,1\}^l$ is selected for each $l \in \mathcal{L}$. Let $\mathrm{Func}(\mathcal{M})$ be the set of all length-preserving functions on $\mathcal{M}$. Write $\mathrm{Perm}(\ell)$ and $\mathrm{Func}(\ell)$ for $\mathrm{Perm}(\{0,1\}^\ell)$ and $\mathrm{Func}(\{0,1\}^\ell)$, respectively.

Let $\mathcal{S} \subseteq \{0,1\}^{\geq 1}$. Then define $\mathcal{S}^2 = \{XY \mid X, Y \in \mathcal{S} \wedge |X| = |Y|\}$. Let $f: \mathcal{S}^2 \to \mathcal{S}^2$ be a length-preserving function. We define the *left projection* of f as the function $f_L: \mathcal{S}^2 \to \mathcal{S}$ where $f_L(X)$ is equal to the first $|X|/2$ bits of $f(X)$. We define the *right projection* of f as the function $f_R: \mathcal{S}^2 \to \mathcal{S}$ where $f_R(X)$ is equal to the last $|X|/2$ bits of $f(X)$. Of course $f(X) = f_L(X) \parallel f_R(X)$.

When we say "Replace the last ℓ bits of M, Last, by $F(\text{Last})$" we mean (1) parse M into $X \parallel \text{Last}$ where $|X| = |M| - \ell$ and $|\text{Last}| = \ell$; (2) let Z be $F(\text{Last})$; and (3) replace M by $X \parallel Z$. We define the semantics of similar uses of "Replace ..." in the natural way.

The notation "$XYZ \leftarrow M$ *of lengths* x, y, z" for any string M with $|M| = x + y + z$ means parse M into three strings of length x, y, and z and assign these values to X, Y, and Z, respectively. The notation is extended to the case of parsing M into two halves in the natural way.

Finally, an *involution* is a permutation g which is its own inverse: $g(g(x)) = x$.

SECURITY NOTIONS. When an adversary $\mathcal{A}$ is run with an oracle $\mathcal{O}$ we let $A^{\mathcal{O}} \Rightarrow 1$ denote the event that $\mathcal{A}$ outputs the bit 1. Let $\mathcal{E}: \mathcal{K} \times \mathcal{M} \to \mathcal{M}$ be a cipher. Then we define the following advantages for an adversary $\mathcal{A}$:

$$\mathbf{Adv}_{\mathcal{E}}^{\pm \mathrm{prp}}(\mathcal{A}) = \Pr\Big[K \xleftarrow{\$} \mathcal{K} : A^{\mathcal{E}_K, \mathcal{D}_K} \Rightarrow 1\Big] - \Pr\Big[\pi \xleftarrow{\$} \mathrm{Perm}(\mathcal{M}) : A^{\pi, \pi^{-1}} \Rightarrow 1\Big]$$

$$\mathbf{Adv}_{\mathcal{E}}^{\pm \mathrm{prf}}(\mathcal{A}) = \Pr\Big[K \xleftarrow{\$} \mathcal{K} : A^{\mathcal{E}_K, \mathcal{D}_K} \Rightarrow 1\Big] - \Pr\Big[\rho, \sigma \xleftarrow{\$} \mathrm{Func}(\mathcal{M}) : A^{\rho, \sigma} \Rightarrow 1\Big]$$

where the probabilities are over the choice of K or choice of π (resp. ρ, σ) and the coins used by $\mathcal{A}$. The first experiment represents distinguishing $\mathcal{E}$ and its inverse from a random length-preserving permutation and its inverse and the second experiment represents distinguishing $\mathcal{E}$ and its inverse from two random length-preserving functions. In both settings, we demand that the adversary $\mathcal{A}$, given oracles f, g, does not repeat any query, does not ask $g(Y)$ after receiving Y in response to some query $f(X)$, and does not ask $f(X)$ after receiving X in response to some query $g(Y)$. Such forbidden queries are termed *pointless*.

While the above formalization allows *variable input length* (VIL) adversaries, we can also restrict adversaries to only query messages of a single length. We call such adversaries *fixed input length* (FIL) adversaries.

Informally, a cipher is called a "strong pseudorandom permutation" if no reasonable adversary $\mathcal{A}$ can distinguish the enciphering and deciphering functions, randomly keyed, from a randomly selected permutation and its inverse: $\mathbf{Adv}_{\mathcal{E}}^{\pm \mathrm{prp}}(\mathcal{A})$ is small. Our theorems make concrete statements about this and so we will not have to formalize "reasonable" or "small." Resources we pay attention to are the adversary's maximum running time (which, by convention, includes the length of the program); the number of queries it asks; and the lengths of the queries. For any cipher $\mathcal{E}$ with inverse $\mathcal{D}$, define $\mathsf{Time}_{\mathcal{E}}(\mu) = \max\{T_{key}, T_{\mathcal{E}}, T_{\mathcal{D}}\}$ where T_{key} is the maximum time required to generate a key L for the scheme, $T_{\mathcal{E}}$ is the maximum time to run $\mathcal{E}_L$ on a message of at most μ bits, and $T_{\mathcal{D}}$ is the maximum time to run $\mathcal{D}_L$ on a ciphertext of at most μ bits.

3 The XLS Construction

Fix a blocksize n. Let $\mathcal{E}: \mathcal{K}_{\mathcal{E}} \times \mathcal{M} \to \mathcal{M}$ be a cipher with $\mathcal{M} \subseteq \{0,1\}^{\geq n}$ and let $E: \mathcal{K}_E \times \{0,1\}^n \to \{0,1\}^n$ be a blockcipher. Finally, define a length-preserving permutation mix: $\mathcal{S}^2 \to \mathcal{S}^2$ where $\mathcal{S} \supseteq \cup_{i=1}^{n-1}\{0,1\}^i$. Then we define a cipher $\mathcal{E}^* = \mathrm{XLS}[\mathrm{mix}, \mathcal{E}, E]$ with key space $\mathcal{K}^* = \mathcal{K}_{\mathcal{E}} \times \mathcal{K}_E$ and message space $\mathcal{M}^* = \mathcal{M} \parallel \{0,1\}^{<n}$. For keys $L \in \mathcal{K}_{\mathcal{E}}$ and $K \in \mathcal{K}_E$ we have $\mathcal{E}^*_{L,K}(\cdot) = \mathcal{E}^*((L,K),\cdot)$. See Fig. 1 for the definition.

Enciphering a message M with $\mathcal{E}^* = \mathrm{XLS}[\mathrm{mix}, \mathcal{E}, E]$ is straightforward. If $M \in \mathcal{M}$, then simply apply $\mathcal{E}$. Otherwise, apply E to the last full n-bit block of M and replace those bits with the result. Then 'mix together' the last $2s$ bits, again replacing the appropriate bits with the resulting mixture. Flip bit $|M| - 2s$, which is the first bit from the right not affected by mix. Apply $\mathcal{E}$ to as many bits as possible, starting from the left. Finally, just repeat the first three steps in reverse order. Deciphering is equally simple, and in fact, if one implements mix with an involution, as we suggest, then the inverse of $\mathcal{E}^*$ is just $\mathcal{D}^* = \mathrm{XLS}[\mathrm{mix}, \mathcal{D}, D]$.

Algorithm $\mathcal{E}^*_{L,K}(M)$

00   If $M \in \mathcal{M}$ then return $\mathcal{E}_L(M)$
01   Let $l < |M|$ be largest number such that $\{0,1\}^l \subseteq \mathcal{M}$; $s \leftarrow |M| - l$
02   If $s \geq n$ or $l < n$ then Return $\perp$
03   Replace the last full n-bit block of M, LastFull, with $E_K(\text{LastFull})$
04   Replace the last $2s$ bits of M, Last, with mix(Last)
05   Replace the $(|M| - 2s)$'th bit of M, b, with $\mathtt{flip}(b)$
06   Replace the first l bits of M, First, with $\mathcal{E}_L(\text{First})$
07   Replace the $(|M| - 2s)$'th bit of M, b, with $\mathtt{flip}(b)$
08   Replace the last $2s$ bits of M, Last, with mix(Last)
09   Replace the last full n-bit block of M, LastFull, with $E_K(\text{LastFull})$
10   Return M

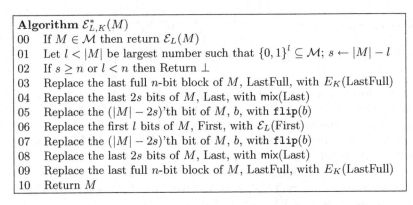

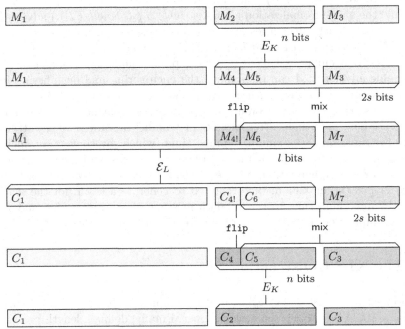

Fig. 1. Top: Enciphering algorithm $\mathcal{E}^* = \text{XLS}[\text{mix}, \mathcal{E}, E]$. **Bottom:** Enciphering $M = M_1 \| M_2 \| M_3$ under $\mathcal{E}^*$ where M is not in $\mathcal{M}$; $l < |M|$ is the largest value such that $\{0,1\}^l \subseteq \mathcal{M}$; $s = |M| - l$; $M_1 \in \{0,1\}^{l-n}$; $M_2 \in \{0,1\}^n$; and $M_3 \in \{0,1\}^s$.

Why, intuitively, should XLS work? "Working" entails that each output bit strongly depends on each input bit. Since $\mathcal{E}$ presumably already does a good job of this we need only worry about mixing in the "leftover" s bits for $M \notin \mathcal{M}$. We mix in these bits utilizing the mixing function mix. But since mix will be a simple combinatorial object—it is unkeyed and will have no "cryptographic" property—we need to "protect" its input with the blockcipher. The "symmetrizing" of the protocol—repeating the blockcipher call and the mixing step in the reverse order so that lines 03 → 09 are identical to lines 09 → 03—helps achieve *strong* PRP-security: each input bit must strongly depend on each output bit, as queries can be made in the forward *or* backward direction. Finally, the bit-flipping step is just a symmetry-breaking technique to ensure that different-length messages are treated differently.

If mix does a "good" job of mixing, then XLS will in fact be secure, as we prove in Section 6. But what is the meaning of "good," and how do we make a mixing function that is simultaneously good, efficient, and easy to implement? We now turn towards answering these questions.

4   The Mixing Function

We now look at several possible ways of implementing mix, to build intuition on what properties are needed for the security of XLS. In the end we formally define, quantitatively, the sufficient condition of interest. For ease of exposition we will often silently parse the input to mix into its two halves, i.e., $\text{mix}(AB)$ means that $A \parallel B \in \mathcal{S}^2$ and that $|A| = |B|$. Also we interchangeably write $\text{mix}(AB)$ and $\text{mix}(A, B)$, which are equivalent.

A NAIVE APPROACH. Let's start with a natural construction that, perhaps surprisingly, does not lead to a secure construction. Suppose we define mixWrong by saying that $\text{mixWrong}(AB) = A \oplus B \parallel B$ for equal-length A, B: the mixing function xors the right half of the input into the left half, outputting the result and the original right half. Clearly mixWrong is a length-preserving permutation. Furthermore, it might seem sufficient for XLS because it will mix the "leftover" bits into those handled by $\mathcal{E}$. But this intuition is flawed: $\mathcal{E}^* = \text{XLS}[\text{mixWrong}, \mathcal{E}, E]$ is easily distinguished from a length-preserving permutation on $\mathcal{M}^*$. An adversary can simply query $0^n \parallel 0^{n-1}$ and $1^n \parallel 0^{n-1}$. As one can easily verify, both $\mathcal{E}^*(0^n \parallel 0^{n-1})$ and $\mathcal{E}^*(1^n \parallel 0^{n-1})$ will have output with the last $n-1$ bits equal to 0^{n-1}. This would be true of a random permutation with probability at most $1/2^{n-1}$, and so the adversary's advantage is close to one. In fact mixWrong does not do a good job of mixing: the right half of the output is only a function of the right half of the input. One can try various fixes, but ultimately it appears that using *just* xors is inherently inadequate.

USING ORTHOGONAL LATIN SQUARES. The failure above suggests that what is needed is a mixing function with symmetry, in the sense that both the left and right halves of the output are dependent on both the left and right halves of the input. To achieve such a goal we can turn to the classical combinatorial objects known as a pair of *orthogonal Latin squares* [7], also called a *multipermutation* [27,29]. This is a permutation mix: $\mathcal{S}^2 \to \mathcal{S}^2$ such that, for any $C \in S$, $\text{mix}_{\text{L}}(C, \cdot)$, $\text{mix}_{\text{L}}(\cdot, C)$, $\text{mix}_{\text{L}}(C, \cdot)$, and $\text{mix}_{\text{R}}(\cdot, C)$ are all permutations, where mix_{L} and mix_{R} denote the projection of mix onto its first and second component. Let us describe a concrete realization. Fix a finite field $\mathbb{F}_{2^s}$ for each s and view each

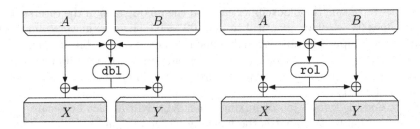

Fig. 2. The mixing functions mix1 (left) and mix2 (right). Here $A, B \in \{0,1\}^{<n}$ with $s = |A| = |B|$. The operation dbl is a multiplication by $\mathbf{2} = 0^{s-2}10 = \mathrm{x}$ in the finite field $\mathbb{F}_{2^s}$ and rol is a circular left rotation by one bit.

s-bit string as an element of this field. Then we can build a mixing function mix1 by saying that

$$\mathrm{mix1}(AB) = (\mathbf{3}A + \mathbf{2}B) \,\|\, (\mathbf{2}A + \mathbf{3}B) = (A + \mathbf{2}(A + B), B + \mathbf{2}(A + B))$$

for equal-length strings A and B (we will assume the length to be at least 2). Here addition and multiplication are over $\mathbb{F}_{2^s}$ and $\mathbf{2} = 0^{s-2}10 = \mathrm{x}$ and $\mathbf{3} = 0^{s-2}11 = \mathrm{x} + 1$. Addition is bitwise xor and multiplication by $\mathbf{2}$, which we also denote dbl, can be implemented by a shift and a conditional xor. The mixing function mix1 has several nice properties. First, it is a permutation and, in fact, an involution (meaning $\mathrm{mix} = \mathrm{mix}^{-1}$). Moreover, for any $C \in \{0,1\}^s$ we have that $\mathrm{mix1}_\mathrm{L}(C, \cdot)$, $\mathrm{mix1}_\mathrm{R}(C, \cdot)$, $\mathrm{mix1}_\mathrm{L}(\cdot, C)$, and $\mathrm{mix1}_\mathrm{R}(\cdot, C)$ are all permutations on $\{0,1\}^s$.

We show later that, when used in XLS, mixing function mix1 leads to a secure construction. Moreover, it is fast and relatively simple. But implementing it in XLS requires a table of constants corresponding to irreducible polynomials, one for each $s \leq n - 1$. As it turns out, we can do better.

THE SIMPLIFIED MIXING FUNCTION. We now simplify the mixing function mix1. Let $\mathrm{rol}(X)$ represent left circular bit-rotation, that is, for any string X of length s let $\mathrm{rol}(X) = X[2]X[3] \cdots X[s]X[1]$. Then define mix2 by

$$\mathrm{mix2}(AB) = (A \oplus \mathrm{rol}(A \oplus B)) \,\|\, (B \oplus \mathrm{rol}(A \oplus B))$$

where A and B are equal-length strings. See Fig. 2. Notice the similarity with mix1: we replaced multiplication by two with a left circular rotation. The bit rotation "approximates" a proper multiplication, eliminating, in an implementation, the conditional xor and the table of constants. As before, mix2 is an involution.

QUANTIFYING THE QUALITY OF MIXING FUNCTIONS. We now formalize the properties of a mixing function that are needed in the proof of XLS.

Definition 1. *Fix a set $\mathcal{S} \subseteq \{0,1\}^{\geq 1}$, let $\mathrm{mix} \colon \mathcal{S}^2 \to \mathcal{S}^2$ be a length-preserving permutation, and let $\epsilon \colon \mathbb{N} \to [0,1]$. We say that mix is an $\epsilon(s)$-**good mixing function** if, for all s such that $\{0,1\}^s \subseteq \mathcal{S}$, we have that*

(1) $\mathsf{mix_L}(A, \cdot)$ *is a permutation for all* $A \in \{0,1\}^s$,

(2) $\mathsf{mix_R}(\cdot, B)$ *is a permutation for all* $B \in \{0,1\}^s$,

(3) $\Pr[R \xleftarrow{\$} \{0,1\}^s : C = \mathsf{mix_L}(R, B)] \leq \epsilon(s)$ *for all* $B, C \in \{0,1\}^s$, *and*

(4) $\Pr[R \xleftarrow{\$} \{0,1\}^s : C = \mathsf{mix_R}(A, R)] \leq \epsilon(s)$ *for all* $A, C \in \{0,1\}^s$. $\square$

The best one can hope for is a 2^{-s}-good mixing function. In fact mix1 is such a function, while the mix2 function is just a factor of two off.

Lemma 1. *The mixing function* mix1 *is a* 2^{-s}-*good mixing function. The mixing function* mix2 *is a* 2^{-s+1}-*good mixing function.* $\square$

Proof: Since for any $s \in [1 .. n-1]$ and any $C \in \{0,1\}^s$ we have that $\mathsf{mix1_L}(C, \cdot)$, $\mathsf{mix1_R}(C, \cdot)$, $\mathsf{mix1_L}(\cdot, C)$, and $\mathsf{mix1_R}(\cdot, C)$ are all permutations, the first half of the lemma is clear.

That mix2 meets parts 1 and 2 of the definition is clear. For the third part, we have that

$$\mathsf{mix2_L}(R, B) = R \oplus \mathtt{rol}(R \oplus B) = R \oplus \mathtt{rol}(R) \oplus \mathtt{rol}(B)$$

and so we bound the number of values R such that $C \oplus \mathtt{rol}(B) = R \oplus \mathtt{rol}(R)$. Let $C' = C \oplus \mathtt{rol}(B)$, which is a constant. Then we have that

$$R[1] \oplus R[s] = C'[1]$$
$$R[2] \oplus R[1] = C'[2]$$
$$\vdots$$
$$R[s-1] \oplus R[s-2] = C'[s-1]$$
$$R[s] \oplus R[s-1] = C'[s]$$

Note that there only exists a string R that satisfies the above equalities if $C'[1] \oplus C'[2] \oplus \cdots \oplus C'[s] = 0$. If there exists a solution, then pick a value for $R[1]$. That choice and the equations above combine to specify $R[2], \ldots, R[s]$. This means that there are at most two possible values of R and so the probability that $C = \mathsf{mix2_L}(R, B)$ is at most $2/2^s$. The proof of the fourth part is symmetric. ∎

We point out that the properties of circular rotations combined with xors as utilized in mix2 have been used before in different settings, such as [23].

We have introduced two mixing functions for the following reason: mix1 is conceptually more elegant, while mix2 is operationally more elegant. In addition, mix2 is in effect an approximation of mix1, making the latter an important conceptual building block. Such mixing functions might prove useful in future provable-security results.

Note that our definition of an $\epsilon(s)$-good mixing function is general, but XLS requires a mixing function for which $\mathcal{S} \supseteq \cup_{i=1}^{n-1}\{0,1\}^i$. This is clearly the case for mix1 and mix2. For the rest of the paper when we refer to an $\epsilon(s)$-good mixing function, we implicitly require that this function is well-defined for such an $\mathcal{S}$.

5   The Bit Flips

In steps 05 and 07 of XLS (see Fig. 1) we flip a single bit. Flipping bits in this manner is unintuitive and might seem unimportant for the security of XLS. However, the bit flips are actually crucial for the security of the scheme when in the VIL setting. Let $\mathcal{E}^\dagger$ be the cipher defined by running the algorithm of Fig. 1 except with lines 05 and 07 omitted. Then the following VIL adversary A easily distinguishes $\mathcal{E}^\dagger$ from a family of random permutations. The adversary A makes two enciphering queries on $M = 0^{n+1}$ and $M' = 0^{n+2}$, getting return values C and C' respectively. If the first n bits of C and C' are equal, then A outputs 1 (the oracles are likely the construction) and otherwise outputs 0 (the oracles are likely a random permutation). We have that $\Pr[K \xleftarrow{\$} \mathcal{K}^* : A^{\mathcal{E}^\dagger(\cdot),\mathcal{D}^\dagger(\cdot)} \Rightarrow 1] = 1$. This is so because for both queries the inputs to $\mathcal{E}$ are necessarily the same (as one can verify quickly by following along in the diagram in Fig. 1; remember to omit the flip steps). Clearly $\Pr[\pi \xleftarrow{\$} \mathrm{Perm}(\mathcal{M}^*) : A^{\pi(\cdot),\pi^{-1}(\cdot)} \Rightarrow 1] = 2^{-n}$ and so A has large advantage.

6   Security of XLS

We are now ready to prove the security of XLS. The proof is broken into two parts: first we show that XLS is secure in an information-theoretic setting (i.e., using actual random permutations as components). Afterwards we pass to a complexity-theoretic setting to get our main result.

Theorem 1. *Fix n and an $\epsilon(s)$-good mixing function* mix. *Let $\mathcal{M} \subseteq \{0,1\}^{\geq n}$ and $\mathcal{E}^* = \mathrm{XLS}[\mathrm{mix}, \mathrm{Perm}(\mathcal{M}), \mathrm{Perm}(n)]$. Then for any adversary A that asks at most q queries we have that $\mathbf{Adv}_{\mathcal{E}^*}^{\pm\mathrm{prf}}(A) \leq 5q^2\,\epsilon(s)/2^{n-s} + 3q^2/2^n$ for any $s \in [1..n-1]$, and so, by Lemma 1,*

$$\mathbf{Adv}_{\mathcal{E}^*}^{\pm\mathrm{prf}}(A) \leq \frac{8q^2}{2^n} \quad \text{and} \quad \mathbf{Adv}_{\mathcal{E}^*}^{\pm\mathrm{prf}}(A) \leq \frac{13q^2}{2^n}$$

for mix = mix1 *and* mix = mix2, *respectively.*   □

Proof. Due to space constraints, we only present a self-contained chunk of the proof together with a sketch of the other portion of the proof. See the full version [25] for the complete proof. Fix n and let mix: $\mathcal{S}^2 \to \mathcal{S}^2$ be an $\epsilon(s)$-good mixing permutation. Let $\mathcal{E}$: $\mathrm{Perm}(\mathcal{M}) \times \mathcal{M} \to \mathcal{M}$ be a cipher with message space $\mathcal{M}$ and let E: $\mathrm{Perm}(n) \times \{0,1\}^n \to \{0,1\}^n$ be a blockcipher. Note that these last two simply implement a family of random length-preserving permutations on $\mathcal{M}$ and a random permutation on $\{0,1\}^n$, respectively. Let A be a $\pm\mathrm{prf}$ adversary against $\mathcal{E}^* = \mathrm{XLS}[\mathrm{mix}, \mathcal{E}, E]$. We therefore must bound

$$\mathbf{Adv}_{\mathcal{E}^*}^{\pm\mathrm{prf}}(A) = \Pr\left[K \xleftarrow{\$} \mathcal{K}^* : A^{\mathcal{E}_K^*, \mathcal{D}_K^*} \Rightarrow 1\right] - \Pr\left[\rho, \sigma \xleftarrow{\$} \mathrm{Func}(\mathcal{M}^*) : A^{\rho,\sigma} \Rightarrow 1\right].$$

Recall that we disallow A from making pointless queries. We utilize a game-playing argument [5] and the first two games are G0 and G1, shown in Fig. 3.

procedure $\mathsf{Choose}\mathcal{E}(X)$
$Y \stackrel{\$}{\leftarrow} \{0,1\}^{|X|}$

If $Y \in \mathcal{RE}$ then $bad \leftarrow \mathsf{true}$, $Y \stackrel{\$}{\leftarrow} \overline{\mathcal{RE}}$

If $X \in \mathcal{DE}$ then $bad \leftarrow \mathsf{true}$, $Y \leftarrow \mathcal{E}(X)$

$\mathcal{E}(X) \leftarrow Y; \mathcal{D}(Y) \leftarrow X$

$\mathcal{RE} \stackrel{\cup}{\leftarrow} Y; \mathcal{DE} \stackrel{\cup}{\leftarrow} X;$ Return Y

procedure $\mathsf{Choose}\mathcal{D}(Y)$ $\boxed{\text{G0}}$ G1
$X \stackrel{\$}{\leftarrow} \{0,1\}^{|Y|}$

If $X \in \mathcal{DE}$ then $bad \leftarrow \mathsf{true}$, $Y \stackrel{\$}{\leftarrow} \overline{\mathcal{DE}}$

If $Y \in \mathcal{RE}$ then $bad \leftarrow \mathsf{true}$, $X \leftarrow \mathcal{D}(Y)$

$\mathcal{E}(X) \leftarrow Y; \mathcal{D}(Y) \leftarrow X$

$\mathcal{RE} \stackrel{\cup}{\leftarrow} Y; \mathcal{DE} \stackrel{\cup}{\leftarrow} X;$ Return X

procedure $\mathsf{Choose}E(X)$
$Y \stackrel{\$}{\leftarrow} \{0,1\}^{|X|}$

If $Y \in RE$ then $bad \leftarrow \mathsf{true}$, $Y \stackrel{\$}{\leftarrow} \overline{RE}$

If $X \in DE$ then $bad \leftarrow \mathsf{true}$, $Y \leftarrow E(X)$

$E(X) \leftarrow Y; D(Y) \leftarrow X$

$RE \stackrel{\cup}{\leftarrow} Y; DE \stackrel{\cup}{\leftarrow} X;$ Return Y

procedure $\mathsf{Choose}D(Y)$
$X \stackrel{\$}{\leftarrow} \{0,1\}^{|Y|}$

If $X \in DE$ then $bad \leftarrow \mathsf{true}$, $Y \stackrel{\$}{\leftarrow} \overline{DE}$

If $Y \in RE$ then $bad \leftarrow \mathsf{true}$, $X \leftarrow D(Y)$

$E(X) \leftarrow Y; D(Y) \leftarrow X$

$RE \stackrel{\cup}{\leftarrow} Y; DE \stackrel{\cup}{\leftarrow} X;$ Return X

procedure $\mathrm{Enc}(M)$
$j \leftarrow j + 1; M^j \leftarrow M;$ If $M^j \in \mathcal{M}$ then Return $C^j \leftarrow \mathsf{Choose}\mathcal{E}(M^j)$
Let s be smallest number s.t. $\{0,1\}^{|M^j|-s} \in \mathcal{M}$
$m \leftarrow |M^j| - n - s; M_1^j \, M_2^j \, M_3^j \leftarrow M^j$ *of lengths* m, n, s
Let $i \in [1..j]$ be smallest index s.t. $M_2^j = M_2^i$
If $i < j$ then $M_4^j \leftarrow M_4^i; M_5^j \leftarrow M_5^i$
Else $M_4^j \, M_5^j \leftarrow \mathsf{Choose}E(M_2^j)$ *of lengths* $n - s, s$
$M_6^j \, M_7^j \leftarrow \mathsf{mix}(M_4^j, M_3^j)$ *of lengths* s, s
$C_1^j \, C_{4!}^j \, C_6^j \leftarrow \mathsf{Choose}\mathcal{E}(M_1^j \,\|\, \mathtt{flip1}(M_4^j) \,\|\, M_6^j)$ *of lengths* $m, n - s, s$
$C_3^j \, C_5^j \leftarrow \mathsf{mix}(C_6^j, M_7^j)$ *of lengths* s, s
$C_2^j \leftarrow \mathsf{Choose}E(\mathtt{flip1}(C_{4!}^j) \,\|\, C_5^j)$
Return $C_1^j \, C_2^j \, C_3^j$

procedure $\mathrm{Dec}(C)$
$j \leftarrow j + 1; C^j \leftarrow C;$ If $C^j \in \mathcal{M}$ then Return $M^j \leftarrow \mathsf{Choose}\mathcal{D}(C^j)$
Let s be smallest number s.t. $\{0,1\}^{|C^j|-s} \in \mathcal{M}$
$m \leftarrow |C^j| - n - s; C_1^j \, C_2^j \, C_3^j \leftarrow C^j$ *of lengths* m, n, s
Let $i \in [1..j]$ be smallest index s.t. $C_2^j = C_2^i$
If $i < j$ then $C_4^j \leftarrow C_4^i, C_5^j \leftarrow C_5^i$
Else $C_4^j \, C_5^j \leftarrow \mathsf{Choose}D(C_2^j)$ *of lengths* $n - s, s$
$C_6^j \, M_7^j \leftarrow \mathsf{mix}(C_4^j, C_3^j)$ *of lengths* s, s
$M_1^j \, M_{4!}^j \, M_6^j \leftarrow \mathsf{Choose}\mathcal{D}(C_1^j \,\|\, \mathtt{flip1}(C_4^j) \,\|\, C_5^j)$ *of lengths* $m, n - s, s$
$M_5^j \, M_3^j \leftarrow \mathsf{mix}(M_6^j \, M_7^j)$ *of lengths* s, s
$M_2^j \leftarrow \mathsf{Choose}D(\mathtt{flip1}(M_{4!}^j) \,\|\, M_5^j)$
Return $M_1^j \, M_2^j \, M_3^j$

Fig. 3. Games G0 (boxed statements included) and G1 (boxed statements dropped) used in the proof of Theorem 1. Initially, $j = 0$ and $\mathcal{DE}, \mathcal{RE}, DE, RE$ are empty sets and the partial functions $\mathcal{E}, \mathcal{D}, E, D$ are everywhere undefined. The function $\mathtt{flip1}(X)$, for any bit string $X = X[1] \cdots X[s]$, outputs the string with last bit complemented: $X[1] \cdots X[s-1]\mathtt{flip}(X[s])$.

In G0 we build $\mathcal{E}$ and E lazily using the appropriate Choose procedures, and so $\mathcal{E}$ and E are partial functions in this context. Note that $D\mathcal{E}, R\mathcal{E}, DE, RE$ are initially empty and the functions $\mathcal{E}, \mathcal{D}, E, D$ are everywhere undefined. As usual, $\mathcal{D}$ and D represent the inverses of $\mathcal{E}$ and E. While game G0, which includes the boxed statements, enforces that $\mathcal{E}$ and E be length-preserving permutations, game G1 dispenses with that requirement (the boxed statements are not included in G1). A flag bad is initially false and set to true when, in the course of building $\mathcal{E}$ and E, a duplicate domain or range point is initially selected. In G0 these points are not used (enforcing that the functions are permutations), but in G1 we use them and thus duplicate points can be added to $D\mathcal{E}, R\mathcal{E}, DE$, and RE. A *collision* is just a pair of equal strings in one of the sets. Note that for $D\mathcal{E}$ and $R\mathcal{E}$, only strings of the same length can collide.

Game G0 exactly simulates $\mathcal{E}^*$ and its inverse while G1 always returns random bits. This second statement needs to be justified for the case of a query $M \notin \mathcal{M}$ (or $C \notin \mathcal{M}$). Particularly, if the jth query is to encipher $M^j \notin \mathcal{M}$, then the last s bits returned are $C_3^j = \mathsf{mix}_\mathrm{R}(C_6^j, M_7^j)$. Here C_6^j is uniformly selected, and by the definition of an $\epsilon(s)$-good mixing function, we have that $\mathsf{mix}_\mathrm{R}(C_6^j, M_7^j)$ is a permutation of C_6^j. So C_3^j inherits its distribution. The same reasoning justifies the distribution of deciphering queries $C \notin \mathcal{M}$. We can therefore replace the oracles $\mathcal{A}$ queries with the two described games and apply the fundamental lemma of game playing [5] to get

$$\mathbf{Adv}_{\mathcal{E}^*}^{\pm\mathrm{prf}}(\mathcal{A}) = \Pr[\mathcal{A}^{\mathrm{G0}} \Rightarrow 1] - \Pr[\mathcal{A}^{\mathrm{G1}} \Rightarrow 1] \leq \Pr[\mathcal{A}^{\mathrm{G1}} \text{ sets } bad]. \quad (1)$$

The following lemma captures the bound on the ability of $\mathcal{A}$ to set bad.

Lemma 2. $\Pr[\mathcal{A}^{\mathrm{G1}} \text{ sets } bad] \leq 5q^2 \epsilon(s)/2^{n-s} + 3q^2/2^n$ *for any* $s \in [1 .. n-1]$. $\square$

Combining Lemma 2 with Equation 1 implies the theorem statement, and a full proof of the lemma appears in [25]. Here we informally sketch one of the more interesting cases for proving the lemma above. In particular, we reason about the probability that $\mathcal{A}$ can set bad by causing a collision in the set $D\mathcal{E}$, which represents the domain of $\mathcal{E}$. Note that in the full proof in [25], we go through several game transitions before reasoning about this case—here we do it in the context of game G1 and thus end up being a bit informal. For simplicity we'll just focus on enciphering queries. Suppose that the ith and j^{th} (with $i < j$) enciphering queries result in applying $\mathcal{E}$ to the same domain point. That is, if we let X^i and X^j be the bit strings added to $D\mathcal{E}$ during queries i and j, then a collision in $D\mathcal{E}$ occurs if $X^i = X^j$ and $l \equiv |X^i| = |X^j|$. If such a collision occurs with high probability, then $\mathcal{A}$ would be able to distinguish easily. There are two main cases to consider, based on the lengths of the queried messages M^i and M^j. The cases are marked by triangles.

$\triangleright$ Suppose that $|M^i| \notin \mathcal{M}$ and $|M^j| \notin \mathcal{M}$. Then the two domain points are $X^i = M_1^i \,\|\, \mathsf{flip1}(M_4^i) \,\|\, M_6^i$ and $X^j = M_1^j \,\|\, \mathsf{flip1}(M_4^j) \,\|\, M_6^j$. Let $s^i = |M_6^i|$ and $s^j = |M_6^j|$ (necessarily we have that $|M^i| - s^i = |M^j| - s^j$). We break down the analysis into two subcases:

- If $M_2^i \neq M_2^j$ then $M_4^j \parallel M_5^j$ are selected uniformly and independently from any random choices made during query i. We have then that $\mathtt{flip1}(M_4^j)$ consists of $n - s^j$ randomly selected bits, which will collide with the appropriate $n - s^j$ bits of X^i with probability 2^{-n+s}. Furthermore, $M_6^j = \mathrm{mix_L}(M_5^j, M_3^j)$ where M_5^j is a string of s^j random bits. We can apply the definition of an $\epsilon(s)$-good mixing function to get that the probability that M_6^j collides with some other value is at most $\epsilon(s)$. Combining the two probabilities, we see that the probability that $X^i = M_1^j \parallel \mathtt{flip1}(M_4^j) \parallel M_6^j$ is at most $\epsilon(s)/2^{n-s}$.

- If $M_2^i = M_2^j$ then we can show that the probability that $X^i = X^j$ is zero. First consider if $s \equiv s^i = s^j$. Then we have that $M_4^i = M_4^j$ and $M_5^i = M_5^j$. For a collision to occur it must be that $M_1^i = M_1^j$ and $M_6^i = M_6^j$. But because of the permutivity of $\mathrm{mix_L}(A, \cdot)$, as given by Definition 1 part 1, this last equivalence implies that $M_3^i = M_3^j$. In turn this means that $M^i = M^j$, which would make query j pointless. But since we disallow $\mathcal{A}$ from making pointless queries, we have a contradiction. Second consider, without loss of generality, that $s^i < s^j$. Then we have that X^i can not equal X^j (recall that $|X^i| = |X^j| = l$) because $\mathtt{flip1}$ ensures that $X^i\left[l - s^j\right] \neq X^j\left[l - s^j\right]$. Thus, in either situation, the probability of a collision is zero.

▷ Now suppose that $|M^i| \in \mathcal{M}$ and $|M^j| \notin \mathcal{M}$. The two domain points are M^i and $X^j = M_1^j \parallel \mathtt{flip1}(M_4^j) \parallel M_6^j$. Let $s^j = |M_6^j|$. We have that M_4^j and M_5^j are uniformly selected after the adversary has specified M^i (since $i < j$). Thus, $\mathtt{flip1}(M_4^j)$ will collide with the appropriate bits of M^i with probability 2^{-n+s}. Since $M_6^j = \mathrm{mix_L}(M_5^j, M_3^j)$ we can apply the definition of a good mixing function. This gives that the probability of M_6^j colliding with the appropriate s^j bits of M^i is at most $\epsilon(s)$. Therefore the probability that $M^i = X^j$ is at most $\epsilon(s)/2^{n-s}$.

If on the other hand $|M^i| \notin \mathcal{M}$ while $|M^j| \in \mathcal{M}$, then we can only apply similar reasoning if we show that $\mathcal{A}$ learns nothing about certain random choices made in the course of answering query i. We do just that rigorously in the full proof.

So in the cases above the probability of a collision is no greater than $\epsilon(s)/2^{n-s}$, where $s \in [1 .. n - 1]$. Because each query adds one string to $\mathcal{DE}$, we have that $|\mathcal{DE}| = q$. Thus, the total probability of *bad* being set due to a collision in the domain of $\mathcal{E}$ is at most

$$\binom{q}{2} \frac{\epsilon(s)}{2^{n-s}} \leq \frac{q^2 \, \epsilon(s)}{2^{n-s+1}} \, .$$

Combining this (via a union bound) with analyses of the other ways in which *bad* can be set yields a sketch of the lemma. ∎

The next theorem captures the security of XLS in a complexity-theoretic setting. It's proof is by a standard hybrid argument, which utilizes as one step Theorem 1. See the full version for details [25].

Theorem 2. *Fix n and an $\epsilon(s)$-good mixing function* mix. *Let $\mathcal{E}: \mathcal{K_E} \times \mathcal{M} \to \mathcal{M}$ be a cipher and $E: \mathcal{K}_E \times \{0,1\}^n \to \{0,1\}^n$ be a blockcipher. Let $\mathcal{E}^* =$*

XLS[mix, $\mathcal{E}, E$] and let $\mathcal{A}$ be an adversary that runs in time t and asks at most q queries, each of at most μ bits. Then there exists adversaries $\mathcal{B}$ and $\mathcal{C}$ and an absolute constant c such that $\mathbf{Adv}_{\mathcal{E}^*}^{\pm\mathrm{prp}}(\mathcal{A}) \leq \mathbf{Adv}_{\mathcal{E}}^{\pm\mathrm{prp}}(\mathcal{B}) + \mathbf{Adv}_E^{\pm\mathrm{prp}}(\mathcal{C}) + 5q^2\,\epsilon(s)/2^{n-s} + 4q^2/2^n$ for any $s \in [1\mathinner{..}n-1]$, and so, by Lemma 1

$$\mathbf{Adv}_{\mathcal{E}^*}^{\pm\mathrm{prp}}(\mathcal{A}) \leq \mathbf{Adv}_{\mathcal{E}}^{\pm\mathrm{prp}}(\mathcal{B}) + \mathbf{Adv}_E^{\pm\mathrm{prp}}(\mathcal{C}) + \frac{9q^2}{2^n}$$

when mix = mix1, and the same expression, but with the 9 replaced by 14, when mix = mix2. Here $\mathcal{B}$ runs in time $t_B = t + c\mu q \log q$ and asks $q_B = q$ queries, each of length at most μ, and $\mathcal{C}$ runs in time $t_C \leq t + (q + 1) \cdot \mathsf{Time}_{\mathcal{E}}(\mu) + c\mu q$ and asks $q_C \leq 2q$ queries, each of length at most n. □

7   Supporting Tweaks

A *tweakable cipher* [16,15] is a function $\widetilde{\mathcal{E}}: \mathcal{K}_{\mathcal{E}} \times \mathcal{T} \times \mathcal{M} \to \mathcal{M}$ where $\mathcal{K}_{\mathcal{E}} \neq \emptyset$ is the key space, $\mathcal{T} \neq \emptyset$ is the tweak space, and $\mathcal{M}$ is the message space. We require that $\widetilde{\mathcal{E}}_K^T(\cdot)$ is a length-preserving permutation for all $K \in \mathcal{K}_{\mathcal{E}}$ and $T \in \mathcal{T}$. We write the inverse of $\widetilde{\mathcal{E}}$ as $\widetilde{\mathcal{D}}$. A tweakable blockcipher is a tweakable cipher with $\mathcal{M} = \{0,1\}^n$ for some fixed n. Tweakable ciphers are useful tools for building higher-level protocols. The tweak of a cipher can be used as, for example, a sector index.

The security of a tweakable cipher is based on indistinguishability of the scheme and a tweakable random permutation. More formally, we define the following advantages

$$\mathbf{Adv}_{\widetilde{\mathcal{E}}}^{\pm\widetilde{\mathrm{prp}}}(\mathcal{A}) = \Pr\left[K \xleftarrow{\$} \mathcal{K}_{\mathcal{E}} : \mathcal{A}^{\widetilde{\mathcal{E}}_K, \widetilde{\mathcal{D}}_K} \Rightarrow 1\right] - \Pr\left[\widetilde{\pi} \xleftarrow{\$} \mathrm{Perm}^{\mathcal{T}}(\mathcal{M}) : \mathcal{A}^{\widetilde{\pi}, \widetilde{\pi}^{-1}} \Rightarrow 1\right]$$

$$\mathbf{Adv}_{\widetilde{\mathcal{E}}}^{\pm\widetilde{\mathrm{prf}}}(\mathcal{A}) = \Pr\left[K \xleftarrow{\$} \mathcal{K}_{\mathcal{E}} : \mathcal{A}^{\widetilde{\mathcal{E}}_K, \widetilde{\mathcal{D}}_K} \Rightarrow 1\right] - \Pr\left[\rho, \sigma \xleftarrow{\$} \mathrm{Func}^{\mathcal{T}}(\mathcal{M}) : \mathcal{A}^{\rho, \sigma} \Rightarrow 1\right]$$

where the probabilities are over the choice of K or $\widetilde{\pi}$ (resp. ρ, σ) and the coins used by $\mathcal{A}$. Here $\mathrm{Perm}^{\mathcal{T}}(\mathcal{M})$ is the set of all functions $\widetilde{\pi}: \mathcal{T} \times \mathcal{M} \to \mathcal{M}$ where $\widetilde{\pi}(T, \cdot)$ is a length-preserving permutation and $\mathrm{Func}^{\mathcal{T}}(\mathcal{M})$ is the set of all functions $\rho: \mathcal{T} \times \mathcal{M} \to \mathcal{M}$ where $\rho(T, \cdot)$ is length-preserving.

XLS AND TWEAKS. The XLS construction works for tweaks. By this we mean that if the cipher $\mathcal{E}$ utilized in XLS is tweakable, then the resulting cipher with the enlarged message space is also tweakable. More specifically, fix an $\epsilon(s)$-good mixing function. If we construct $\widetilde{\mathcal{E}}^* = \mathrm{XLS}[\mathrm{mix}, \widetilde{\mathcal{E}}, E]$ where $\widetilde{\mathcal{E}}: \mathcal{K}_{\mathcal{E}} \times \mathcal{T} \times \mathcal{M} \to \mathcal{M}$ is a tweakable cipher and $E: \mathcal{K}_E \times \{0,1\}^n \to \{0,1\}^n$ is a conventional blockcipher then the resulting scheme $\widetilde{\mathcal{E}}^*$ is tweakable with tweak space $\mathcal{T}$ and enlarged message space $\mathcal{M}^* = \mathcal{M} \,||\, \{0,1\}^{<n}$. Here XLS is implicitly changed by adding the tweak T as a superscript to $\mathcal{E}$ on lines 00 and 06 of Fig. 1. The following theorem statements, which are analogous to Theorem 1 and Theorem 2, establish the security of XLS with tweaks.

Theorem 3. *Fix n and an $\epsilon(s)$-good mixing function* mix. *Let $\mathcal{M} \subset \{0,1\}^{\geq n}$, let $\mathcal{T}$ be a nonempty set, and let $\widetilde{\mathcal{E}}^* = \mathrm{XLS}[\mathrm{mix}, \mathrm{Perm}^{\mathcal{T}}(\mathcal{M}), \mathrm{Perm}(n)]$. Then for any adversary A that asks at most q queries we have that $\mathbf{Adv}^{\pm\widetilde{\mathrm{prf}}}_{\widetilde{\mathcal{E}}^*}(A) \leq 5q^2\,\epsilon(s)/2^{n-s} + 3q^2/2^n$ for any $s \in [1..n-1]$, and so, by Lemma 1*

$$\mathbf{Adv}^{\pm\widetilde{\mathrm{prf}}}_{\widetilde{\mathcal{E}}^*}(A) \leq \frac{8q^2}{2^n} \quad \text{and} \quad \mathbf{Adv}^{\pm\widetilde{\mathrm{prf}}}_{\widetilde{\mathcal{E}}^*}(A) \leq \frac{13q^2}{2^n}$$

for mix $=$ mix1 *and* mix $=$ mix2, *respectively.* □

To prove this theorem, we can adjust the proof of Theorem 1 as follows. Replace $\mathcal{E}$ with $\widetilde{\mathcal{E}}$ throughout. Modify games G0 and G1 so enciphering and deciphering queries take a tweak, and lazily build separate random length-preserving functions $\widetilde{\mathcal{E}}$ for each tweak queried (in G0 they are permutations while in G1 they are just functions). Lemma 2, and its proof, can be modified in the natural way. See [25] for details. Combining Theorem 1 with a standard hybrid argument proves the next theorem.

Theorem 4. *Fix n and an $\epsilon(s)$-good mixing function* mix. *Let $\widetilde{\mathcal{E}}$: $\mathcal{K}_{\mathcal{E}} \times \mathcal{T} \times \mathcal{M} \to \mathcal{M}$ be a tweakable cipher and E: $\mathcal{K}_E \times \{0,1\}^n \to \{0,1\}^n$ be a blockcipher. Let $\mathcal{E}^* = \mathrm{XLS}[\mathrm{mix}, \widetilde{\mathcal{E}}, E]$ and let A be an adversary that runs in time t and asks at most q queries, with maximal query length being μ bits Then there exists adversaries B and C and an absolute constant c such that $\mathbf{Adv}^{\pm\widetilde{\mathrm{prp}}}_{\mathcal{E}^*}(A) \leq \mathbf{Adv}^{\pm\widetilde{\mathrm{prp}}}_{\widetilde{\mathcal{E}}}(B) + \mathbf{Adv}^{\pm\mathrm{prp}}_{E}(C) + 5q^2\,\epsilon(s)/2^{n-s} + 4q^2/2^n$ where $s \in [0..n-1]$, and so, by Lemma 1*

$$\mathbf{Adv}^{\pm\widetilde{\mathrm{prp}}}_{\mathcal{E}^*}(A) \leq \mathbf{Adv}^{\pm\widetilde{\mathrm{prp}}}_{\widetilde{\mathcal{E}}}(B) + \mathbf{Adv}^{\pm\mathrm{prp}}_{E}(C) + \frac{9q^2}{2^n}$$

when mix $=$ mix1, *and the same expression, but with the 9 replaced by 14, when* mix $=$ mix2. *Here B runs in time $t_B = t + c\mu q \log q$ and asks $q_B = q$ queries, none longer than μ, and where C runs in time $t_C \leq t + (q+1) \cdot \mathrm{Time}_{\widetilde{\mathcal{E}}}(\mu) + c\mu q$ and asks $q_C \leq 2q$ queries, each of length at most n.* □

Note that the concrete security bounds are the same in both the tweaked and untweaked settings. Intuitively this is because having a tweakable family of permutations is a stronger primitive than a normal family of permutations, and the adversary might as well just focus its attack on a single tweak.

8   XLS with Ordinary PRPs

If we apply the XLS construction to cipher $\widetilde{\mathcal{E}}$ and blockcipher E that are both secure as (ordinary) pseudorandom permutations (i.e., adversaries are restricted to chosen-plaintext attacks), then the resulting cipher with expanded message space is also secure as a PRP. More formally, we define the following advantage

$$\mathbf{Adv}^{\widetilde{\mathrm{prp}}}_{\widetilde{\mathcal{E}}}(A) = \Pr\left[K \xleftarrow{\$} \mathcal{K}_{\mathcal{E}} : A^{\widetilde{\mathcal{E}}_K} \Rightarrow 1\right] - \Pr\left[\widetilde{\pi} \xleftarrow{\$} \mathrm{Perm}^{\mathcal{T}}(\mathcal{M}) : A^{\widetilde{\pi}} \Rightarrow 1\right]$$

where the probability is over the random choice of K or π and the random coins used by $\mathcal{A}$. While the theorems from Section 7 do not imply the security of XLS using ordinary (tweaked) PRPs, their proofs can be (simplified) in a straightforward manner to derive the PRP security of XLS as captured by the next theorem statement.

Theorem 5. *Fix n and an $\epsilon(s)$-good mixing function* mix. *Let $\widetilde{\mathcal{E}}$: $\mathcal{K}_{\mathcal{E}} \times \mathcal{T} \times \mathcal{M} \to \mathcal{M}$ be a tweakable cipher and E: $\mathcal{K}_E \times \{0,1\}^n \to \{0,1\}^n$ be a blockcipher. Let $\mathcal{E}^* = \mathrm{XLS}[\mathrm{mix}, \mathcal{E}, E]$ and let $\mathcal{A}$ be an adversary that runs in time t and asks at most q queries, with maximal query length being μ bits Then there exists adversaries $\mathcal{B}$ and $\mathcal{C}$ and an absolute constant c such that $\mathbf{Adv}_{\widetilde{\mathcal{E}}^*}^{\widetilde{\mathrm{prp}}}(\mathcal{A}) \le \mathbf{Adv}_{\widetilde{\mathcal{E}}}^{\widetilde{\mathrm{prp}}}(\mathcal{B}) + \mathbf{Adv}_E^{\mathrm{prp}}(\mathcal{C}) + 5q^2\, \epsilon(s)/2^{n-s} + 4q^2/2^n$ where $s \in [0..n-1]$, and so, by Lemma 1*

$$\mathbf{Adv}_{\mathcal{E}^*}^{\widetilde{\mathrm{prp}}}(\mathcal{A}) \le \mathbf{Adv}_{\widetilde{\mathcal{E}}}^{\widetilde{\mathrm{prp}}}(\mathcal{B}) + \mathbf{Adv}_E^{\mathrm{prp}}(\mathcal{C}) + \frac{9q^2}{2^n}$$

when mix = mix1, *and the same expression, but with the 9 replaced by 14, when* mix = mix2. *Here $\mathcal{B}$ runs in time $t_B = t + c\mu q \log q$ and asks $q_B = q$ queries, none longer than μ, and where $\mathcal{C}$ runs in time $t_C \le t + (q+1) \cdot \mathsf{Time}_{\widetilde{\mathcal{E}}}(\mu) + c\mu q$ and asks $q_C \le 2q$ queries, each of length at most n.* $\square$

An immediate corollary of the theorem is security of XLS for ordinary, untweaked PRPs (just set the tweak space $\mathcal{T}$ to a single value).

Acknowledgments

We thank Mihir Bellare, Jesse Walker, and the anonymous referees. This work was supported in part by NSF grants CCR-0208842, CNS-0524765, and a gift from Intel Corp; thanks to the NSF and Intel for their kind support.

References

1. An, J., Bellare, M.: Constructing VIL-MACs from FIL-MACs: Message authentication under weakened assumptions. In: Wiener, M.J. (ed.) CRYPTO 1999. LNCS, vol. 1666, pp. 252–269. Springer, Heidelberg (1999)
2. Anderson, R., Biham, E.: Two practical and provably secure block ciphers: BEAR and LION. In: Gollmann, D. (ed.) Fast Software Encryption. LNCS, vol. 1039, pp. 113–120. Springer, Heidelberg (1996)
3. Bellare, M., Rogaway, P.: On the construction of variable-input-length ciphers. In: Knudsen, L.R. (ed.) FSE 1999. LNCS, vol. 1636, pp. 231–244. Springer, Heidelberg (1999)
4. Bellare, M., Rogaway, P.: Encode-then-encipher encryption: How to exploit nonces or redundancy in plaintexts for efficient cryptography. In: Okamoto, T. (ed.) ASIACRYPT 2000. LNCS, vol. 1976, pp. 317–330. Springer, Heidelberg (2000)
5. Bellare, M., Rogaway, P.: The security of triple encryption and a framework for code-based game-playing proofs. In: Vaudenay, S. (ed.) EUROCRYPT 2006. LNCS, vol. 4004, pp. 409–426. Springer, Heidelberg (2006)

6. Black, J., Rogaway, P.: Ciphers with arbitrary finite domains. In: Preneel, B. (ed.) CT-RSA 2002. LNCS, vol. 2271, pp. 114–130. Springer, Heidelberg (2002)
7. Cayley, A.: On Latin squares. Oxford Cambridge Dublin Messenger Math. 19, 135–137 (1890)
8. Chakraborty, D., Sarkar, P.: A new mode of encryption providing a strong tweakable pseudo-random permutation. In: Robshaw, M. (ed.) FSE 2006. LNCS, vol. 4047, pp. 293–309. Springer, Heidelberg (2006)
9. Chakraborty, D., Sarkar, P.: HCH: A new tweakable enciphering scheme using the Hash-Encrypt-Hash approach. In: Barua, R., Lange, T. (eds.) INDOCRYPT 2006. LNCS, vol. 4329, pp. 287–302. Springer, Heidelberg (2006)
10. Cook, D., Yung, M., Keromytis, A.: Elastic AES. Cryptology ePrint archive, report 2004/141 (2004)
11. Cook, D., Yung, M., Keromytis, A.: Elastic block ciphers. Cryptology ePrint archive, report 2004/128 (2004)
12. Halevi, S.: EME*: Extending EME to handle arbitrary-length messages with associated data. In: Canteaut, A., Viswanathan, K. (eds.) INDOCRYPT 2004. LNCS, vol. 3348, pp. 315–327. Springer, Heidelberg (2004)
13. Halevi, S.: TET: A wide-block tweakable mode based on Naor-Reingold. Cryptology ePrint archive, report 20007/14 (2007)
14. Halevi, S., Rogaway, P.: A parallelizable enciphering mode. In: Okamoto, T. (ed.) CT-RSA 2004. LNCS, vol. 2964, pp. 292–304. Springer, Heidelberg (2004)
15. Halevi, S., Rogaway, P.: A tweakable enciphering mode. In: Boneh, D. (ed.) CRYPTO 2003. LNCS, vol. 2729, pp. 482–499. Springer, Heidelberg (2003)
16. Liskov, M., Rivest, R., Wagner, D.: Tweakable block ciphers. In: Yung, M. (ed.) CRYPTO 2002. LNCS, vol. 2442, pp. 31–46. Springer, Heidelberg (2002)
17. Luby, M., Rackoff, C.: How to construct pseudorandom permutations from pseudorandom functions. SIAM Journal of Computing 17(2), 373–386 (1988)
18. Lucks, S.: BEAST: A fast block cipher for arbitrary blocksizes. In: Communications and Multimedia Security, IFIP, vol. 70, pp. 144–153. Chapman & Hill, Sydney, Australia (1996)
19. Lucks, S.: Faster Luby-Rackoff ciphers. In: Gollmann, D. (ed.) Fast Software Encryption. LNCS, vol. 1039, pp. 189–203. Springer, Heidelberg (1996)
20. McGrew, D., Fluhrer, S.: The extended codebook (XCB) mode of operation. Cryptology ePrint archive, report 2004/278 (2004)
21. Meyer, C., Matyas, M.: Cryptography: A New Dimension in Data Security. John Wiley & Sons, New York (1982)
22. Naor, M., Reingold, O.: On the construction of pseudorandom permutations: Luby-Rackoff revisited. Journal of Cryptology 12(1), 29–66 (1999)
23. Patarin, J.: How to construct pseudorandom and super pseudorandom permutations from one single pseudorandom function. In: Rueppel, R.A. (ed.) EUROCRYPT 1992. LNCS, vol. 658, pp. 256–266. Springer, Heidelberg (1993)
24. Patel, S., Ramzan, Z., Sundaram, G.: Efficient constructions of variable-input-length block ciphers. In: Handschuh, H., Hasan, M.A. (eds.) SAC 2004. LNCS, vol. 3357, pp. 326–340. Springer, Heidelberg (2004)
25. Ristenpart, T., Rogaway, P.: How to enrich the Message Space of a Cipher (full version of this paper), http://www.cse.ucsd.edu/users/tristenp/
26. Schneier, B., Kelsey, J.: Unbalanced Feistel networks and block cipher design. In: Gollmann, D. (ed.) Fast Software Encryption. LNCS, vol. 1039, pp. 121–144. Springer, Heidelberg (1996)

27. Schnorr, C., Vaudenay, S.: Black box cryptanalysis of hash networks based on multipermutations. In: De Santis, A. (ed.) EUROCRYPT 1994. LNCS, vol. 950, pp. 47–57. Springer, Heidelberg (1995)
28. Schroeppel, R.: Hasty pudding cipher specification. First AES Candidate Workshop (1998)
29. Vaudenay, S.: On the need for multipermutations: cryptanalysis of MD4 and SAFER. In: Preneel, B. (ed.) Fast Software Encryption. LNCS, vol. 1008, pp. 286–297. Springer, Heidelberg (1995)
30. Wang, P., Feng, D., Wu, W.: HCTR: a variable-input-length enciphering mode. In: Feng, D., Lin, D., Yung, M. (eds.) CISC 2005. LNCS, vol. 3822, pp. 175–188. Springer, Heidelberg (2005)
31. Zheng, Y., Matsumoto, T., Imai, H.: On the construction of block ciphers provably secure and not relying on any unproved hypotheses. In: Brassard, G. (ed.) CRYPTO 1989. LNCS, vol. 435, pp. 461–480. Springer, Heidelberg (1990)

Security Analysis of Constructions Combining FIL Random Oracles

Yannick Seurin and Thomas Peyrin

France Telecom R&D, 38-40 rue du Général Leclerc, F-92794 Issy-les-Moulineaux, France
Université de Versailles, 45 avenue des Etats-Unis, F-78035 Versailles, France
yannick.seurin@m4x.org, thomas.peyrin@orange-ftgroup.com

Abstract. We consider the security of compression functions built by combining smaller perfectly secure compression functions modeled as fixed input length random oracles. We give tight security bounds and generic attacks for various parameters of these constructions and apply our results to recent proposals of block cipher-based hash functions.

Keywords: block ciphers, compression functions, hash functions, provable security, random oracle.

1 Introduction

Cryptographic hash functions are fundamental primitives in information security [17] used in a variety of applications such as message integrity, authentication schemes or digital signatures. Mathematically speaking, it is a function from $\{0,1\}^*$, the set of all finite length bit strings, to $\{0,1\}^l$ where l is the fixed size of the hash value. Ideally, a cryptographic hash function should possess the following properties:

- *collision resistance*: finding a pair $x \neq x' \in \{0,1\}^*$ such that $H(x) = H(x')$ should require $2^{l/2}$ operations
- *2nd preimage resistance*: for a given $x \in \{0,1\}^*$, finding a $x' \neq x$ such that $H(x) = H(x')$ should require 2^l operations
- *preimage resistance*: for a given $y \in \{0,1\}^l$, finding a $x \in \{0,1\}^*$ such that $H(x) = y$ should require 2^l operations.

All currently used hash functions are so-called *iterated* hash functions which are designed by iterating a *compression function* with a fixed-length input, say $h : \{0,1\}^{l+l'} \to \{0,1\}^l$. The iterated hash function H is then defined thanks to *domain extension* methods. The most popular one is the Merkle-Damgård method [5,18] which consists in first padding the input x so that the length of the padded message is a multiple of l' and outputing, for a padded message consisting of m l'-bit blocks $\mathrm{Pad}(x) = x_1\|\ldots\|x_m$, the value y_m defined by the recurrence $y_i = h(x_i\|y_{i-1})$, where y_0 is a fixed constant of $\{0,1\}^l$. The y_i's are called *chaining variables*. The popularity of the MD method comes from the

A. Biryukov (Ed.): FSE 2007, LNCS 4593, pp. 119–136, 2007.

fact that the hash function obtained is at least as resistant to collision attacks as the compression function. However, recent results have highlighted the intrinsic limitations of the MD approach [8,9] and motivated the study of other domain extension methods [1,4].

Most popular hash functions (e.g. MD5, SHA1) make use of compression functions build "from scratch", not appealing to any lower-level primitive. Another direction of research consists in trying to turn a block cipher into a compression function. This approach has been revived by the recent attacks on hash function using compression functions of dedicated design [28,27]. The question of how to turn a block cipher into a single block length (SBL) compression function (i.e. whose output length is the same as the block length of the block cipher) can be more or less considered as closed since the systematic study of Preneel et al. [23] and Black et al. [2]. However, the block length of the most trusted and standardized block ciphers such as DES and AES is too short to prevent collision attacks by the birthday paradox on SBL hash functions based on them. This is why there has been much effort in order to build a double block length (DBL) or more generally a multiple block length (MBL) compression function whose output is twice (or more) the block length of the block cipher. Most of the earlier proposals [3,15,16,22,24] turned out to have weaknesses [10,15]. Proofs of security for block cipher-based hash functions date back to Winternitz [29], who used the ideal cipher model of Shannon [25] to prove the security of the Davies-Meyer scheme against preimage attacks. Black et al. [2] used the same paradigm to study all the natural ways of building SBL compression functions, a work which had been initiated in [23]. Hirose [6,7] demonstrated the security of a family of DBL compression functions using two independent block ciphers with key length twice the block length, again in the ideal block cipher model. However, no secure DBL scheme using block ciphers with key length equal to the block length has been proposed so far. Nandi et al. [20] proposed DBL schemes with better rates than those of Hirose and claimed to have proved that an adversary must make $\Omega(2^{2n/3})$ oracle queries to get a collision and $\Omega(2^{4n/3})$ oracle queries to get a preimage. However, in light of the attacks presented in [11] (where a preimage attack requiring only $O(2^n)$ queries is described), we spotted a mistake in the security proof of [20]. One of the goal of this paper is to remedy the strategy they adopted.

At Asiacrypt '06 [21], Peyrin et al. presented a general framework to analyse how to combine secure compression functions in order to obtain compression functions with longer output. This approach had already been adopted in a series of papers [12,13,14] where partial answers were given thanks to error-correcting codes theory. Analysing two types of generic attacks, Peyrin et al. derived necessary conditions for the compression functions of their framework to be secure. Nevertheless, no security proofs were given. The aim of this paper is to analyse the constructions of the general framework introduced in [21] in a proof oriented manner. Though we will work in the fixed input length (FIL) random oracle model, this must be understood as a first step in the systematic study of MBL compression functions based on block ciphers.

The paper is organized as follows. In section 2 we establish the notations and some useful lemmas. In section 3 and 4 we carry out the security analysis for preimage resistance and collision resistance respectively. In section 5 we apply our results to previous proposals of block cipher-based hash functions and we draw our conclusions and propose future work in section 6.

2   Definitions and Notations

Basic Notations. In all the following, $\mathcal{I}_n$ will denote the set $\{0,1\}^n$, and $\mathcal{F}(a,b)$ the set of all functions from $\{0,1\}^a$ to $\{0,1\}^b$. We will often consider vectors of elements of $\mathcal{I}_n$ of various length which will be denoted by bold letters. For a binary vector $l = (l_1, \ldots, l_r) \in \{0,1\}^r$ and $\boldsymbol{X} = (X_1, \ldots, X_r) \in (\mathcal{I}_n)^r$, $\boldsymbol{X} \cdot l^T = l_1 X_1 \oplus \cdots \oplus l_r X_r$. Similarly, for a binary matrix $L = [l_1^T, \ldots, l_s^T] \in \mathcal{M}_{r,s}(\{0,1\})$, $\boldsymbol{X} \cdot L$ is the vector $(\boldsymbol{X} \cdot l_1^T, \ldots, \boldsymbol{X} \cdot l_s^T)$. Given two vectors $\boldsymbol{X} = (X_1, \ldots, X_r)$ and $\boldsymbol{Y} = (Y_1, \ldots, Y_s)$, $\boldsymbol{X} \| \boldsymbol{Y}$ will denote the vector $(X_1, \ldots, X_r, Y_1, \ldots, Y_s)$. Finally, $\|\cdot\|_H$ will denote the Hamming weight of a vector and $\mathbb{E}$ the expected value of a random variable.

Generic Constructions. The aim of this paper is to analyse the security of a very general class of compression functions build from smaller secure compression functions. Namely, our building blocks will be t compression functions $f^{(1)}, \ldots, f^{(t)}$ taking each k n-bit blocks as input and outputing one n-bit block. For the security analysis, we will assume that these functions are independent random oracles. The larger compression function will take as input m n-bit message blocks and c n-bit chaining variable blocks, which will be denoted respectively $\boldsymbol{M} = (M_1, \ldots, M_m)$ and $\boldsymbol{H} = (H_1, \ldots, H_c)$. These blocks will be named *external* input blocks to distinguish them from the $t * k$ input blocks to the inner compression functions, which will be named *internal* input blocks. They are obtained as linear combinations of the external input blocks. Namely, for each $i \in [1..t]$, there is a binary matrix $\mathcal{A}_i \in \mathcal{M}_{(m+c,k)}(\{0,1\})$ such that the input to the i-th internal compression function is $\boldsymbol{M} \| \boldsymbol{H} \cdot \mathcal{A}_i$.

The output blocks of the internal compression functions

$$\boldsymbol{F} = (f^{(1)}(\boldsymbol{M} \| \boldsymbol{H} \cdot \mathcal{A}_1), \ldots, f^{(t)}(\boldsymbol{M} \| \boldsymbol{H} \cdot \mathcal{A}_t))$$

are then mixed by a linear output layer $\mathcal{B} \in \mathcal{M}_{(t,c)}(\{0,1\})$ to give the external output blocks $\boldsymbol{H}' = (H_1', \ldots, H_c')$ according to $\boldsymbol{H}' = \boldsymbol{F} \cdot \mathcal{B}$. In all the following, it will be assumed that $\mathcal{B}$ has full rank (otherwise the external output blocks are linearly dependent, which is clearly undesirable).

A *compression function construction* h is thus completely determined by the parameters (c, t, k, m) and the input and output layers $(\mathcal{A}_i)_{i \in [1..t]}$ and $\mathcal{B}$. The compression function obtained once the internal compression functions $f^{(1)}, \ldots, f^{(t)}$ are instantiated will be noted $h^{(f^{(1)}, \ldots, f^{(t)})}$. The construction can be summarized by the formula (see also Fig. 1)

$$h^{(f^{(1)}, \ldots, f^{(t)})}(\boldsymbol{M} \| \boldsymbol{H}) = \left(f^{(1)}(\boldsymbol{M} \| \boldsymbol{H} \cdot \mathcal{A}_1), \ldots, f^{(t)}(\boldsymbol{M} \| \boldsymbol{H} \cdot \mathcal{A}_t) \right) \cdot \mathcal{B}. \quad (1)$$

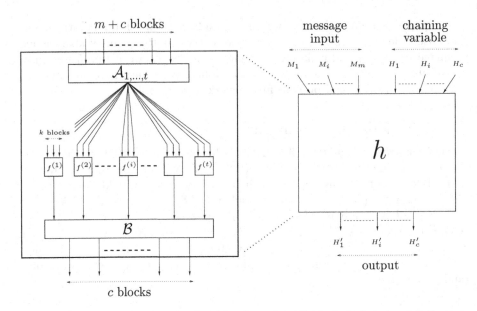

Fig. 1. The compression function h taking $(m+c)$ n-bit blocks in input and delivering c n-bit output blocks. It is build from t compression functions $f^{(i)}$ taking k n-bit input blocks and outputing one n-bit block.

A more general framework could encompass a feedforward of the external input blocks, i.e. allowing to xor some external input blocks with some internal output blocks. Though it would definitely be useful in the ideal block cipher model, we do not believe this would strengthen in any way the constructions in the random oracle model, and thus we consider this feature as out of the scope of the article. In the following, we will often consider linear combinations of the coordinates of the function h. For this, we will use the following notations. For $x \in \{0,1\}^c$, G_x will be the function

$$G_x : \boldsymbol{M} \| \boldsymbol{H} \mapsto h^{(f^{(1)}, \dots, f^{(t)})}(\boldsymbol{M} \| \boldsymbol{H}) \cdot x^T.$$

Alternatively, G_x may be seen as a linear combination of the t functions $\boldsymbol{M} \| \boldsymbol{H} \mapsto f^{(i)}(\boldsymbol{M} \| \boldsymbol{H} \cdot \mathcal{A}_i)$. Indeed, writing $y = x \cdot \mathcal{B}^T \in \{0,1\}^t$, one has

$$G_x(\boldsymbol{M} \| \boldsymbol{H}) = (f^{(1)}(\boldsymbol{M} \| \boldsymbol{H} \cdot \mathcal{A}_1), \dots, f^{(t)}(\boldsymbol{M} \| \boldsymbol{H} \cdot \mathcal{A}_t)) \cdot y^T.$$

It will often be more convenient to define the set S_x of "active" inner compression functions, i.e. the set of integers $j \in [1..c]$ such that the j-th coordinate of $x \cdot \mathcal{B}^T$ is 1. Then G_x can be expressed by

$$G_x(\boldsymbol{M} \| \boldsymbol{H}) = \bigoplus_{j \in S_x} f^{(j)}(\boldsymbol{M} \| \boldsymbol{H} \cdot \mathcal{A}_j).$$

Security Model. In the following we will analyse the resistance of the compression functions we just described against preimage attacks and collision attacks.

An *adversary* will be an algorithm with access to oracles for the inner compression functions $f^{(1)}, \ldots, f^{(t)}$. Given a finite set S, $s \xleftarrow{\$} S$ denotes the operation of selecting s in the probability space S endowed with the uniform distribution. We will work in the random oracle model, meaning that the inner compression functions are uniformly and independently selected in the set $\mathcal{F}(kn, n)$. When these functions are asked queries from an algorithm, their output is uniform and independent from all other outputs, but consistent with answers to queries already asked. We now define the preimage and collision resistance of the compression function constructions.

Definition 1 (Preimage resistance of a compression function). *Let h be a (c, t, k, m)-compression function construction and let $\mathfrak{A}$ be an adversary. Then the advantage of $\mathfrak{A}$ in finding a preimage for h is the real number*

$$\boldsymbol{Adv}_h^{\mathrm{pre}}(\mathfrak{A}) = \Pr\left[(f^{(1)}, \ldots, f^{(t)}) \xleftarrow{\$} \mathcal{F}(kn, n)^t; H' \xleftarrow{\$} (\mathcal{I}_n)^c; \right.$$

$$\left. M\|H \xleftarrow{\$} \mathfrak{A}(H') : h^{(f^{(1)}, \ldots, f^{(t)})}(M\|H) = H'\right].$$

We associate to each compression function construction h the insecurity measure

$$\boldsymbol{Adv}_h^{\mathrm{pre}}(q) = \max_{\mathfrak{A}}\{\boldsymbol{Adv}_h^{\mathrm{pre}}(\mathfrak{A})\}$$

where the maximum is taken over all adversaries making at most q oracle queries to each inner compression function $f^{(1)}, \ldots, f^{(t)}$.

Definition 2 (Collision resistance of a compression function). *Let h be a (c, t, k, m)-compression function construction and let $\mathfrak{A}$ be an adversary. Then the advantage of $\mathfrak{A}$ in finding a collision for h is the real number*

$$\boldsymbol{Adv}_h^{\mathrm{coll}}(\mathfrak{A}) = \Pr\left[(f^{(1)}, \ldots, f^{(t)}) \xleftarrow{\$} \mathcal{F}(kn, n)^t; (M_1\|H_1, M_2\|H_2) \xleftarrow{\$} \mathfrak{A} : \right.$$

$$\left. M_1\|H_1 \neq M_2\|H_2 \wedge h^{(f^{(1)}, \ldots, f^{(t)})}(M_1\|H_1) = h^{(f^{(1)}, \ldots, f^{(t)})}(M_2\|H_2)\right].$$

We associate to each compression function construction h the insecurity measure

$$\boldsymbol{Adv}_h^{\mathrm{coll}}(q) = \max_{\mathfrak{A}}\{\boldsymbol{Adv}_h^{\mathrm{coll}}(\mathfrak{A})\}$$

where the maximum is taken over all adversaries making at most q oracle queries to each inner compression function $f^{(1)}, \ldots, f^{(t)}$.

For the remainder of this paper we will make the following classical assumptions regarding the adversaries. First, they are computationally unbounded, in consequence of what we can restrain ourselves wlog to deterministic adversaries (so that we do not have to take into account any more the randomness coming from the random choices of the algorithm). Second, an adversary does not make the same oracle query more than once. Third, we will restrain ourselves to adversaries making exactly q queries to each inner compression function. These assumptions does not restrict the generality of the analysis in that for any adversary $\mathfrak{A}$ asking at most q queries there exists another adversary $\mathfrak{A}'$ verifying the assumptions that achieves at least the same advantage as $\mathfrak{A}$.

Type I Constructions. A natural requirement for the compression functions studied here would be that the image of two distinct inputs by any linear combination of the output blocks are independent. This is generally not the case, and compressions functions which does not possess this property are subject to devastating attack called DF attacks (*degrees of freedom*) in [21]. This feature is achieved by letting every external output block depend on all external input block, no matter which invertible transformations of the external inputs and outputs are used. Expressing it mathematically yields the following definition.

Definition 3 (Type I (for Independent) compression function construction). *A (c, t, k, m)-compression function construction will be said to be of type I iff for all $x \in \{0, 1\}^c \setminus \{0\}$, $\bigcap_{j \in S_x} \ker \mathcal{A}_j = \{0\}$.*

For such constructions, one can prove the following property (the proof is given in Appendix A).

Lemma 1. *Let h be a (c, t, k, m) compression function construction of type I. Then for all $x \in \{0, 1\}^c \setminus \{0\}$, and for all distinct $M_1 \| H_1$ and $M_2 \| H_2$, $G_x(M_1 \| H_1)$ and $G_x(M_2 \| H_2)$ are uniformly random and independent.*

Not all parameter sets permit to build type I compression function constructions. More precisely, one has the following necessary condition, which was proved in [21].

Lemma 2 ([21]). *Let h be a (c, t, k, m) compression function construction of type I. Then necessarily $\forall x \in \{0, 1\}^c \setminus \{0\}$, $\|x \cdot \mathcal{B}^T\|_H \geq \frac{m+c}{k}$. In other words, $\mathcal{B}$ must have minimal distance at least $\lceil \frac{m+c}{k} \rceil$.*

Computable Inputs. In order to make our explanations more rigorous, we will need the following notions of *computability*, which are generalizations of concepts introduced in [20]. Informally speaking, once an adversary has made certain queries to the inner compression functions, we want to define for each $M \| H$ the number of coordinates of $h(M \| H)$ the adversary is able to compute.

Definition 4 (G_x-computable input). *Let $\mathcal{Q}_1, \ldots, \mathcal{Q}_t \subset (\mathcal{I}_n)^k$ be sets of queries to each of the inner compression functions. For $x \in \{0, 1\}^c$, we will say that an external input $M \| H \in (\mathcal{I}_n)^{m+c}$ is G_x-computable with respect to these sets of queries if $M \| H \cdot \mathcal{A}_i \in \mathcal{Q}_i$, for each $i \in S_x$.*

It is easy to verify that given sets of queries $\mathcal{Q}_1, \ldots, \mathcal{Q}_t$ and $x_1, \ldots, x_r \in \{0, 1\}^c$, $M \| H$ is G_{x_i}-computable for all $i \in [1..r]$ implies that $M \| H$ is G_x-computable for all $x \in \text{Vec}(x_1, \ldots, x_r)$. It is thus natural to give the following definitions.

Definition 5 (V-computable input). *Let V be a subspace of $\{0, 1\}^c$, $V \neq \emptyset$, let $\mathcal{Q}_1, \ldots, \mathcal{Q}_t \subset (\mathcal{I}_n)^k$ be the sets of queries to each of the inner compression functions. We will say that an external input $M \| H \in (\mathcal{I}_n)^{m+c}$ is V-computable with respect to these sets of queries if $M \| H$ is G_x-computable for all $x \in V$.*

Let $V_{M\|H}$ be the biggest subspace such that $M \| H$ is V-computable (possibly reduced to $\{0\}$). If r is the dimension of $V_{M\|H}$, we will say that $M \| H$

is r-computable. *We will also talk of* h-computable *input when* $r = c$ *and of* uncomputable *input when* $r = 0$.

Definition 6 (Maximal number of (at least) r-computable inputs with q queries). *Let* h *be a compression function construction and* $q \geq 1$*. We define the maximal number of (at least) r-computable inputs with q queries $\beta_r(q)$ as being*

$$\beta_r(q) = \max_{\mathcal{Q}_1,\ldots,\mathcal{Q}_t} \#\{M\|H \in (\mathcal{I}_n)^{m+c} \mid M\|H \text{ is at least } r\text{-computable}\}$$

where the maximum is taken over all the possible sets of q queries to the inner compression functions.

We will also need the following slightly different notion for $r = 1$:

$$\beta_1'(q) = \max_{x \in \{0,1\}^c \backslash \{0\}} \max_{\mathcal{Q}_1,\ldots,\mathcal{Q}_t} \#\{M\|H \in (\mathcal{I}_n)^{m+c} \mid M\|H \text{ is } G_x\text{-computable}\}$$

where the maximum is taken over all the non-zero linear combinations of output blocks and over all the possible sets of q queries to the inner compression functions.

The following proposition is rather obvious and given without proof.

Proposition 1

$$q \leq \beta_c(q) \leq \beta_{c-1}(q) \leq \ldots \leq \beta_1(q).$$

$\beta_1(q)$ and $\beta_1'(q)$ capture approximately the same characteristic of the compression function for it is immediate to verify that

$$\beta_1'(q) \leq \beta_1(q) \leq 2^c \beta_1'(q). \tag{2}$$

In our security analysis, we will make an extensive use of the following lemma (see the proof in Appendix A).

Lemma 3 (Independency lemma). *Let* $\mathcal{Q}_1,\ldots,\mathcal{Q}_t \subset (\mathcal{I}_n)^k$ *be sets of queries to the inner compression functions such that* $M\|H$ *is r-computable. Let* $(x_1,\ldots,x_r)$ *be a basis of* $V_{M\|H}$*, and* $X_1,\ldots,X_r, X \in \mathcal{I}_n$*. Then* $\forall x \in \{0,1\}^c \backslash V_{M\|H}$,

$$\Pr[G_x(M\|H) = X \mid G_{x_1}(M\|H) = X_1,\ldots,G_{x_r}(M\|H) = X_r] = \frac{1}{2^n}.$$

More generally, if H' *is such that for all* $i \in [1..r]$*,* $H' \cdot x_i^T = X_i$*, then*

$$\Pr[h(M\|H) = H' \mid G_{x_1}(M\|H) = X_1,\ldots,G_{x_r}(M\|H) = X_r] = \frac{1}{2^{(c-r)n}}.$$

3   Security Analysis for Preimage Resistance

In this section we begin with providing a security bound to preimage attacks for the constructions of the general framework studied in this paper. Then we show that this security bound is tight by analysing an attack whose advantage is close to the security bound.

Theorem 1 (Security bound for preimage resistance). *Let h be a (c, t, k, m)-compression function construction (non necessarily of type I) with parameter $\beta_1(q)$ defined by definition 6. Then*

$$Adv_h^{\mathrm{pre}}(q) \leq \frac{1}{2^n} + \frac{\beta_1(q)}{2^{cn}}.$$

Proof. Let $\mathfrak{A}$ be a preimage-finding adversary attacking the compression function h. We suppose wlog that the random input H' to the adversary is $\mathbf{0}$. We first define Preim as being the set of external inputs $M \| H$ which are h-computable with respect to the final sets of queries of $\mathfrak{A}$ and such that $h^{(f^{(1)}, \ldots, f^{(t)})}(M \| H) = \mathbf{0}$. First of all, if $\mathfrak{A}$ does not find any external input in this set, its probability of success is very low. Indeed, $\mathfrak{A}$ is bound to output an $M \| H$ which is not h-computable. According to Lemma 3, the probability for this output to be a good preimage is $\leq 1/2^n$. Therefore we have $\Pr[\mathfrak{A} \text{ wins}] \leq 1/2^n + \Pr[\text{Preim} \neq \emptyset]$.

We now bound $\Pr[\text{Preim} \neq \emptyset]$. For this, we analyse the behavior of $\mathfrak{A}$ in a sequential manner: $\mathfrak{A}$ makes its queries to the inner functions in a certain order. During this process, each external input $M \| H$ goes through successive states: either it is uncomputable, or it is r-computable and still a potential candidate to be mapped on to $\mathbf{0}$, or it is r-computable and discarded because there exists $x \in V_{M\|H}$ such that $G_x(M \| H) \neq \mathbf{0}$. More precisely, consider partial sets of queries $\mathcal{Q}'_1, \ldots, \mathcal{Q}'_t \subset (\mathcal{I}_n)^k$. We will say that an external input $M \| H$ is *compatible* with these partial sets of queries if $\forall x \in V_{M\|H}, G_x(M \| H) = \mathbf{0}$. Note that an external input which is h-computable with respect to the final sets of queries of $\mathfrak{A}$ was necessarily 1-computable at some stage in the sequential queries of $\mathfrak{A}$. Said differently, an external input cannot "jump" from the state *uncomputable* to a state where it is r-computable for $r > 1$ with one single query because one single query never enables to compute more than one output block or linear combination of output blocks[1]. When it exists, we will note $G^1_{M\|H}$ the linear combination of output blocks associated with the first $x \in \{0, 1\}^c \setminus \{0\}$ such that $M \| H$ is G_x-computable. Let us define the set Pot_1 as being the set of all $M \| H$ such that, at some stage in the sequential queries of $\mathfrak{A}$, $M \| H$ was 1-computable and compatible. Then one clearly has $\text{Pot}_1 \supset \text{Preim}$, so that

$$\Pr[M \| H \in \text{Preim}] = \Pr[M \| H \in \text{Preim} | M \| H \in \text{Pot}_1] \cdot \Pr[M \| H \in \text{Pot}_1].$$

---

[1] Suppose that one single query enables to compute both $G_{x_1}(M \| H)$ and $G_{x_2}(M \| H)$. This means that all the other queries necessary to compute them have been made previously. But this implies that $M \| H$ is already $G_{x_1 \oplus x_2}$-computable, so that in fact the computability of $M \| H$ has only been increased by 1.

The key point in the proof is the fact that according to Lemma 3, one has, for all $M\|H$,

$$\Pr[M\|H \in \text{Preim}|M\|H \in \text{Pot}_1] \leq \Pr[h^{(f^{(1)},\ldots,f^{(t)})}(M\|H) = 0|M\|H \in \text{Pot}_1]$$
$$\leq \frac{1}{2^{(c-1)n}}.$$

In consequence,

$$\Pr[\text{Preim} \neq \emptyset] \leq \frac{1}{2^{(c-1)n}} \sum_{M\|H} \Pr[M\|H \in \text{Pot}_1].$$

Now we want to bound the sum $\sum_{M\|H} \Pr[M\|H \in \text{Pot}_1]$. Recall that by the definition of Pot_1, $M\|H \in \text{Pot}_1$ is the event that $M\|H$ is at least 1-computable with respect to the final sets of queries of $\mathfrak{A}$, and $G^1_{M\|H}(M\|H) = 0$. Now conditioning on the event that $M\|H$ is at least 1-computable with respect to the final sets of queries of $\mathfrak{A}$, the probability that $G^1_{M\|H}(M\|H) = 0$ is $1/2^n$. Summing up this reasoning with formulas yields

$$\sum_{M\|H} \Pr[M\|H \in \text{Pot}_1] \leq \sum_{M\|H} \Pr[G^1_{M\|H}(M\|H) = 0|M\|H \text{ is 1-computable}]\cdot$$
$$\Pr[M\|H \text{ is 1-computable}]$$
$$\leq \frac{1}{2^n} \sum_{M\|H} \Pr[M\|H \text{ is 1-computable}]$$
$$\leq \frac{1}{2^n}\mathbb{E}\left(\#\{M\|H \mid M\|H \text{ is 1-computable}\}\right)$$
$$\leq \frac{\beta_1(q)}{2^n}.$$

The theorem follows immediately. $\qquad\square$

Remark 1. The reasoning used in [20] concludes that preimage resistance is $O(\beta_c(q)/2^{cn})$, which cannot be in view of the generic attack presented hereafter. We reproduce this faulty reasoning and point out the mistake in Appendix C.

Theorem 2 (Preimage attack matching the security bound). *Let h be a (c, t, k, m)-compression function construction of type I with parameter $\beta'_1(q)$ defined by definition 6. Then $\beta'_1(q) = \Omega(2^{cn})$ and $q = \Omega(2^{(c-1)n})$ implies that $Adv_h^{\text{pre}}(q) = \Omega(1)$.*

Proof. Once again we suppose wlog that the random input H' to the adversary is 0. Consider the following adversary: $\mathfrak{A}$ first identifies $x \in \{0,1\}^c \setminus \{0\}$ such that $\beta'_1(q)$ is reached and makes the q queries to the inner compression functions $f^{(i)}$ involved in the calculation of G_x (i.e. such that $i \in S_x$), thus obtaining $\beta'_1(q)$ images by G_x. Let N_{G_x} be the random variable counting among these

$\beta_1'(q)$ G_x-computable inputs the number of them such that $G_x(M\|H) = 0$. The compression function construction considered being of type I, the $\beta_1'(q)$ images by G_x obtained are random and pairwise independent. As $\beta_1'(q) = \Omega(2^{cn})$, with overwhelming probability $N_{G_x} = \Omega(2^{(c-1)n})$. After this first step, $\mathfrak{A}$ selects $\min(N_{G_x}, q)$ $M\|H$ such that $G_x(M\|H) = 0$. If $N_{G_x} > q$, it selects them randomly. $\mathfrak{A}$ then queries the remaining compression functions in order to obtain the full image of the selected external inputs. Restricting the number of selected external inputs to q ensures that $\mathfrak{A}$ is always able to obtain their image by h. The probability for one of these external inputs to be a good preimage is $1/2^{(c-1)n}$. As $q = \Omega(2^{(c-1)n})$ by hypothesis, the adversary finds a preimage of $\mathbf{0}$ with non-negligible probability. Hence the result.   □

Conclusion for Preimage Resistance. The results of this section show that preimage resistance of a (c, t, k, m)-compression function construction of type I is governed by the parameter $\beta_1(q)$. Combining Theorems 1 and 2, and recalling inequality (2) proves that, at least for constructions such that one may have $\beta_1'(q) = \Omega(2^{cn})$ and $q = \Omega(2^{(c-1)n})$ at the same time, preimage resistance is $\Theta(\beta_1(q)/2^{cn})$.

4   Security Analysis for Collision Resistance

Theorem 3 (Security bound for collision resistance). *Let h be a (c, t, k, m) compression function construction (non necessarily of type I) with parameter $\beta_1(q)$ defined by definition 6. Then*

$$Adv_h^{coll}(q) \leq \frac{1}{2^n} + \frac{\beta_1(q)^2}{2 \cdot 2^{cn}}.$$

The proof of this theorem is very similar to the proof of Theorem 1 and is given in Appendix B. We now exhibit two collision attacks matching sometimes the security bound.

First Collision Attack. The first attack presented here is very simple and meets the security bound in some cases. It simply consists in computing the image by h of $\beta_c(q)$ external inputs. For a type I construction, they are random and independent, and a classical calculus tells us that the probability to obtain a collision is $1 - \prod_{i=1}^{\beta_c(q)-1}(1 - i/2^{cn})$. As a consequence we have the following result:

Theorem 4 (First collision attack.). *Let h be a (c, t, k, m)-compression function construction of type I with parameter $\beta_c(q)$ defined by definition 6. Then* $Adv_h^{coll}(q) \geq 0.6 \frac{\beta_c(q)(\beta_c(q)-1)}{2 \cdot 2^{cn}}$.
In consequence, for constructions such that $\beta_c(q) \sim \beta_1(q)$ when $q \to \infty$, the security bound given in Theorem 3 is tight.

Proof. We have the following inequalities:

$$\mathbf{Adv}_h^{\mathrm{coll}}(q) \geq 1 - \prod_{i=1}^{\beta_c(q)-1} \left(1 - \frac{i}{2^{cn}}\right)$$

$$\geq 1 - \exp\left(-\sum_{i=1}^{\beta_c(q)-1} \frac{i}{2^{cn}}\right)$$

$$= 1 - \exp\left(-\frac{\beta_c(q)(\beta_c(q)-1)}{2 \cdot 2^{cn}}\right)$$

$$\geq \left(1 - \frac{1}{e}\right) \frac{\beta_c(q)(\beta_c(q)-1)}{2 \cdot 2^{cn}}.$$

The inequality $(1 - e^{-1}) > 0.6$ completes the proof. $\qquad\square$

Second Collision Attack. The second attack is more similar to the preimage attack presented previously and may achieve or not a better advantage than the first one depending on the input and output mappings. The adversary proceeds as follows. $\mathfrak{A}$ first identifies $x \in \{0,1\}^c \setminus \{0\}$ such that $\beta_1'(q)$ is reached and makes the q queries to the inner compression functions involved in the calculation of G_x, thus obtaining $\beta_1'(q)$ images by G_x. $\mathfrak{A}$ quotients the set of the external inputs which are G_x-computable at this stage by the equivalence relation $G_x(M_1 \| H_1) = G_x(M_2 \| H_2)$ and orders the quotient classes by decreasing cardinal. It then calculates the full image by h of the elements of the quotient classes, looking for a collision on the $(c-1)$ remaining output blocks, and beginning with the quotient class of larger cardinal in order to maximize its probability of success. $\mathfrak{A}$ is able to calculate at least q images. Analysing this adversary enables to enunciate the following result.

Theorem 5 (Second collision attack). *Let h be a (c,t,k,m)-compression function construction of type I with parameter $\beta_1'(q)$ defined by definition 6. Then $q\beta_1'(q) = \Omega(2^{cn})$ and $\beta_1'(q) = \Omega(n2^n)$ implies $\mathbf{Adv}_h^{\mathrm{coll}}(q) = \Omega(1)$.*

Proof. Let us analyse the probability of success of the adversary we just described. The fact that $\beta_1'(q) = \Omega(n2^n)$ implies that with probability $1 - O(1)$, $\mathfrak{A}$ obtains 2^n quotient classes containing $\Theta(\beta_1'(q)/2^n)$ elements each (this is a classical "balls and bins" result, see for example [19]). Though $\mathfrak{A}$ will not always be able to obtain the image by h of all the $\beta_1'(q)$ inputs, we can ensure that it will be able to do so for at least q inputs. So the number C of quotient classes in which it will be able to look for a full collision under h is such that $C \cdot \frac{\beta_1'(q)}{2^n} = q$, i.e. $C = \frac{q2^n}{\beta_1'(q)}$. The events "finding a collision in quotient class i", $i \in [1..C]$ are independent and their probability p_i verify (the proof is analog to the proof of Theorem 4)

$$p_i \geq 0.6 \frac{\#\mathcal{C}_i(\#\mathcal{C}_i - 1)}{2^{(c-1)n}}$$

where $\#\mathcal{C}_i$ is the cardinal of the quotient class being explored for a full collision. As $\#\mathcal{C}_i = \Theta(\beta_1'(q)/2^n)$ with overwhelming probability, we have that the total probability to find a collision is

$$\Omega\left(C \cdot \frac{\left(\frac{\beta_1'(q)}{2^n}\right)^2}{2^{(c-1)n}}\right) = \Omega\left(\frac{q\beta_1'(q)}{2^{cn}}\right).$$

Consequently $q\beta_1'(q) = \Omega(2^{cn})$ implies that the probability of success of the adversary is $\Omega(1)$. This concludes the proof.   □

Conclusion for Collision Resistance. The security analysis of (c,t,k,m)-compression functions for collision resistance is not as tight as for preimage resistance. We proved in this section that collision resistance is $O(\beta_1(q)^2/2^{cn})$, while the attacks we described show that a lower bound for collision resistance is $\Omega(\max(\beta_c(q)^2, q\beta_1(q))/2^{cn})$.

5   Application to Previously Proposed Schemes

Hirose Schemes. We call Hirose schemes the (c,t,k,m)-compression function constructions where $k = m + c$. In this case, using only $t = c$ inner compression functions, setting $M\|H \cdot A_i = M\|H$ for all $i \in [0,t]$ and taking for $\mathcal{B}$ the $c \times c$ identity matrix yields a compression function such that $\beta_1(q) = \beta_c(q) = q$, so that its preimage resistance is $\Theta(\frac{q}{2^{cn}})$ and its collision resistance is $\Theta(\frac{q^2}{2^{cn}})$, which is optimal. This is not a surprising result since it is easy to see that in the random oracle model, the compression function obtained is itself a random function from $\mathcal{F}((m + c)n, cn)$. These type of schemes have been studied by Hirose in [6,7], where it is shown how to construct such an optimally resistant compression function with one ideal block cipher when $c = 2$.

Nandi et al. Schemes. Nandi et al. proposed two schemes in [20] which are depicted in Fig. 2. For these schemes, it was shown in [20] that $\beta_2(q) \le q^{3/2}$ and

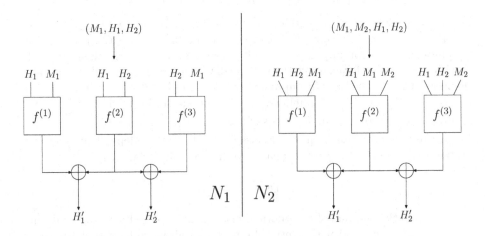

Fig. 2. Nandi et al. schemes [20]

it is not difficult to convince oneself that $\beta_1'(q) \leq q^2$. Conversely, $\beta_2(q) \geq \lfloor q^{1/2} \rfloor^3$ and, for $q \leq 2^n$, $\beta_1(q) \geq q^2$. Consequently their preimage resistance is $\Theta(q^2/2^{2n})$, and an attack requiring $\Theta(2^n)$ operations was described in [11]. For the collision resistance, our security proof shows that it is $O(q^4/2^{2n})$ while the two collision attacks we described achieve advantage $\Omega(q^3/2^{2n})$. The authors of [20] claimed to have proved that collision resistance for their schemes is $O(q^3/2^{2n})$, however we explain in Appendix C why their reasoning is incorrect. Nevertheless, we conjecture that our security proof can be enhanced to prove that collision resistance is indeed $O(q^3/2^{2n})$. But this must not discourage to look in the direction of finding better collision attacks than the one described in [11], which needs $\Theta(2^{2n/3})$ oracle queries.

Peyrin et al. Schemes. Peyrin et al. proposed two schemes in [21] which verified the necessary conditions they established for a scheme to be secure. They are depicted in Fig. 3. For the first scheme, one can prove with techniques similar to the ones used for Nandi et al. schemes that $\beta_1(q) = \Theta(q^{3/2})$ and $\beta_2(q) = \Theta(q^{3/2})$, so that the security analysis is tight in the collision case (collision resistance is $\Theta(q^3/2^{2n})$) as well as in the preimage case (preimage resistance is $\Theta(q^{3/2}/2^{2n})$). For the second scheme, $\beta_1(q) = \Theta(q^{3/2})$ and $\beta_2(q) = \Theta(q^{4/3})$.

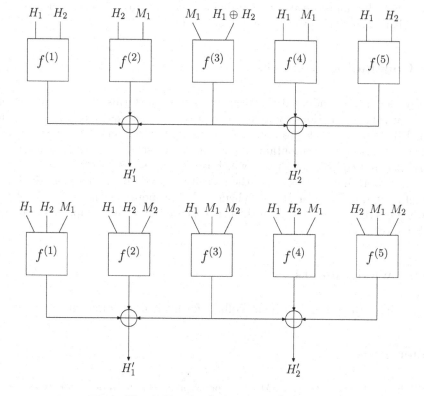

Fig. 3. Peyrin et al. schemes [21]

Here preimage resistance is $\Theta(q^{3/2}/2^{2n})$, and collision resistance is $O(q^3/2^{2n})$, while the first collision attack achieves advantage $\Omega(q^{8/3}/2^{2n})$. Here again it is an open question to close the gap between the security proof and the attack.

Related Algorithmic Problems. We want to emphasize that one must make a clear distinction between security analysis in terms of number of *oracle queries* and number of *operations*. While the number of queries is an obvious lower bound for the number of operations, it is not always clear how an attacker will be able to reach this lower complexity bound. For example for the scheme N1 of Fig. 2, the preimage resistance is $\Theta(q^2/2^{2n})$ so that an adversary must make $\Theta(2^n)$ oracle queries to find a preimage with non negligible probability. The authors of [11] presented an attack also requiring $\Theta(2^n)$ operations. Fundamentally, this is achievable thanks to an efficient algorithm for solving the so-called 2-sum problem which consists in finding, in two lists L_1 and L_2, two elements $x_1 \in L_1$ and $x_2 \in L_2$ such that $x_1 \oplus x_2 = 0$. The generalization to k lists was thoroughly studied by Wagner [26]. In the same way as the (in)security of the schemes of Nandi *et al.* is linked to efficient ways of solving the 2-sum problem [26], we conjecture that the security in terms of operations of the schemes of Peyrin *et al.* is related to the 3-sum problem, for which no good algorithm is known. Giving a reductionist security proof linking the security of these schemes to a 3-sum hard problem would be an elegant result.

6   Concluding Remarks

In this paper we conducted the security analysis in terms of oracle queries of very general constructions combining compression functions modeled as independent FIL random oracles to obtain a compression function with longer output. Using the concept of computable input, we gave a security bound for preimage resistance and collision resistance which is tight for some constructions.

Future work includes carrying the security analysis in the ideal block cipher model as it was done for Hirose schemes [6,7], a more systematic study of the parameters $\beta_i(q)$, and closing in the general case the security gap, especially for collision resistance.

Acknowledgements

The authors are grateful to Henri Gilbert for his helpful comments.

References

1. Bellare, M., Ristenpart, T.: Multi-property-preserving hash domain extension and the EMD transform. In: Lai, X., Chen, K. (eds.) ASIACRYPT 2006. LNCS, vol. 4284, pp. 299–314. Springer, Heidelberg (2006)

2. Black, J.R., Rogaway, P., Shrimpton, T.: Black-box analysis of the block-cipher-based hash-function constructions from PGV. In: Yung, M. (ed.) CRYPTO 2002. LNCS, vol. 2442, pp. 320–335. Springer, Heidelberg (2002)

3. Coppersmith, D., Pilpel, S., Meyer, C.H., Matyas, S.M., Hyden, M.M., Oseas, J., Brachtl, B., Schilling, M.: Data authentication using modification dectection codes based on a public one way encryption function. U.S. Patent No. 4,908,861 (March 13, 1990)

4. Coron, J.-S., Dodis, Y., Malinaud, C., Puniya, P.: Merkle-Damgård revisited: How to construct a hash function. In: Shoup, V. (ed.) CRYPTO 2005. LNCS, vol. 3621, pp. 430–448. Springer, Heidelberg (2005)

5. Damgård, I.: A design principle for hash functions. In: Brassard, G. (ed.) CRYPTO 1989. LNCS, vol. 435, pp. 416–427. Springer, Heidelberg (1990)

6. Hirose, S.: Provably secure double-block-length hash functions in a black-box model. In: Park, C.-s., Chee, S. (eds.) ICISC 2004. LNCS, vol. 3506, pp. 330–342. Springer, Heidelberg (2005)

7. Hirose, S.: In: Robshaw, M. (ed.) FSE 2006. LNCS, vol. 4047, Springer, Heidelberg (2006)

8. Joux, A.: Multicollisions in iterated hash functions. application to cascaded constructions. In: Franklin, M. (ed.) CRYPTO 2004. LNCS, vol. 3152, pp. 306–316. Springer, Heidelberg (2004)

9. Kelsey, J., Schneier, B.: Second preimages on n-bit hash functions for much less than 2^n work. In: Cramer, R.J.F. (ed.) EUROCRYPT 2005. LNCS, vol. 3494, pp. 474–490. Springer, Heidelberg (2005)

10. Knudsen, L.R, Lai, X.: New attacks on all double block length hash functions of hash rate 1, including the parallel-DM. In: De Santis, A. (ed.) EUROCRYPT 1994. LNCS, vol. 950, pp. 410–418. Springer, Heidelberg (1995)

11. Knudsen, L.R., Muller, F.: Some attacks against a double length hash proposal. In: Roy, B. (ed.) ASIACRYPT 2005. LNCS, vol. 3788, pp. 462–473. Springer, Heidelberg (2005)

12. Knudsen, L.R., Preneel, B.: Hash functions based on block ciphers and quaternary codes. In: Kim, K.-c., Matsumoto, T. (eds.) ASIACRYPT 1996. LNCS, vol. 1163, pp. 77–90. Springer, Heidelberg (1996)

13. Knudsen, L.R., Preneel, B.: Fast and secure hashing based on codes. In: Kaliski Jr., B.S. (ed.) CRYPTO 1997. LNCS, vol. 1294, pp. 485–498. Springer, Heidelberg (1997)

14. Knudsen, L.R., Preneel, B.: Construction of secure and fast hash functions using nonbinary error-correcting codes. IEEE Transactions on Information Theory 48(9), 2524–2539 (2002)

15. Lai, X., Massey, J.L.: Hash function based on block ciphers. In: Rueppel, R.A. (ed.) EUROCRYPT 1992. LNCS, vol. 658, pp. 55–70. Springer, Heidelberg (1993)

16. Lai, X., Waldvogel, C., Hohl, W., Meier, T.: Security of iterated hash functions based on block ciphers. In: Stinson, D.R. (ed.) CRYPTO 1993. LNCS, vol. 773, pp. 379–390. Springer, Heidelberg (1994)

17. Menezes, A., van Oorschot, P.C., Vanstone, S.A.: Handbook of Applied Cryptography. CRC Press (1996)

18. Merkle, R.C.: One way hash functions and DES. In: Brassard, G. (ed.) CRYPTO 1989. LNCS, vol. 435, pp. 428–446. Springer, Heidelberg (1990)

19. Motwani, R., Raghavan, P.: Randomized algorithms. Cambridge University Press, Cambridge (1995)

20. Nandi, M., Lee, W., Sakurai, K., Lee, S.: Security analysis of a 2/3-rate double length compression function in black-box model. In: Gilbert, H., Handschuh, H. (eds.) FSE 2005. LNCS, vol. 3557, pp. 243–254. Springer, Heidelberg (2005)

21. Peyrin, T., Gilbert, H., Muller, F., Robshaw, M.J.B.: Combining compression functions and block cipher-based hash functions. In: Lai, X., Chen, K. (eds.) ASIACRYPT 2006. LNCS, vol. 4284, pp. 315–331. Springer, Heidelberg (2006)

22. Preneel, B., Bosselaers, A., Govaerts, R., Vandewalle, J.: Collision-free hash functions based on block cipher algorithms. In: Proceedings, International Carnahan Conference on Security Technology, IEEE 1989, IEEE catalog number 89CH2774-8, pp. 203–210 (1989)

23. Preneel, B., Govaerts, R., Vandewalle, J.: Hash functions based on block ciphers: A synthetic approach. In: Stinson, D.R. (ed.) CRYPTO 1993. LNCS, vol. 773, pp. 368–378. Springer, Heidelberg (1994)

24. Quisquater, J.-J., Girault, M.: 2n-bit hash-functions using n-bit symmetric block cipher algorithms. In: Quisquater, J.-J., Vandewalle, J. (eds.) EUROCRYPT 1989. LNCS, vol. 434, pp. 102–109. Springer, Heidelberg (1990)

25. Shannon, C.: Communication theory of secrecy systems. Bell System Technical Journal 28(4), 656–715 (1949)

26. Wagner, D.: A generalized birthday problem. In: Yung, M. (ed.) CRYPTO 2002. LNCS, vol. 2442, pp. 288–303. Springer, Heidelberg (2002)

27. Wang, X., Yin, Y.L., Yu, H.: Finding collisions in the full SHA-1. In: Shoup, V. (ed.) CRYPTO 2005. LNCS, vol. 3621, pp. 17–36. Springer, Heidelberg (2005)

28. Wang, X., Yu, H.: How to break MD5 and other hash functions. In: Cramer, R.J.F. (ed.) EUROCRYPT 2005. LNCS, vol. 3494, pp. 19–35. Springer, Heidelberg (2005)

29. Winternitz, R.S.: A secure one-way hash function built from DES. In: IEEE Symposium on Security and Privacy, pp. 88–90. IEEE Computer Society Press, Los Alamitos (1984)

A   Proof of Lemmata

Proof of Lemma 1. Consider $x \in \{0,1\}^c \setminus \{0\}$ and two distinct inputs $M_1 \| H_1$ and $M_2 \| H_2$. By definition of S_x, we have that

$$G_x(M \| H) = \bigoplus_{j \in S_x} f^{(j)}(M \| H \cdot \mathcal{A}_j).$$

The fact that $G_x(M_1 \| H_1)$ and $G_x(M_2 \| H_2)$ are uniformly random is obvious because they are linear combination of the uniformly random outputs of the $f^{(j)}$'s. Moreover the internal input blocks of the $f^{(j)}$'s differ in at least one bit for $M_1 \| H_1$ and $M_2 \| H_2$, otherwise, as $M_1 \| H_1 \neq M_2 \| H_2$ it would be possible to construct a non zero element in $\bigcap_{j \in S_x} \ker \mathcal{A}_j$, which is $\{0\}$ by hypothesis. As the output of the $f^{(j)}$'s are random and independent, $G_x(M_1 \| H_1)$ and $G_x(M_2 \| H_2)$ are also independent.

Proof of Lemma 3. As $x \in \{0,1\}^c \setminus \text{Vec}(x_1, \ldots, x_r)$, and as $\mathcal{B}$ has full rank, then necessarily $x \cdot \mathcal{B}^T$ is not a linear combination of the $(x_i \cdot \mathcal{B}^T)_{i \in [1..r]}$. Consequently, there exists $j \in [1..t]$ such that $f^{(j)}$ intervenes in G_x but in none of the $(G_{x_i})_{i \in [1..r]}$. As the outputs of the inner compression functions are independent, $G_x(M \| H)$ is independent from $(G_{x_i}(M \| H))_{i \in [1..r]}$. The generalization follows easily by induction.

B   Proof of Theorem 3

Let $\mathfrak{A}$ be a collision-finding adversary attacking the compression function h. Instead of working on single external inputs as for the proof of Theorem 1, we will work on pairs of distinct external inputs, but the reasoning will be quite similar and readers are recommended to read the preimage proof before this one. Let $\mathcal{P}_2$ be the set of all 2-elements subsets of $\{0,1\}^{m+c}$. External inputs will be noted X instead of $M\|H$ for concision. Let define Coll as being the set of pairs of distinct external inputs $\{X_1, X_2\}$ which are h-computable with respect to the final set of queries of $\mathfrak{A}$ and which collide under h. As for preimage, it is easy to see that $\Pr[\mathfrak{A} \text{ wins}] \leq 1/2^n + \Pr[\text{Coll} \neq \emptyset]$.

We now bound $\Pr[\text{Coll} \neq \emptyset]$. Given partial sets of queries $\mathcal{Q}'_1, \ldots, \mathcal{Q}'_t \subset (\mathcal{I}_n)^k$, we will say that the pair $\{X_1, X_2\} \in \mathcal{P}_2$ is *compatible* if X_1 and X_2 collide on $V_{\{X_1,X_2\}} = V_{X_1} \cap V_{X_2}$, meaning that for all $x \in V_{\{X_1,X_2\}}$, $G_x(X_1) = G_x(X_2)$. Here also it is possible to show that if $V_{\{X_1,X_2\}} \neq \{0\}$ with respect to the final sets of queries of $\mathfrak{A}$, then there is an unique $x \in \{0,1\}^c \setminus \{0\}$ such that $V_{\{X_1,X_2\}} = \text{Vec}(x)$ when it becomes strictly bigger than $\{0\}$. We will note $G^1_{\{X_1,X_2\}}$ the linear combination of output blocks associated with this x. We define Pot_1 as being the set of all $\{X_1, X_2\} \in \mathcal{P}_2$ such that X_1 and X_2 are at least 1-compatible with respect to the final sets of queries of $\mathfrak{A}$ and $G^1_{\{X_1,X_2\}}(X_1) = G^1_{\{X_1,X_2\}}(X_2)$. It is now straightforward to follow the same reasoning as for the preimage proof, which we do without further justification:

$$
\begin{aligned}
\Pr[\text{Coll} \neq \emptyset] &\leq \sum_{\{X_1,X_2\}} \Pr[\{X_1,X_2\} \in \text{Coll}] \\
&\leq \sum_{\{X_1,X_2\}} \Pr[\{X_1,X_2\} \in \text{Coll} | \{X_1,X_2\} \in \text{Pot}_1] \cdot \\
&\qquad\qquad\qquad\qquad \Pr[\{X_1,X_2\} \in \text{Pot}_1] \\
&\leq \frac{1}{2^{(c-1)n}} \sum_{\{X_1,X_2\}} \Pr[\{X_1,X_2\} \in \text{Pot}_i] \\
&\leq \frac{1}{2^{(c-1)n}} \sum_{\{X_1,X_2\}} \Pr[G^1_{\{X_1,X_2\}}(X_1) = G^1_{\{X_1,X_2\}}(X_2) | X_1 \text{ and } X_2 \text{ are} \\
&\qquad\qquad 1-\text{computable}] \cdot \Pr[X_1 \text{ and } X_2 \text{ are } 1-\text{computable}] \\
&\leq \frac{1}{2^{(c-1)n}} \frac{1}{2^n} \sum_{\{X_1,X_2\}} \Pr[X_1 \text{ and } X_2 \text{ are } 1-\text{computable}] \\
&\leq \frac{\beta_1(q)^2}{2 \cdot 2^{cn}}.
\end{aligned}
$$

Hence the result.

C   Why the Reasoning of [20] Was Faulty

We'd like to emphasize why the following reasoning, which was the one used in [20] for their security proof for preimage attacks, is tempting but fallacious.

One can surely write (see the proof of Theorem 1 for the notations)

$$\Pr[\text{Preim} \neq \emptyset] \leq \sum_{\boldsymbol{M}\|\boldsymbol{H}} \Pr[\boldsymbol{M}\|\boldsymbol{H} \in \text{Preim}]$$

$$\leq \sum_{\boldsymbol{M}\|\boldsymbol{H}} \Pr[\boldsymbol{M}\|\boldsymbol{H} \text{ is } h\text{-computable} \wedge h^{(f^{(1)},\ldots,f^{(t)})}(\boldsymbol{M}\|\boldsymbol{H}) = \boldsymbol{0}].$$

At this stage, it is tempting to claim that the events "$\boldsymbol{M}\|\boldsymbol{H}$ is h-computable" and "$h^{(f^{(1)},\ldots,f^{(t)})}(\boldsymbol{M}\|\boldsymbol{H}) = \boldsymbol{0}$" are independent, thus concluding that

$$\Pr[\text{Preim} \neq \emptyset] \leq \frac{1}{2^{cn}} \sum_{\boldsymbol{M}\|\boldsymbol{H}} \Pr[\boldsymbol{M}\|\boldsymbol{H} \text{ is } h\text{-computable}]$$

$$\leq \frac{1}{2^{cn}} \mathbb{E}(\#\{\boldsymbol{M}\|\boldsymbol{H} \mid \boldsymbol{M}\|\boldsymbol{H} \text{ is } h\text{-computable}\})$$

$$\leq \frac{\beta_c(q)}{2^{cn}}.$$

However, this is false because these two events are not independent. Indeed, one can intuitively argue that the fact that $h^{(f^{(1)},\ldots,f^{(t)})}(\boldsymbol{M}\|\boldsymbol{H}) = \boldsymbol{0}$, being detected on one of the output blocks by the adversary, will increase the probability that $\mathfrak{A}$ makes the queries needed to compute $\boldsymbol{M}\|\boldsymbol{H}$ on other output blocks, thus increasing the probability for $\boldsymbol{M}\|\boldsymbol{H}$ to be h-computable.

The same type of problem arises for collision resistance, where an analogue but still hasty reasoning would conclude that $\Pr[\text{Coll} \neq \emptyset] \leq \frac{\beta_c(q)^2}{2\cdot 2^{cn}}$.

Bad and Good Ways of Post-processing Biased Physical Random Numbers

Markus Dichtl

Siemens AG
Corporate Technology
81730 München
Germany
Markus.Dichtl@siemens.com

Abstract. Algorithmic post-processing is used to overcome statistical deficiencies of physical random number generators. We show that the quasigroup based approach for post-processing random numbers described in [MGK05] is ineffective and very easy to attack. We also suggest new algorithms which extract considerably more entropy from their input than the known algorithms with an upper bound for the number of input bits needed before the next output is produced.

Keywords: quasigroup, physical random numbers, post-processing, bias.

1 Introduction

It seems that all physical random number generators show some deviation from the mathematical ideal of statistically independent and uniformly distributed bits.

Algorithmic post-processing is used to eliminate or reduce the imperfections of the output. Clearly, the per bit entropy of the output can only be increased if the post-processing algorithm is compressing, that is, more than one bit of input is used to get one bit of output.

Bias, that is a deviation of the 1-probability from $1/2$, is a very frequent problem. There are various ways to deal with it, when one assumes that bias is the only problem, that is, the bits are statistically independent.

In the rest of the paper, it is assumed that the physical random number generator produces statistically independent bits. It is also assumed that the generator is stationary, that is, the bias is constant.

We show that the post-processing algorithm based on quasigroups suggested in [MGK05] is ineffective. We describe an attack on the post-processed output bits which requires very little computational effort.

We show that post-processing algorithms for biased random numbers which produce completely unbiased output cannot have upper bounds on the number of input bits required until the next output bit is produced. We describe new algorithms for post-processing biased random numbers which have a fixed number of input bits. The new algorithms extract considerably more entropy from the input than the known algorithms with a fixed number of input bits.

A. Biryukov (Ed.): FSE 2007, LNCS 4593, pp. 137–152, 2007.

2   An Ineffective Post-processing Method for Biased Random Numbers

At FSE 2005, S. Markovski. D. Gligoroski, and L. Kocarev suggested in [MGK05] a method based on quasigroups for post-processing biased random numbers. In this paper, we only give results for their E-algorithm, but all results can be transferred trivially to the E'-algorithm.

2.1   The E-Transform

A quasigroup $(A, *)$ of finite order s is a set A of cardinality s with an operation $*$ on A such that the operation table of $*$ is a Latin square (that is, all elements of A appear exactly once in each row and column of the table).

The mapping $e_{b_0,*}$ maps a finite string of elements $a_1, a_2, \ldots, a_n$ of A to a finite string $b_1, b_2, \ldots b_n$ such that $b_{i+1} = a_{i+1} * b_i$ for $i = 0, 1, \ldots, n - 1$. b_0 is called the leader of the mapping $e_{b_0,*}$. The leader must be chosen in such a way that $b_0 * b_0 \neq b_0$ holds.

For a fixed positive integer k, the E-transform of a finite string of elements from A is just the k-fold application of the function $e_{b_0,*}$. Here we deviate slightly from the terminology of [MGK05]. There, the E-transform allows different leaders for the k applications of the function e, but later in the description of the post-processing algorithm the leader is a fixed element.

2.2   True and Claimed Properties of the E-Transform

Why the suggested quasi group approach is ineffective for post-processing biased random numbers. Theorem 1 of [MGK05] states correctly that the E-transform is bijective. As a consequence of this theorem the E-transform is ineffective for post-processing biased random numbers. As a bijection, it cannot extract entropy from its input bits. The entropy of its output bits is just the same as the entropy of the input bits. Of course, applying a bijection to some random bits does not do any harm either.

The output of the E-transform is not uniformly distributed. Theorem 2 of [MGK05] claims incorrectly that the substrings of length $l \leq k$ are uniformly distributed in the E-transform of a sufficiently long arbitrary input string.

Let x be an element of the quasigroup A. The E-transform is a bijection, so for each string length n, there is an input string w which is mapped by the E-transform to the string with n times x in sequence. Clearly, all substrings of the E-transform of w are just repetitions of x. Hence, the distribution of substrings of length $l \leq k$ in the E-transform of w is as far from uniform as possible.

When we look at the proof of the theorem, we see that the authors do not deal with an arbitrary input string, but with one where all elements of the string are chosen randomly and statistically independently according to a fixed distribution. In their proof, the authors do not try to show that the distribution of the output substrings is exactly uniform, but refer to the stationary state of a Markov chain. So their result could hold at most asymptotically. However, the Lemma 1 they use in their proof, is completely wrong. Let x be the input string for the first application of $e_{b_0,*}$. Then the lemma states that for all m the probability of a fixed element of A at the mth position of $e_{b_0,*}(x)$ is approximately $1/s$, s being the order of the quasigroup. To see how wrong this is, we consider a highly biased generator suggested in [MGK05]. We assume that the probability of 0-bits is 999/1000 and the probability of 1-bits is 1/1000. The bits are assumed to be statistically independent. We name this generator HB (highly biased). (The value of the bias is -0.499 .) For post-processing, we use a quasigroup ($\{0,1,2,3\},*$) of order 4. We map each pair of input bits bijectively to a quasigroup element by using the binary value of the pair. Hence the post-processing input 0 has probability 0.998001, 1 and 2 have probability 0.000999, and the probability of 3 is 0.000001. The first element of $e_{b_0,*}(x)$ depends only on the first element of x. So, one element of A appears as first element of the output string with probability 0.998001, two with probability 0.000999, and one with probability 0.000001. All these values are very far away from the value 1/4 suggested by the lemma.

One should note that the generator HB, which we consider several times in this paper, was not constructed to demonstrate the weakness of the quasigroup post-processing, but that the authors of [MGK05] claim explicitly that their method is suited for post-processing the output of HB.

2.3   What Is Really Going on in the E-Transform

The elements at the end of the output string of the E-transform approach the uniform distribution very slowly, when the input string from a strongly biased source is growing longer. This is shown in Figure 1. We used the quasigroup of order 4 from the Example 1 in [MGK05] and the leader 1 with $k = 128$ (as suggested by the authors of [MGK05] for highly biased input) to post-process 10000 samples of 100000 bits (50000 input elements) from the generator HB.

To achieve an approximately uniform distribution, about 50000 input elements, or 100000 bits have to be processed.

Now, it would be wrong to conclude that everything is fine after 100000 bits. Since the E-transform is bijective for all input lengths, the entropy of the later output elements is as low as the entropy of the first ones. How can the entropy of the later output bits be so low if they are approximately uniformly distributed? The low entropy is the result of strong statistical dependencies between the output elements. For the later output elements, the E-transform just replaces one statistical problem, bias, with another one, dependency. Of course one cannot expect anything better from a bijective post-processing function.

2.4   Attacking the Post-processed Output of the E-Transform

Since the E-transform is bijective, it is, from an information theoretical point of view, clear that its output resulting from an input with biased probabilities can be predicted with a probability greater than $1/s$, where s is the cardinality of A.

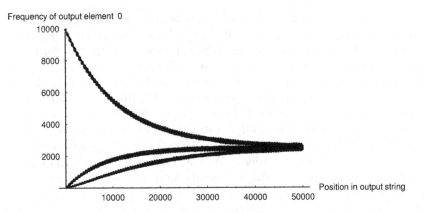

Fig. 1. Frequencies of the element 0 in the post-processed output of the generator HB. For each position in the output string, one dot indicates the frequency of the output element 0 at that position in 10000 experiments. To be visible, the dots have to be so large the individual dots are not discernible and give the impression of three "curves". For subsequent output elements, the output frequency sometimes jumps from one of the three "curves" visible in the figure to another one.

We will show now that the computation of the output element with the highest probability is very easy.

The first output element of the E-transform is the easiest to guess. We just choose the string containing the most frequent input symbol and apply the E-transform to get the most probable first output element. For predicting later output elements, we have to know the previous input elements. They can be easily computed from the output elements. Due to the Latin square property of the operation table of $*$, the equation $b_{i+1} = a_{i+1} * b_i$, which defines $e_{b_0,*}$ can be uniquely resolved for a_{i+1} if b_{i+1} and b_i are given. To the previous input elements we append the most probable input symbol and apply the E-transform. The last element of the output string is the most probable next output element. Our prediction is correct if the most probable input symbol occurred. For the generator HB and a quasigroup of order 4, this means that we have a probability of 0.998001 to predict the next output element correctly. The computational effort for the attack is about the same as for the application of the E-transform.

2.5   Attacking Unknown Quasigroups and Leaders

The authors of [MGK05] do not suggest to keep secret the quasigroup and the leader chosen for post-processing the random numbers. We show here for one set of parameters suggested in [MGK05] that even keeping the parameters secret would not prevent the prediction of output bits. For the attack, we assume that the attacker gets to see all output bits from post-processing, and that she may restart the generator such that the post-processing is reinitialized. Since the post-processing is bijective, an attacker can learn nothing from observing the post-processed output of a perfect random number generator. Instead, we consider again the highly biased generator HB.

We choose the parameters for the post-processing function according to the suggestions of [MGK05]. The order of the quasigroup was chosen to be 4, and the number k of applications of $e_{b_0,*}$ as 128.

Normalised Latin squares of order 4 have 0, 1, 2, 3 as their first row and column. By trying out all possibilities, one sees easily that there are exactly 4 normalised Latin squares of order 4. By permuting all four columns and the last three rows of the 4 normalised Latin squares we find all possible Latin squares of order 4. Hence, there are $4 \cdot 4! \cdot 3! = 576$ quasigroups of order 4. With 4 leaders for each, this would mean 2304 choices of parameters to consider for the E-transform. However, we have to take into account that the leader l may not have the property $l * l = l$. After sorting out those undesired cases, we are left with 1728 cases.

From the observed post-processed bits, we want to uniquely determine the parameters used.

The attack is very simple. We apply the inverse E-transform for all 1728 possible choices of quasigroup and leaders to the first n of the observed postprocessed bits. With high probability, the correct choice of parameters is revealed by an overwhelming majority of 0 bits in the inverse. When choosing $n = 32$, the correct parameters are uniquely identified by an inverse of 32 0 bits in 61 % of all cases. Using more post-processed bits, we can get arbitrarily close to certainty about having found the right parameters. This variant of the attack is due to an anonymous reviewer of FSE 2007, mine was unnecessarily complicated.

When we have determined the parameters of the E-transform, the attack can continue as described in the previous section.

This finishes our treatment of quasigroup post-processing for true random numbers.

3   Two Classes of Random Number Post-processing Functions

We have seen that bijective methods of post-processing random numbers are not useful, as they can not increase entropy. We have to accept that efficient post-processing functions have less output bits than input bits. This paper only deals with the post-processing of random bits which are assumed to be statistically independent, but which are possibly biased. The source of randomness is assumed to be stationary, that is the bias does not change with time. Although the bias is constant, it is unrealistic to assume that its numerical value, that is the deviation of the probability of 1 bits from $1/2$, is exactly known. A useful post-processing function should work for all biases from a not too small range.

Probably the oldest method of post-processing biased random numbers was invented by John von Neumann [vN63]. He partitions the bits from the true random number generator in adjacent, non-overlapping pairs. The result of a 01 pair is a 0 output bit. A 10 pair results in a 1 output bit. Pairs 00 and 11 are just discarded. Since the bits are assumed to be independent and the bias is assumed to be constant, the pairs 01 and 10 have exactly the same probability. The output is therefore completely unbiased.

However, this method has two drawbacks: firstly, even for a perfect true random number generator it results in an average output data rate 4 times slower than the input data rate, and secondly, the waiting times until output bits are available can become arbitrarily large. Although pairs 01 or 10 statistically turn up quite often, it is not certain that they will do so within any fixed number of pairs considered. This is a unsatisfactory situation for the software developer who has to use the random numbers. In many protocols, time-outs occur when reactions to messages take too long. Although the probability for this event can be made arbitrarily low, it can not be reduced to zero when using the von Neumann post-processing method.

The low output date rate may be overcome by using methods like the one described by Peres [Per92]. A very general method for post-processing any stationary, statistcally independent data is given in [JJSH00]. The algorithm described there can asymptotically extract all the entropy from its input. So the data rate is asymptotically optimal. The problem of arbitrarily long waiting times, however, can not be solved. This is proved in the following theorem:

Theorem 1. *Let f be an arbitrary post-processing function which maps n bits to one bit, n being a positive integer. Let $X_1, \ldots, X_n$ be n independent random bits with the same probability p of 1-bits. Let S be an infinite subset of the unit interval. Then $Pr[f(X_1, \ldots, X_n) = 1] \neq 1/2$ holds for infinitely many $p \in S$.*

Proof: $\Pr[f(X_1, \ldots, X_n) = 1]$ is a polynomial of at most n-th degree in p. Its value for $p = 0$ is either 0 or 1. If the function is constant, it is unequal $1/2$ on the whole unit interval. If it is a non-constant polynomial, there are at most n p-values, for which it assumes the value $1/2$. For all other $p \in S$ holds $\Pr[f(X_1, \ldots, X_n) = 1] \neq 1/2$. q. e. d.

So we have two classes of functions for post-processing true random numbers: those with bounded numbers of input bits to produce one bit of output, and those with unbounded. We can reformulate our theorem as

An algorithm for post-processing biased, but statistically independent random bits with a bounded number of input bits for one output bit cannot produce unbiased output bits for an infinite set of biases.

For practical implementations, an unbounded number of input bits means an unbounded waiting time until the next output is produced.

So we have to accept some output bias for bounded waiting time post-processing functions, but we will show subsequently how to make this bias very small.

Although not as well known as the von Neumann post-processing algorithm, some algorithms with a fixed number of input bits are frequently used to improve the statistical quality of the output of true random number generators.

Probably the simplest method is to XOR n bits from the generator in order to get one bit of output where n is a fixed integer greater than 1.

Another popular method is to use a linear feedback shift register where the bits from the true random number generator are XORed to the feedback value computed according to the feedback polynomial. The result of this XOR operation is then fed back to the shift register. After clocking m bits from the physical random number generator into the shift register, one bit of output is taken from a cell of the shift register. This method tries to make use of the good statistical properties of linear feedback shift registers. Even when the physical source of randomness breaks down completely and produces only constant bits, the LFSR computes output which at first sight looks random. When the bits from the physical source are non-constant but biased, some of the bias is removed, but additional dependencies between output bits are introduced by this LFSR method.

4 Improved Random Number Post-processing Functions with a Fixed Number of Input Bits

4.1 The Concrete Problem Considered

We consider the following problem: we have a stationary source of random bits which produces statistically independent, but biased output bits. Let p be the probability for a 1 bit. The post-processing algorithm we are looking for must not depend on the value of p. The post-processing algorithm uses 16 bits from the physical source of randomness in order to produce 8 bits of output. Our aim is to choose an algorithm such that the entropy of the output byte is high.

The number of input bits of the function we are looking for is 16. This is a trivial upper bound on the number of input bits needed before the next output is produced. Clearly, any decent implementation will also produce the output after a bounded waiting time.

4.2 A Solution for Low Area Hardware Implementation

Let $a_0, a_1, \ldots, a_{15}$ be the input bits for post-processing. We define the 8 bits $b_0, b_1, \ldots, b_7$ by $b_i = a_i$ XOR $a_{(i+1) \bmod 8}$. We note that the mapping from $a_0, \ldots, a_7$ to $b_0, \ldots, b_7$ defined in this way is not bijective. Only 128 different values for $b_0, b_1, \ldots, b_7$ are possible. When the input is a perfect random number generator, we destroy one bit of entropy by mapping $a_0, \ldots, a_7$ to $b_0, \ldots, b_7$. The output $c_0, c_1, \ldots, c_7$ of the suggested post-processing function is defined by $c_i = b_i$ XOR a_{i+8}. We call this function, which maps 16 input bits to 8 output bits, H.

If we split the 16 input bits of H into two bytes we name $a1$ and $a2$, we can write H in pseudocode as

```
H(a1,a2)= XOR(XOR(a1,rotateleft(a1,1)),a2)
```

Figure 2 shows the design of this post-processing function.

The entropy of the output of H is shown in Figure 3 as a function of p.

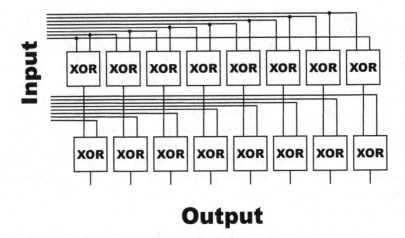

Fig. 2. The post-processing function H

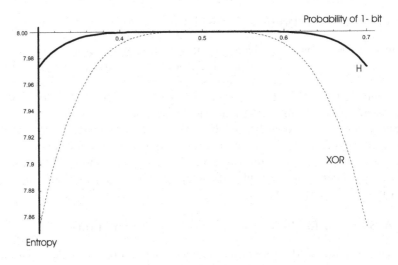

Fig. 3. The entropy of H

The dashed line in the figure shows, for comparison, the entropy one gets when compressing two input bytes to one output byte by bitwise XOR. We see that H extracts significantly more entropy from its input than the XOR.

Figure 3 might suggest that for low biases the entropies of H and XOR are quite comparable. Indeed both are close to 8, but H is considerably closer. Let's look at a generator which produces 1-bits with a probability of 0.51. This is quite good for a physical random number generator. When we use XOR to compress two bytes from this generator to one output byte, the output byte has an entropy of 7.9999990766751 . Using H, we get $e_H = 7.9999999996305$. So in this case the deviation from the ideal value 8 is reduced by a factor more than 2499 if we use H instead of XOR. How good must a generator be to produce output bytes with entropy e_H when using XOR for postprocessing? From 2 input-bytes with a 1-probability of 0.501417 we get an output entropy of e_h if we XOR them. With XOR the 1-probability must be seven times closer to 0.5 than with H to get the output entropy e_H.

The improved entropy of H is achieved with very low hardware costs, just 16 XOR gates with two inputs are needed.

4.3   Analysis of H

What is going on in H? Why is its entropy higher than the entropy of XOR although one first discards one bit of entropy from one input byte? Let us look at the probabilities in more detail.

We define the quantitative value of the bias as

$$\epsilon = p - 1/2.$$

First we compute the probability that the source produces a raw (unprocessed) byte B. Since the bits of B are independent, the probability depends only on the Hamming weight w of B. It is

$$(1/2 - \epsilon)^{8-w} \cdot (1/2 + \epsilon)^{w}.$$

For the different values of w we obtain

w	byte probability for a raw data byte
0	$\frac{1}{256} - \frac{\epsilon}{16} + \frac{7\epsilon^2}{16} - \frac{7\epsilon^3}{4} + \frac{35\epsilon^4}{8} - 7\epsilon^5 + 7\epsilon^6 - 4\epsilon^7 + \epsilon^8$
1	$\frac{1}{256} - \frac{3\epsilon}{64} + \frac{7\epsilon^2}{32} - \frac{7\epsilon^3}{16} + \frac{7\epsilon^5}{4} - \frac{7\epsilon^6}{2} + 3\epsilon^7 - \epsilon^8$
2	$\frac{1}{256} - \frac{\epsilon}{32} + \frac{\epsilon^2}{16} + \frac{\epsilon^3}{8} - \frac{5\epsilon^4}{8} + \frac{\epsilon^5}{2} + \epsilon^6 - 2\epsilon^7 + \epsilon^8$
3	$\frac{1}{256} - \frac{\epsilon}{64} - \frac{\epsilon^2}{32} + \frac{3\epsilon^3}{16} - \frac{3\epsilon^5}{4} + \frac{\epsilon^6}{2} + \epsilon^7 - \epsilon^8$
4	$\frac{1}{256} - \frac{\epsilon^2}{16} + \frac{3\epsilon^4}{8} - \epsilon^6 + \epsilon^8$
5	$\frac{1}{256} + \frac{\epsilon}{64} - \frac{\epsilon^2}{32} - \frac{3\epsilon^3}{16} + \frac{3\epsilon^5}{4} + \frac{\epsilon^6}{2} - \epsilon^7 - \epsilon^8$
6	$\frac{1}{256} + \frac{\epsilon}{32} + \frac{\epsilon^2}{16} - \frac{\epsilon^3}{8} - \frac{5\epsilon^4}{8} - \frac{\epsilon^5}{2} + \epsilon^6 + 2\epsilon^7 + \epsilon^8$
7	$\frac{1}{256} + \frac{3\epsilon}{64} + \frac{7\epsilon^2}{32} + \frac{7\epsilon^3}{16} - \frac{7\epsilon^5}{4} - \frac{7\epsilon^6}{2} - 3\epsilon^7 - \epsilon^8$
8	$\frac{1}{256} + \frac{\epsilon}{16} + \frac{7\epsilon^2}{16} + \frac{7\epsilon^3}{4} + \frac{35\epsilon^4}{8} + 7\epsilon^5 + 7\epsilon^6 + 4\epsilon^7 + \epsilon^8$

For a good generator, ϵ should be small. As a consequence, the terms with the lower powers of ϵ are the more disturbing ones, the linear ones being the worst.

Before we start the analysis of H, we consider the XOR post-processing. The XOR of two raw input bits results in a 1-bit with a probability of $2p(1 - p) = 1/2 - 2\epsilon^2$. That is, the bias of the resulting bit is $-2\epsilon^2$. When we insert this bias term for the ϵ in the table above we obtain the byte probabilities for the bitwise XOR of two raw data bytes:

w	byte probability for the XOR of two raw data bytes
0	$\frac{1}{256} - \frac{\epsilon^2}{8} + \frac{7\epsilon^4}{4} - 14\,\epsilon^6 + 70\,\epsilon^8 - 224\,\epsilon^{10} + 448\,\epsilon^{12} - 512\,\epsilon^{14} + 256\,\epsilon^{16}$
1	$\frac{1}{256} - \frac{3\epsilon^2}{32} + \frac{7\epsilon^4}{8} - \frac{7\epsilon^6}{2} + 56\,\epsilon^{10} - 224\,\epsilon^{12} + 384\,\epsilon^{14} - 256\,\epsilon^{16}$
2	$\frac{1}{256} - \frac{\epsilon^2}{16} + \frac{\epsilon^4}{4} + \epsilon^6 - 10\,\epsilon^8 + 16\,\epsilon^{10} + 64\,\epsilon^{12} - 256\,\epsilon^{14} + 256\,\epsilon^{16}$
3	$\frac{1}{256} - \frac{\epsilon^2}{32} - \frac{\epsilon^4}{8} + \frac{3\epsilon^6}{2} - 24\,\epsilon^{10} + 32\,\epsilon^{12} + 128\,\epsilon^{14} - 256\,\epsilon^{16}$
4	$\frac{1}{256} - \frac{\epsilon^4}{4} + 6\,\epsilon^8 - 64\,\epsilon^{12} + 256\,\epsilon^{16}$
5	$\frac{1}{256} + \frac{\epsilon^2}{32} - \frac{\epsilon^4}{8} - \frac{3\epsilon^6}{2} + 24\,\epsilon^{10} + 32\,\epsilon^{12} - 128\,\epsilon^{14} - 256\,\epsilon^{16}$
6	$\frac{1}{256} + \frac{\epsilon^2}{16} + \frac{\epsilon^4}{4} - \epsilon^6 - 10\,\epsilon^8 - 16\,\epsilon^{10} + 64\,\epsilon^{12} + 256\,\epsilon^{14} + 256\,\epsilon^{16}$
7	$\frac{1}{256} + \frac{3\epsilon^2}{32} + \frac{7\epsilon^4}{8} + \frac{7\epsilon^6}{2} - 56\,\epsilon^{10} - 224\,\epsilon^{12} - 384\,\epsilon^{14} - 256\,\epsilon^{16}$
8	$\frac{1}{256} + \frac{\epsilon^2}{8} + \frac{7\epsilon^4}{4} + 14\,\epsilon^6 + 70\,\epsilon^8 + 224\,\epsilon^{10} + 448\,\epsilon^{12} + 512\,\epsilon^{14} + 256\,\epsilon^{16}$

Indeed, all the linear terms in ϵ are gone.

We also analysed some linear feedback shift registers used for true random number post-processing. In the cases we considered, the ratio of the numbers of input and output bits was 2. We observed that the output probabilities contained no linear powers of ϵ, but squares occurred. So XOR and LFSRs with an input/output ratio of 2 seem to be random number post-processing functions of similar quality.

Now we compute the probabilities of the output bytes of H. One approach is to consider every possible pair of input bytes to determine its probability as the product of the byte probabilities from the table for raw bytes above, and to sum up these probabilities separately for all input byte pairs which lead to the same value of H. We obtain 30 different probabilities. Formulas for theses probabilities are shown in Table 1. In all these probabilities, there are no linear or quadratic terms in ϵ! This explains why H is better than the simple XOR of 2 input bytes.

But now, there is a challenge: Can we eliminate further powers of ϵ?

4.4   Improving

Can we eliminate further powers of ϵ? Surprisingly, slight modifications of H also eleminate the third and forth power of ϵ. Again, we split up the 16 input bits into two bytes $a1$ and $a2$ and define the functions $H2$ and $H3$ by the following pseudocode:

```
H2(a1,a2)= XOR(XOR(XOR(a1,rotateleft(a1,1)),rotateleft(a1,2)),a2)

H4(a1,a2)= XOR(XOR(XOR(XOR(a1,rotateleft(a1,1)),
                    rotateleft(a1,2)),rotateleft(a1,4)),a2)
```

For $H2$ we find that the lowest powers of ϵ in the probabilities of the output bytes are ϵ^4. For $H3$ they are ϵ^5.

Table 1. Probabilities of the outputs of H for bias ϵ

output byte	probability for H
	$\frac{1}{256} + \frac{\epsilon^3}{16} - \frac{\epsilon^4}{4} - \frac{3\epsilon^5}{4} + \frac{\epsilon^6}{2} + 3\epsilon^7 + 3\epsilon^8 - 4\epsilon^9 - 8\epsilon^{10}$
	$\frac{1}{256} - \frac{\epsilon^3}{16} - \frac{\epsilon^4}{4} + \frac{3\epsilon^5}{4} + \frac{\epsilon^6}{2} - 3\epsilon^7 + 3\epsilon^8 + 4\epsilon^9 - 8\epsilon^{10}$
	$\frac{1}{256} - \frac{\epsilon^3}{16} - \frac{\epsilon^4}{4} - \frac{\epsilon^6}{2} + 5\epsilon^7 - \epsilon^8 - 12\epsilon^9 + 8\epsilon^{10}$
	$\frac{1}{256} + \frac{\epsilon^3}{16} + \frac{\epsilon^5}{4} - \frac{\epsilon^6}{2} - 5\epsilon^7 - \epsilon^8 + 12\epsilon^9 + 8\epsilon^{10}$
	$\frac{1}{256} - \frac{\epsilon^3}{16} + \frac{\epsilon^4}{4} - \frac{\epsilon^5}{4} - \frac{3\epsilon^6}{2} + \epsilon^7 - 5\epsilon^8 + 20\epsilon^9 + 24\epsilon^{10} - 64\epsilon^{11}$
	$\frac{1}{256} + \frac{3\epsilon^3}{16} + \frac{\epsilon^4}{4} + \frac{\epsilon^5}{4} + \frac{5\epsilon^6}{2} + 3\epsilon^7 - 5\epsilon^8 - 20\epsilon^9 - 40\epsilon^{10} - 32\epsilon^{11}$
	$\frac{1}{256} + \frac{\epsilon^3}{16} - \frac{\epsilon^4}{4} - \frac{\epsilon^5}{2} + \frac{\epsilon^6}{2} - 3\epsilon^7 + 3\epsilon^8 + 20\epsilon^9 - 8\epsilon^{10} - 32\epsilon^{11}$
	$\frac{1}{256} - \frac{\epsilon^3}{16} + \frac{\epsilon^5}{4} - \frac{\epsilon^6}{2} - \epsilon^7 - \epsilon^8 + 12\epsilon^9 + 8\epsilon^{10} - 32\epsilon^{11}$
	$\frac{1}{256} - \frac{3\epsilon^3}{16} + \frac{\epsilon^4}{4} - \frac{\epsilon^5}{4} + \frac{5\epsilon^6}{2} - 3\epsilon^7 - 5\epsilon^8 + 20\epsilon^9 - 40\epsilon^{10} + 32\epsilon^{11}$
	$\frac{1}{256} - \frac{\epsilon^3}{16} - \frac{\epsilon^4}{4} + \frac{\epsilon^5}{4} + \frac{\epsilon^6}{2} + 3\epsilon^7 + 3\epsilon^8 - 20\epsilon^9 - 8\epsilon^{10} + 32\epsilon^{11}$
	$\frac{1}{256} + \frac{\epsilon^3}{16} - \frac{\epsilon^5}{4} - \frac{\epsilon^6}{2} + \epsilon^7 - \epsilon^8 - 12\epsilon^9 + 8\epsilon^{10} + 32\epsilon^{11}$
	$\frac{1}{256} + \frac{\epsilon^3}{16} + \frac{\epsilon^4}{4} + \frac{\epsilon^5}{4} - \frac{3\epsilon^6}{2} - \epsilon^7 - 5\epsilon^8 - 20\epsilon^9 + 24\epsilon^{10} + 64\epsilon^{11}$
	$\frac{1}{256} - 2\epsilon^6 + 9\epsilon^8 - 32\epsilon^{12}$
	$\frac{1}{256} - \frac{\epsilon^4}{4} + \epsilon^6 - 3\epsilon^8 + 16\epsilon^{10} - 32\epsilon^{12}$
	$\frac{1}{256} - \frac{\epsilon^3}{8} + \frac{\epsilon^5}{2} + 2\epsilon^7 - 7\epsilon^8 - 8\epsilon^9 + 32\epsilon^{10} - 32\epsilon^{12}$
	$\frac{1}{256} + \frac{\epsilon^3}{8} - \frac{\epsilon^5}{2} - 2\epsilon^7 - 7\epsilon^8 + 8\epsilon^9 + 32\epsilon^{10} - 32\epsilon^{12}$
	$\frac{1}{256} + \frac{\epsilon^3}{8} + \frac{\epsilon^4}{4} + \frac{\epsilon^5}{2} + \epsilon^6 + 2\epsilon^7 + 5\epsilon^8 - 8\epsilon^9 - 48\epsilon^{10} - 64\epsilon^{11} - 32\epsilon^{12}$
	$\frac{1}{256} - \frac{\epsilon^3}{8} + \frac{\epsilon^4}{4} - \frac{\epsilon^5}{2} + \epsilon^6 - 2\epsilon^7 + 5\epsilon^8 + 8\epsilon^9 - 48\epsilon^{10} + 64\epsilon^{11} - 32\epsilon^{12}$
	$\frac{1}{256} + \frac{\epsilon^4}{4} - 2\epsilon^6 + \epsilon^8 + 32\epsilon^{12}$
	$\frac{1}{256} - \frac{\epsilon^4}{4} + 9\epsilon^8 - 32\epsilon^{10} + 32\epsilon^{12}$
	$\frac{1}{256} - \frac{\epsilon^4}{2} + 3\epsilon^6 - 3\epsilon^8 - 16\epsilon^{10} + 32\epsilon^{12}$
	$\frac{1}{256} - \frac{\epsilon^3}{8} + \frac{\epsilon^5}{2} - \epsilon^6 + 2\epsilon^7 + 5\epsilon^8 - 8\epsilon^9 - 16\epsilon^{10} + 32\epsilon^{12}$
	$\frac{1}{256} + \frac{\epsilon^3}{8} - \frac{\epsilon^5}{2} - \epsilon^6 - 2\epsilon^7 + 5\epsilon^8 + 8\epsilon^9 - 16\epsilon^{10} + 32\epsilon^{12}$
	$\frac{1}{256} + \epsilon^6 - 11\epsilon^8 + 16\epsilon^{10} + 32\epsilon^{12},$
	$\frac{1}{256} - \frac{\epsilon^3}{4} + \frac{\epsilon^4}{2} - \epsilon^5 + 7\epsilon^6 - 20\epsilon^7 + 45\epsilon^8 - 112\epsilon^9 + 176\epsilon^{10} - 128\epsilon^{11} + 32\epsilon^{12}$
	$\frac{1}{256} - \frac{\epsilon^5}{2} - \epsilon^6 + 2\epsilon^7 + 5\epsilon^8 + 8\epsilon^9 - 16\epsilon^{10} - 32\epsilon^{11} + 32\epsilon^{12}$
	$\frac{1}{256} - \frac{\epsilon^5}{8} + \epsilon^6 + 4\epsilon^7 - 11\epsilon^8 + 16\epsilon^{10} - 32\epsilon^{11} + 32\epsilon^{12}$
	$\frac{1}{256} + \frac{\epsilon^5}{2} - \epsilon^6 - 2\epsilon^7 + 5\epsilon^8 - 8\epsilon^9 - 16\epsilon^{10} + 32\epsilon^{11} + 32\epsilon^{12}$
	$\frac{1}{256} + \frac{\epsilon^3}{8} + \epsilon^6 - 4\epsilon^7 - 11\epsilon^8 + 16\epsilon^{10} + 32\epsilon^{11} + 32\epsilon^{12}$
	$\frac{1}{256} + \frac{\epsilon^3}{4} + \frac{\epsilon^4}{2} + \epsilon^5 + 7\epsilon^6 + 20\epsilon^7 + 45\epsilon^8 + 112\epsilon^9 + 176\epsilon^{10} + 128\epsilon^{11} + 32\epsilon^{12}$

4.5   An Even Better Solution

It seems that the linear methods which could eliminate all powers of ϵ up to the forth do not work for the fifth. Whereas the linear solutions turned up rather wondrously, we have to labour now. What must we do in order to eliminate the first through fifth powers of ϵ? We have to partition the set of 65536 inputs into 256 sets of cardinality 256 in such a way that the first through fifth powers of ϵ disappear in all 256 sums over the 256 input probabilities. Fortunately, we do not have to deal with too many different input probabilities, since they depend

Table 2. Probabilities of 16-bit-vectors with bit bias ϵ

w	Occurrences	Probability of 16 bit input with Hamming weight w
0	1	$\frac{1}{65536} - \frac{\epsilon}{2048} + \frac{15\,\epsilon^2}{2048} - \frac{35\,\epsilon^3}{512} + \frac{455\,\epsilon^4}{1024} - \frac{273\,\epsilon^5}{128} + \frac{1001\,\epsilon^6}{128} - \frac{715\,\epsilon^7}{32} + \frac{6435\,\epsilon^8}{128} - \frac{715\,\epsilon^9}{8} + \frac{1001\,\epsilon^{10}}{8} - \frac{273\,\epsilon^{11}}{2} + \frac{455\,\epsilon^{12}}{4} - 70\,\epsilon^{13} + 30\,\epsilon^{14} - 8\,\epsilon^{15} + \epsilon^{16}$
1	16	$\frac{1}{65536} - \frac{7\,\epsilon}{8192} + \frac{45\,\epsilon^2}{8192} - \frac{175\,\epsilon^3}{4096} + \frac{455\,\epsilon^4}{2048} - \frac{819\,\epsilon^5}{1024} + \frac{1001\,\epsilon^6}{512} - \frac{715\,\epsilon^7}{256} + \frac{715\,\epsilon^9}{64} - \frac{1001\,\epsilon^{10}}{32} + \frac{819\,\epsilon^{11}}{16} - \frac{455\,\epsilon^{12}}{8} + \frac{175\,\epsilon^{13}}{4} - \frac{45\,\epsilon^{14}}{2} + 7\,\epsilon^{15} - \epsilon^{16}$
2	120	$\frac{1}{65536} - \frac{3\,\epsilon}{8192} + \frac{\epsilon^2}{256} - \frac{49\,\epsilon^3}{2048} + \frac{91\,\epsilon^4}{1024} - \frac{91\,\epsilon^5}{512} + \frac{143\,\epsilon^7}{128} - \frac{429\,\epsilon^8}{128} + \frac{143\,\epsilon^9}{32} - \frac{91\,\epsilon^{11}}{8} + \frac{91\,\epsilon^{12}}{4} - \frac{49\,\epsilon^{13}}{2} + 16\,\epsilon^{14} - 6\,\epsilon^{15} + \epsilon^{16}$
3	560	$\frac{1}{65536} - \frac{5\,\epsilon}{16384} + \frac{21\,\epsilon^2}{4096} - \frac{45\,\epsilon^3}{2048} + \frac{39\,\epsilon^4}{1024} - \frac{39\,\epsilon^5}{512} - \frac{143\,\epsilon^6}{256} + \frac{143\,\epsilon^7}{64} - \frac{143\,\epsilon^9}{32} + \frac{39\,\epsilon^{10}}{16} - \frac{39\,\epsilon^{12}}{8} + \frac{45\,\epsilon^{13}}{4} - \frac{21\,\epsilon^{14}}{2} + 5\,\epsilon^{15} - \epsilon^{16}$
4	1820	$\frac{1}{65536} - \frac{3\,\epsilon}{4096} + \frac{3\,\epsilon^2}{2048} - \frac{3\,\epsilon^3}{1024} - \frac{9\,\epsilon^4}{1024} + \frac{15\,\epsilon^5}{256} - \frac{11\,\epsilon^6}{128} - \frac{11\,\epsilon^7}{64} + \frac{99\,\epsilon^8}{128} - \frac{11\,\epsilon^9}{16} - \frac{11\,\epsilon^{11}}{8} + \frac{15\,\epsilon^{11}}{4} - \frac{9\,\epsilon^{12}}{4} - 3\,\epsilon^{13} + 6\,\epsilon^{14} - 4\,\epsilon^{15} + \epsilon^{16}$
5	4368	$\frac{1}{65536} - \frac{3\,\epsilon}{8192} + \frac{5\,\epsilon^2}{4096} + \frac{5\,\epsilon^3}{4096} - \frac{25\,\epsilon^4}{2048} + \frac{17\,\epsilon^5}{1024} + \frac{33\,\epsilon^6}{512} - \frac{55\,\epsilon^7}{256} + \frac{55\,\epsilon^9}{64} - \frac{33\,\epsilon^{10}}{32} + \frac{17\,\epsilon^{11}}{16} + \frac{25\,\epsilon^{12}}{8} - \frac{5\,\epsilon^{13}}{4} - \frac{5\,\epsilon^{14}}{2} + 3\,\epsilon^{15} - \epsilon^{16}$
6	8008	$\frac{1}{65536} - \frac{\epsilon}{8192} + \frac{5\,\epsilon^3}{2048} - \frac{5\,\epsilon^4}{1024} - \frac{9\,\epsilon^5}{512} + \frac{\epsilon^6}{16} + \frac{5\,\epsilon^7}{128} - \frac{45\,\epsilon^8}{128} + \frac{5\,\epsilon^9}{32} + \epsilon^{10} - \frac{9\,\epsilon^{11}}{8} - \frac{5\,\epsilon^{12}}{4} + \frac{5\,\epsilon^{13}}{2} - 2\,\epsilon^{15} + \epsilon^{16}$
7	11440	$\frac{1}{65536} - \frac{\epsilon}{16384} - \frac{3\,\epsilon^2}{8192} + \frac{7\,\epsilon^3}{4096} + \frac{7\,\epsilon^4}{2048} - \frac{21\,\epsilon^5}{1024} - \frac{7\,\epsilon^6}{512} + \frac{35\,\epsilon^7}{256} - \frac{35\,\epsilon^9}{64} + \frac{7\,\epsilon^{10}}{32} + \frac{21\,\epsilon^{11}}{16} - \frac{7\,\epsilon^{12}}{8} - \frac{7\,\epsilon^{13}}{4} + \frac{3\,\epsilon^{14}}{2} + \epsilon^{15} - \epsilon^{16}$
8	12870	$\frac{1}{65536} - \frac{\epsilon^2}{2048} + \frac{7\,\epsilon^4}{1024} - \frac{7\,\epsilon^6}{128} + \frac{35\,\epsilon^8}{128} - \frac{7\,\epsilon^{10}}{8} + \frac{7\,\epsilon^{12}}{4} - 2\,\epsilon^{14} + \epsilon^{16}$
9	11440	$\frac{1}{65536} + \frac{\epsilon}{16384} - \frac{3\,\epsilon^2}{8192} - \frac{7\,\epsilon^3}{4096} + \frac{7\,\epsilon^4}{2048} + \frac{21\,\epsilon^5}{1024} - \frac{7\,\epsilon^6}{512} - \frac{35\,\epsilon^7}{256} + \frac{35\,\epsilon^9}{64} + \frac{7\,\epsilon^{10}}{32} - \frac{21\,\epsilon^{11}}{16} - \frac{7\,\epsilon^{12}}{8} + \frac{7\,\epsilon^{13}}{4} + \frac{3\,\epsilon^{14}}{2} - \epsilon^{15} - \epsilon^{16}$
10	8008	$\frac{1}{65536} + \frac{\epsilon}{8192} - \frac{5\,\epsilon^3}{2048} - \frac{5\,\epsilon^4}{1024} + \frac{9\,\epsilon^5}{512} + \frac{\epsilon^6}{16} - \frac{5\,\epsilon^7}{128} - \frac{45\,\epsilon^8}{128} - \frac{5\,\epsilon^9}{32} + \epsilon^{10} + \frac{9\,\epsilon^{11}}{8} - \frac{5\,\epsilon^{12}}{4} - \frac{5\,\epsilon^{13}}{2} + 2\,\epsilon^{15} + \epsilon^{16}$
11	4368	$\frac{1}{65536} + \frac{3\,\epsilon}{16384} + \frac{5\,\epsilon^2}{4096} - \frac{5\,\epsilon^3}{4096} - \frac{25\,\epsilon^4}{2048} - \frac{17\,\epsilon^5}{1024} + \frac{33\,\epsilon^6}{512} + \frac{55\,\epsilon^7}{256} - \frac{55\,\epsilon^9}{64} - \frac{33\,\epsilon^{10}}{32} + \frac{17\,\epsilon^{11}}{16} + \frac{25\,\epsilon^{12}}{8} + \frac{5\,\epsilon^{13}}{4} - \frac{5\,\epsilon^{14}}{2} - 3\,\epsilon^{15} - \epsilon^{16}$
12	1820	$\frac{1}{65536} + \frac{\epsilon}{4096} + \frac{3\,\epsilon^2}{2048} + \frac{3\,\epsilon^3}{1024} - \frac{9\,\epsilon^4}{1024} - \frac{15\,\epsilon^5}{256} - \frac{11\,\epsilon^6}{128} + \frac{11\,\epsilon^7}{64} + \frac{99\,\epsilon^8}{128} + \frac{11\,\epsilon^9}{16} - \frac{11\,\epsilon^{10}}{8} - \frac{15\,\epsilon^{11}}{4} - \frac{9\,\epsilon^{12}}{4} + 3\,\epsilon^{13} + 6\,\epsilon^{14} + 4\,\epsilon^{15} + \epsilon^{16}$
13	560	$\frac{1}{65536} + \frac{5\,\epsilon}{16384} + \frac{21\,\epsilon^2}{8192} + \frac{45\,\epsilon^3}{4096} + \frac{39\,\epsilon^4}{2048} - \frac{39\,\epsilon^5}{1024} - \frac{143\,\epsilon^6}{512} - \frac{143\,\epsilon^7}{256} + \frac{143\,\epsilon^9}{64} + \frac{143\,\epsilon^{10}}{32} + \frac{39\,\epsilon^{11}}{16} - \frac{39\,\epsilon^{12}}{8} - \frac{45\,\epsilon^{13}}{4} - \frac{21\,\epsilon^{14}}{2} - 5\,\epsilon^{15} - \epsilon^{16}$
14	120	$\frac{1}{65536} + \frac{3\,\epsilon}{8192} + \frac{\epsilon^2}{256} + \frac{49\,\epsilon^3}{2048} + \frac{91\,\epsilon^4}{1024} + \frac{91\,\epsilon^5}{512} - \frac{143\,\epsilon^7}{128} - \frac{429\,\epsilon^8}{128} - \frac{143\,\epsilon^9}{32} + \frac{91\,\epsilon^{11}}{8} + \frac{91\,\epsilon^{12}}{4} + \frac{49\,\epsilon^{13}}{2} + 16\,\epsilon^{14} + 6\,\epsilon^{15} + \epsilon^{16}$
15	16	$\frac{1}{65536} + \frac{7\,\epsilon}{16384} + \frac{45\,\epsilon^2}{4096} + \frac{175\,\epsilon^3}{4096} + \frac{455\,\epsilon^4}{2048} + \frac{819\,\epsilon^5}{1024} + \frac{1001\,\epsilon^6}{512} + \frac{715\,\epsilon^7}{256} - \frac{715\,\epsilon^9}{64} - \frac{1001\,\epsilon^{10}}{32} - \frac{819\,\epsilon^{11}}{16} - \frac{455\,\epsilon^{12}}{8} - \frac{175\,\epsilon^{13}}{4} - \frac{45\,\epsilon^{14}}{2} - 7\,\epsilon^{15} - \epsilon^{16}$
16	1	$\frac{1}{65536} + \frac{\epsilon}{2048} + \frac{15\,\epsilon^2}{2048} + \frac{35\,\epsilon^3}{512} + \frac{455\,\epsilon^4}{1024} + \frac{273\,\epsilon^5}{128} + \frac{1001\,\epsilon^6}{128} + \frac{715\,\epsilon^7}{32} + \frac{6435\,\epsilon^8}{128} + \frac{715\,\epsilon^9}{8} + \frac{1001\,\epsilon^{10}}{8} + \frac{273\,\epsilon^{11}}{2} + \frac{455\,\epsilon^{12}}{4} + 70\,\epsilon^{13} + 30\,\epsilon^{14} + 8\,\epsilon^{15} + \epsilon^{16}$

only on the Hamming weight w of the 16 bit input. The probabilities and their numbers of occurrences are given in Table 2.

A careful look at this table shows that all odd powers of ϵ disappear in the probabilities, if we partition the inputs in such a way that 16 bit input values and their bitwise complements are always in the same partition. This makes our problem significantly easier and leads to the much simpler Table 3.

Table 3. Probabilities for combining 16 bit inputs and their complements (bit bias ϵ)

w	Occurrences	Probability of input + probability of complement
0	1	$\frac{1}{32768} + \frac{15\epsilon^2}{1024} + \frac{455\epsilon^4}{512} + \frac{1001\epsilon^6}{64} + \frac{6435\epsilon^8}{64} + \frac{1001\epsilon^{10}}{4} + \frac{455\epsilon^{12}}{2} + 60\epsilon^{14} + 2\epsilon^{16}$
1	16	$\frac{1}{32768} + \frac{45\epsilon^2}{4096} + \frac{455\epsilon^4}{1024} + \frac{1001\epsilon^6}{256} - \frac{1001\epsilon^{10}}{16} - \frac{455\epsilon^{12}}{4} - 45\epsilon^{14} - 2\epsilon^{16}$
2	120	$\frac{1}{32768} + \frac{\epsilon^2}{128} + \frac{91\epsilon^4}{512} - \frac{429\epsilon^6}{64} + \frac{91\epsilon^{12}}{2} + 32\epsilon^{14} + 2\epsilon^{16}$
3	560	$\frac{1}{32768} + \frac{21\epsilon^2}{4096} + \frac{39\epsilon^4}{1024} - \frac{143\epsilon^6}{256} + \frac{143\epsilon^{10}}{16} - \frac{39\epsilon^{12}}{4} - 21\epsilon^{14} - 2\epsilon^{16}$
4	1820	$\frac{1}{32768} + \frac{3\epsilon^2}{1024} - \frac{9\epsilon^4}{512} - \frac{11\epsilon^6}{64} + \frac{99\epsilon^8}{64} - \frac{11\epsilon^{10}}{4} - \frac{9\epsilon^{12}}{2} + 12\epsilon^{14} + 2\epsilon^{16}$
5	4368	$\frac{1}{32768} + \frac{5\epsilon^2}{4096} - \frac{25\epsilon^4}{1024} + \frac{33\epsilon^6}{256} - \frac{33\epsilon^{10}}{16} + \frac{25\epsilon^{12}}{4} - 5\epsilon^{14} - 2\epsilon^{16}$
6	8008	$\frac{1}{32768} - \frac{5\epsilon^4}{512} + \frac{\epsilon^6}{8} - \frac{45\epsilon^8}{64} + 2\epsilon^{10} - \frac{5\epsilon^{12}}{2} + 2\epsilon^{16}$
7	11440	$\frac{1}{32768} - \frac{3\epsilon^2}{4096} + \frac{7\epsilon^4}{1024} - \frac{7\epsilon^6}{256} + \frac{7\epsilon^{10}}{16} - \frac{7\epsilon^{12}}{4} + 3\epsilon^{14} - 2\epsilon^{16}$
8	6435	$\frac{1}{32768} - \frac{\epsilon^2}{1024} + \frac{7\epsilon^4}{512} - \frac{7\epsilon^6}{64} + \frac{35\epsilon^8}{64} - \frac{7\epsilon^{10}}{4} + \frac{7\epsilon^{12}}{2} - 4\epsilon^{14} + 2\epsilon^{16}$

For the Hamming weight 8, the number of occurrences in Table 3 is only half of the number in Table 2, because the Hamming weight of complements of inputs with Hamming weight 8 is also 8.

Now, we have to partition the 32768 values from this table into 256 sets with 128 elements, such that the first through fifth powers of ϵ eliminate each other. Here is the solution we found:

Among the 256 sets, we distinguish 7 different types:

Type A consists of once the Hamming weight 0, 112 times the Hamming weight 6, and 15 times the Hamming weight 8. There is only one set of type A.

Type B consists of once the Hamming weight 1, 42 times the Hamming weight 5, and 85 times the Hamming weight 7. There are 16 sets of type B.

Type C consists of 14 times the Hamming weight 4, 28 times the Hamming weight 5, 36 times the Hamming weight 7, and 50 times the Hamming weight 8. There are 46 sets of type C.

Type D consists of twice the Hamming weight 2, 37 times the Hamming weight 5, 16 times the Hamming weight 6, 43 times the Hamming weight 7, and 30 times the Hamming weight 8. There are 60 sets of type D.

Type E consists of 5 times the Hamming weight 3, 7 times the Hamming weight 4, 58 times the Hamming weight 6, 43 times the Hamming weight 7, and 15 times the Hamming weight 8. There are 112 sets of type E.

Type F consists of 13 times the Hamming weight 4, 30 times the Hamming weight 5, 8 times the Hamming weight 6, 2 times the Hamming weight 7, and 75 times the Hamming weight 8. There are 4 sets of type F.

Type G consists of 20 times the Hamming weight 4, 4 times the Hamming weight 5, 24 times the Hamming weight 6, 60 times the Hamming weight 7, and 20 times the Hamming weight 8. There are 17 sets of type G.

The following table shows the output probability for each type:

Type	Probability of output byte
A	$\frac{1}{256} + 28\,\epsilon^6 + 30\,\epsilon^8 + 448\,\epsilon^{10} + 256\,\epsilon^{16}$
B	$\frac{1}{256} + 7\,\epsilon^6 - 112\,\epsilon^{10} - 256\,\epsilon^{16}$
C	$\frac{1}{256} - \frac{21\,\epsilon^6}{4} + 49\,\epsilon^8 - 168\,\epsilon^{10} + 224\,\epsilon^{12} - 64\,\epsilon^{14}$
D	$\frac{1}{256} + \frac{37\,\epsilon^6}{16} - \frac{33\,\epsilon^8}{4} - 78\,\epsilon^{10} + 312\,\epsilon^{12} - 112\,\epsilon^{14} - 64\,\epsilon^{16}$
E	$\frac{1}{256} + \frac{7\,\epsilon^6}{16} - \frac{87\,\epsilon^8}{4} + 134\,\epsilon^{10} - 248\,\epsilon^{12} + 48\,\epsilon^{14} + 64\,\epsilon^{16}$
F	$\frac{1}{256} - \frac{45\,\epsilon^6}{8} + \frac{111\,\epsilon^8}{2} - 212\,\epsilon^{10} + 368\,\epsilon^{12} - 288\,\epsilon^{14} + 128\,\epsilon^{16}$
G	$\frac{1}{256} - \frac{15\,\epsilon^6}{4} + 25\,\epsilon^8 - 24\,\epsilon^{10} - 160\,\epsilon^{12} + 320\,\epsilon^{14}$

All ϵ powers up to the fifth are gone!

In order to define a concrete post-processing function, we have to fix which inputs are mapped to which outputs. We can do this arbitrarily, taking into account the rules indicated above. The many choices we have for defining the concrete function do not influence the statistical quality of the function, but they can be used to ease the implementation of the function. Let S be a fixed function constructed according to the principles described above.

4.6   What About Going Further?

Naturally, we want to go on to eliminate the sixth powers of ϵ. Linear programming shows that the probability for Hamming weight 1 can be used at most $296/891$ times when we want to arrange the probabilities in such a way that for sets of 256 probabilities the ϵ-powers cancel out up to the sixth. As a fraction does not make sense here, we have shown that it is impossible to eliminate the sixth powers of ϵ. We do not claim that S is optimal, as there might be solutions with sixth powers of ϵ with smaller absolute values of the coefficients than S.

We used the linear programming approach to prove another negative result: When considering postprocessing with 32 input and 16 output bits, the probability of the outputs contains nineth or lower powers of ϵ.

4.7   The Entropy of S

Figure 4 shows, as expected, that S extracts even more entropy from its input than H.

In the evaluation directive AIS 31 for true random number generators of the German BSI [KS01], a bias of 0.02 is still considered acceptable. This bias corresponds to an entropy of 0.998846 per bit, or 7.99076 per byte. This entropy is achieved by a source with bias 0.1, if each output bit is the XOR of two input bits. With the post-processing function H, the same entropy is achieved with a source bias of 0.16835. And for the post-processing function S, we obtain this entropy even for a source bias of 0.23106.

4.8   On the Implementation of S

A hardware implementation of the post-processing function S would probably require a considerable amount of chip area. The easiest way to implement S is

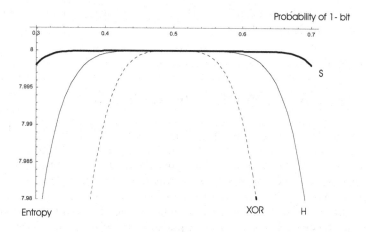

Fig. 4. The entropy of S

just a lookup table, which assigns output bytes to all possible 16 bit inputs. Such a table requires 64 kBytes of ROM. This should be no problem on a modern PC, but could be prohibitive for some smart-card applications. We can halve this storage requirement, if we take into account that inputs and their bitwise complements lead to the same output. If the leading bit of the input is 1, we use the output for its bitwise complement from the table. Therefore, we can use a table of 32 kBytes.

We can implement really compact software for S, if we make use of the freedom we have for fixing the values for the function S. We will only sketch the principle of such a software implementation. The main idea is to use the lexicographically first possible solution for all situations, where one has a choice. Following this principle, the only instance of type A leads to the output of 0, the 16 instances of type B to the output values $1, \ldots, 16$. The 46 instances of type C are assigned the output values $17, \ldots, 62$, and so on. The assignment of input values to instances of types also follows this lexicographic principle. As an example, we consider inputs of Hamming weight 8. The 15 lexicographically lowest 16 bit values with Hamming weight 8 are assigned to the only instance of type A. Type B does not use inputs of Hamming weight 8, so the lexicographically next inputs with this Hamming weight are assigned to instances of type C. The lexicographically next 50 inputs with Hamming weight 8 are assigned to the first instance of type C, the next 50 to the second instance of type C, and so on. When we follow this lexicographic design principle, the tables needed for the computation of S can be stored very compactly in about 50 bytes.

This lexicographic principle can only be used profitably for a software implementation, if we are able to determine the rank of a given 16 bit value of Hamming weight w in the lexicographic order of the inputs of this Hamming weight. Let $p_0, p_1, \ldots, p_{w-1}$ with $p_0 < p_1 < \cdots < p_{w-2} < p_{w-1}$ be the bit positions of the 1-bits of the input. Let the bit position of the least significant bit be 0, and let the bit position of the most significant bit be 15. Then the lexicographic rank is given by

$$\sum_{i=0}^{w-1} \binom{p_i}{i+1}.$$

5 Conclusion and Further Research Topics

We have shown that the quasigroup post-processing method for physical random numbers described in [MGK05] is ineffective and very easy to attack.

We have also shown that there are post-processing functions with a fixed number of input bits for biased physical random number generators which are much better than the ones usually used up to now.

The results of section 4 of this paper can be extended in many directions: e. g., compression rates greater than 2, other input sizes, systematic construction of good post-processing functions.

Acknowledgment

The author would like to thank Bernd Meyer (Siemens AG) for his help and for many valuable discussions.

References

[JJSH00] Juels, A., Jakobsson, M., Shriver, E., Hillyer, B.K.: How to turn loaded dice into fair coins. IEEE Trans. Information Theory 46(3), 911–921 (2000)

[KS01]   Killmann, W., Schindler, W.: A proposal for: Functionality classes and evaluation methodology for true (physical) random number generators (2001), http://www.bsi.bund.de/zertifiz/zert/interpr/trngk31e.pdf

[MGK05]   Markovski, S., Gligoroski, D., Kocarev, L.: Unbiased random sequences from quasigroup string transformations. In: Gilbert, H., Handschuh, H. (eds.) FSE 2005. LNCS, vol. 3557, pp. 163–180. Springer, Heidelberg (2005)

[Per92]   Peres, Y.: Iterating von Neumann's procedure for extracting random bits. The Annals of Statistics 20(3), 590–597 (1992)

[vN63]   von Neumann, J.: Various techniques for use in connection with random digits. Von Neumann's Collected Works, pp. 768–770. Pergamon, London (1963)

Improved Slide Attacks

Eli Biham[1,*], Orr Dunkelman[2,*], and Nathan Keller[3,**]

[1]Computer Science Department, Technion
Haifa 32000, Israel
biham@cs.technion.ac.il
[2]Katholieke Universiteit Leuven,
Dept. of Electrical Engineering ESAT/SCD-COSIC
Kasteelpark Arenberg 10, B-3001 Leuven-Heverlee, Belgium
orr.dunkelman@esat.kuleuven.be
[3]Einstein Institute of Mathematics, Hebrew University
Jerusalem 91904, Israel
nkeller@math.huji.ac.il

Abstract. The slide attack is applicable to ciphers that can be represented as an iterative application of the same keyed permutation. The slide attack leverages simple attacks on the keyed permutation to more complicated (and time consuming) attacks on the entire cipher.

In this paper we extend the slide attack by examining the cycle structures of the entire cipher and of the underlying keyed permutation. Our method allows to find slid pairs much faster than was previously known, and hence reduces the time complexity of the entire slide attack significantly. In addition, since our attack finds as many slid pairs as the attacker requires, it allows to leverage all types of attacks on the underlying permutation (and not only simple attacks) to an attack on the entire cipher.

We demonstrate the strength of our technique by presenting an attack on 24-round reduced GOST whose S-boxes are unknown. Our attack retrieves the unknown S-boxes as well as the secret key with a time complexity of about 2^{63} encryptions. Thus, this attack allows an easier attack on other instances of GOST that use the same S-boxes. When the S-boxes are known to the attacker, our attack can retrieve the secret key of 30-round GOST (out of the 32 rounds).

1 Introduction

Most of the modern block ciphers are constructed as a cascade of repeated keyed components, called rounds (or round functions). The security of such ciphers relies on applying the round function sufficiently many times. This is mainly due

---

* This work was supported in part by the Israel MOD Research and Technology Unit.
** The research presented in this paper was supported by the Adams fellowship.

to the fact that most attacks on block ciphers, e.g., differential cryptanalysis [3] and linear cryptanalysis [16], are statistical in nature and their effectiveness reduces as the number of rounds increases. Even non-statistical techniques, such as the SQUARE attack [7], are also affected by increasing the number of rounds of the cipher.

There are only few attacks on block ciphers that are independent of the number of rounds. One of them is the *related key attack* [1] presented by Biham in 1993. The attack uses encryption under unknown but related keys to derive the actual values of the keys.

In 1999, Biryukov and Wagner explored the framework of related key attacks and introduced the *slide attack* [4]. The slide attack is applied against ciphers that are a cascade of some identical "simple" functions, e.g., $E_k = f_k^l = f_k \circ f_k \circ \cdots \circ f_k$, where f_k is a relatively weak keyed permutation. The attacker uses the birthday paradox to find *slid pairs* (P_1, P_2) such that $P_2 = f_k(P_1)$. Due to the structure of E_k, the corresponding ciphertexts (C_1, C_2) satisfy that $C_2 = f_k(C_1)$. Once a slid pair is found, the attacker uses the "simplicity" of f_k and the input/output pairs (P_1, P_2) and (C_1, C_2) to attack f_k, and thus, to attack E_k. The attack can be used only if f_k is "simple", i.e., can be broken using only two known input/output pairs.

In 2000, Biryukov and Wagner presented new variants of the slide attack, called *complementation slide* and *sliding with a twist* [5]. These variants allow for treating more complex functions in the slide framework. Nevertheless, there are no widely used ciphers that can be attacked using these techniques.

In addition to these new variants, the authors of [5] presented several techniques aimed at finding several slid pairs simultaneously, thus enabling to use the attack even if several input/output pairs are required to break f_k. One of these techniques, fully explored by Furuya [9], uses the fact that if (P_1, P_2) is a slid pair then $(E_k^t(P_1), E_k^t(P_2))$ are also slid pairs for all values of t. This technique allows the attacker to transform any known plaintext attack on f_k that requires m known plaintexts to an attack on E_k. The data complexity of the attack is $O(m \cdot 2^{n/2})$ adaptively chosen plaintexts, where n is the block size, and the time complexity of $O(2^n)$ applications of the known plaintext attack. Other attacks on the underlying function, e.g., chosen plaintext attacks, can be also leveraged to an attack on the entire cipher using the standard transformation to a known plaintext attack. However, in this case both the data complexity and the time complexity become very large, since the attack on the underlying function is applied $O(2^n)$ times during the attack on the whole cipher.

In this paper we present a new variant of the slide attack which enables to find slid pairs much more quickly. In our attack, instead of constructing slid pairs by the birthday paradox, we use the cycle structures of E_k and f_k for the detection of a large set of slid pairs. Then, we can use these pairs in any attack on f_k, whether it is a known plaintext attack, or if enough pairs are found, a chosen plaintext attack (or even an adaptive chosen plaintext and ciphertext attack). Unlike previous attacks, in our attack, the attack on f_k is repeated only once (as the slid pairs are given), no matter what f_k is.

The data complexity of our attack is high — it requires almost the entire codebook. On the other hand, the time complexity is surprisingly low — the first stage of the attack (finding the slid pairs) takes no more than 2^n encryptions, where n is the block size, while the second stage is a (much faster) attack on f_k. Despite the large data requirements, the attack can be useful, as we demonstrate in our attack on 24-round reduced GOST with unknown S-boxes.

The block cipher GOST [10] is the Russian Encryption Standard. Since its publication in 1989 it withstood extensive cryptanalytic efforts. GOST has 64-bit blocks and keys of 256 bits. The main special feature of GOST is the S-boxes that were not published. Every industry was issued a different set of secret S-boxes. An example of a set that leaked is the set used in the Russian banking industry [19], and few attacks have been published on variants of GOST with these S-boxes [12,15,20]. There are also attacks on several variants of GOST with unknown S-boxes, but these attacks are either applicable to only a small number of rounds [20] or to relatively small classes of weak keys [4]. Another attack retrieves the unknown S-boxes by choosing a key from a set of weak keys for which it is possible to apply the slide attack [18].

In this paper we apply our technique to devise an attack on GOST reduced to its 24 first rounds with unknown S-boxes that allows to retrieve both the S-boxes and the secret key, a total of 768 key bits. The data complexity of our attack is 2^{63} adaptively chosen plaintexts and the time complexity is 2^{63} encryptions. A known plaintext variant of the attack has data and time complexities of a little less than 2^{64} encryptions.

A possible application of our attack is by an authorized user that has legitimate access to an instance of 24-round GOST that cannot be reverse engineered. Such a user can apply our attack to find the unknown S-boxes. As these S-boxes are shared by an entire industry, such an attack compromises the security of the entire industry.

When the S-boxes are known, it is possible to apply our attack to 30-round GOST (out of the 32 rounds), such that it is still faster than exhaustive key search. In addition, we present a new class of weak keys consisting of 2^{128} keys, for which our attack can break the entire GOST with unknown S-boxes with the same data and time complexities as the 24-round attack.

This paper is organized as follows: In Section 2 we give a brief description of the related key attacks and the slide attacks. In Section 3 we present our new technique. In Section 4 we present an attack on 24-round GOST with unknown S-boxes and on 30-round GOST with known S-boxes. Appendix A outlines a 7-round truncated differential of GOST with probability 0.494. Finally, Section 5 summarizes this paper.

2   Related-Key Attacks and Slide Attacks

In this section we survey the previous results on related-key attacks. We start with describing the related-key attacks suggested by Biham [1] and Knudsen [14].

Then, we describe the slide attacks by Biryukov and Wagner [4]. We also present some of the extensions of the slide attack.

2.1   Related-Key Attacks

Related-key attacks, introduced in [1,14], are attacks exploiting related-key plaintext pairs. The main idea behind the attack is to find instances of keys for which the encryption processes deploy the same permutation (or almost the same permutation). To illustrate the technique, we shortly present the attack from [1].

Consider a variant of DES in which all the rotate left operations in the key schedule algorithm are by a fixed number of bits.[1] For any key K there exists another key $\hat{K}$ such that the round subkeys $KR_i, \widehat{KR_i}$ produced by K and $\hat{K}$, respectively, satisfy:

$$KR_{i+1} = \widehat{KR_i}, \qquad \text{for } i = 1, \ldots, 15.$$

For such pair of keys, if a pair of plaintexts (P_1, P_2) satisfies $P_2 = f_{KR_1}(P_1)$, where $f_{sk}(P)$ denotes one round DES encryption of P under the subkey sk, then the corresponding ciphertexts, C_1 and C_2, respectively, satisfy $C_2 = f_{\widehat{KR_{16}}}(C_1)$. Given such a pair of plaintexts (called in the sequel "a related-key plaintext pair"), the subkeys KR_1 and $\widehat{KR_{16}}$ can be easily extracted [1]. Due to the Feistel structure of DES, the related-key plaintext pairs can be detected easily. Therefore, this attack can be applied using only 2^{16} chosen plaintexts encrypted under K and 2^{16} chosen plaintexts encrypted under $\hat{K}$, with a time complexity of 2^{17} encryptions.

In the more general case, the related-key attack has three parts: Obtaining related-key plaintexts, identifying the related-key plaintext pairs, and using them to deduce the key. In many cases, identifying the related-key plaintext pairs is best achieved by assuming for each candidate pair that it is a related-key plaintext pair, and then using it as an input for the key recovery phase of the attack. In other cases, the round functions' weaknesses allow the attacker to identify these pairs easily.

2.2   Slide Attacks

When a cipher has self-related keys, i.e., it can be written as $E_k = f_k^l = f_k \circ f_k \circ \cdots \circ f_k$, then it is susceptible to a variant of the related-key attack called the *slide attack* [4]. In this case, it is possible to apply the related-key attack to the cipher with $K = \hat{K}$, thus eliminating the key requirement of having two keys. The attacker looks for a slid pair, i.e., two plaintexts (P_1, P_2) such that $P_2 = f_k(P_1)$. In this case, the pair satisfies $C_2 = f_k(C_1)$ as well. When the round function f_k is simple enough, it is possible to use these two pairs in order to deduce information about the key.

---

[1] Such a variant was proposed by Brown and Seberry [6].

In the slide attack the attacker obtains enough plaintext/ciphertext pairs to contain a slid pair, and has to check for each possible pair of plaintexts whether it is a slid pair by applying the attack on f_k. When dealing with a general block cipher this approach requires $O(2^{n/2})$ known plaintexts and $O(2^n)$ applications of the attack on f_k, where n is the block size. For Feistel block ciphers, the attack can be optimized using $O(2^{n/4})$ chosen plaintexts and $O(1)$ applications of the attack. Note that as in the original related-key attacks, the main drawback of this approach is that the attack can be used only if f_k can be broken using only two known input/output pairs, i.e., given one slid pair.

In 2000, Biryukov and Wagner [5] presented two variants of the slide attack, named *complementation slide* and *sliding with a twist*. These variants allow for treating more complex functions in a slide attack. Nevertheless, there are no widely used ciphers that can be attacked using these techniques.

The authors of [5] also presented several techniques aimed at finding several slid pairs simultaneously, enabling to use the attack even if several input/output pairs are needed for attacking f_k. One of these techniques, fully explored by Furuya [9], uses the fact that when (P_1, P_2) is a slid pair then $(E_k^t(P_1), E_k^t(P_2))$ are also slid pairs for all values of t. This allows the attacker to transform any known plaintext attack on f_k that requires m known plaintexts to an attack on E_k with a data complexity of $O(m \cdot 2^{n/2})$ adaptively chosen plaintexts. The time complexity of this approach is $O(2^n)$ applications of the known plaintext attack on f_k. It is worth mentioning that the technique can be easily improved when f_k is one Feistel round, for which $O(2^{n/4})$ chosen plaintexts and $O(m)$ adaptive chosen plaintexts are sufficient to achieve m slid pairs, which are easily identified.

This approach can be also used if there is only a chosen plaintext attack on f_k. The attacker can repeatedly generate slid pairs until the "plaintext" parts of the slid pairs contain a set of plaintexts satisfying the data requirements of the attack on f_k. However, the attack still requires $O(2^n)$ applications of the attack on f_k (unless there is a special property that allows easy identification of the slid pairs).

3   Our New Technique

Before we start our discussion, let us recall the notations used in the previous section. In our discussion, $E_k = f_k^l$ is a block cipher composed of l applications of the same keyed permutation f_k. The attacker looks for a slid pair, i.e., a pair of plaintexts (P_1, P_2) such that $P_2 = f_k(P_1)$.

3.1   Studying the Cycle Structure

As noted earlier, in the previous variants of the slide attack there was no immediate indication which of the possible pairs is a slid pair (for a general cipher). Thus, an attacker had to try all possible pairs as candidates.

In our attack we use the cycle structure of E_k and of f_k to find a large amount of slid pairs. We emphasize that the attack uses only the cycle structure and is independent of any other properties of the functions E_k and f_k. As a result, the attacker can detect the slid pairs even if there is no efficient attack on f_k. After finding the slid pairs, the attacker can use them to mount any attack she wishes on f_k, in order to retrieve the key material used in f_k.

First we recall several facts regarding the cycle structure of permutations that are used in our attack. Let $g : GF(2^n) \to GF(2^n)$ be a random permutation. For every $x \in GF(2^n)$ we denote $CycleLength(x) = \min\{k > 0 | g^k(x) = x\}$. The lengths of the cycles of g are close to be uniformly distributed [8]. The expectation of the cycle length is $E[CycleLength(x)] = 2^{n-1}$. Therefore, we expect that the largest cycle $Cycle_{max}$ has a size of $O(2^{n-1})$ and for every $x \in GF(2^n)$, $\Pr[x \in Cycle_{max}] \approx 1/2$. Moreover, we expect that the second largest cycle has a size of $O(2^{n-2})$, and generally, the i-th largest cycle has a size of $O(2^{n-i})$. We note that for most of the permutations there are about n cycles that are distributed according to a Poisson distribution (confirming the above claims for most of the cases) [11].

The main observation used in our attack is the following: Let P be a randomly chosen plaintext. We consider the cycles of P with respect to f_k and E_k, denoted by $Cycle_{f_k}(P)$ and $Cycle_{E_k}(P)$, respectively. Denote the length of $Cycle_{f_k}(P)$ by m_1, i.e., $m_1 = \min\{t > 0 | f_k^t(P) = P\}$. Similarly, denote the length of $Cycle_{E_k}(P)$ by m_2. As $E_k = f_k^l$ the relation $E_k^{m_2}(P) = P$ can be written as $f_k^{l \cdot m_2}(P) = P$. Thus, from the definition of cycle lengths $m_1 | l \cdot m_2$. On the other hand, $m_2 = \min\{t > 0 | E_k^t(P) = P\} = \min\{t > 0 | f_k^{(t \cdot l) \bmod m_1}(P) = P\}$. Combining these two statements, we get $m_2 = \min\{t > 0 | tl = 0 \bmod m_1\}$. We conclude that

$$m_2 = m_1/gcd(m_1, l).$$

In particular, if $gcd(m_1, l)=1$, then $m_1 = m_2$.

If $gcd(m_1, l) = 1$ we can find $1 \le d_1 \le l - 1$ and d_2 such that $d_1 \cdot m_1 = (-1) \pmod{l} = d_2 \cdot l - 1$ using Euclid's extended algorithm. These two values d_1 and d_2 satisfy both $f_k^{d_1 \cdot m_1 + 1}(P) = f_k(P)$ and $f_k^{d_1 \cdot m_1 + 1}(P) = f_k^{d_2 \cdot l}(P) = E_k^{d_2}(P)$. Therefore, the pair $(P, E_k^{d_2}(P))$ is a slid pair for E. Moreover, as already observed in [5], the pairs $(E_k^t(P), E_k^{d_2 + t}(P))$ are also slid pairs for every t.

When $gcd(m_1, l) \ne 1$, there is no general way to identify slid pairs as this case may be the product of several reasons. For example, when there is only one cycle (of size 2^n), then the slid counterpart of any given plaintext P is found in the cycle, but in an unknown location (or more precisely, any information that can be used to find the slid pair in the cycle, can be used without the need to construct the cycle).

In other cases, using the cycle structure can help a little bit. Assume that there are many cycles of different lengths, but a small group of cycles with the same size. It is very likely that this group contains slid pairs with one element in one cycle, and its slid counterpart in the other cycle. However, even if this is the case, the order between the different cycles, or even which is the counterpart of a given plaintext, is not necessarily disclosed by studying the cycle structure.

3.2   Using the Cycles in the Slide Attack

Our attack starts at some random plaintext P_0. We apply E_k to it repeatedly until we get P_0 once again. When this happens, we have identified a cycle and the value of $CycleLength_{E_k}(P_0)$. Note that the attack cannot obtain the cycle of f_k, nor its length.

However, when the corresponding values m_1 and l satisfy $gcd(m_1, l) = 1$ our attack can find the cycle length of f_k, as $m_1 = m_2$. Under this assumption we can calculate d_2, and then we get m_2 slid pairs in the cycle of E_k. These slid pairs can be used in any attack on f_k.

For a random value of x, $E[CycleLength(x)] = 2^{n-1}$, thus, we expect to get about 2^{n-1} slid pairs. This amount is sufficient for most possible attacks, including adaptive chosen plaintext and ciphertext attacks. In case the block ciphers (f_k or E_k) have a different behavior, they can be easily distinguished from random permutations using this property.

In our attack, we assume that $m_1 = m_2$. In case this assumption is wrong, our attack fails. The probability of a failure for a random permutation f_k is $1 - \varphi(l)/l$, where $\varphi(l) = |\{1 \leq d \leq (l-1) : gcd(d, l) = 1\}|$ is the Euler's function. In this case, we start with another plaintext $P_1 \notin Cycle_{E_k}(P_0)$ and repeat the attack. If the attack fails again, we take a new plaintext P_2 etc. Assuming that all the cycles lengths are independent of each other, after t applications of the attack, the total success probability is $1 - (1 - \varphi(l)/l)^t$. This assumption is quite appropriate for random permutations, as long as t is smaller than the total number of cycles.

The data complexity of the attack is very large — the attack requires $O(2^{n-1})$ adaptively chosen plaintexts. However, as we shall see in the attack on GOST, there are scenarios in which such an attack can be used despite the large data requirements.

We note that the attack can be transformed into a known plaintext attack that requires almost the entire code book. In that case, using birthday arguments we expect that few long cycles are found, suggesting a large amount of slid pairs.

4   Several Attacks on Reduced Round GOST

In this section we present a slide attack on 24-round GOST. The data complexity of the attack is about 2^{63} adaptive chosen plaintexts or 2^{64} known plaintexts (as required by our technique described in Section 3). The time complexity of the best attack we suggest is dominated by the time required to encrypt the plaintexts.

4.1   A Short Description of GOST

GOST [10] is a 64-bit block and 256-bit key cipher with a Feistel structure of 32 rounds. The round function accepts an input and a subkey of 32 bits each. The input and the subkey are added (modulo 2^{32}), and the outcome is divided into eight groups of four bits each. Each such group enters a different S-box, where the least significant group enters $S1$, and the most significant group enters $S8$. The actual S-boxes that are used are kept secret, and are assigned by the government to a given set of users in each industry.[2] The outputs of all S-boxes are combined to 32 bits, which are then rotated to the left by 11 bits.

The key schedule algorithm takes the 256-bit key and treats it as eight 32-bit words, i.e., $K = K_1, \ldots, K_8$. The subkey SK_r of round r is

$$SK_r = \begin{cases} K_{(r-1) \bmod 8+1} & r \in \{1, \ldots, 24\}; \\ K_{33-r} & r \in \{25, \ldots, 32\}. \end{cases}$$

Thus, the 24 rounds of GOST we attack can be described as f_k^3, where f_k is the first eight rounds of GOST.

There are few results on GOST with the S-boxes used in the banking industry: A differential attack on 13-round GOST requiring 2^{51} chosen plaintexts is described in [20] along with a related-key attack on 21-round GOST that requires 2^{56} related-key chosen plaintexts. A 24-round related-key attack (whose actual complexity and success rate depend on the S-boxes used) is described in [12]. A related-key attack on the full GOST with a data complexity of 2^{35} related-key chosen plaintexts and a time complexity of 2^{36} encryptions is presented in [15]. Only few results on GOST when the S-boxes are unknown were published before: A related-key distinguisher for the entire cipher that requires two related-key chosen plaintexts is presented in [15]. In [5] a slide attack on a 20-round variant of GOST in which the key additions are replaced by XOR with the key is described, along with a weak key class of 2^{128} keys that exists for that variant. A weak key class consisting of 2^{32} keys identified using a slide attack is presented in [9]. A chosen-key attack with a key in the weak key class that retrieves the unknown S-boxes is described in [18]. The attack uses $16 \cdot 2^{32}$ chosen plaintexts and has a running time of 2^{11} encryptions. The attack is a slide attack (that can be improved to an attack that requires 2^{17} chosen plaintexts), and the key can be any of the 2^{32} values $K = (K^*, K^*, K^*, K^*, K^*, K^*, K^*, K^*)$. However, we note that it is very easy to protect GOST against this attack, by preventing the usage of such keys.

4.2   Description of the Attack

As noted earlier, we attack 24 rounds of GOST for which $E_k = f_k^3$. Our attack first finds a large set of slid pairs (P_a, P_b) and their corresponding ciphertexts (C_a, C_b) such that $f_k(P_a) = P_b$ and $f_k(C_a) = C_b$. These slid pairs are found using the algorithm suggested in Section 3. Once these slid pairs are found, we apply a differential attack to f_k. We attack eight rounds of GOST using the following seven round differential that is independent of the actual S-boxes used:

$$P' = 00\ 01\ 00\ 00\ \ 00\ 00\ 00\ 00_x \rightarrow T' = ??\ ??\ ??\ ??\ \ ?U_8\ L_9?\ ??\ ??$$

---

[2] We note that according to the published documentation [10], the S-boxes are not necessarily permutations. Hence, an S-box can be modeled as an unknown 64-bit key.

where ? denotes an unknown value, $L_9 \in \{0, 1_x, \ldots, 7_x\}$, and $U_8 \in \{0_x, 8_x\}$. The complete description of the differential is presented in Figure 1 in Appendix A. We note that as the differential is independent of the actual S-boxes used, we can apply the attack to any instance of GOST.

Roughly speaking, the attack takes pairs of slid pairs (P_a, P_b) and (P_c, P_d) for which the difference $P_a \oplus P_c$ equals the input difference of the differential, i.e., pairs for which $P_a \oplus P_c = P'$. As P_b is the "encryption" of P_a through f_k and P_d is the "encryption" of P_c through f_k, then we treat the two plaintexts P_b and P_d, as the ciphertexts in a differential attack on f_k. Once enough pairs of slid pairs are encountered, we partially decrypt the "ciphertexts" to find whether they satisfy the differential.

For the right guess of S-boxes and subkeys, the probability that a pair with input difference P' has an output difference T' is 0.494. This is a lower bound on the probability that the difference in the four bits 23–26 of the right half (left half after the swap operation) is zero. For a randomly selected pair of values this probability is 2^{-4}. The way to check whether the differential holds, is to partially decrypt the pairs one round, and check whether the difference in these bits is 0 or not.

Our attack on GOST uses the following observations:

Observation 1. *Each cycle suggests many pairs that can be used. Thus, we can impose more conditions on the ciphertexts we analyze, in order to reduce the time complexity of the attack.*

Observation 2. *In order to determine the difference in bits 23–26 of the right half in round 7, it is sufficient to determine the output of S4 in round 8 (bits 12–15).*

Observation 3. *It is possible to consider only ciphertexts that have predetermined inputs to the relevant S-box.*

Based on these observations, the attack algorithm is the following:

1. Obtain about $2^{43.5}$ slid pairs (P_a, P_b) such that $f_k(P_a) = P_b$.
2. Identify 2^{22} pairs of slid pairs (P_a, P_b) and (P_c, P_d) such that $P_a \oplus P_c = P'$.
3. Consider the pairs of slid pairs $((P_a, P_b), (P_c, P_d))$ such that
   (a) The value of the right half of P_b is $X0$ in bits 8–15, (X is the value that enters S4 in round 8), where $X \in \{0, \ldots, F_x\}$ is a predetermined value.
   (b) The value of the right half of P_d is $Z0$ in bits 8–15, (Z is the value that enters S4 in round 8), where $Z \in \{0, \ldots, F_x\} \backslash \{X\}$ is a predetermined value.
4. For each remaining pair $((P_a, P_b), (P_c, P_d))$ of slid pairs:
   (a) Guess the four bits leaving S4 in round 8 for the pair (P_a, P_b) (assuming that the value is the same for all the examined pairs), and guess the four bits leaving S4 in round 8 for the pairs (P_c, P_d) (assuming that the value is the same for all the examined pairs).
   (b) Partially decrypt P_b and P_d through S4, and check whether the difference in round 7 in bits 23–26 of the left half is 0.

   (c) If the difference in these bits after the partial decryption is 0, increment a
       counter that corresponds to the specific guesses made (a total of 4+4=8
       bits).
 5. Output all the guesses whose corresponding counter has a value greater than
    19.

4.3   Analysis of the Attack

For a wrong guess of the S-boxes outputs there is probability of 2^{-4} that a
partially decrypted pair has zero difference in bits 23–26, thus incrementing the
counter. For the correct guess, this probability is about 0.494.

Of the $2^{43.5}$ slid pairs, we expect that $(2^{43.5})^2/2 \cdot 2^{-64} = 2^{22}$ pairs of slid pairs
have "plaintext" difference of P'. Out of these pairs, about $2^{22} \cdot (2^{-8})^2 = 2^6$ pairs
are analyzed in Step 4 (due to the 8-bit condition on each of the ciphertexts).

In Step 4(a) and Step 4(b) the ciphertexts are partially decrypted through
round 8. If the actual inputs to $S4$ in round 8 are fixed, then so do their outputs
that compose the four output bits that are needed for the partial decryption.
Thus, we choose a value for bits 12–15 (those entering S-box $S4$) for each of
the two pairs, i.e., we require that one ciphertext has some arbitrary value X
in the S-box and that the second ciphertext has some other arbitrary value Z,
where $X, Z \in \{0, \ldots, F_x\}$. Setting these conditions does not necessarily imply
that in all pairs the ciphertexts have the same four output bits (four from each
ciphertext) due to the carry that the key addition may cause. Thus, we also
require that in all the ciphertexts bits 8–11 have the value zero, and this reduces
the probability that a carry changes the values entering S-box $S4$.[3]

For a wrong guess, the expected value of the counter is $2^6 \cdot 2^{-4} = 4$, while for
the right guess it is $2^6 \cdot 0.494 = 31.6$. There are 2^8 guesses, and the probability
that one of them has a counter whose value is greater than 19 is less than 2^{-22},
while the correct guess is suggested with probability of $1 - 2^{-7.6}$. We note that
the attack returns two outputs of the S-box $S4$ in round 8. Using more pairs
with different inputs to $S4$, enables mapping all the outputs of $S4$, up to the 16
combinations of key and S-box that are all equivalent (as the key bits are used to
decide on the actual entries that lead to the output). The exact combination can
be identified using different entries to the S-box and using auxiliary techniques.

We conclude that our attack uses $2^{43.5}$ slid pairs. The time complexity of
Step 4 in the attack is less than 2^8 24-round GOST encryptions for the retrieval
of 8 bits of information about the key and $S4$. In this case, the time complexity
of the attack is dominated mostly by the first step of the attack, i.e., finding the
slid pairs, a step that requires 2^{63} encryptions.

After finding the entire $S4$ (up to the exact order, due to the addition with
the key), we can use a rotated version of the differential (that may have a slightly

---

[3] There is a difference in the carry when the four key bits that enter $S3$ are all 1's,
   and in addition there is a difference in the carry operation in the most significant
   bit that enters $S2$. If the attack fails, we deduce that this is the case, and we can
   continue to impose conditions on the ciphertexts, until either the attack succeeds, or
   we obtain the 12 least significant bits of K_8.

smaller probability) to retrieve other S-boxes. The remaining key words can be found using auxiliary techniques (e.g., using the differential in the decryption direction).

As the attack finds the S-boxes along with the key it can be used by an authorized user who wishes to find the S-boxes used in a given implementation of GOST. In this case, the attacker can even know the encryption key (which can help in the process of the attack), and retrieve the unknown S-boxes in time complexity of 2^{63} encryptions (which as a valid user of GOST he can achieve). We note that the work of Saarinen in [18] addresses the same problem for a fixed key for which the slide properties are easily applied. However, it is easy to protect GOST from this attack by preventing the usage of keys of the form (k, k, k, k, k, k, k, k).

We note that the data complexity is about 2^{63} adaptive chosen plaintexts, or about $2^{64} - 2^{18}$ known plaintexts. In the latter case, it is expected that three cycles of expected lengths of at least $2^{43.5}$ values are encountered. There is a high probability that one of these three cycles can be used in our attack (if the gcd of the cycle length with 3 is 1).

4.4   Other Results on GOST

Our attack can be extended for two cases: a weak key class of the full GOST that can be detected even when the S-boxes are unknown, and a 30-round attack when the S-boxes are known.

Our weak key class has 2^{128} keys of the form $K_1, K_2, K_3, K_4, K_4, K_3, K_2, K_1$, i.e., $K_{9-i} = K_i$. Thus, the last eight rounds define the same permutation as the first eight rounds. Hence, it is possible to treat the full GOST of this form as $GOST_K = f_K^4$, and apply our attack (even when the S-boxes are unknown). We note that this weak key class was suggested in [5] for a weakened variant of GOST where the key is XORed and not added.

The attack on 30-round GOST with known S-boxes is of the following nature:

- Ask for almost the entire code book
- For each guess of $K_3, K_4, \ldots, K_8$:
  - Partially decrypt all the ciphertexts through rounds 30–25
  - Apply a variant of the 24-round attack described earlier
  - If the attack succeeds, output the key guess of $K_3, \ldots, K_8$
- Exhaustively search over all possible values of K_1, K_2

The time complexity of this attack is equivalent to partially decrypting each ciphertext 2^{192} times. As the attack requires almost the entire code book, then the actual time complexity of the attack is almost 2^{256} partial decryptions which are equivalent to $2^{253.7}$ 30-round GOST encryptions.

5   Summary and Conclusions

In this paper we have presented a new variant of the slide attack. Our attack uses the relation between the cycle structure of the entire cipher and that of the underlying permutation, and allows to detect a large amount of slid pairs in an efficient way. These pairs are then used to mount various attacks on the underlying permutation.

The new technique allows us to attack 24-round GOST, even when the S-boxes are unknown, and to retrieve both the key and the unknown S-boxes. When the S-boxes are known, this attack can be extended to up to 30-round GOST. In addition, for a weak key class of GOST containing 2^{128} weak keys, the attack is applicable against the full GOST with unknown S-boxes. All the attacks have a data complexity of a little less than 2^{64} known plaintexts (or 2^{63} adaptively chosen plaintexts) with time complexity of 2^{64} (besides the 30-round attack whose time complexity is $2^{253.7}$). The 24-round attack reveals the S-boxes used in GOST, and thus it can be used by an authorized user who wishes to declassify the S-boxes he was given.

References

1. Biham, E.: New Types of Cryptanalytic Attacks Using Related Keys. Journal of Cryptology 7(4), 229–246 (1994)
2. Biham, E., Biryukov, A., Shamir, A.: Miss in the Middle Attacks on IDEA and Khufu. In: Knudsen, L.R. (ed.) FSE 1999. LNCS, vol. 1636, pp. 124–138. Springer, Heidelberg (1999)
3. Biham, E., Shamir, A.: Differential Cryptanalysis of the Data Encryption Standard. Springer, Heidelberg (1993)
4. Biryukov, A., Wagner, D.: Slide Attacks. In: Knudsen, L.R. (ed.) FSE 1999. LNCS, vol. 1636, pp. 245–259. Springer, Heidelberg (1999)
5. Biryukov, A., Wagner, D.: Advanced Slide Attacks. In: Preneel, B. (ed.) EURO-CRYPT 2000. LNCS, vol. 1807, pp. 586–606. Springer, Heidelberg (2000)
6. Brown, L., Seberry, J.: Key Scheduling in DES Type Cryptosystems. In: Seberry, J., Pieprzyk, J.P. (eds.) AUSCRYPT 1990. LNCS, vol. 453, pp. 221–228. Springer, Heidelberg (1990)
7. Daemen, J., Knudsen, L.R., Rijmen, V.: The Block Cipher Square. In: Biham, E. (ed.) FSE 1997. LNCS, vol. 1267, pp. 149–165. Springer, Heidelberg (1997)
8. Davies, D.W., Parkin, G.I.P.: The Average Cycle Size of the Key Stream in Output Feedback Encipherment (Abstract). In: McCurley, K.S., Ziegler, C.D. (eds.) CRYPTO 1982. LNCS, vol. 1440, pp. 97–98. Springer, Heidelberg (1982)
9. Furuya, S.: Slide Attacks with a Known-Plaintext Cryptanalysis. In: Kim, K.-c. (ed.) ICISC 2001. LNCS, vol. 2288, pp. 214–225. Springer, Heidelberg (2002)
10. GOST: Gosudarstvennei Standard 28147-89, Cryptographic Protection for Data Processing Systems, Government Committee of the USSR for Standards (1989)
11. Granville, A.: Cycle lengths in a permutation are typically Poisson distributed. Electronic Journal of Combinatorics 13(1), 107 (2006), http://www.dms.umontreal.ca/ andrew/PDF/CycleLengths.pdf
12. Kelsey, J., Schneier, B., Wagner, D.: Key-Schedule Cryptoanalysis of IDEA, G-DES, GOST, SAFER, and Triple-DES. In: Koblitz, N. (ed.) CRYPTO 1996. LNCS, vol. 1109, pp. 237–251. Springer, Heidelberg (1996)
13. Kelsey, J., Schneier, B., Wagner, D.: Related-Key Cryptanalysis of 3-WAY, Biham-DES, CAST, DES-X, NewDES, RC2, and TEA. In: Han, Y., Quing, S. (eds.) ICICS 1997. LNCS, vol. 1334, pp. 233–246. Springer, Heidelberg (1997)

14. Knudsen, L.R.: Cryptanalysis of LOKI91. In: Zheng, Y., Seberry, J. (eds.) AUSCRYPT 1992. LNCS, vol. 718, pp. 196–208. Springer, Heidelberg (1993)
15. Ko, Y., Hong, S., Lee, W., Lee, S., Kang, J.-S.: Related Key Differential Attacks on 27 Rounds of XTEA and Full-Round GOST. In: Roy, B., Meier, W. (eds.) FSE 2004. LNCS, vol. 3017, pp. 299–316. Springer, Heidelberg (2004)
16. Matsui, M.: Linear Cryptanalysis Method for DES Cipher. In: Helleseth, T. (ed.) EUROCRYPT 1993. LNCS, vol. 765, pp. 386–397. Springer, Heidelberg (1994)
17. National Bureau of Standards: Data Encryption Standard, Federal Information Processing Standards Publications No. 46 (1977)
18. Saarinen, M.-J.: A Chosen Key Attack against the Secret S-boxes of GOST (1998), http://citeseer.ist.psu.edu/saarinen98chosen.html
19. Schneier, B.: Applied Cryptography, 2nd edn. John Wiley & Sons, Chichester (1996)
20. Seki, H., Kaneko, T.: Differential Cryptanalysis of Reduced Rounds of GOST. In: Stinson, D.R., Tavares, S. (eds.) SAC 2000. LNCS, vol. 2012, pp. 315–323. Springer, Heidelberg (2001)
21. Yuval, G.: Reinventing the wheel Travois: Encryption/MAC in 30 ROM Bytes. In: Biham, E. (ed.) FSE 1997. LNCS, vol. 1267, pp. 205–209. Springer, Heidelberg (1997)

A   A 7-Round Differential of GOST with Unknown S-Boxes

The input difference of the differential is of the form $P' = 00\ 01\ 00\ 00\ 00\ 00\ 00\ 00_x$. The zero difference enters the first round and has a zero output difference. In the second round a difference of $00\ 01\ 00\ 00_x$ enters the round function. As there is an addition, with probability 15/16 the differences in the carry of the addition operation (if there are such) do not affect other S-boxes. Thus, there is a non-zero input difference to one of the S-boxes of round 2, while all the rest have a zero input difference. The differential evolves, and after seven rounds there are four bits whose difference is known to be zero.

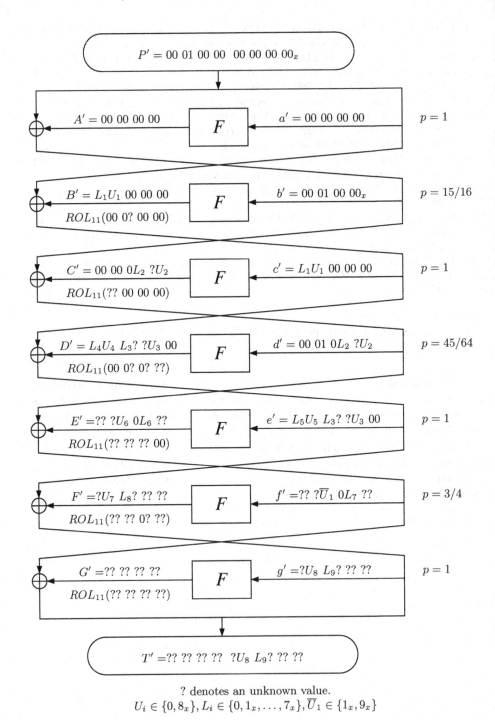

Fig. 1. A 7-Round Differential of GOST with Probability $\frac{15}{16} \cdot \frac{45}{64} \cdot \frac{3}{4} = 0.494$

A New Class of Weak Keys for Blowfish

Orhun Kara and Cevat Manap

TÜBİTAK UEKAE, Gebze, Kocaeli, Turkey
{orhun,cmanap}@uekae.tubitak.gov.tr

Abstract. The reflection attack is a recently discovered self similarity analysis which is usually mounted on ciphers with many fixed points. In this paper, we describe two reflection attacks on r-round Blowfish which is a fast, software oriented encryption algorithm with a variable key length k. The attacks work successfully on approximately $2^{k+32-16r}$ number of keys which we call *reflectively weak keys*. We give an almost precise characterization of these keys. One interesting result is that 2^{34} known plaintexts are enough to determine if the unknown key is a reflectively weak key, for any key length and any number of rounds. Once a reflectively weak key is identified, a large amount of subkey information is revealed with no cost. Then, we recover the key in roughly $r \cdot 2^{16r+22}$ steps. Furthermore, it is possible to improve the attack for some key lengths by using memory to store all reflectively weak keys in a table in advance. The pre-computation phase costs roughly $r \cdot 2^{k-11}$ steps. Then the unknown key can be recovered in $2^{(k+32-16r)/64}$ steps. As an independent result, we improve Vaudenay's analysis on Blowfish for reflectively weak keys. Moreover, we propose a new success criterion for an attack working on some subset of the key space when the key generator is random.

Keywords: Blowfish, cryptanalysis, reflection attack, fixed point, key dependent S-Box, self similarity analysis, weak key.

1   Introduction

Self similarity attacks, such as slide attack [4,5], related key attack [2], and a very recently discovered attack, reflection attack [10], generally work on ciphers with very simple key schedules. In this paper, we propose reflection attacks on full-round Blowfish which has a very complicated key schedule.

Blowfish is a widely used, unpatented, license-free, fast block cipher designed by Schneier in 1994 [16]. Blowfish is a 16-round Feistel network and uses a large number of subkeys. Security of the algorithm is particularly based on the key dependent S-boxes and the difficulty of recovering the key from a partial knowledge of some subkeys. No attacks have been published on a full version of Blowfish so far. Nevertheless, there have been a few studies on cryptanalysis of Blowfish. The analysis by Rijmen [14] is a second order differential attack against 4-round Blowfish. Another differential cryptanalysis is by Vaudenay [18] and uses $3 \cdot 2^{51}$ chosen plaintexts with the assumption that the round function

A. Biryukov (Ed.): FSE 2007, LNCS 4593, pp. 167–180, 2007.

F is known and weak in the sense that some of its S-boxes are not one to one. The number of the keys producing the weak F functions is approximately 2^{k-15}. The slide attack by Biryukov and Wagner [4] uses only 2^{27} chosen plaintexts and works under the powerful assumption that all the P subkeys are equal to zero which happens with a probability of roughly 2^{-576}.

1.1 Our Contributions and Organization of the Paper

The notion of fixed points of weak DES keys is well known[9,6,12,13][1]. These works focus on algebraic properties of DES permutations and their short cycles. In this paper, we also exploit permutations with many fixed points. However, we aim to identify weak keys and recover these keys in Blowfish.

We give two new models of description of Blowfish and deduce some reflection properties of Blowfish by utilizing these models. In particular, we show that certain keys produce $(r-2)$-round Blowfish encryption function with many fixed points. We call these keys as *reflectively weak keys*. The number of reflectively weak keys is approximately $2^{k+32-16r}$. We propose two reflection attacks on Blowfish with variable number of rounds. These attacks work on reflectively weak keys.

We identify a reflectively weak key using roughly 2^{34} known plaintexts for any number of rounds and any key length. Moreover, we characterize a reflectively weak key by certain equalities among its subkeys. The characterization is not precise, but it is true with a probability almost 1. Theorem 1 states the characterization in detail. The first attack is a guess-and-determine type attack utilizing this characterization. First, we determine whether the unknown key is a reflectively weak key. Once a reflectively weak key is identified, we obtain a large amount of subkey information with no cost. This information leads to a guess-and-determine attack. The time complexity of the attack is roughly $r \cdot 2^{16r+22}$ steps where each step is equal to one step of exhaustive search. In the second attack, we improve the time complexity of the first attack by using memory for some key lengths. We detect all reflectively weak keys and save them in a table in advance by checking all keys. The check mechanism deduced from the characterization of reflectively weak keys reduces the workload since we do not have to implement the whole key schedule, even though we check all the keys. The pre-computation phase costs roughly $r \cdot 2^{k-11}$ steps and we recover the key in $2^{(k+32-16r)/64}$ steps using $2^{k+32-16r}$ memory. Note that this is an improvement of the first attack when $k < 16r + 32$. Some interesting examples are given in Table 1 for $r = 8$ and $r = 16$.

Another result is a straightforward improvement of Vaudenay's attack. We reduce the number of chosen plaintexts from $3 \cdot 2^{51}$ to $3 \cdot 2^{44}$ on a set of keys of size roughly 2^{k-271}.

In addition, we propose a new success criterion for an attack working on some subset of the key space. We argue that such an attack is successful if the workload

---

[1] We would like to thank the anonymous referees for pointing these references.

Table 1. Some Complexity Examples of The Attack. w is the average number of weak keys; PC is Pre-computation steps; M is Memory Space Used; T is Time Steps. Each step is equal to one step in exhaustive search. Data complexity is roughly 2^{34} known plaintexts. Complexities without memory are the examples of the first attack.

r	k	w	PC	M	T
	128	2^{32}	$2^{120.6}$	2^{32}	-
	160	2^{64}	$2^{152.6}$	2^{64}	1
8	192	2^{96}	$2^{184.6}$	2^{96}	2^{32}
	192	2^{96}	-	-	$2^{153.3}$
	256	2^{32}	$2^{249.3}$	2^{32}	-
	288	2^{64}	$2^{281.3}$	2^{64}	1
16	320	2^{96}	$2^{313.3}$	2^{96}	2^{32}
	384	2^{160}	-	-	$2^{282.1}$
	448	2^{224}	-	-	$2^{282.1}$

of determining that the unknown key is in the subset, is less than the number of keys in the subset, and recovering it is less than that of exhaustive search.

The paper is organized as follows. Blowfish is described briefly in Section 2. Moreover, we give two new descriptions of the Blowfish algorithm. We state reflection properties of Blowfish in Section 3 as a preparation phase for the statements of the attacks. The notion of reflectively weak key and its characterization is introduced in this section. Then, we give the details of the attack and its improvement in Section 4. The improvement of Vaudenay's analysis for a subset of keys is given in Section 5. In the last section, we discuss the similarity degrees of the functions producing subkeys and the parameters of Blowfish, and give the argument about the success criterion for attacks working on some keys.

1.2 Notation

We use the following notation throughout the paper: k is the bit length of the key; r is the number of rounds; F is the key dependent round function of Blowfish; P is the array of 32 bit subkeys of Blowfish and P_i is its i-th component for $i = 1, .., r + 2$; I is the array of hexadecimal digits of π XORed with the key bits; and $\oplus$ is the XOR operator.

2 High Level Descriptions of Blowfish

Blowfish is a 16-round Feistel network with 64 bit block length and a variable key length of at most 448 bits. It is also specified as an 8-round cipher with a key length of at most 192 bits [17]. Blowfish uses a large number of subkeys. The set of subkeys consists of two parts: The P array which contains $r + 2$ number of 32-bit subkeys, $P_1, ..., P_{r+2}$, and four 8×32 key dependent S-boxes used in the F function. The encryption process starts after the P array and the S-boxes are generated. Let (x_1, y_1) be a plaintext for $x_1, y_1 \in GF(2)^{32}$. Then, i-th round of the encryption process is given as $(x_{i+1}, y_{i+1}) = (F(P_i \oplus x_i) \oplus y_i, P_i \oplus x_i)$. The corresponding ciphertext is defined as $(y_{r+1} \oplus P_{r+2}, x_{r+1} \oplus P_{r+1})$. We do not give the details of F function since we do not use it in the analyses. The high level description of Blowfish is depicted as Type I in Figure 1.

Generating the P array and the S-boxes is as follows:

- Initialize the I array and the S-boxes with hexadecimal digits of π.
- XOR I_1 with the first 32 bits of the key, I_2 with the second 32 bits of the key and so on for all bits of the key. Repeatedly cycle through the key bits until the entire I array has been XORed with key bits. Then copy the contents of the I array to the P array.
- Encrypt the all-zero string with the Blowfish algorithm, using the P array as subkeys. Replace P_1 and P_2 with the output.
- Repeat the last process, replacing all entries of P array, four S-boxes in order, with the output of continuously changing Blowfish algorithm.

2.1   New Models for Description

The XOR operator is commutative. Hence, an XOR operator of a subkey in Blowfish can be pushed through other XOR operators until a non-commutative operation such as an F operation is obtained. So, by moving certain subkeys we can obtain various descriptions of Blowfish. Two descriptions are depicted in Figure 1. We call them *the Type II description of Blowfish* and *the Type III description of Blowfish*. For example, in the Type II description we move the third round key through the second round to the first round and the fourth round key through the third round to the second round. So we consider the third round key as a part of the first round and the fourth round key as a part of the second round.

Repeating this process we obtain a new description of Blowfish where half of the rounds use two subkeys and the other half of the rounds use no subkey. The type III description can be obtained similarly. These descriptions facilitate the attack idea. Particularly, we treat two-round Blowfish as keyed or unkeyed:

Definition 1. *Two-rounds of Blowfish given as*

$$x' = F(P_{i_1} \oplus x) \oplus y \oplus P_{i_3} \oplus P_{i_4} \text{ and}$$
$$y' = F(F(P_{i_1} \oplus x) \oplus y \oplus P_{i_3}) \oplus P_{i_1} \oplus P_{i_2} \oplus x$$

is called a two-round keyed Blowfish function ($\mathcal{K}_2$ in short) and

$$x' = F(x) \oplus y \text{ and}$$
$$y' = F(F(x) \oplus y) \oplus x$$

is called a two-round unkeyed Blowfish function ($\mathcal{U}_2$ in short). Here (x, y) is an input and (x', y') is the corresponding output.

Two-round keyed/unkeyed Blowfish functions are depicted in Figure 2.

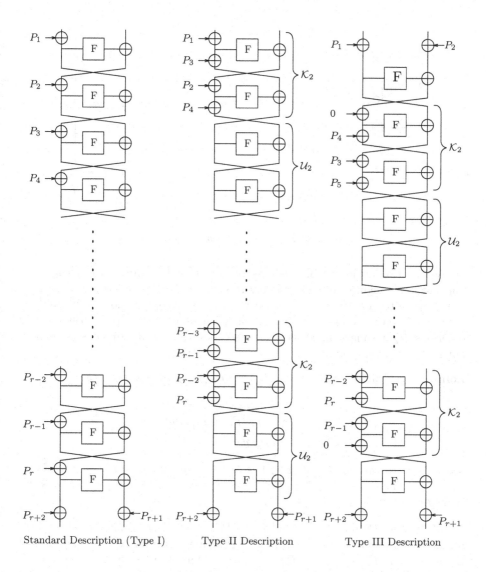

Fig. 1. Three different descriptions of r-round Blowfish

3   Reflection Properties of Blowfish

One of the very recent cryptanalysis methods is the reflection attack [10]. This attack exploits the similarities between the round functions of the encryption and the round functions of the decryption. It can be very powerful especially against product ciphers using involutions, such as Feistel networks. In general, self-similarity attacks are mounted on ciphers with simple key schedules. We

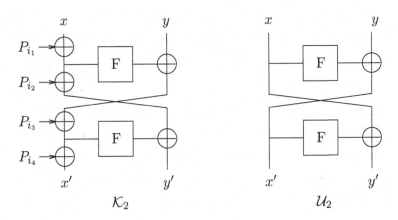

Fig. 2. General Description of $\mathcal{K}_2$ and $\mathcal{U}_2$

apply reflection attack on Blowfish, as an exceptional example with a very complicated key schedule, and successfully recover the key in some special cases.

Let us note that Blowfish can be written as a composition of $\mathcal{K}_2$ and $\mathcal{U}_2$ functions (see Figure 1). The reflection attack exploits certain properties of these functions. $\mathcal{U}_2$ has many fixed points and so does $\mathcal{K}_2$ for some subkeys. Moreover, any $\mathcal{U}_2$ is an involution and $\mathcal{K}_2^{-1}$ has the same structure as $\mathcal{K}_2$.

Lemma 1. *Consider a two-round keyed Blowfish function given as*

$$x' = F(P_{i_1} \oplus x) \oplus y \oplus P_{i_3} \oplus P_{i_4} \text{ and}$$
$$y' = F(F(P_{i_1} \oplus x) \oplus y \oplus P_{i_3}) \oplus P_{i_1} \oplus P_{i_2} \oplus x$$

where (x, y) is the input and (x', y') is the corresponding output. If the subkeys $P_{i_1} = P_{i_4}$ and $P_{i_2} = P_{i_3}$, then the two-round keyed Blowfish function has 2^{32} fixed points.

Proof. Assume that $P_{i_1} = P_{i_4}$ and $P_{i_2} = P_{i_3}$. Then, (x, y) is encrypted to (x, y) if and only if $y = F(P_{i_1} \oplus x) \oplus x \oplus P_{i_1} \oplus P_{i_2}$. However, we have 2^{32} plaintexts of the form $(x, F(P_{i_1} \oplus x) \oplus x \oplus P_{i_1} \oplus P_{i_2})$. Therefore, the $\mathcal{K}_2$ has 2^{32} fixed points.
□

A straightforward corollary of Lemma 1 is obtained when two-round Blowfish is unkeyed, i.e., $P_1 = P_2 = P_3 = P_4 = 0$. Hence, any two-round unkeyed Blowfish function ($\mathcal{U}_2$) has 2^{32} fixed points of the form $(x, F(x) \oplus x)$.

Definition 2. *A $(4r' - 2)$-round function $\mathcal{K}_2 S \mathcal{U}_2 S \mathcal{K}_2 S \mathcal{U}_2 S \cdots S \mathcal{U}_2 S \mathcal{K}_2$ is called a $(4r' - 2)$-round Blowfish function, where S is the swap operation of Feistel network.*

Note that, we have r' number of $\mathcal{K}_2$ and $r' - 1$ number of $\mathcal{U}_2$ operations in a $(4r' - 2)$-round Blowfish function.

We show that when there are certain relations among the round subkeys used in a $(4r' - 2)$-round Blowfish function, then $(4r' - 2)$-round Blowfish function has many fixed points.

Proposition 1. *Let $\mathcal{B}$ be a $(4r' - 2)$-round Blowfish function. Assume that the i-th $\mathcal{K}_2$ of $\mathcal{B}$ is the inverse of the $(r' - i + 1)$-th $\mathcal{K}_2$ of $\mathcal{B}$ for $i = 1, ..., \lfloor \frac{r'}{2} \rfloor$ where $\lfloor \frac{r'}{2} \rfloor$ is the integer part of $\frac{r'}{2}$. If r' is odd, then assume also that the $(\frac{r'+1}{2})$-th $\mathcal{K}_2$ has 2^{32} fixed points. Then the function $\mathcal{B}$ has 2^{32} fixed points.*

Proof. Let x be an input of $\mathcal{B}$. The intermediate two-round in the center of encryption is $(\frac{r'+1}{2})$-th $\mathcal{K}_2$ if r' is odd and $(\frac{r'}{2})$-th $\mathcal{U}_2$ if r' is even. $\mathcal{U}_2$ has 2^{32} fixed points by Lemma 1 and the $(\frac{r'+1}{2})$-th $\mathcal{K}_2$ also has 2^{32} fixed points by the assumption. Moreover, the i-th $\mathcal{K}_2$ is the inverse of $(r' - i + 1)$-th $\mathcal{K}_2$ by the assumption. Then, the input of the i-th $\mathcal{K}_2$ is equal to the output of the $(r' - i + 1)$-th $\mathcal{K}_2$ for any $i = 1, ..., \lfloor \frac{r'}{2} \rfloor$ corresponding to x means x is a fixed of $\mathcal{B}$. Hence, any fixed point of the central round produces a fixed point of $\mathcal{B}$ and any fixed point of $\mathcal{B}$ gives a fixed point of the central round. Therefore, $\mathcal{B}$ has 2^{32} fixed points.

□

Observe that, the i-th $\mathcal{K}_2$ is the inverse of the $(r' - i + 1)$-th $\mathcal{K}_2$ if and only if their subkeys are equal in twist order. That is, if $P_{i_1}, P_{i_2}, P_{i_3}, P_{i_4}$ are the subkeys of i-th $\mathcal{K}_2$ and $P'_{i_1}, P'_{i_2}, P'_{i_3}, P'_{i_4}$ are the subkeys of $(r' - i + 1)$-th $\mathcal{K}_2$ then we have $P_{i_1} = P'_{i_4}$, $P_{i_2} = P'_{i_3}$, $P_{i_3} = P'_{i_2}$ and $P_{i_4} = P'_{i_1}$. On the other hand, a sufficient condition for having 2^{32} fixed points of $(\frac{r'+1}{2})$-th $\mathcal{K}_2$ is given by Lemma 1.

Definition 3. *A key is called a reflectively weak key with respect to a $(4r' - 2)$-round Blowfish function $\mathcal{B}$ if $\mathcal{B}$ has 2^{32} fixed points.*

Assumptions of Proposition 1 and Lemma 1 are satisfied when certain equalities among subkeys hold. In this case, we have many fixed points. In the following theorem, we prove that this is the only case resulting in many fixed points, by assuming that Blowfish is a random permutation when the assumptions do not hold.

Theorem 1. *Assume that a given key is reflectively weak with respect to an $(r - 2)$-round Blowfish function $\mathcal{B}$ and also assume that $\mathcal{B}$ is a random permutation when the assumptions of Proposition 1 and Lemma 1 do not hold. Then, the subkeys of $\mathcal{B}$ satisfy the assumptions of Proposition 1 and Lemma 1 with a probability approximately*

$$1 - \frac{2^{16 \cdot r} - 1}{2^{32}! \cdot e + 2^{16 \cdot r} - 1}$$

where e is the Euler constant.

Proof. Let $X_{\mathcal{A}}$ be the event that $\mathcal{B}$ satisfies the assumptions of Proposition 1 and Lemma 1 and $X_{\mathcal{F}}$ be the event that $\mathcal{B}$ has 2^{32} fixed points. Assumptions of

Proposition 1 and Lemma 1 imply that we have $\frac{r}{2}$ equalities between r subkeys of $\mathcal{B}$. Each equality holds with a probability approximately 2^{-32}. Hence, $\Pr(X_{\mathcal{A}}) = 2^{-16 \cdot r}$. We have $\Pr(X_{\mathcal{F}} \mid X_{\mathcal{A}}) = 1$ by Proposition 1. *The rencontres numbers,* i.e., the number $D(n, m)$ of permutations of n containing m fixed points, is given as

$$D(n, m) = \frac{n!}{m!} \sum_{k=0}^{n-m} \frac{(-1)^k}{k!}$$

which immediately implies that $\frac{D(n,m)}{n!} \approx \frac{e^{-1}}{m!}$ for large n (see [15] for details). Since $\mathcal{B}$ is a random permutation when the assumptions of Proposition 1 and Lemma 1 do not hold, we have $\Pr(X_{\mathcal{F}} \mid \overline{X_{\mathcal{A}}}) = \frac{e^{-1}}{2^{32}!}$. Then,

$$\Pr(X_{\mathcal{F}}) = \Pr(X_{\mathcal{A}}) \Pr(X_{\mathcal{F}} \mid X_{\mathcal{A}}) + \Pr(\overline{X_{\mathcal{A}}}) \Pr(X_{\mathcal{F}} \mid \overline{X_{\mathcal{A}}})$$
$$= 2^{-16 \cdot r} + \frac{e^{-1}}{2^{32}!}(1 - 2^{-16 \cdot r}).$$

Now, applying Bayes Rule we get,

$$\Pr(X_{\mathcal{A}} \mid X_{\mathcal{F}}) = \frac{\Pr(X_{\mathcal{F}} \mid X_{\mathcal{A}}) \cdot \Pr(X_{\mathcal{A}})}{\Pr(X_{\mathcal{F}})} = 1 - \frac{2^{16r} - 1}{2^{32}! \cdot e + 2^{16r} - 1}$$

which is the probability that the assumptions of Proposition 1 and Lemma 1 hold, given that $\mathcal{B}$ has 2^{32} fixed points.   □

Remark 1. Using Stirling's approximation, we have $2^{32}! \approx (2^{32})^{2^{32}} = 2^{2^{37}}$ and the probability given in Theorem 1 is almost 1 (roughly $1 - 2^{-2^{37}}$). Therefore, the converses of both Proposition 1 and Lemma 1 are true with a probability almost 1. This probability affects the success rate of the attacks. However, Theorem 1 implies that the false alarm probability is negligible. Hence, we can assume that we have $\frac{r}{2}$ equalities between r subkeys of $\mathcal{B}$ if $\mathcal{B}$ has 2^{32} fixed points.

Example 1. Let $r = 16$ and let $\mathcal{B}$ be the 14-round Blowfish function obtained by removing the first and the last rounds in Type III description of Blowfish. Then, loosely speaking, a key is a reflectively weak key with respect to $\mathcal{B}$ means that the following 7 equalities are satisfied: $P_4 = P_{15}$, $P_3 = P_{16}$, $P_5 = P_{14}$, $P_6 = P_{13}$, $P_7 = P_{12}$, $P_8 = P_{11}$ and $P_9 = P_{10}$. The eighth equality already holds (0=0) by the definition of Type III description of Blowfish.

4   Two Reflection Attacks

In this section, we introduce an attack and its improvement for some key lengths. The improvement has two phases: Reflectively weak keys are collected in the precomputation phase, and the reflectively weak key is recovered during the on-line phase.

4.1   First Attack

The attack consists of two parts. In the first part, we identify if the key is reflectively weak. Subkeys of a reflectively weak key satisfies certain equalities and we utilize these equalities to recover the key in the second part.

We use several known plaintext-ciphertext pairs to identify a reflectively weak key . Let (x, y) denote plaintext and (x', y') denote the corresponding ciphertext. Assume that a reflectively weak key is used with respect to the $(r - 2)$-round Blowfish function $\mathcal{B}$, obtained by removing the first and the last rounds in Type III description of Blowfish. Then, $\mathcal{B}$ will have 2^{32} fixed points by Proposition 1. Observe that, if $\mathcal{B}$ has a fixed point for (x, y), then we have $P_1 \oplus P_{r+2} = x \oplus x'$ and $P_2 \oplus P_{r+1} = y \oplus y'$. Hence, we expect the value $(P_1 \oplus P_{r+2}, P_2 \oplus P_{r+1})$ to occur with probability 2^{-32} and other values to occur with probability 2^{-64} when a reflectively weak key is used. On the other hand, we expect that each vector, $(x \oplus x', y \oplus y')$, occurs with probability 2^{-64} if the key is not a reflectively weak key. Consequently, we identify a reflectively weak key and obtain $(P_1 \oplus P_{r+2}, P_2 \oplus P_{r+1})$ with 2^{34} known plaintexts.

Once we identify a reflectively weak key, we can recover information on $\frac{r}{2} + 1$ subkeys of P array since we have $\frac{r}{2} + 1$ equalities between $r + 2$ subkeys. $\frac{r}{2} - 1$ of the equalities are deduced by Theorem 1 and two equalities are obtained while identifying the reflectively weak key. By guessing $\frac{r}{2} + 1$ subkeys we can determine remaining $\frac{r}{2} + 1$ subkeys and obtain the whole P array. One can recover the I array and the key by reversing the key schedule by the following Lemma.

Lemma 2. *Assume that the P array of a key is known. Then it is possible to recover the key by $\frac{r}{2} + 1$ encryptions.*

Proof. For any $i = 1, ..., \frac{r}{2} + 1$, (P_{2i-1}, P_{2i}) is the encryption of the all-zero string with the Blowfish algorithm with the subkeys

$$(P_1, P_2, \cdots, P_{2i-3}, P_{2i-2}, I_{2i-1}, I_{2i}, \cdots, I_{r+1}, I_{r+2})$$

and a publicly known F function.

We encrypt the all-zero string up to r rounds with subkeys $(P_1, P_2, \cdots, P_{r-1}, P_r)$ and obtain $(P_{r+1} \oplus I_{r+2}, P_{r+2} \oplus I_{r+1})$. Then, we can easily recover the subkeys (I_{r+1}, I_{r+2}). We need to recover (I_{r-1}, I_r) by using the P array and (I_{r+1}, I_{r+2}), and recover (I_{r-3}, I_{r-2}) by using the P array and $(I_{r-1}, I_r, I_{r+1}, I_{r+2})$. We repeat this process until the whole I array is obtained. So, the problem of reversing the key schedule and recovering the I array from the P array is reduced to the problem of recovering the value of (I_{2i-1}, I_{2i}) using $(P_1, P_2, \cdots, P_{2i-3}, P_{2i-2}, I_{2i+1}, I_{2i+2}, \cdots, I_{r+1}, I_{r+2})$ and (P_{2i-1}, P_{2i}). Observe that this problem is recovering the subkeys for two-round Blowfish given an input-output pair. The second subkey can be moved up to the first round and both subkeys can be considered as an initial whitening. Hence we can decrypt the output for two rounds, XOR the result with the input and obtain the subkeys (I_{2i-1}, I_{2i}). This process costs only one Blowfish encryption. Therefore we recover the whole I array with a cost of $\frac{r}{2} + 1$ encryptions. Then the key is extracted from I with no cost.   □

Observe that repetition of 32 bit key words in the construction of the I array allows one to check whether the obtained I array is a valid array. Although several candidates may turn out to give valid I arrays, complexity of exhaustively searching these candidates is dominated by the complexity of the attack.

We guess $16r + 32$ bits in total and each guess is checked in $\frac{r}{2} + 1$ encryptions by Lemma 2. On the other hand, each step of exhaustive search costs $\frac{r}{2} + 514$ encryptions. Therefore, the time complexity is $\frac{2^{16r+32} \cdot (r+2)}{r+1028}$ exhaustive search steps. Hence we have the following theorem.

Theorem 2. *Let $\mathcal{B}$ be the $(r-2)$-round Blowfish function obtained by removing the first and the last rounds in the type III description of Blowfish. Then, one may determine if an unknown key is a reflectively weak key with respect to $\mathcal{B}$ by using approximately 2^{34} known plaintexts and then recover the key in $\frac{2^{16r+32}(r+2)}{r+1028}$ steps if it is reflectively weak, where each step is a key loading time plus one encryption.*

Let us note that the time complexity of the attack is independent of the key length. On the other hand, a key is a reflectively weak key with respect to the $(r-2)$-round Blowfish function $\mathcal{B}$ in Type III description of Blowfish if $\frac{r}{2} - 1$ equalities are satisfied among $r - 2$ subkeys of $\mathcal{B}$ by Theorem 1. Therefore, assuming the Blowfish key schedule is random, the probability that a key is a reflectively weak key is 2^{32-16r}.

4.2   Second Attack

For some key lengths, the previous attack can be improved using memory. We search all the keys and collect reflectively weak keys in a table with their subkeys $(P_1 \oplus P_{r+2}, P_2 \oplus P_{r+1})$, sorted with respect to $(P_1 \oplus P_{r+2}, P_2 \oplus P_{r+1})$. A key is loaded to the key schedule and the P array is generated. Then, we check whether the subkeys $P_2, ..., P_r$ satisfy the equalities so as to satisfy the assumptions of Proposition 1. Note that the first check is the equality $P_{r/2+1} = P_{r/2+2}$ and most of the keys are eliminated in this step where it is enough to produce P arrays up to $P_{r/2+2}$ which costs $\frac{r}{4} + 1$ encryptions. We need to produce P subkeys up to $P_{r/2+4}$ for only one in 2^{32} keys, up to $P_{r/2+6}$ for only one in 2^{64} keys and so on. Hence, the total time complexity is given as

$$\frac{2^{k-1}(r+4)}{r+1028} + \frac{2^{k-33}(r+8)}{r+1028} + \frac{2^{k-65}(r+12)}{r+1028} + \cdots \approx \frac{2^{k-1}(r+4)}{r+1028}$$

which seems to cost almost that of exhaustive search. However, it is done once and faster than exhaustive search.

The table of weak keys occupies $2^{k+32-16r}$ spaces in memory. The attack is now straightforward. First, we determine if the unknown key is reflectively weak with 2^{34} known plaintexts as in the first attack. If the key is a reflectively weak key, then we obtain $(P_1 \oplus P_{r+2}, P_2 \oplus P_{r+1})$. By searching the table sorted with respect to $(P_1 \oplus P_{r+2}, P_2 \oplus P_{r+1})$, we get approximately $2^{(k+32-16r)/64}$ candidates. The correct key can be recovered by searching these candidates. Therefore, the time complexity will be $2^{(k+32-16r)/64}$ steps.

5   Improvement of Vaudenay's Cryptanalysis on a Subset of Keys

In [18], Vaudenay proposes a differential attack on 16-round Blowfish with $3 \cdot 2^{51}$ chosen plaintexts with the assumption that the F function is known and weak in the sense that some of the S-boxes are not one to one. The attack works for approximately 2^{k-15} keys. The number of chosen plaintexts required for the attack can be generalized as $2^{2+7 \cdot \lceil \frac{r-2}{2} \rceil}$ for $r \leq 10$ and $3 \cdot 2^{2+7 \cdot \lceil \frac{r-2}{2} \rceil}$ for $r \geq 11$.

We improve Vaudenay's attack on 16-round Blowfish in a certain subset of Vaudenay's weak key class by reducing the amount of chosen plaintext required for the attack. The improved version has two steps: In the first step, we recover both whitening keys P_{17} and P_{18}, by a reflection attack to reduce the algorithm to 14 rounds. In the second step, we apply Vaudenay's differential attack to the 14-round algorithm.

Assume that F is known and weak. Assume also that a reflectively weak key with respect to the first 14 rounds of the Type II description of Blowfish is used. The latter assumption can be checked by collecting roughly 2^{34} vectors $(F(x) \oplus y \oplus x', F(F(x) \oplus y) \oplus x \oplus y')$ where (x, y) is a plaintext and (x', y') is the corresponding ciphertext. If a plaintext (x, y) is a fixed point of the first 14 rounds, then

$$(P_{18}, P_{17}) = (F(x) \oplus y \oplus x', F(F(x) \oplus y) \oplus x \oplus y').$$

Hence, we expect one of the vectors to occur approximately four times in the collection of approximately 2^{34} plaintexts, encrypted by a reflectively weak key. This vector is (P_{18}, P_{17}) since the probability that (P_{18}, P_{17}) occurs is slightly more than 2^{-32} whereas the probability that any arbitrary vector occurs is approximately 2^{-64}. Therefore, if a reflectively weak key with respect to the first 14 rounds of the Type II description is used, then we can identify it and recover the whitening keys (P_{18}, P_{17}) by using approximately 2^{34} known plaintexts. Then, we peel the last two rounds, obtaining 14-round Blowfish. Applying the differential attack in [18] to 14 round Blowfish requires $3 \cdot 2^{44}$ chosen plaintexts. Therefore, we reduce the number of chosen plaintexts from $3 \cdot 2^{51}$ to $3 \cdot 2^{44}$.

The attack works if the key is weak both in terms of Vaudenay's attack and reflectively. F is weak for 2^{-15} of key space. On the other hand, by Proposition 1 and Theorem 1, if a key is reflectively weak with respect to the first 14 rounds of the Type II description of Blowfish, then certain eight equalities between subkeys in these 14 rounds hold. Thus, the probability that a reflectively weak key is used is approximately 2^{-256} since each equation holds with probability 2^{-32}. Therefore, the attack works for a subset of key space of size 2^{k-271}.

6   Discussion of Attacks

A new definition of similarity degree between two functions is given in [10]. The definition is as follows:

Definition 4. *Let $F_1, F_2 : GF(2)^n \to GF(2)^m$ be two functions. Then, F_1 and F_2 are called similar of degree (d_1, d_2) with probability p if the number of $(x, x') \in GF(2)^n \times GF(2)^n$ satisfying*

$$HW(x \oplus x') \le n - d_1 \Rightarrow HW(F_1(x) \oplus F_2(x')) \le m - d_2$$

is $p \cdot 2^n \cdot \sum_{i=0}^{n-d_1} \binom{n}{i}$ where $HW()$ is the Hamming Weight of binary vectors.

It is argued in [10] that round functions should not be similar of large degrees with large probabilities. In Blowfish the functions producing round keys are given as

$$\phi_i : GF(2)^k \longrightarrow GF(2)^{32800}, \phi_i(K) = (P_i, F), \text{ for } i = 1, ..., r.$$

Hence, any two functions ϕ_i and ϕ_j are similar of degree $(k, 32768)$ with probability 1. In other words, they are similar of the full degree, $(k, 32800)$, with probability 2^{-32}. That is, the functions producing round keys are highly similar even though they are one way functions by themselves. We exploited this property to mount a reflection attack on Blowfish. This example indicates that ciphers having round functions which are similar of high degree with high probability, may be vulnerable to self similarity attacks even though they have very complicated key schedules. Moreover, we argue that the attacks presented here can be improved further by taking different degrees of similarity or by decreasing the number of pairs of the functions compared.

Another property that we exploited is the relatively short block length of Blowfish. Its key length can be as long as 448 bits whereas its block length is always 64 bits. We propose that block length of a block cipher should not be smaller than key length. This is also necessary to provide resistance to tradeoff attacks [8,1,7] when the cipher is operated in a stream mode.

By Lemma 2, it is easy to take inverse of the key schedule of Blowfish and deduce the key if one knows the P array . For example, if the P array were updated once more after the F function were constructed in the key schedule, then we could not recover the key from the P array. For 16-round Blowfish, this change makes the key schedule only 1.7% slower than the original key schedule.

Unlike typical Feistel structure, P_i's are XORed outside F. Adding the subkeys to left half of the Feistel network seems to hinder the self similarity attacks since it destroys the typical symmetry of Feistel network. However, we reconstructed the symmetry by describing Blowfish in different manners (see Figure 1). Moreover, one can push some of the subkey XORs to the whitening which allows to increase the number of weak keys.

A recent phenomenon is how to evaluate an attack mounted on some so called weak keys. Most conventional attacks such as differential cryptanalysis [3] or linear cryptanalysis [11] work generally on any key. On the other hand, the attacks working only on a subset of the keys form a new class. These attacks have two important parameters: The number of weak keys and the workload to identify that the unknown key is weak. For a given attack, let W be the workload of identifying a weak key and w be the number weak keys. Given a set of $\frac{2^k}{w}$ randomly generated keys, we expect one weak key on the average. To identify the weak key, we run the identification process on all $\frac{2^k}{w}$ keys with a cost of $W\frac{2^k}{w}$. So, a necessary condition for the success of the attack is $W\frac{2^k}{w} < 2^k$, i.e., $W < w$. This leads to the following criteria: Assuming that the keys are produced randomly, we propose that an attack on weak keys should be considered successful, if the workload of identification of a weak key is less than the number of weak keys, in addition to the widely adopted phenomenon that the workload of key recovery is less than that of exhaustive search.

Acknowledgments

We thank Hüseyin Demirci, Nezih Geçkinli, Atilla Hasekioğlu and Ali Aydın Selçuk for their comments.

References

1. Babbage, S.: Improved Exhaustive Search Attacks on Stream Ciphers. European Convention on Security and Detection, IEE Conference publication No. 408, pp. 161-166, IEE (1995)
2. Biham, E.: New Types of Cryptanalytic Attacks Using Related Keys. J. of Cryptology 7, 229–246 (1994)
3. Biham, E., Shamir, A.: Differential Cryptanalysis of Data Encryption Standard. Springer, Heidelberg (1993)
4. Biryukov, A., Wagner, D.: Slide Attacks. In: Knudsen, L.R. (ed.) FSE 1999. LNCS, vol. 1636, pp. 245–259. Springer, Heidelberg (1999)
5. Biryukov, A., Wagner, D.: Advanced Slide Attacks. In: Preneel, B. (ed.) EURO-CRYPT 2000. LNCS, vol. 1807, pp. 589–606. Springer, Heidelberg (2000)
6. Coppersmith, D.: The Real Reason for Rivest's Phenomenon. In: Williams, H.C. (ed.) CRYPTO 1985. LNCS, vol. 218, pp. 535. Springer, Heidelberg (1986)
7. Golić, J.: Cryptanalysis of Alleged A5 Stream Cipher, In: Fumy, W. (ed.) EURO-CRYPT 1997. LNCS, vol. 1233, pp. 239–255. Springer, Heidelberg (1997)
8. Hong, J., Sarkar, P.: Rediscovery of the Time Memory Tradeoff, Cryptology ePrint Archive, Report 2005/090 (2005)
9. Kaliski, B.S., Rivest, R.L., Sherman, A.T.: Is DES a pure Cipher(Results of More Cycling Experiments on DES). In: Williams, H.C. (ed.) CRYPTO 1985. LNCS, vol. 218, p. 212. Springer, Heidelberg (1986)
10. Kara, O.: Reflection Attacks on Product Ciphers, Cryptology ePrint Archive, Report 2007/043 (2007)
11. Matsui, M.: Linear Cryptanalysis Method of DES Cipher. In: Helleseth, T. (ed.) EUROCRYPT 1993. LNCS, vol. 765, pp. 386–397. Springer, Heidelberg (1994)
12. Moore, J.H., Simmons, G.J.: Cycle Structures of the DES with Weak and Semi-Weak Keys, In: Odlyzko, A.M. (ed.) CRYPTO 1986. LNCS, vol. 263, pp. 9–32. Springer, Heidelberg (1987)
13. Moore, J.H., Simmons, G.J.: Cycle Structure of the DES for Keys Having Palindromic (or Antipalindromic) Sequences of Round Keys. In: IEEE Transactions on Software Engineering, vol. SE-13, pp. 262–273. IEEE Computer Society Press, Los Alamitos (1987)
14. Rijmen, V.: Cryptanalysis and Design of Iterated Block Ciphers, Doctoral Dissertation, K.U. Leuven (1997)

15. Riordan, J.: An Introduction to Combinatorial Analysis. Wiley, New York (1958)
16. Schneier, B.: Description of a New Variable - Length Key, 64 Bit Block Cipher (Blowfish). In: Anderson, R. (ed.) Fast Software Encryption. LNCS, vol. 809, pp. 191–204. Springer, Heidelberg (1994)
17. http://www.schneier.com/blowfish.html
18. Vaudenay, S.: On the Weak Keys of Blowfish. In: Gollmann, D. (ed.) Fast Software Encryption. LNCS, vol. 1039, pp. 27–32. Springer, Heidelberg (1996)

The 128-Bit Blockcipher CLEFIA
(Extended Abstract)

Taizo Shirai[1], Kyoji Shibutani[1], Toru Akishita[1],
Shiho Moriai[1], and Tetsu Iwata[2]

[1] Sony Corporation
1-7-1 Konan, Minato-ku, Tokyo 108-0075, Japan
{taizo.shirai,kyoji.shibutani,toru.akishita,shiho.moriai}@jp.sony.com
[2] Nagoya University
Furo-cho, Chikusa-ku, Nagoya 464-8603, Japan
iwata@cse.nagoya-u.ac.jp

Abstract. We propose a new 128-bit blockcipher CLEFIA supporting key lengths of 128, 192 and 256 bits, which is compatible with AES. CLEFIA achieves enough immunity against known attacks and flexibility for efficient implementation in both hardware and software by adopting several novel and state-of-the-art design techniques. CLEFIA achieves a good performance profile both in hardware and software. In hardware using a 0.09 μm CMOS ASIC library, about 1.60 Gbps with less than 6 Kgates, and in software, about 13 cycles/byte, 1.48 Gbps on 2.4 GHz AMD Athlon 64 is achieved. CLEFIA is a highly efficient blockcipher, especially in hardware.

Keywords: blockcipher, generalized Feistel structure, DSM, CLEFIA.

1 Introduction

A lot of secure and high performance blockciphers have been designed benefited from advancing research which started since the development of DES [11]. For example, IDEA, MISTY1, AES, and Camellia are fruits of such research activities [1, 10, 15, 19]. New design and evaluation techniques are evolved day by day; topics on algebraic immunity and related-key attacks are paid attention recently [8, 6]. Moreover, light-weight ciphers suitable for a very limited resource environment are still active research fields. FOX and HIGHT are examples of such newly developed blockciphers [12, 13].

We think it is good timing to show a new blockcipher design based on current state-of-the-art techniques. In this paper, we propose a new 128-bit blockcipher CLEFIA supporting key lengths of 128, 192 and 256 bits, which are compatible with AES. CLEFIA achieves enough immunity against known cryptanalyses and flexibility for very efficient implementation in hardware and software. The fundamental structure of CLEFIA is a generalized Feistel structure consisting of 4 data lines, in which there are two 32-bit F-functions per one round.

A. Biryukov (Ed.): FSE 2007, LNCS 4593, pp. 181–195, 2007.

One of novel design approaches of CLEFIA is that these F-functions employ the Diffusion Switching Mechanism (DSM) [22, 23]: they use different diffusion matrices, and two different S-boxes are used to obtain stronger immunity against a certain class of attacks. Consequently, the required number of rounds can be reduced. Moreover, the two S-boxes are based on different algebraic structures, which is expected to increase algebraic immunity. Other novel ideas include the secure and compact key scheduling design and the *DoubleSwap* function used in it. The key scheduling part uses a generalized Feistel structure, and it is possible to share it with the data processing part. The *DoubleSwap* function can be compactly implemented to enable efficient round key generation in encryption and decryption.

CLEFIA achieves about 1.60 Gbps with less than 6 Kgates in hardware using a 0.09 μm CMOS ASIC library, and about 13 cycles/byte, 1.48 Gbps on 2.4 GHz AMD Athlon 64 processor in software. We consider CLEFIA is a well-balanced blockcipher in security and performance, and the performance is advantageous among other blockciphers especially in hardware.

This paper is organized as follows: in Sect. 2, notations are first introduced. In Sect. 3, we give the specification of CLEFIA. Then design rationale is shown in Sect. 4. Sect. 5 describes the evaluation results on both of security and performance aspects. Finally Sect. 6 concludes the paper.

2   Notations

This section describes mathematical notations, conventions and symbols used throughout this paper.

0x	: A prefix for a binary string in a hexadecimal form
$a_{(b)}$	: b denotes the bit length of a
$a\|b$ or $(a\|b)$	: Concatenation
$(a,\ b)$ or $(a\ b)$	: Vector style representation of $a\|b$
$a \leftarrow b$	: Updating a value of a by a value of b
${}^t a$	: Transposition of a vector or a matrix a
$a \oplus b$	: Bitwise exclusive-OR. Addition in $GF(2^n)$
$a \cdot b$	: Multiplication in $GF(2^n)$
$\overline{a}$	: Logical negation
$a \lll b$	: b-bit left cyclic shift operation
$\mathbf{w_b}(a)$	: For an $8n$-bit string $a = a_0\|a_1\|\ldots\|a_{n-1}$, $a_i \in \{0,1\}^8$, $\mathbf{w_b}(a)$ denotes the number of non-zero a_is.

3   Specification

This section describes the specification of CLEFIA. We first define a function $GFN_{d,r}$ which is a fundamental structure for CLEFIA, followed by definitions of a data processing part and a key scheduling part.

3.1   Definition of $GFN_{d,r}$

CLEFIA uses a 4-branch and an 8-branch Type-2 generalized Feistel network [24]. We denote d-branch r-round generalized Feistel network employed in CLEFIAas $GFN_{d,r}$. $GFN_{d,r}$ employs two different 32-bit F-functions F_0 and F_1 whose input/output are defined as follows.

$$F_0, F_1 : \begin{cases} \{0,1\}^{32} \times \{0,1\}^{32} \to \{0,1\}^{32} \\ (RK_{(32)}, x_{(32)}) \mapsto y_{(32)} \end{cases}$$

For d 32-bit input X_i and output Y_i $(0 \le i < d)$, and $dr/2$ 32-bit round keys RK_i $(0 \le i < dr/2)$, $GFN_{d,r}$ $(d = 4, 8)$ are defined as follows.

$$GFN_{4,r} : \begin{cases} \{\{0,1\}^{32}\}^{2r} \times \{\{0,1\}^{32}\}^4 \to \{\{0,1\}^{32}\}^4 \\ (RK_{0(32)}, \dots, RK_{2r-1(32)}, X_{0(32)}, \dots, X_{3(32)}) \mapsto Y_{0(32)}, \dots, Y_{3(32)} \end{cases}$$

Step 1. $T_0 \mid T_1 \mid T_2 \mid T_3 \leftarrow X_0 \mid X_1 \mid X_2 \mid X_3$
Step 2. For $i = 0$ to $r - 1$ do the following:
    Step 2.1 $T_1 \leftarrow T_1 \oplus F_0(RK_{2i}, T_0)$, $T_3 \leftarrow T_3 \oplus F_1(RK_{2i+1}, T_2)$
    Step 2.2 $T_0 \mid T_1 \mid T_2 \mid T_3 \leftarrow T_1 \mid T_2 \mid T_3 \mid T_0$
Step 3. $Y_0 \mid Y_1 \mid Y_2 \mid Y_3 \leftarrow T_3 \mid T_0 \mid T_1 \mid T_2$

$$GFN_{8,r} : \begin{cases} \{\{0,1\}^{32}\}^{4r} \times \{\{0,1\}^{32}\}^8 \to \{\{0,1\}^{32}\}^8 \\ (RK_{0(32)}, \dots, RK_{4r-1(32)}, X_{0(32)}, \dots, X_{7(32)}) \mapsto Y_{0(32)}, \dots, Y_{7(32)} \end{cases}$$

Step 1. $T_0 \mid T_1 \mid \ \dots \ \mid T_7 \leftarrow X_0 \mid X_1 \mid \ \dots \ \mid X_7$
Step 2. For $i = 0$ to $r - 1$ do the following:
    Step 2.1 $T_1 \leftarrow T_1 \oplus F_0(RK_{4i}, T_0)$, $T_3 \leftarrow T_3 \oplus F_1(RK_{4i+1}, T_2)$
            $T_5 \leftarrow T_5 \oplus F_0(RK_{4i+2}, T_4)$, $T_7 \leftarrow T_7 \oplus F_1(RK_{4i+3}, T_6)$
    Step 2.2 $T_0 \mid T_1 \mid \ \dots \ \mid T_6 \mid T_7 \leftarrow T_1 \mid T_2 \mid \ \dots \ \mid T_7 \mid T_0$
Step 3. $Y_0 \mid Y_1 \mid \ \dots \ \mid Y_6 \mid Y_7 \leftarrow T_7 \mid T_0 \mid \ \dots \ \mid T_5 \mid T_6$

The inverse function $GFN_{d,r}^{-1}$ are realized by changing the order of RK_i and the direction of word rotation at Step 2.2 and Step 3.

3.2   Data Processing Part

The data processing part of CLEFIA consists of ENC_r for encryption and DEC_r for decryption. ENC_r and DEC_r use a 4-branch generalized Feistel structure $GFN_{4,r}$. Let $P, C \in \{0,1\}^{128}$ be a plaintext and a ciphertext, and let $P_i, C_i \in \{0,1\}^{32}$ $(0 \le i < 4)$ be divided plaintext and ciphertext where $P = P_0|P_1|P_2|P_3$ and $C = C_0|C_1|C_2|C_3$, and let $WK_0, WK_1, WK_2, WK_3 \in \{0,1\}^{32}$ be whitening keys and $RK_i \in \{0,1\}^{32}$ $(0 \le i < 2r)$ be round keys provided by the key scheduling part. Then, r-round encryption function ENC_r is defined as follows:

$$ENC_r : \begin{cases} \{\{0,1\}^{32}\}^4 \times \{\{0,1\}^{32}\}^{2r} \times \{\{0,1\}^{32}\}^4 \to \{\{0,1\}^{32}\}^4 \\ (WK_{0(32)}, \dots, WK_{3(32)}, RK_{0(32)}, \dots, RK_{2r-1(32)}, P_{0(32)}, \dots, P_{3(32)}) \\ \qquad \mapsto C_{0(32)}, \dots, C_{3(32)} \end{cases}$$

> Step 1. $T_0 \mid T_1 \mid T_2 \mid T_3 \leftarrow P_0 \mid (P_1 \oplus WK_0) \mid P_2 \mid (P_3 \oplus WK_1)$
> Step 2. $T_0 \mid T_1 \mid T_2 \mid T_3 \leftarrow GFN_{4,r}(RK_0, \ldots, RK_{2r-1}, T_0, T_1, T_2, T_3)$
> Step 3. $C_0 \mid C_1 \mid C_2 \mid C_3 \leftarrow T_0 \mid (T_1 \oplus WK_2) \mid T_2 \mid (T_3 \oplus WK_3)$

The decryption function DEC_r is the inverse function of ENC_r which is defined by using $GFN_{d,r}^{-1}$. Fig. 1 in Appendix C illustrates the ENC_r function. The number of rounds, r, is 18, 22 and 26 for 128-bit, 192-bit and 256-bit keys, respectively.

3.3   Key Scheduling Part

The key scheduling part of CLEFIA supports 128, 192 and 256-bit keys and outputs whitening keys WK_i $(0 \le i < 4)$ and round keys RK_j $(0 \le j < 2r)$ for the data processing part. We first define the *DoubleSwap* function which is used in the key scheduling part.

Definition 1 (The *DoubleSwap* Function Σ).
The DoubleSwap function $\Sigma : \{0,1\}^{128} \rightarrow \{0,1\}^{128}$ is defined as follows:

$$X_{(128)} \mapsto Y_{(128)}$$
$$Y = X[7-63] \mid X[121-127] \mid X[0-6] \mid X[64-120] \, ,$$

where $X[a-b]$ denotes a bit string cut from the a-th bit to the b-th bit of X. 0-th bit is the most significant bit (See Fig. 2 in Appendix C).

Let K be a k-bit key, where k is 128, 192 or 256. The key scheduling part is divided into the following two sub-parts. (1) Generating an intermediate key L from K, and (2) Expanding K and L to generate WK_i and RK_j. The key scheduling is explained according to the sub-parts.

Key Scheduling for a 128-bit Key. The 128-bit intermediate key L is generated by applying $GFN_{4,12}$ which takes twenty-four 32-bit constant values $CON_i^{(128)}$ $(0 \le i < 24)$ as round keys and $K = K_0|K_1|K_2|K_3$ as an input. Then K and L are used to generate WK_i $(0 \le i < 4)$ and RK_j $(0 \le j < 36)$ in the following steps. In the latter part, thirty-six 32-bit constant values $CON_i^{(128)}$ $(24 \le i < 60)$ are used. The generation steps of CON are explained in Sect. 3.5.

> (Generating L from K)
> Step 1. $L \leftarrow GFN_{4,12}(CON_0^{(128)}, \ldots, CON_{23}^{(128)}, K_0, \ldots, K_3)$
>
> (Expanding K and L)
> Step 2. $WK_0|WK_1|WK_2|WK_3 \leftarrow K$
> Step 3. For $i = 0$ to 8 do the following:
> $\quad T \leftarrow L \oplus (CON_{24+4i}^{(128)} \mid CON_{24+4i+1}^{(128)} \mid CON_{24+4i+2}^{(128)} \mid CON_{24+4i+3}^{(128)})$
> $\quad L \leftarrow \Sigma(L)$
> $\quad$ if i is odd: $T \leftarrow T \oplus K$
> $\quad RK_{4i}|RK_{4i+1}|RK_{4i+2}|RK_{4i+3} \leftarrow T$

Key Scheduling for a 192-bit Key. Two 128-bit values K_L, K_R are generated from a 192-bit key $K = K_0|K_1|K_2|K_3|K_4|K_5$, $K_i \in \{0,1\}^{32}$. Then two 128-bit values L_L, L_R are generated by applying $GFN_{8,10}$ which takes $CON_i^{(192)}$ ($0 \leq i < 40$) as round keys and $K_L|K_R$ as a 256-bit input. Then K_L, K_R and L_L, L_R are used to generate WK_i ($0 \leq i < 4$) and RK_j ($0 \leq j < 44$) in the following steps. In the latter part, forty-four 32-bit constant values $CON_i^{(192)}$ ($40 \leq i < 84$) are used.

The following steps show the 192-bit/256-bit key scheduling. For the 192-bit key scheduling, the value of k is set as 192.

(Generating L_L, L_R from K_L, K_R for a k-bit key)

Step 1. Set $k = 192$ or $k = 256$

Step 2. If $k = 192$: $K_L \leftarrow K_0|K_1|K_2|K_3$, $K_R \leftarrow K_4|K_5|\overline{K_0}|\overline{K_1}$
        else if $k = 256$: $K_L \leftarrow K_0|K_1|K_2|K_3$, $K_R \leftarrow K_4|K_5|K_6|K_7$

Step 3. Let $K_L = K_{L0}|K_{L1}|K_{L2}|K_{L3}$, $K_R = K_{R0}|K_{R1}|K_{R2}|K_{R3}$
     $L_L|L_R \leftarrow GFN_{8,10}(CON_0^{(k)}, \ldots, CON_{39}^{(k)}, K_{L0}, \ldots, K_{L3}, K_{R0}, \ldots, K_{R3})$

(Expanding K_L, K_R and L_L, L_R for a k-bit key)

Step 4. $WK_0|WK_1|WK_2|WK_3 \leftarrow K_L \oplus K_R$

Step 5. For $i = 0$ to 10 (if $k = 192$), or 12 (if $k = 256$) do the following:
     If ($i \bmod 4$) = 0 or 1:
$$T \leftarrow L_L \oplus (CON_{40+4i}^{(k)} \mid CON_{40+4i+1}^{(k)} \mid CON_{40+4i+2}^{(k)} \mid CON_{40+4i+3}^{(k)})$$
$$L_L \leftarrow \Sigma(L_L)$$
        if i is odd: $T \leftarrow T \oplus K_R$
     else:
$$T \leftarrow L_R \oplus (CON_{40+4i}^{(k)} \mid CON_{40+4i+1}^{(k)} \mid CON_{40+4i+2}^{(k)} \mid CON_{40+4i+3}^{(k)})$$
$$L_R \leftarrow \Sigma(L_R)$$
        if i is odd: $T \leftarrow T \oplus K_L$
     $RK_{4i}|RK_{4i+1}|RK_{4i+2}|RK_{4i+3} \leftarrow T$

Key Scheduling for a 256-bit Key. For a 256-bit key, the value of k is set as 256, and the steps are almost the same as in the 192-bit key case. The difference is that we use $CON_i^{(256)}$ ($0 \leq i < 40$) as round keys to generate L_L and L_R, and then to generate RK_j ($0 \leq j < 52$), we use fifty-two 32-bit constant values $CON_i^{(256)}$ ($40 \leq i < 92$).

3.4 F-Functions

F-functions $F_0, F_1 : (RK_{(32)}, x_{(32)}) \mapsto y_{(32)}$ are defined as follows:

[F-function F_0]	[F-function F_1]
Step 1. $T \leftarrow RK \oplus x$	*Step 1.* $T \leftarrow RK \oplus x$
Step 2. Let $T = T_0\|T_1\|T_2\|T_3$, $T_i \in \{0,1\}^8$	*Step 2.* Let $T = T_0\|T_1\|T_2\|T_3$, $T_i \in \{0,1\}^8$
$T_0 \leftarrow S_0(T_0)$, $T_1 \leftarrow S_1(T_1)$	$T_0 \leftarrow S_1(T_0)$, $T_1 \leftarrow S_0(T_1)$
$T_2 \leftarrow S_0(T_2)$, $T_3 \leftarrow S_1(T_3)$	$T_2 \leftarrow S_1(T_2)$, $T_3 \leftarrow S_0(T_3)$
Step 3. Let $y = y_0\|y_1\|y_2\|y_3$, $y_i \in \{0,1\}^8$	*Step 3.* Let $y = y_0\|y_1\|y_2\|y_3$, $y_i \in \{0,1\}^8$
${}^t(y_0, y_1, y_2, y_3) = M_0 \, {}^t(T_0, T_1, T_2, T_3)$	${}^t(y_0, y_1, y_2, y_3) = M_1 \, {}^t(T_0, T_1, T_2, T_3)$

S_0 and S_1 are nonlinear 8-bit S-boxes. The orders of these S-boxes are different in F_0 and F_1. Tables in Appendix B show the input/output values of each S-box. In these tables all values are expressed in hexadecimal form, suffixes '0x' are omitted. For an 8-bit input of an S-box, the upper 4-bit indicates a row and the lower 4-bit indicates a column .

Two matrices M_0 and M_1 in Step 3 are defined as follows.

$$
M_0 = \begin{pmatrix} \text{0x01} & \text{0x02} & \text{0x04} & \text{0x06} \\ \text{0x02} & \text{0x01} & \text{0x06} & \text{0x04} \\ \text{0x04} & \text{0x06} & \text{0x01} & \text{0x02} \\ \text{0x06} & \text{0x04} & \text{0x02} & \text{0x01} \end{pmatrix}, \quad M_1 = \begin{pmatrix} \text{0x01} & \text{0x08} & \text{0x02} & \text{0x0a} \\ \text{0x08} & \text{0x01} & \text{0x0a} & \text{0x02} \\ \text{0x02} & \text{0x0a} & \text{0x01} & \text{0x08} \\ \text{0x0a} & \text{0x02} & \text{0x08} & \text{0x01} \end{pmatrix}.
$$

These are 4×4 Hadamard-type matrices with elements $h_{ij} = a_{i \oplus j}$ for a certain set $\{a_0, a_1, a_2, a_3\}$[1].

The multiplications between matrices and vectors are performed in $\mathrm{GF}(2^8)$ defined by the lexicographically first primitive polynomial $z^8 + z^4 + z^3 + z^2 + 1$. Fig. 3 in Appendix C illustrates the construction of F_0 and F_1.

3.5   Constant Values

32-bit constant values $CON_i^{(k)}$ are used in the key scheduling algorithm. We need 60, 84 and 92 constant values for 128, 192 and 256-bit keys, respectively. Let $\mathbf{P}_{(16)} = \text{0xb7e1}$ ($= (e - 2) \cdot 2^{16}$) and $\mathbf{Q}_{(16)} = \text{0x243f}$ ($= (\pi - 3) \cdot 2^{16}$), where e is the base of the natural logarithm (2.71828...) and π is the circle ratio (3.14159...). $CON_i^{(k)}$, for $k = 128, 192, 256$, are generated by the following way (See Table 1 for the repetition numbers $l^{(k)}$ and the initial values $IV^{(k)}$).

> Step 1. $T \leftarrow IV^{(k)}$
> Step 2. For $i = 0$ to $l^{(k)} - 1$ do the following:
>   Step 2.1. $CON_{2i}^{(k)} \leftarrow (T \oplus \mathbf{P}) \mid (\overline{T} \lll 1)$
>   Step 2.2. $CON_{2i+1}^{(k)} \leftarrow (\overline{T} \oplus \mathbf{Q}) \mid (T \lll 8)$
>   Step 2.3. $T \leftarrow T \cdot \text{0x0002}^{-1}$

In Step 2.3, multiplications are performed in $\mathrm{GF}(2^{16})$ defined by a primitive polynomial $z^{16} + z^{15} + z^{13} + z^{11} + z^5 + z^4 + 1$ ($=\text{0x1a831}$)[2].

Table 1. Required Numbers of Constant Values

k	# of CON	$l^{(k)}$	$IV^{(k)}$	
128	60	30	0x428a	($= (\sqrt[3]{2} - 1) \cdot 2^{16}$)
192	84	42	0x7137	($= (\sqrt[3]{3} - 1) \cdot 2^{16}$)
256	92	46	0xb5c0	($= (\sqrt[3]{5} - 1) \cdot 2^{16}$)

---

[1] An Hadamard-type matrix is used in the blockcipher Anubis [2].

[2] The lower 16-bit value is defined as $\text{0xa831} = (\sqrt[3]{101} - 4) \cdot 2^{16}$. '101' is the smallest prime number satisfying the primitive polynomial condition in this form.

4   Design Rationale

CLEFIA is designed to realize good balance on three fundamental directions for practical ciphers: (1) security, (2) speed, and (3) cost for implementations. This section describes the design rationale of several aspects of CLEFIA.

Structure. CLEFIA employs a 4-branch generalized Feistel structure which is considered as an extension of a 2-branch traditional Feistel structure. There are many types of generalized Feistel structures. We choose one instance which is known as "Generalized type-2 transformations" defined by Zheng et al. [24]. The type-2 structure has two F-functions in one round for the four data lines case. The type-2 structure has the following features:

- F-functions are smaller than that of the traditional Feistel structure
- Plural F-functions can be processed simultaneously
- Tends to require more rounds than the traditional Feistel structure

The first feature is a great advantage for software and hardware implementations, and the second one is suitable for efficient implementation especially in hardware. We conclude that the advantages of the type-2 structure surpass the disadvantage of the third one for our blockcipher design. Moreover, the new design technique, which is explained in the next, enables to reduce the number of rounds effectively.

Diffusion Switching Mechanism. CLEFIA employs two different diffusion matrices to enhance the immunity against differential attacks and linear attacks by using the Diffusion Switching Mechanism (DSM). This design technique was originally developed for the traditional Feistel structures [22,23]. We customized this technique suitable for $GFN_{d,r}$, which is one of the unique propositions of this cipher. Due to the DSM, we can prevent difference cancellations and linear mask cancellations in the neighborhood rounds in the cipher. As a result the guaranteed number of active S-boxes is increased.

Let the branch number of a function P be $\mathcal{B}(P) = \min_{a \neq 0}\{\mathbf{w_b}(a) + \mathbf{w_b}(P(a))\}$. The two matrices M_0 and M_1 used in CLEFIA satisfy the following branch number conditions of the DSM.

$$\mathcal{B}(M_0) = \mathcal{B}(M_1) = 5, \quad \mathcal{B}(M_0|M_1) = \mathcal{B}({}^tM_0^{-1}|{}^tM_1^{-1}) = 5 \ .$$

Table 2 shows the guaranteed numbers of active S-boxes of CLEFIA. The guaranteed data are obtained from computer simulations using a exhaustive-type search algorithm. Now we focus on the columns indexed by '$GFN_{4,r}$'. The columns of 'D' and 'L' in the table show the guaranteed number of differential and linear active S-boxes, respectively. The 'DSM' denotes that the DSM is used, and the 'w/o DSM' denotes that DSM is not used, where only one matrix with branch number 5 is employed. From this table we can confirm the effects of the DSM when $r \geq 3$, and these guaranteed numbers increase about $20\% - 40\%$ than the structure without DSM. Consequently, the numbers of rounds can be reduced, which implies that the performance is improved.

Table 2. Guaranteed Numbers of Active S-boxes

	$GFN_{4,r}$			$GFN_{8,r}$		$GFN_{4,r}$			$GFN_{8,r}$
	D & L	D	L	D		D & L	D	L	D
r	w/o DSM	DSM	DSM	DSM	r	w/o DSM	DSM	DSM	DSM
1	0	0	0	0	14	25	34	34	48
2	1	1	1	1	15	26	36	36	50
3	2	2	5	2	16	30	38	39	53
4	6	6	6	6	17	32	40	42	56
5	8	8	10	8	18	36	44	46	59
6	12	12	15	12	19	36	46	48	62
7	12	14	16	14	20	37	50	50	66
8	13	18	18	21	21	38	52	52	71
9	14	20	20	24	22	42	55	55	76
10	18	22	23	29	23	44	56	58	81
11	20	24	26	34	24	48	59	62	86
12	24	28	30	39	25	48	62	64	91
13	24	30	32	44	26	49	65	66	94

Table 3. Tables of SS_i $(0 \le i < 4)$

x	0 1 2 3 4 5 6 7 8 9 a b c d e f	x	0 1 2 3 4 5 6 7 8 9 a b c d e f
$SS_0(x)$	e 6 c a 8 7 2 f b 1 4 0 5 9 d 3	$SS_1(x)$	6 4 0 d 2 b a 3 9 c e f 8 7 5 1
$SS_2(x)$	b 8 5 e a 6 4 c f 7 2 3 1 0 d 9	$SS_3(x)$	a 2 6 d 3 4 5 e 0 7 8 9 b f c 1

Choices of two S-boxes. CLEFIA employs two different types of 8-bit S-boxes: one is based on four 4-bit random S-boxes, and the other is based on the inverse function over $GF(2^8)$ which has the best possible maximum differential probability DP_{max} and linear probability LP_{max}. The both S-boxes are selected to be implemented efficiently especially in hardware. The two 8-bit S-boxes S_0 and S_1 are defined as:

$$S_0, S_1 : \begin{cases} \{0,1\}^8 \to \{0,1\}^8 \\ x_{(8)} \mapsto y_{(8)} \end{cases}$$

S_0 is generated by combining four 4-bit S-boxes SS_0, SS_1, SS_2 and SS_3 in the following way. The values of these S-boxes are defined as Table 3.

Step 1. $t_0 \leftarrow SS_0(x_0)$, $\quad t_1 \leftarrow SS_1(x_1)$, where $x = x_0\|x_1$, $x_i \in \{0,1\}^4$
Step 2. $u_0 \leftarrow t_0 \oplus \text{0x2} \cdot t_1$, $\quad u_1 \leftarrow \text{0x2} \cdot t_0 \oplus t_1$
Step 3. $y_0 \leftarrow SS_2(u_0)$, $\quad y_1 \leftarrow SS_3(u_1)$, where $y = y_0\|y_1$, $y_i \in \{0,1\}^4$

The multiplication in $\text{0x2} \cdot t_i$ is performed in $GF(2^4)$ defined by the lexicographically first primitive polynomial $z^4 + z + 1$.

S_1 is defined as follows:

$$y = \begin{cases} g(f(x)^{-1}) \text{ if } f(x) \ne 0 \\ g(0) \quad\quad\; \text{if } f(x) = 0 \end{cases} .$$

The inverse function is performed in $GF(2^8)$ defined by a primitive polynomial $z^8 + z^4 + z^3 + z^2 + 1$. $f(\cdot)$ and $g(\cdot)$ are affine transformations over $GF(2)$, which are defined as follows.

$$
f : x_{(8)} \mapsto y_{(8)} \qquad\qquad g : x_{(8)} \mapsto y_{(8)}
$$

$$
\begin{pmatrix} y_0 \\ y_1 \\ y_2 \\ y_3 \\ y_4 \\ y_5 \\ y_6 \\ y_7 \end{pmatrix}
=
\begin{pmatrix}
0&0&0&1&1&0&0&0 \\
0&1&0&1&0&0&0&1 \\
0&0&0&0&0&0&0&1 \\
0&0&0&0&0&1&1&0 \\
0&1&1&0&0&1&0&1 \\
0&1&0&1&1&1&0&0 \\
0&1&1&0&0&0&0&0 \\
1&0&0&0&0&0&0&1
\end{pmatrix}
\begin{pmatrix} x_0 \\ x_1 \\ x_2 \\ x_3 \\ x_4 \\ x_5 \\ x_6 \\ x_7 \end{pmatrix}
+
\begin{pmatrix} 0 \\ 0 \\ 0 \\ 1 \\ 1 \\ 1 \\ 1 \\ 0 \end{pmatrix}
$$

$$
\begin{pmatrix} y_0 \\ y_1 \\ y_2 \\ y_3 \\ y_4 \\ y_5 \\ y_6 \\ y_7 \end{pmatrix}
=
\begin{pmatrix}
0&0&0&0&1&0&1&0 \\
0&1&0&0&0&0&0&1 \\
0&1&0&1&1&0&0&0 \\
0&0&1&0&0&0&0&0 \\
0&0&1&1&0&0&0&0 \\
0&0&0&0&0&0&1&0 \\
1&0&0&1&0&0&0&0 \\
0&1&0&0&0&1&0&0
\end{pmatrix}
\begin{pmatrix} x_0 \\ x_1 \\ x_2 \\ x_3 \\ x_4 \\ x_5 \\ x_6 \\ x_7 \end{pmatrix}
+
\begin{pmatrix} 0 \\ 1 \\ 1 \\ 0 \\ 1 \\ 0 \\ 0 \\ 1 \end{pmatrix}
$$

Here, $x = x_0|x_1|x_2|x_3|x_4|x_5|x_6|x_7$ and $y = y_0|y_1|y_2|y_3|y_4|y_5|y_6|y_7$, $x_i, y_i \in \{0,1\}$. The constants in f and g can be represented as 0x1e and 0x69, respectively. If we apply isomorphic mapping ϕ from $GF(2^8)$ to $GF((2^4)^2)$ defined by an irreducible polynomial $z^2 + z + \omega^3$, where ω is a root of $z^4 + z + 1 = 0$, the merged transformations $\phi \circ f$ and $g \circ \phi^{-1}$ require only few XOR operations.

For security parameters, DP_{max} of S_0 is $2^{-4.67}$ and its LP_{max} is $2^{-4.38}$, the minimum Boolean degree is 6, and the minimum number of terms over $GF(2^8)$ is 244. For S_1, DP_{max} and LP_{max} are both $2^{-6.00}$, the minimum Boolean degree is 7 and it has at least 252 terms over $GF(2^8)$.

Designs for Efficient Implementations. CLEFIA can be implemented efficiently both in hardware and software. In Table 4, we summarize the design aspects for efficient implementations.

Table 4. Design Aspects for Efficient Implementations

GFN	· Small size F-functions (32-bit in/out)
	· No need for the inverse F-functions
SP-type F-function	· Enabling the fast table implementation in software
DSM	· Reducing the numbers of rounds
S-boxes	· Very small footprint of S_0 and S_1 in hardware
Matrices	· Using elements with low hamming weights only
Key Schedule	· Sharing the structure with the data processing part
	· Requiring only a 128-bit register for a 128-bit key
	· Small footprint of *DoubleSwap*

5   Evaluations

5.1   Security

As a result of our security evaluation, full-round CLEFIA is considered as a secure blockcipher against known attacks. Here, we mention the cryptanalytic results of several attacks which are considered effective for reduced-round CLEFIA.

Differential Cryptanalysis [7]. For differential attack, we adopt an approach to count the number of active S-boxes of differential characteristics. This method was adopted by AES, Camellia and other blockciphers [10, 1]. We found the guaranteed number of differential active S-boxes of CLEFIA by computer search as shown in Table 2. Using 28 active S-boxes for 12-round CLEFIA and $DP_{max}^{S_0} = 2^{-4.67}$, it is shown that $DCP_{max}^{12r} \leq 2^{28 \times (-4.67)} = 2^{-130.76}$. This means there is no useful 12-round differential characteristic for an attacker. Moreover, since S_1 has lower DP_{max}, the actual upper-bound of DCP is expected to be lower than the above estimation. As a result, we believe that full-round CLEFIA is secure against differential cryptanalysis.

Linear Cryptanalysis [17]. We also apply active S-boxes based approach for the evaluation of linear cryptanalysis. Since $LP_{max}^{S_0} = 2^{-4.38}$, combining 30 active S-boxes for 12-round CLEFIA, $LCP_{max}^{12r} \leq 2^{30 \times (-4.38)} = 2^{-131.40}$. We conclude that it is difficult for an attacker to find 12-round linear-hulls which can be used to distinguish CLEFIA from a random permutation. As a result, full-round CLEFIA is secure enough against linear cryptanalysis.

Impossible Differential Cryptanalysis [4]. We consider that impossible differential attack is one of the most powerful attacks against CLEFIA. The following two impossible differential paths are found.

$$(0, \alpha, 0, 0) \overset{9r}{\not\to} (0, \alpha, 0, 0) \text{ and } (0, 0, 0, \alpha) \overset{9r}{\not\to} (0, 0, 0, \alpha) \quad p = 1$$

where $\alpha \in \{0, 1\}^{32}$ is any non-zero difference. These paths are confirmed by the check algorithm proposed by Kim *et al.* [14]. Using the above distinguisher, we can mount actual key-recovery attacks for each key length. Table 5 shows the summary of the complexity required for the impossible differential attacks. According to Table 5, it is expected that full-round CLEFIA has enough security margin against this attack.

Table 5. Summary of Impossible Differential Cryptanalysis

# of rounds	key length	key whitening	# of chosen plaintexts	time complexity
10	128, 192, 256	yes	$2^{101.7}$	2^{102}
11	192, 256	yes	$2^{103.5}$	2^{188}
12	256	no	$2^{103.8}$	2^{252}

Saturation Cryptanalysis [9]. We also consider that saturation attack is one of the most powerful attacks against CLEFIA. In this analysis, we consider a 32-bit word based saturation attack. Let $X_i \in \{0, 1\}^{32}$ $(0 \leq i < 2^{32})$ be 2^{32} 32-bit variables. Now we classify X_i into four states depending on the conditions satisfied.

Const (C) : $\forall i, j \ X_i = X_j$,   All (A) : $\forall i, j \ i \neq j \Leftrightarrow X_i \neq X_j$,
Balance (B) : $\bigoplus_{i=0}^{2^{32}-1} X_i = 0$,   Unknown (U) : unknown .

Using the above notation, the following 6-round distinguishers are found.

$$(C, A, C, C) \xrightarrow{6r} (B, U, U, U) \quad and \quad (C, C, C, A) \xrightarrow{6r} (U, U, B, U) \quad p = 1$$

These distinguishers can be extended to an 8-round distinguisher by adding two more rounds before the above 6 round path. Let $A_{(96)}$ be an "All" state of 96-bit word, and we divide it into 3 segments as $A_{(96)} = A_0|A_1|A_2$. Then the 8-round distinguishers are given as follows.

$$(A_0, C, A_1, A_2) \xrightarrow{8r} (B, U, U, U) \quad and \quad (A_0, A_1, A_2, C) \xrightarrow{8r} (U, U, B, U) \quad p = 1$$

Using the above 8-round distinguisher, it turns out that 10-round 128-bit key CLEFIA can be attacked with complexity slightly less than 2^{128} F-function calculations. From the above observations, we conclude that full-round CLEFIA has enough security margin against this attack.

Algebraic Attack [8]. Let CLEFIA-I be a modified version of CLEFIA by replacing all 4-bit S-boxes by the identity function I, $I : \{0, 1\}^4 \rightarrow \{0, 1\}^4$, where $I(x) = x$. Based on the estimation method by Courtois and Pieprzyk the total number of terms can be estimated as follows [8]. $T = 81^8 \binom{144}{8} > 2^{50+41} = 2^{91}$ for CLEFIA-I with $r = 18$, which gives the complexity $T^{2.376} = 2^{216}$ and $T^3 = 2^{273}$. For CLEFIA-I with $r = 22$, we have $T = 81^8 \binom{352}{8} > 2^{50+52} = 2^{102}$, and thus $T^{2.376} = 2^{242}$ and $T^3 = 2^{306}$. Finally, for CLEFIA-I with $r = 26$, we have $T = 81^8 \binom{416}{8} > 2^{50+54} = 2^{104}$, $T^{2.376} = 2^{247}$, and $T^3 = 2^{312}$. Although we give the results of the estimation, we should interpret these estimations with an extreme care: the real complexity of the XSL attacks is by no means clear at the time of writing and is the subject of much controversy [20, 16].

Related-Key Attack [3]. As for CLEFIA with a 128-bit key, L is generated by $L = GFN_{4,12}(CON^{(128)}, K)$. As in Table 2, $GFN_{4,12}$ has at least 28 active S-boxes, and we have $DCP_{max} \leq 2^{-130.76}$. Therefore, for any ΔK and ΔL, a differential probability of $(\Delta K \rightarrow \Delta L)$ is expected to be less than 2^{-128}, i.e., no useful differential $(\Delta K \rightarrow \Delta L)$ exists.

Also for 192 and 256-bit keys, (L_L, L_R) is generated by applying $GFN_{8,10}$ to K_L, K_R. From Table 2, it has at least 29 differential active S-boxes, which implies there are no differential characteristics with probability more than 2^{-128}.

The above observations imply the probability of any related-key differential $(\Delta P, \Delta C, \Delta K)$ is less than 2^{-128}, if all the information on ΔL is needed for differential. Because all the bits in L are used in 2 or 6 consecutive rounds. As a result, we believe that CLEFIA holds strong immunity against related-key cryptanalysis. We also expect that CLEFIA holds enough immunity against other related-key type attacks including related-key boomerang and related-key rectangle attacks [5].

Security against Other Attacks. Due to the page limitation, the details of the security evaluation against known general attacks are omitted. Immunity against some of the known attacks can be estimated by the evaluation results of similar type attacks already mentioned in this section. We consider any attack does not threat full-round CLEFIA.

Table 6. Results on Hardware Performance of CLEFIA

	Key Length	Enc/Dec (cycles)	Key Setup (cycles)	Optimization	Area (gates)	Freq. (MHz)	Speed (Mbps)	Speed/Area (Kbps/gate)
CLEFIA (0.09 μm)	128	18	12	area	5,979	225.83	1,605.94	268.63
				speed	12,009	422.29	3,003.00	250.06
		36	24	area	4,950	201.28	715.69	144.59
				speed	9,377	389.55	1,385.10	147.71
	192	22	20	area	8,536	206.56	1,201.85	140.81
	256	26	20	area	8,482	206.56	1,016.95	119.89
AES [21] (0.13 μm)	128	11	N/A	area	12,454	145.35	1,691.35	135.81
		54	N/A	area	5,398	131.24	311.09	57.63

Table 7. Results on Software Performance of CLEFIA (assembly language)

	Type of implementation	Key Length	Encryption (cycles/byte)	Decryption (cycles/byte)	Key Setup (cycles)	Table size (KB)
CLEFIA	single-block	128	12.9	13.3	217	8
		192	15.8	16.2	272	
		256	18.3	18.4	328	
	two-block parallel encryption	128	11.1	11.1	217	16
		192	13.3	13.3	272	
		256	15.6	15.6	328	
AES [18]	single-block	128	10.6	N/A	N/A	8

5.2 Performance

Table 6 shows evaluation results of hardware performance of CLEFIA using a 0.09μm CMOS ASIC library, where one gate is equivalent to a 2-way NAND and the speed is evaluated under the worst-case condition. The Verilog-HDL models were synthesized by specifying area or speed optimization to a logic synthesis tool. The synthesized circuit of CLEFIA with 128-bit key by area optimization occupies only 5,979 gates at a throughput of about 1.60 Gbps. Although we take into account the difference of ASIC libraries, these figures indicate high efficiency of CLEFIA in hardware implementation compared to the best known results of hardware performance of AES [21].

Table 7 shows software performance results on Athlon 64 (AMD64) 4000+ 2.4 GHz processor running Windows XP 64-bit Edition. We measured software processing speed of enc/dec and key setup using the rdtsc instruction. In the single-block (common) implementation, 12.9 cycle/byte (1.48 Gbps on the processor) is achieved. The two-block parallel encryption is suitable for CTR mode, because two blocks can be processed simultaneously [18].

6   Conclusion

We proposed a 128-bit blockcipher CLEFIA, which supports 128-bit, 192-bit, and 256-bit keys. CLEFIA employs several new design approaches, including the DSM technique. As a result, enough immunity against known attacks is achieved. Moreover, the design of CLEFIA allows very efficient implementation in a variety of environments. Some results of highly efficient implementation are exemplified.

Acknowledgment

The authors thank Lars Knudsen, Bart Preneel, Vincent Rijmen, and Serge Vaudenay for their helpful comments. The authors also thank the anonymous referees for their valuable comments.

References

1. Aoki, K., Ichikawa, T., Kanda, M., Matsui, M., Moriai, S., Nakajima, J., Tokita, T.: Camellia: A 128-bit block cipher suitable for multiple platforms. In: Stinson, D.R., Tavares, S. (eds.) SAC 2000. LNCS, vol. 2012, pp. 41–54. Springer, Heidelberg (2001)
2. Barreto, P.S.L.M., Rijmen, V.: The Anubis block cipher. NESSIE, September 2000 (primitive submitted), Available at http://www.cryptonessie.org/
3. Biham, E.: New types of cryptanalytic attacks using related keys. Journal of Cryptology 7(4), 229–246 (1994)
4. Biham, E., Biryukov, A., Shamir, A.: Cryptanalysis of Skipjack reduced to 31 rounds using impossible differentials. In: Stern, J. (ed.) EUROCRYPT 1999. LNCS, vol. 1592, pp. 12–23. Springer, Heidelberg (1999)
5. Biham, E., Dunkelman, O., Keller, N.: Related-key boomerang and rectangle attacks. In: Cramer, R.J.F. (ed.) EUROCRYPT 2005. LNCS, vol. 3494, pp. 507–525. Springer, Heidelberg (2005)
6. Biham, E., Dunkelman, O., Keller, N.: Related-key impossible differential attacks on 8-round AES-192. In: Pointcheval, D. (ed.) CT-RSA 2006. LNCS, vol. 3860, pp. 21–33. Springer, Heidelberg (2006)
7. Biham, E., Shamir, A.: Differential Cryptanalysis of the Data Encryption Standard. Springer, Heidelberg (1993)
8. Courtois, N., Pieprzyk, J.: Cryptanalysis of block ciphers with overdefined systems of equations. In: Zheng, Y. (ed.) ASIACRYPT 2002. LNCS, vol. 2501, pp. 267–287. Springer, Heidelberg (2002)
9. Daemen, J., Knudsen, L.R., Rijmen, V.: The block cipher SQUARE. In: Biham, E. (ed.) FSE 1997. LNCS, vol. 1267, pp. 149–165. Springer, Heidelberg (1997)
10. Daemen, J., Rijmen, V.: The Design of Rijndael: AES - The Advanced Encryption Standard (Information Security and Cryptography). Springer, Heidelberg (2002)
11. Data Encryption Standard: Federal Information Processing Standard (FIPS), Publication 46, National Bureau of Standards, U.S. Department of Commerce, Washington, DC (January 1977)
12. Hong, D., Sung, J., Hong, S., Lim, J., Lee, S., Koo, B., Lee, C., Chang, D., Lee, J., Jeong, K., Kim, H., Kim, J., Chee, S.: Hight: A new block cipher suitable for low-resource device. In: Goubin, L., Matsui, M. (eds.) CHES 2006. LNCS, vol. 4249, pp. 46–59. Springer, Heidelberg (2006)

13. Junod, P., Vaudenay, S.: FOX: A new family of block ciphers. In: Handschuh, H., Hasan, M.A. (eds.) SAC 2004. LNCS, vol. 3357, pp. 114–129. Springer, Heidelberg (2004)
14. Kim, J., Hong, S., Sung, J., Lee, C., Lee, S.: Impossible differential cryptanalysis for block cipher structure. In: Johansson, T., Maitra, S. (eds.) INDOCRYPT 2003. LNCS, vol. 2904, pp. 82–96. Springer, Heidelberg (2003)
15. Lai, X., Massey, J.L., Murphy, S.: Markov ciphers and differential cryptanalysis. In: Davies, D.W. (ed.) EUROCRYPT 1991. LNCS, vol. 547, pp. 17–38. Springer, Heidelberg (1991)
16. Lim, C., Khoo, K.: An Analysis of XSL Applied to BES. In: Pre-proceedings of Fast Software Encryption '07, FSE'07, pp. 253–265 (2007)
17. Matsui, M.: Linear cryptanalysis of the data encryption standard. In: Helleseth, T. (ed.) EUROCRYPT 1993. LNCS, vol. 765, pp. 386–397. Springer, Heidelberg (1994)
18. Matsui, M.: How far can we go on the x64 processors? In: Robshaw, M. (ed.) FSE 2006. LNCS, vol. 4047, pp. 341–358. Springer, Heidelberg (2006)
19. Matsui, M.: New block encryption algorithm MISTY. In: Biham, E. (ed.) FSE 1997. LNCS, vol. 1267, pp. 54–68. Springer, Heidelberg (1997)
20. Murphy, S., Robshaw, M.: Comments on the security of the AES and the XSL technique. Electronic Letters 39(1), 36–38 (2003)
21. Satoh, A., Morioka, S.: Hardware-focused performance comparison for the standard block ciphers AES, Camellia, and Triple-DES. In: Boyd, C., Mao, W. (eds.) ISC 2003. LNCS, vol. 2851, pp. 252–266. Springer, Heidelberg (2003)
22. Shirai, T., Preneel, B.: On Feistel ciphers using optimal diffusion mappings across multiple rounds. In: Lee, P.J. (ed.) ASIACRYPT 2004. LNCS, vol. 3329, pp. 1–15. Springer, Heidelberg (2004)
23. Shirai, T., Shibutani, K.: On Feistel structures using a diffusion switching mechanism. In: Robshaw, M. (ed.) FSE 2006. LNCS, vol. 4047, pp. 41–56. Springer, Heidelberg (2006)
24. Zheng, Y., Matsumoto, T., Imai, H.: On the construction of block ciphers provably secure and not relying on any unproved hypotheses. In: Brassard, G. (ed.) CRYPTO 1989. LNCS, vol. 435, pp. 461–480. Springer, Heidelberg (1990)

A   Test Vectors

We give test vectors of CLEFIA for each key length. The data are expressed in hexadecimal form.

128-bit key:
```
key          ffeeddcc bbaa9988 77665544 33221100
plaintext    00010203 04050607 08090a0b 0c0d0e0f
ciphertext   de2bf2fd 9b74aacd f1298555 459494fd
```
192-bit key:
```
key          ffeeddcc bbaa9988 77665544 33221100 f0e0d0c0 b0a09080
plaintext    00010203 04050607 08090a0b 0c0d0e0f
ciphertext   e2482f64 9f028dc4 80dda184 fde181ad
```
256-bit key:
```
key          ffeeddcc bbaa9988 77665544 33221100
             f0e0d0c0 b0a09080 70605040 30201000
plaintext    00010203 04050607 08090a0b 0c0d0e0f
ciphertext   a1397814 289de80c 10da46d1 fa48b38a
```

B Tables of S_0 and S_1

	.0	.1	.2	.3	.4	.5	.6	.7	.8	.9	.a	.b	.c	.d	.e	.f

S_0

	.0	.1	.2	.3	.4	.5	.6	.7	.8	.9	.a	.b	.c	.d	.e	.f
0.	57	49	d1	c6	2f	33	74	fb	95	6d	82	ea	0e	b0	a8	1c
1.	28	d0	4b	92	5c	ee	85	b1	c4	0a	76	3d	63	f9	17	af
2.	bf	a1	19	65	f7	7a	32	20	06	ce	e4	83	9d	5b	4c	d8
3.	42	5d	2e	e8	d4	9b	0f	13	3c	89	67	c0	71	aa	b6	f5
4.	a4	be	fd	8c	12	00	97	da	78	e1	cf	6b	39	43	55	26
5.	30	98	cc	dd	eb	54	b3	8f	4e	16	fa	22	a5	77	09	61
6.	d6	2a	53	37	45	c1	6c	ae	ef	70	08	99	8b	1d	f2	b4
7.	e9	c7	9f	4a	31	25	fe	7c	d3	a2	bd	56	14	88	60	0b
8.	cd	e2	34	50	9e	dc	11	05	2b	b7	a9	48	ff	66	8a	73
9.	03	75	86	f1	6a	a7	40	c2	b9	2c	db	1f	58	94	3e	ed
a.	fc	1b	a0	04	b8	8d	e6	59	62	93	35	7e	ca	21	df	47
b.	15	f3	ba	7f	a6	69	c8	4d	87	3b	9c	01	e0	de	24	52
c.	7b	0c	68	1e	80	b2	5a	e7	ad	d5	23	f4	46	3f	91	c9
d.	6e	84	72	bb	0d	18	d9	96	f0	5f	41	ac	27	c5	e3	3a
e.	81	6f	07	a3	79	f6	2d	38	1a	44	5e	b5	d2	ec	cb	90
f.	9a	36	e5	29	c3	4f	ab	64	51	f8	10	d7	bc	02	7d	8e

S_1

	.0	.1	.2	.3	.4	.5	.6	.7	.8	.9	.a	.b	.c	.d	.e	.f
0.	6c	da	c3	e9	4e	9d	0a	3d	b8	36	b4	38	13	34	0c	d9
1.	bf	74	94	8f	b7	9c	e5	dc	9e	07	49	4f	98	2c	b0	93
2.	12	eb	cd	b3	92	e7	41	60	e3	21	27	3b	e6	19	d2	0e
3.	91	11	c7	3f	2a	8e	a1	bc	2b	c8	c5	0f	5b	f3	87	8b
4.	fb	f5	de	20	c6	a7	84	ce	d8	65	51	c9	a4	ef	43	53
5.	25	5d	9b	31	e8	3e	0d	d7	80	ff	69	8a	ba	0b	73	5c
6.	6e	54	15	62	f6	35	30	52	a3	16	d3	28	32	fa	aa	5e
7.	cf	ea	ed	78	33	58	09	7b	63	c0	c1	46	1e	df	a9	99
8.	55	04	c4	86	39	77	82	ec	40	18	90	97	59	dd	83	1f
9.	9a	37	06	24	64	7c	a5	56	48	08	85	d0	61	26	ca	6f
a.	7e	6a	b6	71	a0	70	05	d1	45	8c	23	1c	f0	ee	89	ad
b.	7a	4b	c2	2f	db	5a	4d	76	67	17	2d	f4	cb	b1	4a	a8
c.	b5	22	47	3a	d5	10	4c	72	cc	00	f9	e0	fd	e2	fe	ae
d.	f8	5f	ab	f1	1b	42	81	d6	be	44	29	a6	57	b9	af	f2
e.	d4	75	66	bb	68	9f	50	02	01	3c	7f	8d	1a	88	bd	ac
f.	f7	e4	79	96	a2	fc	6d	b2	6b	03	e1	2e	7d	14	95	1d

C Figures

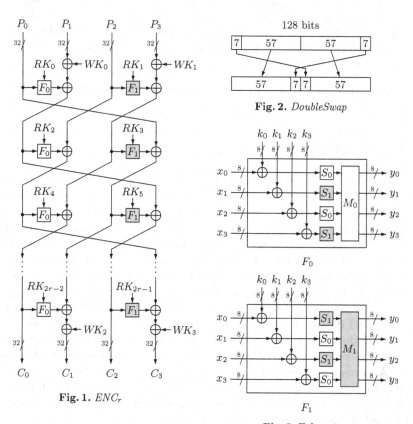

Fig. 1. ENC_r

Fig. 2. *DoubleSwap*

Fig. 3. F-functions

New Lightweight DES Variants

Gregor Leander, Christof Paar, Axel Poschmann, and Kai Schramm

Horst Görtz Institute for IT Security
Communication Security Group (COSY)
Ruhr-University Bochum, Germany
{cpaar,poschmann,schramm}@crypto.rub.de,
leander@rub.de

Abstract. In this paper we propose a new block cipher, DESL (DES Lightweight), which is based on the classical DES (Data Encryption Standard) design, but unlike DES it uses a single S-box repeated eight times.[1] On this account we adapt well-known DES S-box design criteria, such that they can be applied to the special case of a single S-box. Furthermore, we show that DESL is resistant against certain types of the most common attacks, i.e., linear and differential cryptanalyses, and the Davies-Murphy attack. Our hardware implementation results of DESL are very promising (1848 GE), therefore DESL is well suited for ultra-constrained devices such as RFID tags.

Keywords: RFID, DES, DESL, lightweight cryptography, S-box design criteria.

1 Introduction

A flawless and remote identification of products, people or animals plays an important role in many areas of daily life. For example, farmers often have to keep track of the fertility rate of their cattle and hence, identify calf-bearing cows. Other examples are the permanent identification of industrial goods, which improves the supply chain in factories and countersteers thievery, or the need of reliable access control devices, e.g. in form of ski passes or train tickets.

An automatic identification can be achieved with RFID (Radio Frequency Identification) tags. Basically, RFID tags consist of a transponder and an antenna and are able to remotely receive data from an RFID host or reader device. In general, RFID tags can be divided into passive and active devices: active tags provide their own power supply (i.e. in form of a battery), whereas passive tags solely rely on the energy of the carrier signal transmitted by the reader device. As a result, passive RFID devices are not only much less expensive, but also require less chip size and have a longer life cycle [Fin03]. Our proposed DESL algorithm and its low-power, size-optimized implementation aims at very constrained devices such as passive RFID tags.

---

[1] Part of this work has already been presented at RFIDSec '06, a non-proceeding workshop.

A. Biryukov (Ed.): FSE 2007, LNCS 4593, pp. 196–210, 2007.

Very often it is desired to use RFID tags as cryptographic tokens, e.g. in a challenge response protocol. In this case the tag must be able to execute a secure cryptographic primitive. Contactless microprocessor cards [RE02], which are capable to execute cryptographic algorithms, are not only expensive and, hence, not necessarily suited for mass production, but also draw a lot of current. The high, non-optimal power consumption of a microprocessor can usually only be provided by close coupling systems, i.e. a short distance between reader and RFID device has to be ensured [Fin03]. A better approach is to use a custom made RFID chip, which consists of a receiver circuit, a control unit[2], some kind of volatile and/or non-volatile memory and a cryptographic primitive. In [FWR05], Feldhofer et al. propose a very small AES implementation with 3400 gates, which draws a maximum current of $3.0\mu A$ @ $100kHz$. Their AES design is based on a byte-per-byte serialization, which only requires the implementation of a single S-box [DR02] and achieves an encryption within 1032 clock cycles ($= 10.32ms$ @ $100kHz$). Unfortunately, the ISO/IEC 18000 standard requires that the latency of a response of an RFID tag does not exceed $320\mu s$, which is why Feldhofer et al. propose a slightly modified challenge-response protocol based on interleaving.

The remainder of this work is organized as follows: In Section 2 we stress why DES was chosen as the basis for our new family of lightweight algorithms, which we present in Section 3. Subsequently, in Section 4 we show the design criteria for DESL and its lightweight implementation in Section 5. Finally, in Section 6 we summarize our results and conclude this work.

2   Design Considerations for Lightweight Block Ciphers

The pretext of our algorithm family is the desire to find a design for a cipher for extremely lightweight applications such as passive RFIDs. Thus far, there have been two approaches for providing cryptographic primitives for such situations: 1) Optimized low-cost implementations for standardized and trusted algorithms, which means in practice in essence block ciphers such as AES, see e.g., [FDW04]. And 2) design new ciphers with the goal of having low hardware implementation costs (see, e.g., the profile 2 algorithms of the eStream project). Even though both approaches are valid and yielding results, we believe both are not optimum. The problem with the first approach is that most modern block ciphers were primarily designed with good software implementation properties in mind, and not necessarily with hardware-friendly properties. We strongly believe that this was the right approach for today's block ciphers since on the one hand the vast majority of algorithms run in software on PCs or embedded devices, and on the other hand silicon area has become so inexpensive that very high performance hardware implementations (achieved through large chip area) are not a problem any more. However, if the goal is to provide extremely low-cost security on devices where both of those assumptions do not hold, one has to wonder whether modern block ciphers are the best solution.

---

[2] i.e. a finite state machine.

There are also problems with the second approach, designing low cost ciphers anew. First it is well known that it is painfully difficult to design new ciphers without security flaws. Furthermore, as can be seen from the eStream profile 2 algorithms (which have low-cost hardware properties as a main objective), it is far from straightforward to design a new cipher which has a lower hardware complexity than standard DES.

It is fair to claim that an optimum approach would be to have a well investigated cipher, the design of which was driven by low hardware costs. The only known cipher to this respect is the Data Encryption Standard, DES. (The obvious drawback of DES is that its key length is not adequate for many of today's applications, but this will be addressed in Section 3 below.) The promise that DES holds for lightweight hardware implementation can easily be seen by the following observation. If we compare a standard, one-round implementation[3] of AES and DES, the latter consumes about 6% (!) of the logic resources of AES, while having a shorter critical path [VHVM88], [ASM01]. Of course, DES uses a much shorter key so that a direct comparison is not completely accurate, but the time-area advantage of more than one order of magnitude gives an indication. We would also like to stress that it is not a coincidence that DES is so efficient in hardware. DES was designed in the first half of the 1970s and the targeted implementation platform was hardware. However, by today's standard, digital technology was extremely limited in the early 1970s. Hence, virtually all components of DES were heavily driven by low hardware complexity: bit permutation and small S-Boxes.

3   DESL and DESXL: Design Ideas and Security Consideration

The main design ideas of the new cipher family, which are either original DES efficiently implemented or a variant of DES, are:

1. Use of a serial hardware architecture which reduces the gate complexity.
2. Optionally apply key-whitening in order to render brute-force attacks impossible.
3. Optionally replace the 8 original S-Boxes by a single one which further reduces the gate complexity.

If we make use of the first idea, we obtain a lightweight implementation of the original DES algorithm which consumes about 35% less gates than the best known AES implementation [FWR05]. To our knowledge, this is the smallest reported DES implementation, trading area for throughput. The implementation requires also about 86% fewer clock cycles for encrypting of one block than the serialized AES implementation in [FWR05](1032 cycles vs. 144) which makes it easier to use in standardized RFID protocols. However, the security provided is limited by the 56 bit key. Brute forcing this key space takes a few months

---

[3] i.e. one plaintext block is encrypted in one clock cycle.

and hundreds of PCs in software, and only a few days with a special-purpose machine such as COPACOBANA [KPP+06]. Hence, this implementation is only relevant for application where short-term security is needed, or where the values protected are relatively low. However, we can imagine that in certain low cost applications such a security level is adequate.

In situation where a higher security level is needed key whitening, which we define here as follows:

$$DESX_{k.k1.k2}(x) = k2 \oplus DES_k(k1 \oplus x)$$

can be added to standard DES, yielding DESX. The bank of XOR gates and registers increase the gate count by about 14%[4]. The best known key search attack uses a time-memory trade-off and requires 2^{120} time steps and 2^{64} memory locations, which renders this attack entirely out of reach. The best known mathematical attack is linear cryptanalysis [Mat94]. Linear cryptanalysis requires about 2^{43} chosen ciphertext blocks together with the corresponding plaintexts. At a clock speed of 500 kHz, our DESX implementation will take more than 80 years, so that analytical attacks do not pose a realistic threat. Please note that parallelization is only an option if devices with identical keys are available.

In situations where extremely lightweight cryptography is needed, we can further decrease the gate complexity of DES by replacing the eight original S-Boxes by a single new one. This lightweight variant of DES is named DESL and has a brute-force resistance of 2^{56}. In order to strengthen the cipher, key whitening can be applied yielding the cipher DESXL. The crucial question is what the strength of DESL and DESXL is with respect to analytical attacks. We are fully aware that any changes to a cipher might open the door to new attacks, even if the changes have been done very carefully and checked against known attacks. Hence, we believe that DESL (or DESXL) should primarily not be viewed as competitors to AES, but should be used in applications where established algorithms are too costly. In such applications which have to trade security (really: trust in an algorithm) for cost, we argue that it is a cryptographically sounder approach to modestly modify a well studied cipher (in fact, the world's best studied crypto algorithm), rather than designing a new algorithm altogether.

4   Design Criteria of DESL

In this section we describe how a variant of DES with a single S-box can be made resistant against the differential, linear, and Davis-Murphy attack. The work is based on the original design criteria for DES as published by Coppersmith [Cop94] and the work of Kim et al. [KPL93], [KLPL94], [KLPL95] where several criteria for DES type S-boxes are presented to strengthen the resistance against the above mentioned attacks.

Coppersmith states the following eight criteria as the "only cryptographically relevant" ones for the DES S-boxes (see [Cop94]).

---

[4] This number only includes additional XOR gates, because we assume that all keys have to be stored at different memory locations anyway.

(S-1) Each S-box has six bits of input and four bits of output.

(S-2) No output bit of an S-box should be too close to a linear function of the input bits.

(S-3) If we fix the leftmost and rightmost input bits of the S-box and vary the four middle bits, each possible 4-bit output is attained exactly once as the middle input bits range over their 16 possibilities.

(S-4) If two inputs to an S-box differ in exactly one bit, the outputs must differ in at least two bits.

(S-5) If two inputs to an S-box differ in the two middle bits exactly, the outputs must differ in at least two bits.

(S-6) If two inputs to an S-box differ in their first two bits and are identical in their last two bits, the two outputs must not be the same.

(S-7) For any nonzero 6-bit-difference between inputs, ΔI, no more than eight of the 32 pairs of inputs exhibiting ΔI may result in the same output difference ΔO.

(S-8) Minimize the probability that a non zero input difference to three adjacent S-boxes yield a zero output difference.

4.1   Improved Resistance Against Differential Cryptanalysis and Davis Murphy Attack

The criteria (S-1) to (S-7) refer to one single S-box. The only criterion which deals with the combination of S-boxes is criterion (S-8). The designers' goal was to minimize the probability of collisions at the output of the S-boxes and thus at the output of the *f-function*. As a matter of fact, it is only possible to cause a collision, i.e. two different inputs are mapped to the same output, in three adjacent S-boxes, but not in a single S-box or a pair of S-boxes due to the diffusion caused by the expansion permutation. The possibility to have a collision in three adjacent S-boxes leads to the most successful differential attack based on a 2-round iterative characteristic with probability $\frac{1}{234}$.

Clearly better than minimizing the probability for collisions in three or more adjacent S-boxes, is to eliminate them. This was the approach used in [KPL93], [KLPL94], [KLPL95] and can easily be reached by improving one of the design criteria.

We replace (S-6) and (S-8) by an improved design criterion similar to the one given in [KPL93].

Condition 1. *If two inputs to an S-box differ in their first bit and are identical in their last two bits, the two outputs must not be the same.*

This criterion ensures that differential attacks using 2-round iterative characteristics, as the one presented by Biham and Shamir in [BS92], will have all eight S-boxes active and therefore will not be more efficient than exhaustive search anymore.

Moreover, the only criterion that refers to more than one S-box, i.e. (S-8), is now replaced by a condition that refers to one S-box, only. Thus, most of the security analysis remains unchanged when we replace the eight different S-boxes by one S-box repeated eight times.

Note that as described by Biham in [BB97] and by Kim et at. in [KLPL95] this condition also ensures resistance against the Davis Murphy attack [DM95].

4.2 Improved Resistance Against Linear Cryptanalysis

To improve the resistance of our variant of DES with only one S-Box against linear cryptanalysis (LC) is more complex than the protection against the differential cryptanalysis. Kim et.al presented a number of conditions that, when fulfilled by a set of S-boxes, ensure the resistance of DES variants against LC. However several of these conditions focus on different S-boxes and this implies that if one wants to replace all eight S-boxes by just one S-box, there are very tight restrictions to the choice of the S-box. This one S-box has to fulfill *all* conditions given in [KLPL95] referring to *any* S-box.

Let $S_b = \langle b, S(x) \rangle$ denote a combination of output bits that is determined by $b \in \mathrm{GF}(2)^4$. Then, the *Walsh-coefficient* $S_b^{\mathcal{W}}(a)$ for an element $a \in \mathrm{GF}(2)^6$ is defined by

$$S_b^{\mathcal{W}}(a) = \sum_{x \in \mathrm{GF}(2)^6} (-1)^{\langle b, S(x) \rangle + \langle a, x \rangle}. \tag{1}$$

The probability of a linear approximation of a combination of output bits S_b by a linear combination a of input bits can be written as

$$p = \frac{\#\{x | S_b(x) = \langle a, x \rangle\}}{2^6}. \tag{2}$$

Combining equations 1 and 2 leads to

$$p = \frac{S_b^{\mathcal{W}}(a)}{2^7} + \frac{1}{2}.$$

The *linear probability bias* ε is a correlation measure for this deviation from probability $\frac{1}{2}$ for which it is entirely uncorrelated. We have

$$\varepsilon = \left| p - \frac{1}{2} \right| = \left| \frac{S_b^{\mathcal{W}}(a)}{2^7} \right|.$$

Let us denote the maximum absolute value of the Walsh-Transformation by $S_{max}^{\mathcal{W}}$. Then clearly

$$\varepsilon \le \left| \frac{S_{max}^{\mathcal{W}}(a)}{2^7} \right|$$

The smaller the linear probability bias ε is, the more secure the S-box is against linear cryptanalysis. We defined our criterion (S-2") by setting the threshold for $S_{max}^{\mathcal{W}}$ to 28.

Condition 2. $|S_b^{\mathcal{W}}(a)| \le 28$ *for all* $a \in \mathrm{GF}(2)^6$, $b \in \mathrm{GF}(2)^4$.

Note that this is a tightened version of Condition 2 given in [KLPL95] where the threshold was set to 32. In the original DES the best linear approximation has a maximum absolute Walsh coefficient of 40 for S-box S5.

If an LC attack is based on an approximation that involves n S-boxes, under the standard assumption that the round keys are statistically independent, the overall bias ε is (see [Mat94])

$$\varepsilon = 2^{n-1} \prod_{i=1}^{n} \varepsilon_i$$

where the values ε_i are the biases for each of the involved S-box.

A rough approximation of the effort of a linear attack based on a linear approximation with bias ε is ε^{-2}, thus if we require that such an attack is no more efficient than exhaustive search we need $\varepsilon < 2^{-28}$.

It can be easily seen that any linear approximation for 15 round DES involves at least 7 approximations for S-boxes. But as

$$2^6 \prod_{i=1}^{7} \varepsilon_i \leq 2^6 \prod_{i=1}^{7} \frac{7}{32} \approx 2^{-9.35}$$

this bound is clearly insufficient.

Thus in order to prove the resistance against linear attack, we have to make sure that either enough S-boxes are active, i.e. enough S-Boxes are involved in the linear approximation, or, if fewer S-boxes are active, the bound on the probabilities can be tightened. In the first case we need more than 23 active S-boxes as

$$2^{21} \left(\frac{S_{max}^{W}}{128} \right)^{22} > 2^{-28} > 2^{22} \left(\frac{S_{max}^{W}}{128} \right)^{23} \tag{3}$$

For the second case several conditions have been developed in [KLPL94], [KLPL95]. Due to our special constraints we have to slightly modify these conditions. Following [KLPL95] we discuss several cases of iterative linear approximations. We denote a linear approximation of the F function of DES by

$$\langle I, Z_1 \rangle + \langle K, Z_3 \rangle = \langle O, Z_2 \rangle$$

where $Z_1, Z_2, Z_3 \in \mathrm{GF}(2)^{32}$ specify the input, output and key bits used in the linear approximation.

An n round iterative linear approximation is of the form

$$\langle I_1, \cdot \rangle + \langle I_n, \cdot \rangle = \langle K_2, \cdot \rangle + \cdots + \langle K_{n-1}, \cdot \rangle$$

and consists of linear approximations for the rounds 2 until $n - 1$.

Similar as it was done in [KLPL94] it can be shown that a three round (3R) iterative linear approximation is not possible with a non zero bias, due to condition 1.

We therefore focus on the case of a 4 and 5 round iterative approximation only.

4.3   4R Iterative Linear Approximation

A four round iterative linear approximation consists of two linear approximations for the F function of the second and third round. We denote these approximations as

$$A : \langle I_2, Z_1 \rangle + \langle K_2, Z_3 \rangle = \langle O_2, Z_2 \rangle$$
$$B : \langle I_3, Y_1 \rangle + \langle K_3, Y_3 \rangle = \langle O_3, Y_2 \rangle$$

In order to get a linear approximation of the form

$$\langle I_1, \cdot \rangle + \langle I_4, \cdot \rangle = \langle K_2, \cdot \rangle + \langle K_3, \cdot \rangle$$

Using $O_2 = I_1 + I_3$ and $O_3 = I_2 + I_4$ it must hold that

$$Z_2 = Y_1 \text{ and } Z_1 = Y_2$$

The 15 round approximation is

$$-AB - BA - AB - BA - AB$$

If the number of S-boxes involved in the approximation of A is a and for B is b we denote by $\mathcal{A} = (a, b)$. First assume that $\mathcal{A} = (1, 1)$. Due to $Z_2 = Y_1$ and the property of the P-permutation, which distributes the output bits of one S-box to 6 different S-Boxes in the next round, it must hold that $|Y_1| = |Z_2| = 1$. For the same reason we get $|Z_1| = |Y_2| = 1$. To minimize the probability of such an approximation we stipulate the following condition

Condition 3. *The S-box has to fulfill $S_b^{\mathcal{W}}(a) \leq 4$ for all $a \in \mathrm{GF}(2)^6, b \in \mathrm{GF}(2)^4$ with $\mathrm{wt}(a) = \mathrm{wt}(b) = 1$.*

This condition is comparable to Condition 4 in [KLPL95], however, as we only have a single S-box, we could not find a single S-box fulfilling all the restrictions from condition 4 in [KLPL95]. If the S-box fulfils condition 3 the overall bias for the linear approximation described above is bounded by

$$\varepsilon \leq 2^9 \left(\frac{4}{128} \right)^{10} < 2^{-40}.$$

As this is (much) smaller than 2^{-28} this does not yield to a useful approximation.

Assume now that $\mathcal{A} = (1, 2)$ (the case $\mathcal{A} = (2, 1)$ is very similar). If B involves two S-boxes we have $|Y_1| = |Y_2| = 2$ and thus $|Y_2| = |Z_1| = 2$. In particular for both S-boxes involved in B Condition 3 applies which results in a threshold

$$\varepsilon \leq 2^{14} \left(\frac{4}{128} \right)^{10} \left(\frac{28}{128} \right)^5 < 2^{-46}$$

for the overall linear bias.

Next we assume that $\mathcal{A} = (2, 2)$. In this case we get (through the properties of the P function) that each S-box involved in A and B has at most two input and output bits involved in the linear approximation. In order to avoid this kind of approximation we add another condition.

Condition 4. *The S-box has to fulfill $S_b^{\mathcal{W}}(a) \leq 16$ for all $a \in \mathrm{GF}(2)^6, b \in \mathrm{GF}(2)^4$ with $\mathrm{wt}(a), \mathrm{wt}(b) \leq 2$.*

This condition is a tightened version of Condition 5 in [KLPL95] where the threshold was set to 20 . In this case (remember that we now have 20 S-boxes involved) we get

$$\varepsilon \leq 2^{19} \left(\frac{16}{128} \right)^{20} < 2^{-40}.$$

In all other cases, more than 23 S-boxes involved and thus the general upper bound (3) can be applied.

4.4 5R Iterative Linear Approximation

A five round iterative linear approximation consists of three linear approximations for the F function of the second, third and fourth round. We denote these approximations as

$$A : \langle I_2, Z_1 \rangle + \langle K_2, Z_3 \rangle = \langle O_2, Z_2 \rangle$$
$$B : \langle I_3, Y_1 \rangle + \langle K_3, Y_3 \rangle = \langle O_3, Y_2 \rangle$$
$$C : \langle I_4, X_1 \rangle + \langle K_4, X_3 \rangle = \langle O_4, X_2 \rangle.$$

In order to get a linear approximation of the form

$$\langle I_1, \cdot \rangle + \langle I_5, \cdot \rangle = \langle K_2, \cdot \rangle + \langle K_3, \cdot \rangle + \langle K_4, \cdot \rangle$$

it must hold that

$$Z_1 = Y_2 = X_1 \text{ and } Y_1 + Z_2 + X_2 = 0$$

The 15 round approximation is

$$-ABC - CBA - ABC - DE$$

for some linear approximations D and E each involving at least one S-box. Clearly, as the inputs of A and C are the same we have $\mathcal{A} = (a, b, a)$, i.e. the number of involved S-boxes in A and C are the same.

Case $b = 1$: Assume that $b = 1$, i.e. only one S-box is involved in the linear approximation B. If $|Z_1| \geq 3$ than we must have $a \geq 3$ and so the number of S-boxes involved is at least 23, which makes the approximation useless. If $|Z_1| = 2$ we have two active S-boxes for A and B. Moreover as $b = 1$ we must have $|Y_1| = |Z_2 + X_2| = 1$. Due to properties of the P function, the S-boxes involved in A and B are never adjacent S-boxes, therefore exactly one input bit is involved in the approximation for each of the two S-boxes. In order to minimize the probability for such an approximation, we stipulate the following condition

Condition 5. *The S-box has to fulfill*

$$|S_{b_1}^{\mathcal{W}}(a) S_{b_2}^{\mathcal{W}}(a)| \leq 240$$

for all $a \in GF(2)^6, b_1, b_2 \in GF(2)^4$ with $\mathrm{wt}(a) = 1, \mathrm{wt}(b_1 + b_2) = 1$.

This is a modified version of Condition 7 in [KLPL95]. With an S-box fulfilling this condition we derive an upper bound for the overall bias

$$\varepsilon \leq 2^{16} \left(\frac{240}{128^2} \right)^6 \left(\frac{16}{128} \right)^3 \left(\frac{28}{128} \right)^2 < 2^{-33}$$

If $|Z_1| = 1$ then $a = 1$ and we have $|Y_1| = |Z_2 + X_2| = 1$ and $|Z_1| = 1$. We stipulate one more condition.

Condition 6. *The S-box has to fulfill*

$$S_b^{\mathcal{W}}(a) = 0$$

for $a \in \{(010000), (000010)\}, b \in \mathrm{GF}(2)^4$ with $\mathrm{wt}(b) = 1$.

This implies that the input to B is such that a middle bit is affected. Due to the properties of the P function this implies that in the input of A and C a non-middle bit is affected. As for any DES type S-box it holds that $S_b^{\mathcal{W}}(100000) = S_b^{\mathcal{W}}(000001) = 0$ for all b the only possible input values for the S-box involved in A and C are (010000) and (000010). To avoid the second one we define the next condition.

Condition 7

$$|S_{b_1}^{\mathcal{W}}(000010) S_{b_2}^{\mathcal{W}}(000010)| = 0$$

for all $b_1, b_2 \in \mathrm{GF}(2)^4$ with $\mathrm{wt}(b_1 + b_2) = 1$.

The other possible input value, i.e. 01000 occurs only when S-box 1 is active in B and S-box 5 is active in A and C. In this case the input values for the S-box in B is (000100) and the output value is (0100). The next condition makes this approximation impossible.

Condition 8. *The S-box has to fulfill*

$$S_{(0100)}^{\mathcal{W}}(000100) = 0$$

Case $b = 2$: Assume that $b = 2$, i.e. exactly two S-boxes are involved for B. If $a > 2$ then at least 23 S-boxes are involved in total. If $a = 2$ we have for each S-box involved in B at most 2 input bits and at most 2 output bits. Therefore we can apply the bound from condition 4 to the two S-boxes from B. Applying the general bound for all the other S-boxes we get

$$\varepsilon \leq 2^{19} \left(\frac{16}{128} \right)^6 \left(\frac{28}{128} \right)^{14} < 2^{-29}.$$

In the case where $a = 1$ the two S-boxes involved in B have one input and one output bit involved each, thus we can apply the strong bound from condition 3 for these S-boxes (6 in total) and the general bound for the other S-boxes to get

$$\varepsilon \leq 2^{13} \left(\frac{4}{128} \right)^6 \left(\frac{28}{128} \right)^8 < 2^{-34}.$$

Case b > 2: In this case we must have $a, b \geq 2$ and thus at least 29 S-boxes are involved in total.

4.5   nR Iterative Linear Approximation

For an n round iterative linear approximation with only one S-box involved in each round (denoted as Type-I by Matsui) our condition 3 ensures that if more than 7 S-boxes are involved in total the approximation will not be useful for an attack as

$$\varepsilon \leq 2^6 \left(\frac{4}{128}\right)^7 = 2^{-29} \tag{4}$$

4.6   Resistance Against Algebraic Attacks

There is no structural reason why algebraic attacks should pose a greater threat to DESL than to DES. The DESL S-box has been randomly generated in the set of all S-boxes fulfilling the design criteria described above. Therefore we do not expect any special weakness of the chosen S-box. Indeed we computed the number of low degree equations between the input and output bits of the S-box. There exist one quadratic equation and 88 equations of degree 3. Note that for each 6 to 4 Bit S-box, there exist at least 88 equations of degree 3. Given the comparison with the corresponding results for the original DES S-boxes in Table 1 we anticipate that DESL is as secure as DES with respect to algebraic attacks.

Table 1. Number of Degree two and Degree three Equations

DES S-box	S1	S2	S3	S4	S5	S6	S7	S8	DESL
# deg 2	1	0	0	5	1	0	0	0	1
# deg 3	88	88	88	88	88	88	88	88	88

4.7   Improved S-Box

We randomly generated S-boxes, which fulfill the original DES criteria (S-1), (S-3), (S-4), (S-5), (S-7), the condition 1 and our modified conditions 2 to 8. Our goal was to find one single S-box, which is significantly more resistant against differential and linear cryptanalyses than the original eight S-boxes of DES. In our DESL algorithm this S-box is repeated eight times and replaces all eight S-boxes in DES.

5   Lightweight Implementation of DESL

We implemented DES in the hardware description language VHDL, where we sacrificed time for area wherever possible. In this serialized DES ASIC design, registers take up the main part of chip size (33.78%), followed by the S-boxes (32.11%), and multiplexors (31.19%). Chip size of registers and multiplexors can

Table 2. Improved DESL S-box

S
14 5 7 2 11 8 1 15 0 10 9 4 6 13 12 3
5 0 8 15 14 3 2 12 11 7 6 9 13 4 1 10
4 9 2 14 8 7 13 0 10 12 15 1 5 11 3 6
9 6 15 5 3 8 4 11 7 1 12 2 0 14 10 13

not be minimized any further, hence we thought about further possibilities to optimize the chip size of the S-boxes.

While it does not seem to be possible to find better logic minimizations of the original DES S-boxes, there have been other approaches to alter the S-boxes, e.g. key-dependent S-boxes [BB94], [BS92] or the so-called $s^i DES$ [KLPL94], [KLPL95], [KPL93]. While all these approaches, despite the fact that some of them have worse cryptographic properties than DES [Knu92], just change the content and not the **number** of S-boxes. To the best of our knowledge, no DES variant has been proposed in the past which uses a single S-box, repeated eight times.

The main difference between DESL and DES lies in the f-function. We substituted the eight original DES S-boxes by a single but cryptographically stronger S-box (see Table 2), which is repeated eight times. Furthermore, we omitted the initial permutation (IP) and its inverse (IP^{-1}), because they do not provide additional cryptographic strength, but at the same time require area for wiring. The design of our DESL algorithm is exactly the same as for the DES algorithm, except for the (IP) and (IP^{-1}) wiring and the *sbox* module.

We used Synopsys Design Vision V-2004.06-SP2 to map our DESL design to the Artisan UMC 0.18μm L180 Process 1.8-Volt Sage-X Standard Cell Library and Cadence Silicon Ensemble 5.4 for the Placement & Routing-step. Synopsys NanoSim was used to simulate the power consumption of the back-annotated verilog netlist of the ASIC.

Our serialized DESL ASIC implementation has an area requirement of 1848 GE (gate equivalences) and it takes 144 clock cycles to encrypt one 64-bit block of plaintext. For one encryption at 100 kHz the average current consumption is 0.89 μA and the throughput reaches 5.55 KB/s. For further details on the implementational aspects of our DES and DESL architecture we refer to [PLSP07].

6   Results and Conclusion

In Section 2 we stated eight conditions which a single S-box has to fulfill in order to be resistant against certain types of linear and differential cryptanalyses, and the Davies-Murphy attack. We presented a strengthened S-box, which is used in the single S-box DES variants DESL and DESXL. Furthermore, we showed, that a differential cryptanalysis with characteristics similar to the characteristics used by Biham and Shamir in [BS91] is not feasible anymore. We also showed, that

DESL is more resistant against the most promising types of linear cryptanalysis than DES due to the improved non-linearity of the S-box.

Table 3 shows, that our DESL cipher needs 20% less gate equivalences and uses 25% less average current than our DES implementation. In comparison with the AES design presented by Feldhofer et al. [FWR05], our design needs 45% less gate equivalents and 86% less clock cycles. Note that the AES design by Feldhofer et al. was implemented in a $0.35\mu m$ standard cell technology, whereas our design was implemented in a $0.18\mu m$ standard cell technology. Therefore a fair comparison is only possible with regard to the gate equivalences. Regarding area consumption, our DESL is competitive even to stream ciphers recently proposed within the eSTREAM project [GB07]. More interesting, DESL would be the second smallest stream cipher in terms of gate count compared to all eSTREAM candidates (see Table 3). Due to the low current consumption and the small chip size required for our DESL design, it is especially suited for resource limited applications, for example RFID tags and wireless sensor nodes.

Table 3. Comparison of Efficient Ciphers based on Gate Count, Clock Cycles, and Current Consumption

	gate equiv.		cycles /	μA at	Process
	total	rel.	block	100 kHz	μm
DESL	**1848**	**1**	**144**	**0.89**	0.18
DES	2309	1.25	144	1.19	0.18
DESX	2629	1.42	144	–	0.18
DESXL	2168	1.17	144	–	0.18
AES-128 [FWR05]	3400	1.84	1032	3.0	0.35
HIGHT [HSH+06]	3048	1.65	1	–	0.25
Trivium [GB07]	2599	1.41	–	–	0.13
Grain-80 [GB07]	1294	0.70	–	–	0.13

Finally, we can conclude, that DESL is more secure against certain types of linear and differential Cryptanalyses and the Davies-Murphy attack, more size-optimized, and more power efficient than DES. Furthermore, DESL is worth to be considered as an alternative for stream ciphers.

Acknowledgments

The authors would like to thank Matt Robshaw for his insights and valuable comments on various aspects of the S-boxes. The work presented in this paper was supported in part by the European Commission within the STREP UbiSec&Sens of the EU Framework Programme 6 for Research and Development (www.ist-ubisecsens.org). The views and conclusions contained herein are those of the authors and should not be interpreted as necessarily representing the official policies or endorsements, either expressed or implied, of the UbiSecSens project or the European Commission.

References

[ASM01] Satoh, K.T.A., Morioka, S., Munetoh, S.: A Compact Rijndael Hardware Architecture with S-Box Optimization. In: Boyd, C. (ed.) ASIACRYPT 2001. LNCS, vol. 2248, pp. 239–254. Springer, Heidelberg (2001)

[BB94] Biham, Biryukov,: How to Strengthen DES Using Existing Hardware. In: Safavi-Naini, R., Pieprzyk, J.P. (eds.) ASIACRYPT 1994. LNCS, vol. 917, Springer, Heidelberg (1995), available for download at `citeseer.ist.psu.edu/biham94how.html`

[BB97] Biham, E., Biryukov, A.: An Improvement of Davies' Attack on DES. Journal of Cryptology: the journal of the International Association for Cryptologic Research 10(3), 195–205 (1997), available for download at `citeseer.ist.psu.edu/467934.html`

[BS91] Biham, E., Shamir, A.: Differential Cryptanalysis of DES-like Cryptosystems. In: Menezes, A.J., Vanstone, S.A. (eds.) CRYPTO 1990. LNCS, vol. 537, pp. 2–21. Springer, Heidelberg (1991)

[BS92] Biham, E., Shamir, A.: Differential Cryptanalysis of the Full 16-Round DES. In: Brickell, E.F. (ed.) CRYPTO 1992. LNCS, vol. 740, pp. 487–496. Springer, Heidelberg (1993), available for download at `citeseer.ist.psu.edu/biham93differential.html`

[Cop94] Coppersmith, D.: The Data Encryption Standard (DES) and its Strength Against Attacks. Technical report rc 186131994, IBM Thomas J. Watson Research Center (December 1994)

[DM95] Davies, D., Murphy, S.: Pairs and Triplets of DES S-Boxes. Journal of Cryptology 8(1), 1–25 (1995)

[DR02] Daemen, J., Rijmen, V.: The Design of Rijndael. Springer Verlag, Berlin, Heidelberg (2002)

[FDW04] Feldhofer, M., Dominikus, S., Wolkerstorfer, J.: Strong Authentication for RFID Systems. In: Joye, M., Quisquater, J.-J. (eds.) CHES 2004. LNCS, vol. 3156, pp. 357–370. Springer, Heidelberg (2004)

[Fin03] Finkenzeller, K.: RFID-Handbook: Fundamentals and Applications in Contactless Smart Cards and Identification. John Wiley and Sons, Chichester (2003)

[FWR05] Feldhofer, M., Wolkerstorfer, J., Rijmen, V.: AES Implementation on a Grain of Sand. Information Security, IEE Proceedings 152(1), 13–20 (2005)

[GB07] Good, T., Benaissa, M.: Hardware Results for selected Stream Cipher Candidates. State of the Art of Stream Ciphers 2007 (SASC 2007), Workshop Record (February 2007)

[HSH+06] Hong, D., Sung, J., Hong, S., Lim, J., Lee, S., Koo, B.-S., Lee, C., Chang, D., Lee, J., Jeong, K., Kim, H., Kim, J., Chee, S.: HIGHT: A New Block Cipher Suitable for Low-Resource Device. In: Goubin, L., Matsui, M. (eds.) CHES 2006. LNCS, vol. 4249, pp. 46–59. Springer, Heidelberg (2006)

[KLPL94] Kim, K., Lee, S., Park, S., Lee, D.: DES Can Be Immune to Linear Cryptanalysis. In: Proceedings of the Workshop on Selected Areas in Cryptography SAC'94, pp. 70–81 (May 1994), available for download at `citeseer.csail.mit.edu/kim94des.html`

[KLPL95] Kim, K., Lee, S., Park, S., Lee, D.: Securing DES S-boxes Against Three Robust Cryptanalysis. In: Proceedings of the Workshop on Selected Areas in Cryptography SAC'95, pp. 145–157 (1995), available for download at citeseer.ist.psu.edu/kim95securing.html

[Knu92]   Knudsen, L.R.: Iterative Characteristics of DES and s^2-DES. In: Brickell, E.F. (ed.) CRYPTO 1992. LNCS, vol. 740, pp. 497–511. Springer, Heidelberg (1993)

[KPL93]   Kim, K., Park, S., Lee, S.: Reconstruction of s^2-DES S-Boxes and their Immunity to Differential Cryptanalysis. In: Proceedings of 1993 Korea-Japan Joint Workshop on Information Security and Cryptology (JW-ISC'93) (October 1993), available for download at citeseer.csail.mit.edu/kim93reconstruction.html

[KPP+06]   Kumar, S., Paar, C., Pelzl, J., Pfeiffer, G., Schimmler, M.: Breaking Ciphers with COPACOBANA - A Cost-Optimized Parallel Code Breaker. In: Goubin, L., Matsui, M. (eds.) CHES 2006. LNCS, vol. 4249, Springer, Heidelberg (2006)

[Mat94]   Matsui, M.: Linear Cryptanalysis of DES Cipher. In: Helleseth, T. (ed.) EUROCRYPT 1993. LNCS, vol. 765, pp. 286–397. Springer, Heidelberg (1994)

[PLSP07]   Poschmann, A., Leander, G., Schramm, K., Paar, C.: New Ligh-Weight Crypto Algorithms for RFID. In: Proceedings of The IEEE International Symposium on Circuits and Systems 2007 – ISCAS 2007, IEEE Computer Society Press, Los Alamitos, 2007 (to appear)

[RE02]   Rankl, W., Effing, W.: Smart Card Handbook. Carl Hanser Verlag, München, Germany, 2nd edn. (2002)

[VHVM88]   Verbauwhede, I., Hoornaert, F., Vandewalle, J., De Man, H.: Security and Performance Optimization of a New DES Data Encryption Chip. IEEE Journal of Solid-State Circuits 23(3), 647–656 (1988)

A New Attack on 6-Round IDEA

Eli Biham[1,*], Orr Dunkelman[2,*], and Nathan Keller[3,**]

[1] Computer Science Department, Technion
Haifa 32000, Israel
biham@cs.technion.ac.il
[2] Katholieke Universiteit Leuven, Dept. of Electrical Engineering ESAT/SCD-COSIC
Kasteelpark Arenberg 10, B-3001 Leuven-Heverlee, Belgium
orr.dunkelman@esat.kuleuven.be
[3] Einstein Institute of Mathematics, Hebrew University
Jerusalem 91904, Israel
nkeller@math.huji.ac.il

Abstract. IDEA is a 64-bit block cipher with 128-bit keys introduced by Lai and Massey in 1991. IDEA is one of the most widely used block ciphers, due to its inclusion in several cryptographic packages, such as PGP. Since its introduction in 1991, IDEA has withstood extensive cryptanalytic effort, but no attack was found on the full (8.5-round) variant of the cipher.

In this paper we present the first known attack on 6-round IDEA faster than exhaustive key search. The attack exploits the weak key-schedule algorithm of IDEA, and combines Square-like techniques with linear cryptanalysis to increase the number of rounds that can be attacked. The attack is the best known attack on IDEA. We also improve previous attacks on 5-round IDEA and introduce a 5-round attack which uses only 16 known plaintexts.

1 Introduction

The International Data Encryption Algorithm (IDEA) is a 64-bit, 8.5-round block cipher with 128-bit keys proposed by Lai and Massey in 1991 [19]. Due to its inclusion in several cryptographic packages, such as PGP , IDEA is one of the most widely used block ciphers. Since its introduction, IDEA resisted intensive cryptanalytic efforts [1,2,3,5,6,9,10,11,13,14,15,16,17,21,22,23]. The best published chosen-plaintext attack on IDEA is an attack on 5-round IDEA that requires 2^{19} chosen plaintexts, and has time complexity of 2^{103} encryptions [3]. The best published related-key attack is an attack on 7.5-round IDEA that requires $2^{43.5}$ known plaintexts and has a time complexity of $2^{115.1}$ encryptions [3]. Along with the attacks on reduced-round variants, several weak-key

---

* This work was supported in part by the Israel MOD Research and Technology Unit.
** The research presented in this paper was supported by the Adams fellowship.

A. Biryukov (Ed.): FSE 2007, LNCS 4593, pp. 211–224, 2007.

classes for the entire IDEA were found. The largest weak key class (identified by a boomerang technique) contains 2^{64} keys, and the membership test requires 2^{16} adaptive chosen plaintexts and ciphertexts and has a time complexity of 2^{16} encryptions [6].

In this paper we present the first known attacks against 5.5-round and 6-round IDEA. These higher-order differential-linear [4] attacks consist of three components:

1. Constructing linear equations involving the least significant bits of the intermediate values of the cipher. We note that this idea was proposed and used in [3,17,23].
2. Using a higher-order differential (or a Square property) to simplify the linear equations. We note that this modification was proposed in [17]. However, as we show later, the Square distinguisher used in [17] is incorrect, and hence, we replace it by another distinguisher.
3. Taking advantage of the weak key schedule of IDEA — we observe that in some cases, guessing only 103 of the 128 key bits of IDEA is sufficient for encrypting two full rounds of the cipher, and even more than that.

The 5.5-round attack requires 2^{32} chosen plaintexts and has a time complexity of about 2^{127} encryptions, about twice faster than exhaustive key search. The 6-round attack requires almost the entire code book and has a time complexity similar to that of the 5.5-round attack. We note that the time complexity of the attacks could be improved significantly if the Square distinguisher could be replaced by a better one, like the one presented in [17]. However, we were not able to find such a distinguisher at this stage.

We then show two improvements to the 5-round attack presented in [3]. The first improvement reduces the data complexity of the attack by a factor of $\sqrt{2}$ to $2^{18.5}$ known plaintexts, without affecting the time complexity of the attack of 2^{103} encryptions. The second improvement reduces the data complexity to 16 known plaintexts, while raising the data complexity to 2^{114} encryptions.

The complexities of the new attacks, along with selected previously known attacks, are summarized in Table 1.

The paper is organized as follows: In Section 2 we briefly describe the structure of IDEA. In Section 3 we present the new attack on 5.5-round IDEA. In Section 4 we extend the 5.5-round attack to an attack on 6-round IDEA. We present improved attacks on 5-round IDEA in Section 5. Finally, Section 6 summarizes the paper.

2 Description of IDEA and the Notations Used in the Paper

IDEA [19] is a 64-bit, 8.5-round block cipher with 128-bit keys. It uses a composition of XOR operations, additions modulo 2^{16}, and multiplications over $GF(2^{16} + 1)$.

Table 1. Selected Known Attacks on IDEA and Our New Results

Rounds	Attack Type	Complexity		Source
		Data	Time	
4	Impossible Differential	2^{37} CP	2^{70}	[2]
4	Linear	114 KP	2^{114}	[23]
4	Square	2^{32} CP	2^{114}	[13]
4	Square	2^{23} CP	2^{98}	[17][†]
4.5	Impossible Differential	2^{64} CP	2^{112}	[2]
4.5	Linear	16 CP	2^{103}	[3]
5	Meet-in-the-Middle Attack	2^{24} CP	2^{126}	[14]
5	Meet-in-the-Middle Attack	$2^{24.6}$ CP	2^{124}	[1]
5	Linear	2^{19} KP	2^{103}	[3]
5	Linear	$2^{18.5}$ KP	2^{103}	Section 5
5	Linear	16 KP	2^{114}	Section 5
5.5	Higher-Order Differential-Linear	2^{32} CP	$2^{126.85}$	Section 3
6	Higher-Order Differential-Linear	$2^{64} - 2^{52}$ KP	$2^{126.8}$	Section 4

KP – Known plaintexts, CP – Chosen plaintexts
Time complexity is measured in encryption units
[†] – As we show in Section 3.2, this attack does not work

Every round of IDEA is composed of two layers. The round input of round i is composed of four 16-bit words denoted by $(X_1^i, X_2^i, X_3^i, X_4^i)$. In the first layer, denoted by KA, the first and the fourth words are multiplied by subkey words (mod $2^{16} + 1$) where 0 is replaced by 2^{16}, and the second and the third words are added to subkey words in (mod 2^{16}). The intermediate values after this half-round are denoted by $(Y_1^i, Y_2^i, Y_3^i, Y_4^i)$. Formally, let Z_1^i, Z_2^i, Z_3^i, and Z_4^i be the four subkey words, then

$$Y_1^i = Z_1^i \odot X_1^i; \quad Y_2^i = Z_2^i \boxplus X_2^i; \quad Y_3^i = Z_3^i \boxplus X_3^i; \quad Y_4^i = Z_4^i \odot X_4^i$$

Then, $(p^i, q^i) = (Y_1^i \oplus Y_3^i, Y_2^i \oplus Y_4^i)$ enters to the second layer, a structure composed of multiplications and additions denoted by MA. We denote the two output words of the MA transformation by (u^i, t^i). Denoting the subkey words that enter the MA function by Z_5^i and Z_6^i,

$$u^i = (p^i \odot Z_5^i) \boxplus t^i; \quad t^i = (q^i \boxplus (p^i \odot Z_5^i)) \odot Z_6^i$$

Another notation we use in the attack refers to the intermediate value in the MA layer: we denote the value $p^i \odot Z_5^i$ by s^i.

The output of the i-th round is $(Y_1^i \oplus t^i, Y_3^i \oplus t^i, Y_2^i \oplus u^i, Y_4^i \oplus u^i)$. In the last round (round 9) the MA layer is omitted. Thus, the ciphertext is $(Y_1^9 || Y_2^9 || Y_3^9 || Y_4^9)$. The structure of a single round of IDEA is shown in Figure 1.

IDEA's key schedule is linear. Each subkey is composed of bits selected from the key. However, the exact structure of the key schedule is crucial for our attacks and hence the entire key schedule is described in Table 2.

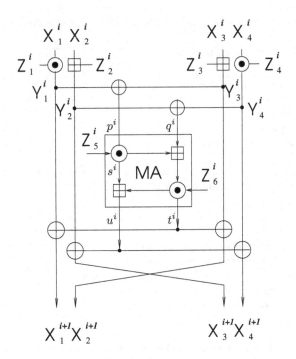

Fig. 1. One Round of IDEA

Table 2. The Key Schedule Algorithm of IDEA

Round	Z_1^i	Z_2^i	Z_3^i	Z_4^i	Z_5^i	Z_6^i
$i = 1$	0–15	16–31	32–47	48–63	64–79	80–95
$i = 2$	96–111	112–127	25–40	41–56	57–72	73–88
$i = 3$	89–104	105–120	121–8	9–24	50–65	66–81
$i = 4$	82–97	98–113	114–1	2–17	18–33	34–49
$i = 5$	75–90	91–106	107–122	123–10	11–26	27–42
$i = 6$	43–58	59–74	100–115	116–3	4–19	20–35
$i = 7$	36–51	52–67	68–83	84–99	125–12	13–28
$i = 8$	29–44	45–60	61–76	77–92	93–108	109–124
$i = 9$	22–37	38–53	54–69	70–85		

3   A New Attack on 5.5-Round IDEA

In this section we present the new attack on 5.5-round IDEA. First we present the three components of the attack.

3.1   The First Component — A Linear Equation Involving the LSBs of the Intermediate Encryption Values

We start with an observation due to Biryukov (according to [23]) and Demirci [14]. Let us examine the second and the third words in all the intermediate stages of the encryption. There is a relation between the values of these words and the outputs of the MA layer in the intermediate rounds that uses only XOR and modular addition, but not multiplication. Let $P = (P_1, P_2, P_3, P_4)$ be a plaintext and let $C = (C_1, C_2, C_3, C_4)$ be its corresponding ciphertext, then

$$((((((((((((((P_2 \boxplus Z_2^1) \oplus u^1) \boxplus Z_3^2) \oplus t^2) \boxplus Z_3^3) \oplus u^3) \boxplus Z_3^4) \oplus t^4) \boxplus Z_2^5) \oplus u^5)$$
$$\boxplus Z_3^6) \oplus t^6) \boxplus Z_2^7) \oplus u^7) \boxplus Z_3^8) \oplus t^8) \boxplus Z_2^9) = C_2.$$

$$(1)$$

Similarly,

$$(((((((((((((((P_3 \boxplus Z_3^1) \oplus t^1) \boxplus Z_2^2) \oplus u^2) \boxplus Z_3^3) \oplus t^3) \boxplus Z_2^4) \oplus u^4) \boxplus Z_3^5) \oplus t^5)$$
$$\boxplus Z_2^6) \oplus u^6) \boxplus Z_2^7) \oplus t^7) \boxplus Z_2^8) \oplus u^8) \boxplus Z_3^9) = C_3.$$

$$(2)$$

When we consider the value of the least significant bit (LSB) of the words, modular addition is equivalent to XOR and we can simplify the above equations into:

$$LSB(P_2 \oplus Z_2^1 \oplus u^1 \oplus Z_3^2 \oplus t^2 \oplus Z_3^3 \oplus u^3 \oplus Z_3^4 \oplus t^4 \oplus Z_2^5 \oplus u^5 \oplus Z_3^6 \oplus t^6 \oplus Z_2^7$$
$$\oplus u^7 \oplus Z_3^8 \oplus t^8 \oplus Z_2^9) = LSB(C_2),$$

$$(3)$$

and

$$LSB(P_3 \oplus Z_3^1 \oplus t^1 \oplus Z_2^2 \oplus u^2 \oplus Z_3^3 \oplus t^3 \oplus Z_2^4 \oplus u^4 \oplus Z_3^5 \oplus t^5 \oplus Z_2^6 \oplus u^6 \oplus Z_3^7$$
$$\oplus t^7 \oplus Z_2^8 \oplus u^8 \oplus Z_3^9) = LSB(C_3).$$

$$(4)$$

Since $u^i = t^i \boxplus s^i$ then $LSB(u^i) = LSB(t^i \boxplus s^i)$, thus, $LSB(u^i \oplus t^i) = LSB(s_i)$. Taking this into consideration and XORing the above two equations we obtain

$$LSB(P_2 \oplus P_3 \oplus Z_2^1 \oplus Z_3^1 \oplus s^1 \oplus Z_2^2 \oplus Z_3^2 \oplus s^2 \oplus Z_2^3 \oplus Z_3^3 \oplus s^3 \oplus Z_2^4 \oplus Z_3^4 \oplus s^4$$
$$\oplus Z_2^5 \oplus Z_3^5 \oplus s^5 \oplus Z_2^6 \oplus Z_3^6 \oplus s^6 \oplus Z_2^7 \oplus Z_3^7 \oplus s^7 \oplus Z_2^8 \oplus Z_3^8 \oplus s^8 \oplus Z_2^9 \oplus Z_3^9)$$
$$= LSB(C_2 \oplus C_3).$$

$$(5)$$

This equation is called in [17] "the Biryukov-Demirci relation".

In order to simplify this equation, we consider the XOR of the intermediate values of several encrypted plaintexts. In [3] the XOR difference between two plaintexts is used; in our attack we use the XOR value of larger sets of plaintexts.

Consider a structure of plaintexts $P^1, \ldots, P^m$, where m is an even integer. Then the XOR of the equations of the form (5) given by $P^1, \ldots, P^m$ gives

$$LSB \left(\bigoplus_{j=1}^{m} (P_2^j \oplus P_3^j) \oplus \bigoplus_{i=1}^{8} S^i \right) = LSB \left(\bigoplus_{j=1}^{m} (C_2^j \oplus C_3^j) \right), \qquad (6)$$

where $S^i = \oplus_{j=1}^m s^i(P^j)$.

Equation (6) is the basic equation used in our attack, where m and the exact structure of plaintexts are specified later.

3.2 The Second Component — A Square-Like Structure

In order to further simplify Equation (6) we want to use special structures of plaintexts, for which we will get $S^1 = S^2 = 0$, independently of the key. Our structures are higher-order differentials, a special case of Square-like structures that were used in [7,12,18,20]. The structures we use and their properties are described in the following proposition.

Proposition 1. *Let T be a structure of $m = 2^{16}$ plaintexts, such that the intermediate values after the first KA layer, denoted by $Y^{1,j}$, where $1 \leq j \leq m$, satisfy the following requirements:*

1. *$Y_2^{1,j}$ and $Y_4^{1,j}$ are fixed for all j.*
2. *The 2^{16} values $Y_1^{1,j}$ (for $1 \leq j \leq 2^{16}$) are different.*
3. *$Y_3^{1,j} = Y_1^{1,j} \oplus C$ for some fixed C.*

Then for $T = \{P^1, \ldots, P^m\}$ the relation $S^1 = S^2 = 0$ holds, independently of the key.

Proof. First, note that by the assumption, (p^1, q^1) is fixed for all the 2^{16} elements of the structure. Hence, s^1 is fixed as well, leading to $S^1 = \bigoplus_{j=1}^{2^{16}} s^1(P^j) = 0$. The output values of the MA structure, (u^1, t^1), are also fixed. Thus, the values X_3^2 are constant for all the elements of the structure, as well as the values X_4^2. In addition, all the values X_1^2 are different, as well as the values X_2^2. As a result, all the values Y_1^2 are different and all the values Y_3^2 are constant. Hence, all the values $p^2 = Y_1^2 \oplus Y_3^2$ are different. Thus, all the values $s^2 = Z_5^2 \odot p^2$ are different. However, since there are only 2^{16} possible values of s^2, it means that s^2 assumes each possible value once and only once. Hence, $S^2 = \bigoplus_{j=1}^{2^{16}} s^2(P^j) = 0$, and this completes the proof.

In the sequel of the paper we call structures which satisfy the above conditions, "right structures".

It follows from the proposition that if we take a right structure as $\{P^1, \ldots, P^m\}$, Equation (6) is simplified to

$$LSB\left(\bigoplus_{j=1}^{m}(P_2^j \oplus P_3^j) \oplus \bigoplus_{i=3}^{8} s^i\right) = LSB\left(\bigoplus_{j=1}^{m}(C_2^j \oplus C_3^j)\right). \tag{7}$$

We note that [17] makes use of a seemingly better Square-like structure that is described in the following statement [17, Section 3.7, Lemma 2]:

Statement 1. *Let L be a structure of $m = 2^{16}$ plaintexts, denoted by $P^1, \ldots, P^m$, having the following properties:*

1. *P_1^j, P_2^j, P_3^j are fixed for all j.*
2. *The 2^{16} values P_4^j (for $1 \leq j \leq 2^{16}$) are different.*

Then for $T = \{P^1, \ldots, P^m\}$ *we have* $S^1 = S^2 = 0$, *independently of the key.*

If this statement was correct, it could be used to improve significantly the time complexity of our attack. However, it appears that the statement is incorrect. For the structure described in the statement we indeed have $S^1 = 0$, but we do not have $S^2 = 0$, since nothing can be said about the values $s^2(P^j)$. The following set of constants is a counterexample for Statement 1:

$$P_1^j = F78b_x; \quad P_2^j = 245_x; \quad P_3^j = ABCD_x;$$
$$Z_1^1 = 8_x; \quad Z_2^1 = 567A_x; \quad Z_3^1 = 2C68_x; \quad Z_4^1 = 4_x;$$
$$Z_5^1 = 5_x; \quad Z_6^1 = 6_x; \quad Z_1^2 = 1238_x; \quad Z_2^2 = 999_x;$$

We note that Z_2^2 is not relevant to the approximation, but it is required for defining the key for which the above statement fails. Using the key schedule, we derive the following subkeys as well: $Z_3^2 = F458_x$, $Z_4^2 = D000_x$, $Z_5^2 = 800_x$, and $Z_6^2 = A00_x$.

3.3   The Third Component — Exploiting the Weak Key Schedule

The fact that the key schedule of IDEA is relatively weak has been known for a long time, and was already used to devise related-key attacks on reduced-round IDEA (e.g.,[5]) and to find large weak key classes for the entire cipher (e.g.,[6]). However, other key recovery attacks (e.g.,[2]) usually exploited other properties of IDEA and took only a small advantage of the key schedule.

In our attack, we use the weakness of the key schedule in order to calculate all the remaining S^i values, while guessing a relatively small number of key bits.

Consider a 5.5-round variant of IDEA starting with the third round. We observe that in order to compute the values S^5, S^6, S^7 from the ciphertexts, it is sufficient to know only 103 key bits. The values S^5, S^6, S^7 are determined by the values of the ciphertext and of the subkeys $Z_4^8, Z_3^8, Z_2^8, Z_1^8, Z_6^7, Z_5^7, Z_4^7, Z_3^7, Z_2^7, Z_1^7,$ $Z_6^6, Z_5^6, Z_1^6, Z_2^6, Z_5^5$. These subkeys are sufficient in order to partially decrypt the ciphertexts through the last two rounds and to find the value of S^5. Note that S^5 is independent of the values of Z_3^6 and Z_4^6. All the required 15 subkeys use only 103 bits of the master key, whereas bits 100–124 of the master key remain unused.

Since in our case (for 5.5 rounds), Equation (7) is reduced to

$$LSB\left(\bigoplus_{j=1}^{m}(P_2^j \oplus P_3^j) \oplus \bigoplus_{i=5}^{7} S^i\right) = LSB\left(\bigoplus_{j=1}^{m}(C_2^j \oplus C_3^j)\right), \qquad (8)$$

we can guess 103 key bits and check whether the equation holds.

3.4   The Basic 5.5-Round Attack

In this subsection we combine the components presented in the previous sub-sections to devise an attack on 5.5-round IDEA. As was shown in Section 3.3, once we have a right structure, it is possible to check the guess of 103 key bits. This is done by partially decrypting the entire set under the subkey guess, and checking whether Equation (8) holds.

The problem in finding a right structure is the key addition layer, which prevents constructing the right structures immediately. The solution to this problem is to use more plaintexts, which increases the data complexity, but in exchange, the set is ensured to contain (several) right structures. We use a set of 2^{32} plaintexts, such that the second and the fourth words are fixed to some arbitrary constant, and the first and the third words obtain all possible values. In such a structure, for any given subkeys, it is possible to find 2^{16} right structures of 2^{16} plaintexts each.

The basic algorithm of the attack is the following:

1. **Data collection phase:** Ask for the encryption of a structure of chosen plaintexts, of the form (x, B, y, D) for two randomly selected constants (B, D) and all possible values of x and y.
2. **Constructing a right structure:** For each possible value of bits 89–104 and 121–8 of the key, perform the following:
   (a) Choose an arbitrary 16-bit value F_1 (a candidate for p^3).
   (b) For all $0 \leq j \leq 2^{16} - 1$, partially decrypt the words j under the guess of Z_1^3 and $F_1 \oplus j$ under the guess of Z_3^3. Denote the list of resulting values by (A_1^j, C_1^j).
3. **Checking the linear equation:** Choose all plaintexts of the form (A_1^j, B, C_1^j, D). For each possible value of key bits 0–104 and 121–127 (note that from these bits we already guessed key bits 89–104 and 121–8) partially decrypt the ciphertexts corresponding to the plaintexts of the right structure to get the values S^5, S^6, S^7. Check whether Equation (8) holds. If not, discard the partial key guess. If the equation holds, pass the key guess for further analysis.
4. **Filtering the remaining key guesses:** For the remaining key guesses, take another right structure of plaintexts, i.e., pick a different value for F, and repeat Steps 2–3, with three different selections. If a partial key guess passed all the four tests, perform exhaustive key search on the remaining key bits.

3.5   Analysis and Improvement of the Basic Attack

The attack requires one structure of 2^{32} plaintexts, and thus the data complexity of the attack is 2^{32} chosen plaintexts. The time complexity of Step (1) is 2^{32} encryptions.

The time complexity of Step (2) is 2^{48} partial decryptions. We note that this step can be performed as a precomputation, but there is no need to do it since the time complexity of this step is negligible with respect to the total time complexity of the attack.

The most time consuming part of the attack is Step (3) which is repeated 2^{32} times (for each guess of bits 89–104 and 121–8). In this step, the attacker guesses 80 key bits and performs a partial decryption of 2^{16} ciphertexts. The partial decryption includes two full rounds and three out of the eight operations of an additional round. Hence, the time complexity of this stage is about $0.43 \cdot 2^{128}$ encryptions in total. About half of the keys are expected to pass to Step (4). For these keys, the attacker performs three additional filterings in Step (4), with time complexity of $(0.22 + 0.11 + 0.05) \cdot 2^{128}$. The 2^{124} remaining keys are checked by exhaustive key search. Hence, the total time complexity of the attack is $(0.43 + 0.22 + 0.11 + 0.05 + 0.06) \cdot 2^{128} = 0.87 \cdot 2^{128}$ encryptions, which is only slightly better than exhaustive key search.

In order to reduce the time complexity of the attack, we introduce a small change in Step (2). We observe, that there is no need to fix a concrete value of F such that $p^3 = F$ for all the values of the structure; it is sufficient to have for all the plaintexts in the considered structure a fixed value for p^3. We exploit this observation by eliminating the need to guess bit 121 of the key in Step (2). We do not guess the value of this bit, but rather assume that its value is zero. As a result, when we decrypt the values $j \oplus F$ through the addition with Z_3^3, all the bits except for the MSB are correct, but the MSB might be wrong. However, if our assumption was incorrect and the values of the MSBs are wrong, this happens to all the elements of the structure simultaneously, since

$$x \boxplus (y \boxplus 8000_x) = (x \boxplus y) \boxplus 8000_x = (x \boxplus y) \oplus 8000_x. \tag{9}$$

Hence, if we take the plaintext structure (A_1^j, B, C_1^j, D) as in the basic attack, we have two possibilities: If our assumption was correct, we get a right structure as in the basic attack. If our assumption was incorrect, then for all the elements of the structure we have $p^3 = F \oplus 8000_x$. However, p^3 still assumes the same value for all the elements of the structure, and hence, the structure is a right structure.

Therefore, we can obtain right structures without guessing the value of bit 121 of the key (by making sure to choose the F values without using two values which differ only in the MSB). This improvement reduces the time complexity of all the steps of the attack (except for the final exhaustive key search) by a factor of two. Hence, the time complexity of the improved attack is $(0.22 + 0.11 + 0.05 + 0.03 + 0.06) \cdot 2^{128} = 0.45 \cdot 2^{128} = 2^{126.85}$, about twice faster than exhaustive key search.

It is also possible to use up to 2^{15} right structures from a given plaintext structure, and use them to filter wrong subkey guesses. When using k right structures with the improved attack, the data complexity is 2^{32} chosen plaintexts, and the time complexity is

$$2^{111} \cdot 2^{16} \cdot \frac{2\frac{3}{8}}{5\frac{1}{2}} \cdot \left(1 + \frac{1}{2} + \cdots + \frac{1}{2^{k-1}} \right) + 2^{128-k}$$

encryptions. When using 16 right structures, the time complexity is $2^{126.8}$ encryptions. The memory requirements of the attack are mostly for storing the data, i.e., 2^{32} blocks of 128-bit each (or 2^{36} bytes).

4   The 6-Round Attack

In this section we extend the 5.5-round attack to an attack on 6-round IDEA. The variant we attack starts before the MA layer of round 2 and ends before the MA layer of round 8.

We observe that the key bits used in the MA layer of round 2 (bits 57–88) are included in the bits that are guessed in the 5.5-round attack. Hence, we can add this half round in the beginning of the attack without enlarging the time complexity.

However, in this case it is much harder to construct right structures, and the data complexity is significantly increased. Our best method is to ask for the encryption of almost the entire code book, and look for the right structures after a partial encryption. We note that if a way to construct right structures is found, then the data complexity of the attack can be reduced. However, we did not find appropriate structures at this stage.

Assume that we start with $2^{64} - a$ known plaintexts, where a is a constant we determine later. For a fixed value of the subkeys $Z_5^2, Z_6^2, Z_1^3, Z_3^3$, we can divide the plaintexts into 2^{48} classes according to the value of the triplet (X_2^3, X_4^3, p^3) (obtained from the plaintexts using the fixed subkeys). Each class consists of 2^{16} plaintexts, and forms a right structure. Hence, a has to be small enough, such that for almost every value of the subkeys, there will be at least four "full" classes in the pool of known plaintexts. In order to estimate the number of full classes, we look at the plaintexts that are not known to the attacker. Assuming that these plaintexts are uniformly distributed over the 2^{48} classes, the probability that none of them falls in some prescribed class is $(1 - 2^{-48})^a \approx e^{-a/2^{48}}$. Hence, the expected number of classes that are not ruined by missing plaintexts, i.e., the full classes, is $2^{48} e^{-a/2^{48}}$. If we take $a = 2^{53}$, this expected number is close to four. However, we want that for most of the possible values of the subkeys there will be four right structures. Hence, we take $a = 2^{52}$. In this case, the expected number of right structures is about 2^{25}, and for most of the possible values of the subkeys, there will be enough right structures. If for some subkey guess, there are not enough right structures, the attacker has to check this guess separately by exhaustive key search over the remaining key bits.[1] We note that this improvement can also reduce the data complexity of the 5.5-round attack by about 2^{18} chosen plaintexts.

Therefore, our 6-round attack requires $2^{64} - 2^{52}$ known plaintexts. The first stage of the attack is slightly changed to the following:

1. Ask for the encryption of $2^{64} - 2^{52}$ arbitrary plaintexts.
2. For every possible value of key bits 57—104 and 121—8:
   (a) Repeat until a candidate right structure is found in the data set — Choose (X_2^3, X_4^3, p^3) at random. Partially decrypt the set $(j, X_2^3, j \oplus p^3, X_4^3)$ for $j = 0, \ldots 2^{16} - 1$ under the subkey guess, and check that all

---

[1] The probability that for $2^{64} - 2^{52}$ random plaintexts, there exists such a key is about $e^{-33554367.2}$.

the resulting plaintexts are in the data set.[2] Stop once at least four such sets are found.

(b) Apply Steps (2) and (3) of the 5.5-round attack given the four right structures using the guessed key bits.

This algorithm allows to detect four right structures that are used in the same way as in the 5.5-round attack. We note that since the expected number of right structures is about 2^{25}, we expect that after about 2^{25} checked triples (X_2^3, X_4^3, p^3), four right structures are found with a high probability. Hence, the time complexity of this stage is $2^{64} \cdot 2^{25} \cdot 2^{16} = 2^{105}$ partial decryptions, which is negligible compared to Step (3) of the 5.5-round attack.

The rest of the attack is the same as in the 5.5-round attack. However, the time complexity measured in encryption units is reduced, since we still decrypt only 2.375 rounds, while the total number of rounds is increased to six. Hence, the time complexity of the attack is $(0.2+0.1+0.05+0.02+0.06) \cdot 2^{128} = 0.43 \cdot 2^{128} = 2^{126.8}$ six-round encryptions. Like in the 5.5-round attack, the memory complexity of the 6-round attack is dominated mostly by the memory required for the data itself, i.e., $2^{64} - 2^{52}$ blocks of 128 bits each. We note that as this value is very close to the entire code book, it can be improved by a factor of 2, by not storing the plaintexts themselves (and keeping only the ciphertexts).

5   An Improved 5-Round Attack

In [3] a 5-round attack on IDEA with data complexity of 2^{19} known plaintexts and time complexity of 2^{103} operations is presented. The attack is based on the Biryukov-Demirci relation, when two plaintexts (P^1, P^2) are used. The relation is used for the four and a half rounds from the beginning of round 4 till after the key addition layer of round 8. For this case the Biryukov-Demirci relation is reduced to:

$$LSB(P_2^1 \oplus P_3^1 \oplus P_2^2 \oplus P_3^2 \oplus \Delta s^4 \oplus \Delta s^5 \oplus \Delta s^6 \oplus \Delta s^7) = LSB(C_2^1 \oplus C_3^1 \oplus C_2^2 \oplus C_3^2),$$
(10)

where P_2^1, P_3^1, P_2^2, and P_3^2 are taken at the beginning of round 4.

By requiring that $\Delta(X_1^4, X_2^4, X_3^4, X_4^4) = (0, \beta, 0, \gamma)$, the attacker ensures that $\Delta s^4 = 0$. Due to the key schedule of IDEA, it is sufficient to guess 103 bits of the key in order to compute $\Delta s^5, \Delta s^6$, and Δs^7. The attack is then quite a straightforward filtering of wrong subkey guesses which suggest that the Biryukov-Demirci relation does not hold.

In [3], the relation $\Delta(X_1^4, X_2^4, X_3^4, X_4^4) = (0, \beta, 0, \gamma)$ is achieved by using 2^{19} known plaintexts, which compose about 2^{37} pairs. On average, about 32 satisfy the requirement on $\Delta(X_1^4, X_2^4, X_3^4, X_4^4)$.

---

[2] We alert the reader to the abuse we have taken in the notations. While j and $j \oplus p^3$ are given after the KA layer of round 3, X_2^3 and X_4^3 are given before it. Thus, we partially decrypt two words through the KA layer of round 3, and then continue to partially decrypt all four words under the MA layer of round 2.

Our first improvement reduces the data complexity of the attack. We note that $\Delta s^4 = 0$ is satisfied whenever $\Delta p^4 = 0$ holds. In addition, we observe that the subkey Z_1^4 is covered by the 103 key bits guessed during the attack. We also note that a difference in the MSB is preserved by addition with an unknown subkey. Using these three observations, we can enlarge the number of plaintext pairs that can be used in the attack. Instead of using only pairs for which $\Delta(X_1^4, X_3^4) = (0, 0)$, we can use also pairs for which $\Delta Y_1^4 = 8000_x$ and $\Delta X_3^4 = 8000_x$, since for such pairs we also have $\Delta s^4 = 0$. As a result, we can start with $2^{18.5}$ known plaintexts, and out of the 2^{36} possible pairs we can still find 32 pairs satisfying $\Delta s^4 = 0$. The rest of the attack is similar to the attack algorithm of [3].

5.1   A 5-Round Attack Using Only 16 Known Plaintexts

Our second improvement has much in common with the first improvement. We note that guessing the subkey Z_3^4 adds only eleven bits to the total number of guessed key bits. On the other hand, after guessing this subkey, we are able to compute the exact value of p^4 for all the plaintexts. Moreover, since the subkey Z_5^4 is also covered by the guessed key bits, we are able also to compute the exact values of s^4 for all the plaintexts, and hence to compute the value of Δs^4 for any pair of plaintexts. As a result, all the plaintext pairs, and not only a part of them, can be used to filter subkey candidates. Since each plaintext pair filters about half of the subkey candidates, 16 pairs are sufficient to reduce the number of possible keys to 2^{113}, and these candidates can be checked by exhaustive key search.

We note that the 16 known plaintexts compose $16 \cdot 15/2 = 120$ pairs. However, only 15 of these pairs can be linearly independent, and hence only 15 pairs can be used for filtering key candidates. Using a smart ordering of the operations, it can be shown that the time complexity of this attack is 2^{114} encryptions.

6   Summary and Conclusions

In this paper we presented the first known attacks on 5.5-round and 6-round variants of IDEA. Our attack on 6-round IDEA has a time complexity of $2^{126.8}$ encryptions and data complexity of $2^{64} - 2^{52}$ known plaintexts. The attacks exploit three techniques: Constructing a linear equation involving the LSBs of several intermediate encryption values, using Square-like structures, and exploiting the weak key schedule. Each of these techniques was already used to attack reduced variants of IDEA, but the novel combination of the techniques allows to improve the previously known attacks significantly.

We also showed that it possible to attack 5-round IDEA using only 16 known plaintexts with time complexity of 2^{114} encryptions. The 5-round attack exploits the weakness of the key schedule of IDEA, that allows to recover many subkeys while guessing only a subset of the key bits.

References

1. Ayaz, E.S., Selçuk, A.A.: Improved DST Cryptanalysis of IDEA. In: Proceedings of Selected Areas in Cryptography 2006, SAC 2006, LNCS, vol. 4356, Springer, Heidelberg (to appear, 2007)
2. Biham, E., Biryukov, A., Shamir, A.: Miss in the Middle Attacks on IDEA and Khufu. In: Knudsen, L.R. (ed.) FSE 1999. LNCS, vol. 1636, pp. 124–138. Springer, Heidelberg (1999)
3. Biham, E., Dunkelman, O., Keller, N.: New Cryptanalytic Results on IDEA. In: Lai, X., Chen, K. (eds.) ASIACRYPT 2006. LNCS, vol. 4284, pp. 412–427. Springer, Heidelberg (2006)
4. Biham, E., Dunkelman, O., Keller, N.: New Combined Attacks on Block Ciphers. In: Gilbert, H., Handschuh, H. (eds.) FSE 2005. LNCS, vol. 3557, pp. 126–144. Springer, Heidelberg (2005)
5. Biham, E., Dunkelman, O., Keller, N.: Related-Key Boomerang and Rectangle Attacks. In: Cramer, R.J.F. (ed.) EUROCRYPT 2005. LNCS, vol. 3494, pp. 507–525. Springer, Heidelberg (2005)
6. Biryukov, A., Nakahara Jr., J., Preneel, B., Vandewalle, J.: New Weak-Key Classes of IDEA. In: Deng, R.H., Qing, S., Bao, F., Zhou, J. (eds.) ICICS 2002. LNCS, vol. 2513, pp. 315–326. Springer, Heidelberg (2002)
7. Biryukov, A., Shamir, A.: Structural Cryptanalysis of SASAS. In: Pfitzmann, B. (ed.) EUROCRYPT 2001. LNCS, vol. 2045, pp. 394–405. Springer, Heidelberg (2001)
8. Borisov, N., Chew, M., Johnson, R., Wagner, D.: Multiplicative Differentials. In: Daemen, J., Rijmen, V. (eds.) FSE 2002. LNCS, vol. 2365, pp. 17–33. Springer, Heidelberg (2002)
9. Borst, J., Knudsen, L.R., Rijmen, V.: Two Attacks on Reduced Round IDEA. In: Fumy, W. (ed.) EUROCRYPT 1997. LNCS, vol. 1233, pp. 1–13. Springer, Heidelberg (1997)
10. Daemen, J., Govaerts, R., Vandewalle, J.: Cryptanalysis of 2.5 Rounds of IDEA (Extended Abstract), technical report 93/1, Department of Electrical Engineering, ESAT–COSIC, Belgium (1993)
11. Daemen, J., Govaerts, R., Vandewalle, J.: Weak Keys for IDEA. In: Stinson, D.R. (ed.) CRYPTO 1993. LNCS, vol. 773, pp. 224–231. Springer, Heidelberg (1994)
12. Daemen, J., Knudsen, L.R., Rijmen, V.: The Block Cipher Square. In: Biham, E. (ed.) FSE 1997. LNCS, vol. 1267, pp. 149–165. Springer, Heidelberg (1997)
13. Demirci, H.: Square-like Attacks on Reduced Rounds of IDEA. In: Nyberg, K., Heys, H.M. (eds.) SAC 2002. LNCS, vol. 2595, pp. 147–159. Springer, Heidelberg (2003)
14. Demirci, H., Selçuk, A.A., Türe, E.: A New Meet-in-the-Middle Attack on the IDEA Block Cipher. In: Matsui, M., Zuccherato, R.J. (eds.) SAC 2003. LNCS, vol. 3006, pp. 117–129. Springer, Heidelberg (2004)
15. Hawkes, P.: Differential-Linear Weak Keys Classes of IDEA. In: Nyberg, K. (ed.) EUROCRYPT 1998. LNCS, vol. 1403, pp. 112–126. Springer, Heidelberg (1998)
16. Hawkes, P., O'Connor, L.: On Applying Linear Cryptanalysis to IDEA. In: Kim, K.-c., Matsumoto, T. (eds.) ASIACRYPT 1996. LNCS, vol. 1163, pp. 105–115. Springer, Heidelberg (1996)
17. Junod, P.: New Attacks Against Reduced-Round Versions of IDEA. In: Gilbert, H., Handschuh, H. (eds.) FSE 2005. LNCS, vol. 3557, pp. 384–397. Springer, Heidelberg (2005)

18. Knudsen, L.R., Wagner, D.: Integral Cryptanalysis. In: Daemen, J., Rijmen, V. (eds.) FSE 2002. LNCS, vol. 2365, pp. 112–127. Springer, Heidelberg (2002)
19. Lai, X., Massey, J.L., Murphy, S.: Markov Ciphers and Differential Cryptanalysis. In: Davies, D.W. (ed.) EUROCRYPT 1991. LNCS, vol. 547, pp. 17–38. Springer, Heidelberg (1991)
20. Lucks, S.: The Saturation Attack — A Bait for Twofish. In: Matsui, M. (ed.) FSE 2001. LNCS, vol. 2355, pp. 1–15. Springer, Heidelberg (2002)
21. Meier, W.: On the Security of the IDEA Block Cipher. In: Helleseth, T. (ed.) EUROCRYPT 1993. LNCS, vol. 765, pp. 371–385. Springer, Heidelberg (1994)
22. Nakahara Jr., J., Barreto, P.S.L.M., Preneel, B., Vandewalle, J., Kim, H.Y.: SQUARE Attacks Against Reduced-Round PES and IDEA Block Ciphers, IACR Cryptology ePrint Archive, Report 2001/068 (2001)
23. Nakahara. Jr., J., Preneel, B., Vandewalle, J.: The Biryukov-Demirci Attack on Reduced-Round Versions of IDEA and MESH Ciphers. In: Wang, H., Pieprzyk, J., Varadharajan, V. (eds.) ACISP 2004. LNCS, vol. 3108, pp. 98–109. Springer, Heidelberg (2004)
24. NESSIE, Performance of Optimized Implementations of the NESSIE Primitives, NES/DOC/TEC/WP6/D21/a, available on-line at http://www.nessie.eu.org/nessie
25. Raddum, H.: Cryptanalysis of IDEA-X/2. In: Johansson, T. (ed.) FSE 2003. LNCS, vol. 2887, pp. 1–8. Springer, Heidelberg (2003)

Related-Key Rectangle Attacks on Reduced AES-192 and AES-256[*]

Jongsung Kim[1], Seokhie Hong[1], and Bart Preneel[2]

[1] Center for Information Security Technologies (CIST),
Korea University, Anam Dong, Sungbuk Gu, Seoul, Korea
{joshep,hsh}@cist.korea.ac.kr
[2] Katholieke Universiteit Leuven, Dept. ESAT/SCD-COSIC,
Kasteelpark Arenberg 10, B-3001 Heverlee, Belgium
Bart.Preneel@esat.kuleuven.be

Abstract. This paper examines the security of AES-192 and AES-256 against a related-key rectangle attack. We find the following new attacks: 8-round reduced AES-192 with 2 related keys, 10-round reduced AES-192 with 64 or 256 related keys and 9-round reduced AES-256 with 4 related keys. Our attacks reduce the complexity of earlier attacks presented at FSE 2005 and Eurocrypt 2005: for reduced AES-192 with 8 rounds, we decrease the required number of related keys from 4 to 2 at the cost of a higher data and time complexity; we present the first shortcut attack on AES-192 reduced to 10 rounds; for reduced AES-256 with 9 rounds, we decrease the required number of related keys from 256 to 4 and both the data and time complexity at the cost of a smaller number of attacked rounds. Furthermore, we point out some flaw in the 9-round AES-192 attack presented at Eurocrypt 2005, show how to fix it and enhance the attack in terms of the number of related keys.

Keywords: Block Ciphers, Cryptanalysis, AES, Related-Key Rectangle Attack.

1 Introduction

The Advanced Encryption Standard (AES), the successor to the Data Encryption Standard (DES), is a block cipher adopted as mandatory encryption standard by the US government. Since NIST announced that the block cipher Rijndael, designed by Daemen and Rijmen [12], was selected for the AES in 2000, it has gradually become one of the most widely used encryption algorithms in the world. Therefore, it is important to continue studying the security of AES.

[*] This work was supported in part by the Concerted Research Action (GOA) Ambiorics 2005/11 of the Flemish Government, by the IUAP P6/26 BCRYPT of the Belgian Federal Science Policy Office and by the European Commission through the IST Programme under Contract IST 2002-507932 ECRYPT and in part by the MIC (Ministry of Information and Communication), Korea, under the ITRC (Information Technology Research Center) support program supervised by the IITA (Institute of Information Technology Assessment).

A. Biryukov (Ed.): FSE 2007, LNCS 4593, pp. 225–241, 2007.

The AES algorithm is a 128-bit SP-network block cipher, which uses 128-bit, 192-bit or 256-bit keys. According to the length of the keys, the AES performs different key scheduling algorithms, different numbers of rounds, but the same round function which is made up of SubBytes (SB), ShiftRows (SR), MixColumns (MC) and AddRoundKey (ARK). These different versions of AES are referred to as AES-128, AES-192 and AES-256.

One of the most powerful known attacks on block ciphers is differential cryptanalysis [1] introduced by Biham and Shamir in 1990. It uses a differential with a non-trivial probability to retrieve subkeys for the first or last few rounds. After this attack was introduced, it has been applied effectively to many known block ciphers and various variants of this attack have been proposed such as the truncated differential attack [24], the higher order differential attack [24], the differential-linear attack [26], the impossible differential attack [3], the boomerang attack [32] and the rectangle attack [5]. Unlike differential cryptanalysis, in the boomerang and rectangle attacks [32,5], two consecutive differentials are used, which are independent of each other, in order to retrieve subkeys for the first or last few rounds.

In 1992 and 1993, Knudsen [23] and Biham [2] independently introduced a cryptanalytic method using related keys, called the related-key attack [2], which applies differential cryptanalysis to the cipher with different, but related unknown keys. This attack is based on the key scheduling algorithm and on the encryption/decryption algorithms, hence a block cipher with a weak key scheduling algorithm may be vulnerable to this kind of attack. Several cryptanalytic results of this attack were reported in [18,19,10,20].

The related-key rectangle attack introduced in [21,16,6] combines the rectangle and related-key attacks by applying the rectangle attack to the cipher with different, but related unknown keys: [21,16,6] show how to apply the rectangle attack with 2, 4, and more than 4 related keys, and show that this kind of attack can be applied to 8-round reduced AES-192 with 4 related keys [16], 9-round reduced AES-192 with 256 related keys [6], and 10-round reduced AES-256 with 256 related keys [6]. Several other articles have been published that demonstrate the power of this attack [7,13,27,28].

In this paper we examine the security of AES-192 and AES-256 against a related-key rectangle attack in other related-key settings. We show that a related-key rectangle attack is applicable to 8-round reduced AES-192 with 2 related keys, 10-round reduced AES-192 with 64 or 256 related keys and 9-round reduced AES-256 with 4 related keys. Our 10-round AES-192 attack leads to the best known attack on AES-192 and our 8-round AES-192, 9-round AES-256 attacks are both better than the previously best known attacks on AES-192 with 2 related keys and AES-256 with 4 related keys in terms of the number of attacked rounds and a data or time complexity. We also demonstrate a flaw in the 9-round AES-192 attack presented at Eurocrypt 2005 [6], show how to fix it and enhance the attack in terms of the number of related keys (from 256 to 64 related keys). See Table 1 for the comparison of our results and the previous ones on AES.

The outline of this paper is as follows: in Sect. 2, we give a brief description of AES and in Sect. 3, we describe a general method of the related-key rectangle

Table 1. Summary of the previous attacks and our attacks on AES

Block Cipher	Type of Attack	Number of Rounds	Number of keys	Complexity Data / Time
AES-128 (10 rounds)	Imp. Diff.	5	1	$2^{29.5}$CP / 2^{31} [4]
		6	1	$2^{91.5}$CP / 2^{122} [11]
	Boomerang	6	1	2^{71}ACPC / 2^{71} [9]
	Partial Sums	6	1	$6 \cdot 2^{32}$CP / 2^{44} [14]
		7	1	$2^{128} - 2^{119}$CP / 2^{120} [14]
AES-192 (12 rounds)	Imp. Diff.	7	1	2^{92}CP / 2^{186} [31]
	Square	7	1	2^{32}CP / 2^{184} [29]
	Partial Sums	7	1	$19 \cdot 2^{32}$CP / 2^{155} [14]
		7	1	$2^{128} - 2^{119}$CP / 2^{120} [14]
		8	1	$2^{128} - 2^{119}$CP / 2^{188} [14]
	RK Imp. Diff.	7	2	2^{111}RK-CP / 2^{116} [17]
		7	32	2^{56}CP / 2^{94} [8]
		8	2	2^{88}RK-CP / 2^{183} [17]
		8	32	2^{116}CP / 2^{134} [8]
		8	32	2^{92}CP / 2^{159} [8]
		8	32	$2^{68.5}$CP / 2^{184} [8]
	RK Rectangle	8	4	$2^{86.5}$RK-CP / $2^{86.5}$ [16]
		8	2	2^{94}RK-CP / 2^{120} (New)
		9†	256	2^{86}RK-CP / 2^{125} [6]
		9‡	64	2^{85}RK-CP / 2^{182} (New)
		10	256	2^{125}RK-CP / 2^{182} (New)
		10	64	2^{124}RK-CP / 2^{183} (New)
AES-256 (14 rounds)	Partial Sums	8	1	$2^{128} - 2^{119}$CP / 2^{204} [14]
		9	256	2^{85}CP / $5 \cdot 2^{224}$ [14]
	RK Rectangle	9	4	2^{99}RK-CP / 2^{120} (New)
		10	256	$2^{114.9}$RK-CP / $2^{171.8}$ [6]
		10	64	$2^{113.9}$RK-CP / $2^{172.8}$ (New)

CP: Chosen Plaintexts, ACPC: Adaptive Chosen Plaintexts and Ciphertexts.
RK: Related-Key, Time: Encryption units.
†: the attack with some flaw, ‡: the attack correcting the flaw in †.

attack. Sections 4 and 5 present our related-key rectangle attacks on reduced AES-192 and AES-256. Section 6 gives some comments on the previous 9-round AES-192 attack. Finally, we conclude the paper in Sect. 7.

2   Description of AES

AES encrypts data blocks of 128 bits with 128, 192 or 256-bit keys. According to the length of the keys, AES uses a different number of rounds, i.e., it has 10, 12 and 14 rounds when used with 128, 192 and 256-bit keys, respectively. The round function of AES consists of the following four basic transformations:

- SubBytes (SB) is a nonlinear byte-wise substitution that applies the same 8×8 S-box to every byte.
- ShiftRows (SR) is a cyclic shift of the i-th row by i bytes to the left.
- MixColumns (MC) is a matrix multiplication applied to each column.
- AddRoundKey (ARK) is an exclusive-or with the round key.

Each round function of AES applies the SB, SR, MC and ARK steps in order, but MC is omitted in the last round. Before the first round, an extra ARK step is applied. We call the key used in this step a whitening key. For more details of the above four transformations, we refer to [12].

AES uses different key scheduling algorithms according to the length of the supplied keys. The key schedule of AES-128 accepts a 128-bit key (W_0, W_1, W_2, W_3) and generates subkeys $W_4, W_5, \cdots, W_{43}$, where each W_i is a 32-bit word composed of 4 bytes in column. The subkeys are generated by the following procedure:

- For $i = s$ till $i = t$ do the following (for AES-128, $s = 4$ and $t = 43$),
  - If $i \equiv 0 \bmod s$, then $W_i = W_{i-s} \oplus SB(RotByte(W_{i-1})) \oplus Rcon(i/s)$,
  - else $W_i = W_{i-s} \oplus W_{i-1}$,

where RotByte represents one byte rotation and Rcon denotes fixed constants depending on its input. In AES-128, the whitening key is (W_0, W_1, W_2, W_3) and the subkey of round i is $(W_{4i+4}, W_{4i+5}, W_{4i+6}, W_{4i+7})$, where $0 \leq i \leq 9$.

Similarly, the key schedules of AES-192 and AES-256 accept 192- and 256-bit keys, and generate as many subkeys as required. The key schedule of AES-192 is exactly the same as that of AES-128 except for the use of $s = 6$ and $t = 51$. The subkeys of AES-256 are derived from the following procedure:

- For $i = 8$ till $i = 59$ do the following,
  - If $i \equiv 0 \bmod 8$ then $W_i = W_{i-8} \oplus SB(RotByte(W_{i-1})) \oplus Rcon(i/8)$,
  - If $i \equiv 4 \bmod 8$ then $W_i = W_{i-8} \oplus SB(W_{i-1})$,
  - else $W_i = W_{i-8} \oplus W_{i-1}$.

In this paper a 128-bit block of AES is represented by a 4×4 byte matrix as in Fig. 1 or by $((X_0, X_1, X_2, X_3), (X_4, X_5, X_6, X_7), (X_8, X_9, X_{10}, X_{11}), (X_{12}, X_{13}, X_{14}, X_{15}))$.

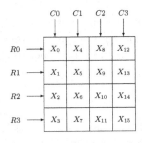

Fig. 1. Byte coordinate of a 128-bit block of AES (Ri: Row i, Ci: Column i, X_i: Byte i)

3   The Related-Key Rectangle Attack

The related-key rectangle attack is based on two consecutive related-key differentials with relatively high probabilities which are independent of each other: the underlying cipher $E : \{0,1\}^n \times \{0,1\}^k \to \{0,1\}^n$ is treated as a cascade of two sub-ciphers $E = E^1 \circ E^0$, where $\{0,1\}^k$ and $\{0,1\}^n$ are the key space and the plaintext/ciphertext space, respectively. We assume that for E^0 there exists a related-key differential $\alpha \to \beta$ with probability p and for E^1 there exists a related-key differential $\gamma \to \delta$ with probability q. More precisely,

$$p = Pr[E_K^0(X) \oplus E_{K^*}^0(X \oplus \alpha) = \beta] = Pr[E_{K'}^0(X) \oplus E_{K'^*}^0(X \oplus \alpha) = \beta],$$

$$q = Pr[E_K^1(X) \oplus E_{K'}^1(X \oplus \gamma) = \delta] = Pr[E_{K^*}^1(X) \oplus E_{K'^*}^1(X \oplus \gamma) = \delta],$$

where $K^* = K \oplus \Delta K$, $K' = K \oplus \Delta K'$ and $K'^* = K \oplus \Delta K \oplus \Delta K'$, i.e., $K \oplus K^* = K' \oplus K'^* = \Delta K$ and $K \oplus K' = K^* \oplus K'^* = \Delta K'$ for known key differences ΔK and $\Delta K'$. Then, these consecutive related-key differentials can be used efficiently to the following related-key rectangle distinguisher:

- Choose two random n-bit plaintexts P and P' and compute two other plaintexts $P^* = P \oplus \alpha$ and $P'^* = P' \oplus \alpha$.
- With a chosen plaintext attack scenario, obtain the corresponding ciphertexts $C = E_K(P)$, $C^* = E_{K^*}(P^*)$, $C' = E_{K'}(P')$ and $C'^* = E_{K'^*}(P'^*)$.
- Check if $C \oplus C' = C^* \oplus C'^* = \delta$.

The probability that the ciphertext quartet $(C, C^*, C'C'^*)$ satisfies the last δ test is computed as follows: let X, X^*, X' and X'^* denote the encrypted values of P, P^*, P' and P'^* under E^0 with K, K^*, K' and K'^*, respectively. Then, the probability that $X \oplus X^* = X' \oplus X'^* = \beta$ is about p^2 by the related-key differential for E^0. In the above process, we randomly choose two plaintexts P and P', so we expect $X \oplus X' = \gamma$ with probability 2^{-n}. Once the two above events occur, $X^* \oplus X'^* = (X \oplus X^*) \oplus (X' \oplus X'^*) \oplus (X \oplus X') = \beta \oplus \beta \oplus \gamma = \gamma$ with probability 1. Since the probability of the related-key differential $\gamma \to \delta$ for E^1 is q, $X \oplus X' = X^* \oplus X'^* = \gamma$ goes to $C \oplus C' = C^* \oplus C'^* = \delta$ with a probability of about q^2. Therefore, the total probability that the last δ test in the above process is satisfied is about

$$\sum_{\beta,\gamma} p^2 \cdot 2^{-n} \cdot q^2 = \widehat{p}^2 \cdot \widehat{q}^2 \cdot 2^{-n}, \text{where } \widehat{p} = \sqrt{\sum_\beta p^2} \text{ and } \widehat{q} = \sqrt{\sum_\gamma q^2}.$$

On the other hand, for a random cipher, the δ test holds with probability 2^{-2n} and thus if the above probability is larger than 2^{-2n} for any 4-tuple $(\alpha, \delta, \Delta K, \Delta K')$, i.e., if $\widehat{p} \cdot \widehat{q} > 2^{-n/2}$, the related-key rectangle distinguisher can be used to distinguish E from a random cipher. Similarly, two consecutive related-key truncated differentials can be used to form a related-key rectangle distinguisher.

According to the condition of the key differences ΔK and $\Delta K'$, the above related-key rectangle distinguisher is used in different ways. If $\Delta K \neq 0$ and $\Delta K' = 0$, then the distinguisher works with two related keys; a related-key differential for E^0 and a regular (non-related-key) differential for E^1. If $\Delta K = 0$ and $\Delta K' \neq 0$, then the distinguisher also works with two related keys; however, a regular differential for E^0 and a related-key differential for E^1 are used. If

$\Delta K \neq 0$, $\Delta K' \neq 0$ and $\Delta K \neq \Delta K'$, then the distinguisher works with four related keys, in which related-key differentials for both E^0 and E^1 are used. Further, more than four related keys can be used in the related-key rectangle distinguisher as in [6,7]; the basic idea is the same as that of the distinguisher with two or four related keys.

4   Related-Key Rectangle Attack on 10-Round AES-192

This section shows how to exploit the related-key rectangle attack to devise key recovery attacks on 10-round AES-192 with 64 or 256 related keys. We first focus on 10-round AES-192 with 256 related keys.

Denote the 10 rounds of AES-192 by $E = E^f \circ E^1 \circ E^0 \circ E^b$, where E^b is round 0 including the whitening key addition step and excluding the key addition step of round 0, E^0 is rounds 1-4 including the key addition step of round 0, E^1 is rounds 5-8 and E^f is round 9. In our 10-round AES-192 attack, we use a related-key truncated differential for E^0 depicted in Fig. 2 and another related-key truncated differential for E^1 depicted in Fig. 3. After we convert these related-key truncated differentials for E^0 and E^1 into a related-key rectangle distinguisher for $E^1 \circ E^0$, we apply it to recover some portions of the keys in E^b and E^f. Before describing our attack, we define some notation which is used in our attacks on AES.

- $K_w, K_w^*, K_w', K_w'^*$: whitening keys generated from master keys K, K^*, K', K'^*, respectively.
- $K_i, K_i^*, K_i', K_i'^*$: subkeys of round i generated from K, K^*, K', K'^*, respectively.
- P, P^*, P', P'^*: plaintexts encrypted under K, K^*, K', K'^*, respectively.
- $I_i, I_i^*, I_i', I_i'^*$: input values to round i caused by plaintexts P, P^*, P', P'^* under K, K^*, K', K'^*, respectively.
- a: a fixed nonzero byte value.
- b, c: output differences of S-box for the fixed nonzero input difference a.
- $*$: a variable and unknown byte.

4.1   8-Round Related-Key Rectangle Distinguisher

Our related-key truncated differentials depicted in Figs. 2 and 3 exploit the slow difference propagation of the key schedule of AES-192, which makes it possible that 3-round key differences $\Delta K_0 || \Delta K_1 || \Delta K_2$ and $\Delta K_5' || \Delta K_6' || \Delta K_7'$ satisfy $HW_b(\Delta K_0) = HW_b(\Delta K_5') = 2$, $HW_b(\Delta K_1) = HW_b(\Delta K_6') = 0$ and $HW_b(\Delta K_2) = HW_b(\Delta K_7') = 1$, where $HW_b(X)$ is the byte Hamming weight of X. Using these key differences with small byte Hamming weights we make $\Delta I_1 = \Delta I_6' = 0$ in our related-key truncated differentials which induce $HW_b(\Delta I_3) = HW_b(\Delta I_8') = 1$ (see Figs. 2 and 3 for ΔK_i, $\Delta K_i'$, ΔI_i and $\Delta I_i'$) and we add one or two more rounds to get longer related-key differentials. Note that our related-key truncated differential for E^0 is the same as the one for rounds 0-3 (including the whitening key addition step) used in [16].

In order to convert the two 4-round related-key truncated differentials into an 8-round related-key rectangle distinguisher, we first make the following Assumptions 1, 2 and 3.

Assumption 1. The key quartet (K, K^*, K', K'^*) is related as follows;

$$K \oplus K^* = K' \oplus K'^* = \Delta K, \ \ K \oplus K' = K^* \oplus K'^* = \Delta K'.$$

Assumption 2. A plaintext quartet (P, P^*, P', P'^*) is related as follows;

$$P \oplus P^*, \ P' \oplus P'^* \in \Delta P.$$

Assumption 3. $E_K^b(P) \oplus E_{K^*}^b(P^*) = E_{K'}^b(P') \oplus E_{K'^*}^b(P'^*) = \Delta K_0.$

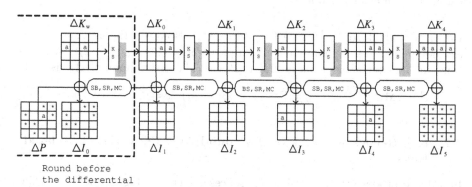

Round before
the differential

Fig. 2. The first related-key truncated differential for rounds 1-4 including the key addition step of round 0 (E^0), and the preceding differential for round 0 including the whitening key addition step and excluding the key addition step of round 0 (E^b)

Note that ΔK is the same as the first six columns of $\Delta K_w \| \Delta K_0$ in Fig. 2. As stated in our notation, $I_5 = E_K^0(E_K^b(P))$, $I_5^* = E_{K^*}^0(E_{K^*}^b(P^*))$, $I_5' = E_{K'}^0(E_{K'}^b(P'))$ and $I_5'^* = E_{K'^*}^0(E_{K'^*}^b(P'^*))$. By the related-key truncated differential for E^0, $I_5 \oplus I_5^*$ is equal to $I_5' \oplus I_5'^*$ with a probability of about $(2^{-32} \cdot 2^{-7})^2 \cdot (2^7 - 2) \cdot 2^{32} + (2^{-32} \cdot 2^{-6})^2 \cdot 2^{32} \approx 2^{-39}$ (this probability has been used for the 8-round AES-192 attack presented in [16]). It follows from counting over all the differentials that can be generated by the active S-box with input difference a and the other four active S-boxes in round 4. Since ShiftRows and MixColumns are linear layers, they can be ignored in round 4 when computing the probability (see Fig. 2). Moreover, the probability that $I_5 \oplus I_5'$, $I_5^* \oplus I_5'^* \in \Delta I_5'$ is about 2^{-64} under the condition $I_5 \oplus I_5^* = I_5' \oplus I_5'^*$ (see Fig. 3 for $\Delta I_5'$). Hence the probability that $I_5 \oplus I_5'$, $I_5^* \oplus I_5'^* \in \Delta I_5'$ is about $2^{-39} \cdot 2^{-64} = 2^{-103}$. Since $e_K(I_5) \oplus e_{K'}(I_5') = 0$ with probability 2^{-64} and $e_{K^*}(I_5^*) \oplus e_{K'^*}(I_5'^*) = 0$ with a probability of about 2^{-64} under the condition $I_5 \oplus I_5'$, $I_5^* \oplus I_5'^* \in \Delta I_5'$, where e is the encryption for round 5,

$$E_K^1(I_5) \oplus E_{K'}^1(I_5'), \ E_{K^*}^1(I_5^*) \oplus E_{K'^*}^1(I_5'^*) \in \Delta I_9'$$

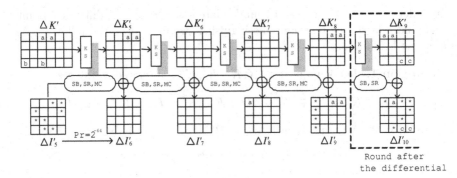

Fig. 3. The second related key truncated differential for rounds 5-8 (E^1) and the following differential for round 9 (E^f)

with a probability of 2^{-231} (see Fig. 3 for $\Delta I_9'$). However, the same statement can be applied to a random cipher with probability $(2^{-128} \cdot (2^7 - 1))^2 \approx 2^{-242}$, since the number of elements in $\Delta I_9'$ is $2^7 - 1$. The first column of $\Delta I_9'$ is

$$\mathcal{B} = \{\text{MC}(y, 0, 0, 0) \mid y = \text{BS}(x) \oplus \text{BS}(x \oplus a), \ x = 0, 1, 2, \cdots, 255\}. \tag{1}$$

4.2 Key Recovery Attack on 10-Round AES-192 with 256 Related Keys

In order to produce the round-key differences depicted in Fig. 3, the 8-bit difference a should satisfy the 8-bit difference b after S-box during the key generation for the third column of $\Delta K_3'$. Given the 8-bit difference a there are 127 possible candidates for the b difference, hence the attack starts by gathering all possible key quartets $(K, K^*, \widetilde{K}', \widetilde{K}'^*)$ of which one satisfies the desired key condition. Note that the keys $K^* = K \oplus \Delta K$, $\widetilde{K}' = K \oplus \Delta \widetilde{K}'$ and $\widetilde{K}'^* = K \oplus \Delta K \oplus \Delta \widetilde{K}'$ where ΔK is fixed as ΔK_w and the first two columns of ΔK_0 in Fig. 2 and $\Delta \widetilde{K}'$ is one of the 127 possible differences; bytes 8 and 12 are both a, bytes 3 and 11 are both b' and other bytes are all zeros, where b' is one of the 127 possible candidates for the b difference. So the total number of required related keys is 256. We apply the related-key rectangle attack to 10-round AES-192 for each key quartet. During this procedure, we stop our attack when we have found a key quartet $(K, K^*, \widetilde{K}', \widetilde{K}'^*)$ that satisfies the desired key condition $b' = b$, i.e., $\Delta \widetilde{K}' = \Delta K'$, $(K, K^*, \widetilde{K}', \widetilde{K}'^*) = (K, K^*, K', K'^*)$.

The aim of our attack is to recover bytes 1, 2, 6, 7, 8, 11, 12, 13 of the whitening key quartet $(K_w, K_w^*, K_w', K_w'^*)$ and bytes 0, 7, 8, 10, 12, 13 of the subkey quartet $(K_9, K_9^*, K_9', K_9'^*)$, for which the byte positions are marked as $*$ on ΔP and $\Delta I_{10}'$ depicted in Fig. 2 and Fig. 3. This attack distinguishes a right key quartet from wrong ones by analyzing enough plaintext quartets with each guessed key quartet. In this attack, we need 2^{64} guesses for the whitening key quartet and 2^{72} guesses for the subkey quartet in round 9, since bytes 0, 7, 8, 10, 12, 13 of ΔK_9 are d, 0, d, e, 0, f, respectively, where d, e and f

are unknown 8-bit values (note that bytes 0, 7, 8, 10, 12, 13 of $\Delta K'_9$ are fixed by a, 0, 0, 0, 0, 0, respectively). Thus, taking into account the guessing of candidates for the difference b, we need about 2^{143} key guesses in total (in our attack it can be reduced by a factor of two on average).

The attack algorithm goes as follows:

1. Choose 2^{54} structures S_1, S_2, $\cdots$, $S_{2^{54}}$ of 2^{64} plaintexts each, where in each structure the 64 bits of bytes 0, 3, 4, 5, 9, 10, 14, 15 are fixed. With a chosen plaintext attack scenario, obtain all their corresponding ciphertexts under the key K. (Step 1 takes 2^{118} chosen plaintexts and about 2^{118} encryptions. Note that n encrpytions mean n 10-round AES-192 encryptions.)

2. Compute 2^{54} structures S_1^*, S_2^*, $\cdots$, $S_{2^{54}}^*$ of 2^{64} plaintexts each by XORing the plaintexts in S_1, S_2, $\cdots$, $S_{2^{54}}$ with a 128-bit value M of which byte 9 is a and all the other bytes are 0. With a chosen plaintext attack scenario, obtain all their corresponding ciphertexts under the key K^*, where $K^* = K \oplus \Delta K$. (Similarly, Step 2 takes 2^{118} chosen plaintexts and about 2^{118} encryptions.)

3. Guess a candidate for the difference b and compute $\Delta \widetilde{K}'$. For the key difference $\Delta \widetilde{K}'$, do the following:

   3.1 Choose 2^{54} structures S'_1, S'_2, $\cdots$, $S'_{2^{54}}$ of 2^{64} plaintexts each, where in each structure the 64 bits of bytes 0, 3, 4, 5, 9, 10, 14, 15 are fixed. With a chosen plaintext attack scenario, obtain all their corresponding ciphertexts under the key $\widetilde{K}'$, where $\widetilde{K}' = K \oplus \Delta \widetilde{K}'$. (For each guess of $\Delta \widetilde{K}'$, Step 3.1 takes 2^{118} chosen plaintexts and about 2^{118} encryptions.)

   3.2 Compute 2^{54} structures $S_1'^*$, $S_2'^*$, $\cdots$, $S_{2^{54}}'^*$ of 2^{64} plaintexts each by XORing the plaintexts in S'_1, S'_2, $\cdots$, $S'_{2^{54}}$ with M. With a chosen plaintext attack scenario, obtain all their corresponding ciphertexts under the key $\widetilde{K}'^*$, where $\widetilde{K}'^* = K \oplus \Delta K \oplus \Delta \widetilde{K}'$. Go to Step 4. (For each guess of $\Delta \widetilde{K}'$, this step also takes 2^{118} chosen plaintexts and about 2^{118} encryptions.)

4. Guess a 64-bit subkey k_w in the position of bytes 1, 2, 6, 7, 8, 11, 12, 13 of the whitening key and compute $k_w^* = k_w \oplus \Delta k_w$, $k'_w = k_w \oplus \Delta \widetilde{k}'_w$, $k_w'^* = k_w \oplus \Delta k_w \oplus \Delta \widetilde{k}'_w$, where Δk_w and $\Delta \widetilde{k}'_w$ are the fixed 64-bit key differences in the position of bytes 1, 2, 6, 7, 8, 11, 12, 13 of ΔK_w (depicted in Fig. 2) and $\Delta \widetilde{K}'_w$, respectively. For the subkey quartet $(k_w, k_w^*, k'_w, k_w'^*)$, do the following:

   4.1 Partially encrypt each plaintext P_{i,l_0} in S_i through E^b under k_w, $i = 1, 2, \cdots, 2^{54}$, $l_0 = 1, 2, \cdots, 2^{64}$. We denote the partially encrypted value by x_{i,l_0}. Partially decrypt each $x_{i,l_0} \oplus \Delta K_{0,R}$ through E^b under k_w^*, and find the corresponding plaintext in S_i^*, denoted P_{i,l_0}^*. We denote the corresponding ciphertexts of P_{i,l_0} and P_{i,l_0}^* by C_{i,l_0} and C_{i,l_0}^*, respectively. (For each guess of 64-bit k_w, Step 4.1 takes about $2^{64+1} \cdot (8/16) \cdot (1/10) = 2^{60.7}$ encryptions. Note that this step is independent of $\Delta \widetilde{K}'$ in Step 3, so there is no need to run this step for every iteration of Step 3.)

4.2 Partially encrypt each plaintext P'_{j,l_1} in S'_j through E^b under k'_w, $j = 1, 2, \cdots, 2^{54}$, $l_1 = 1, 2, \cdots, 2^{64}$. We denote the partially encrypted value by x'_{j,l_1}. Partially decrypt each $x'_{j,l_1} \oplus \Delta K_{0,R}$ through E^b under k'^*_w, and find the corresponding plaintext in S'^*_j, denoted P'^*_{j,l_1}. We denote the corresponding ciphertexts of P'_{j,l_1} and P'^*_{j,l_1} by C'_{j,l_1} and C'^*_{j,l_1}, respectively. (For each guess of 71-bit $(k_w, \Delta \widetilde{K}')$, Step 4.2 takes about $2^{64+1} \cdot (8/16) \cdot (1/10) = 2^{60.7}$ encryptions.)

4.3 Insert $C_{i,l_0} \| C^*_{i,l_0}$ in a hash table (indexed by bytes 1, 2, 3, 4, 5, 6, 9, 14) and then check that $(C_{i,l_0} \| C^*_{i,l_0}) \oplus (C'_{j,l_1} \| C'^*_{j,l_1}) \in \Delta I'_{10}(1) \| \Delta I'_{10}(2)$ for all i, j, l_0 and l_1, where $\Delta I'_{10}(1) = \{((*, 0, 0, 0), (a, 0, 0, *), (b_1, 0, *, b_2), (b_3, *, 0, b_2))\}$, $\Delta I'_{10}(2) = \{((*, 0, 0, 0), (a, 0, 0, *), (b_4, 0, *, b_2), (b_5, *, 0, b_2))\}$, $*$ is any 8-bit value, and b_i is one of the output differences caused by the input difference a to the S-box. Note that $\Delta I'_{10}(1)$ and $\Delta I'_{10}(2)$ are both candidates for $\Delta I'_{10}$, i.e., b_2 is a candidate for c (see Fig. 3). Keep in a table all the ciphertext quartets $(C_{i,l_0}, C'_{j,l_1}, C^*_{i,l_0}, C'^*_{j,l_1})$ passing the both tests and go to Step 5 with this table. Since $\Delta I'_{10}(1)$ in the first test has 2^{53} out of 2^{128} values and $\Delta I'_{10}(2)$ in the second test has 2^{46} out of 2^{128} values, the expected number of quartets kept in the table is about $2^{(54+64)\cdot 2} \cdot 2^{-128+53} \cdot 2^{-128+46} = 2^{79}$. (For each guess of 71-bit $(k_w, \Delta \widetilde{K}')$, Step 4.3 takes about 2^{119} memory accesses that are equivalent to approximately 2^{112} encryptions according to the implementations of NESSIE primitives [30].)

5. Guess an 8-bit subkey $k_{9,v}$ in the position of byte 12 in round 9 and set $k^*_{9,v} = k'_{9,v} = k'^*_{9,v} = k_{9,v}$. For the 8-bit subkey quartet $(k_{9,v}, k^*_{9,v}, k'_{9,v}, k'^*_{9,v})$, do the following:

   5.1 For all the remaining ciphertext quartets $(C_{i,l_0}, C'_{j,l_1}, C^*_{i,l_0}, C'^*_{j,l_1})$, partially decrypt C_{i,l_0} and C'_{j,l_1} under $k_{9,v}$ and $k'_{9,v}$ through E^f, respectively. If the partially decrypted pairs do not have the difference a, then discard the corresponding ciphertext quartets. Since it has approximately a 7-bit filtering, the number of remaining quartets after this step is about 2^{72}. (The partial decryptions can be done after the remaining ciphertext quartets have been sorted by byte 12 of (C_{i,l_0}, C'_{j,l_1}) or this step can use a pre-computed table, so Step 5.1 takes a relatively small time complexity.)

   5.2 For all the remaining ciphertext quartets $(C_{i,l_0}, C'_{j,l_1}, C^*_{i,l_0}, C'^*_{j,l_1})$, partially decrypt C^*_{i,l_0} and C'^*_{j,l_1} under $k^*_{9,v}$ and $k'^*_{9,v}$ through E^f, respectively. If the partially decrypted pairs do not have the difference a, discard the corresponding ciphertext quartets and then go to Step 6. It also imposes approximately a 7-bit filtering, hence the number of remaining quartets after this step is about 2^{65}. (Similarly, Step 5.2 can be performed efficiently.)

6. Guess an 8-bit subkey $k_{9,w}$ in the position of byte 8 in round 9 and set $k'_{9,w} = k_{9,w}$. For the 8-bit subkey pair $(k_{9,w}, k'_{9,w})$, do the following:

6.1 For all the remaining ciphertext quartets $(C_{i,l_0}, C'_{j,l_1}, C^*_{i,l_0}, C'^*_{j,l_1})$, partially decrypt C_{i,l_0} and C'_{j,l_1} under $k_{9,w}$ and $k'_{9,w}$ through E^f, respectively. If the partially decrypted pairs do not have the difference a, then discard the corresponding ciphertext quartets. Since this imposes approximately a 7-bit filtering, the number of remaining quartets after this step is about 2^{58}.

6.2 Guess an 8-bit value d to form an 8-bit subkey pair $(k^*_{9,w} = k_{9,w} \oplus d$, $k'^*_{9,w} = k_{9,w} \oplus d)$ in the position of byte 8 in round 9. For the 8-bit subkey pair $(k^*_{9,w}, k'^*_{9,w})$, do the following:

    6.2.1 For all the remaining ciphertext quartets $(C_{i,l_0}, C'_{j,l_1}, C^*_{i,l_0}, C'^*_{j,l_1})$, partially decrypt C^*_{i,l_0} and C'^*_{j,l_1} under $k^*_{9,w}$ and $k'^*_{9,w}$ through E^f, respectively. If the partially decrypted pairs do not have the difference a, discard the corresponding ciphertext quartets and then go to Step 7. It also induces approximately a 7-bit filtering, hence the number of remaining quartets after this step is about 2^{51}. (Similarly, Step 6 can be performed efficiently.)

7. Guess a 32-bit subkey $k_{9,y}$ in the position of bytes 0, 7, 10, 13 in round 9 and compute $k'_{9,y} = k_{9,y} \oplus (a, 0, 0, 0)$. For the 32-bit subkey pair $(k_{9,y}, k'_{9,y})$, do the following:

7.1 For all the remaining ciphertext quartets $(C_{i,l_0}, C'_{j,l_1}, C^*_{i,l_0}, C'^*_{j,l_1})$, partially decrypt C_{i,l_0} and C'_{j,l_1} under $k_{9,y}$ and $k'_{9,y}$ through E^f, respectively. If the differences of the partially decrypted pairs are not in $\mathcal{B}$ (see Eq. (1)), then discard the corresponding ciphertext quartets. Since $\mathcal{B}$ has $2^7 - 1$ out of 2^{32} values, the remaining quartets after this step is about 2^{26}. (For each guess of 127-bit $(k_{9,y}, d, k_{9,w}, k_{9,v}, k_w, \Delta \widetilde{K}')$, Step 7.1 takes $2^{51+1} \cdot (4/16) \cdot (1/10) = 2^{46.7}$ encryptions.)

7.2 Guess two 8-bit values e, f to form a 32-bit subkey pair $(k^*_{9,y} = k_{9,y} \oplus (d, 0, e, f)$, $k'^*_{9,y} = k_{9,y} \oplus (d \oplus a, 0, e, f))$ in the position of bytes 0, 7, 10, 13 in round 9. For the 32-bit subkey pair $(k^*_{9,y}, k'^*_{9,y})$, do the following:

    7.2.1 For all the remaining ciphertext quartets $(C_{i,l_0}, C'_{j,l_1}, C^*_{i,l_0}, C'^*_{j,l_1})$, partially decrypt C^*_{i,l_0} and C'^*_{j,l_1} under $k^*_{9,y}$ and $k'^*_{9,y}$ through E^f, respectively. If the differences of the partially decrypted pairs are not in $\mathcal{B}$, discard the corresponding ciphertext quartets and then go to Step 8. This also induces approximately about a 25-bit filtering, hence the number of remaining quartets after this step is about 2 for each wrong key guess. (For each guess of 143-bit $(e, f, k_{9,y}, d, k_{9,w}, k_{9,v}, k_w, \Delta \widetilde{K}')$, this step takes $2^{26+1} \cdot (4/16) \cdot (1/10) = 2^{21.7}$ encryptions.)

8. For the remaining ciphertext quartets $(C_{i,l_0}, C'_{j,l_1}, C^*_{i,l_0}, C'^*_{j,l_1})$, classify the quartets according to the differences of C_{i,l_0} and C'_{j,l_1} by byte 11. Discard all the ciphertext quartets except for the group with the largest number of quartets and then go to Step 9. Since this results in approximately a 7-bit filtering for each pair of quartets, the remaining quartets after this step is expected to be about 2^{-6} for each wrong key guess. (It takes a relatively small time complexity.)

9. If there are more than 16 quartets in the table, then output the guessed subkey quartet as the right one. Otherwise, run the above steps with another guess for the subkey quartet, i.e., $(e, f, k_{9,y}, d, k_{9,w}, k_{9,v}, k_w, \Delta \tilde{K}')$.

About 2^{125} chosen plaintexts in Steps 1, 2 and 3 are encrypted on average, hence the data complexity of this attack is about 2^{125} related-key chosen plaintexts and the time complexity of Steps 1, 2 and 3 is about 2^{125} encryptions. Step 4 runs about 2^{70} times, so the time complexity of Step 4.2 is about $2^{60.7+70} = 2^{130.7}$ encryptions (it can be improved by a factor of about 2^4 by using a pre-computed table[1]) and the time complexity of Step 4.3 is about $2^{112+70} = 2^{182}$ encryptions. As stated above, Steps 5, 6 and 8 take relatively small time complexities compared to other steps.

The time complexity for Step 7 depends on how many times this step runs, which can be measured by the number of guessed subkeys (including d, e and f). Since Steps 7.1 and 7.2 run in this attack 2^{126} and 2^{142} times on average, these steps take $2^{172.7}$ and $2^{163.7}$ encryptions, respectively. However, the time complexities of these steps can be improved by using a divide and conquer technique. In Step 7.1, two of the four bytes of the remaining ciphertext quartets are first decrypted (these partial decryptions can be performed after the remaining ciphertext quartets are sorted by these two bytes) and discard the ciphertext quartets of which the decrypted two bytes do not have a difference in B with respect to the two-byte position, and then do this test with other two bytes of the remaining ciphertext quartets byte by byte. With this divide and conquer technique, we can also run Step 7.2. This method allows Steps 7.1 and 7.2 to decrease their time complexities down to about $2^{135.7}$ and $2^{146.7}$ encryptions, respectively.

We can calculate the success rate of the attack by using the Poisson distribution. Since the expected number of remaining quartets for each wrong subkey quartet is 2^{-6}, the probability that the number of remaining quartets for each wrong subkey quartet is larger than 16 is 2^{-150} by the Poisson distribution, $X \sim Poi(\lambda = 2^{-6})$, $Pr_X[X > 16] \approx 2^{-150}$. It follows that the probability that the attack outputs a wrong subkey quartet is quite low, since the total number of guessed wrong subkey quartets is about 2^{142}. On the other hand, the expected number of remaining quartets for the right subkey quartet is about $2^5 = 2^{236} \cdot 2^{-231}$ due to our 8-round related-key rectangle distinguisher. Thus, the probability that the number of remaining quartets for the right key quartet is larger than 16 is 0.99 by the Poisson distribution, $Y \sim Poi(\lambda = 2^5)$, $Pr_Y[Y > 16] \approx 0.99$.

Therefore, this attack works with a data complexity of about 2^{125} related-key chosen plaintexts and with a time complexity of about 2^{182} encryptions and with a success rate of 0.99.

---

[1] Before running this attack, we can pre-compute a table which keeps 2^{64} input pairs (I_0, I_0^*) to round 0, where $I_0^* = BS^{-1}(SR^{-1}(MC^{-1}(MC(SR(BS(I_0))) \oplus \Delta K_{0,R})))$. If Step 4.1 has access to this table for each guessed subkey (k_w, k_w^*), it can find plaintext pairs $(P_{i,l}, P_{i,l}^*)$ by XORing (k_w, k_w^*) with (I_0, I_0^*).

4.3   Reducing the Number of Related Keys from 256 to 64

If we take more delicate related keys in our attack, we can reduce the number of related keys from 256 to 64 (note that the basic idea of this method has been introduced in [8]). The following 64 related keys can be used in our attack:

- 16 key candidates K^i ($i = 0, 1, \cdots, 15$) for the key K such that all the bytes of the 16 K^i have the same values in each byte position except that byte 3 is the same as byte 11 in each K^i, say s_i, but s_0, $s_1, \cdots$, s_{15} are all pairwise distinct.
- 16 key candidates K^{*i} for the key K^* such that bytes 1 and 9 of $K^i \oplus K^{*i}$ are both a, and the other bytes of $K^i \oplus K^{*i}$ are all 0.
- 16 key candidates K'^j ($j = 0, 1, \cdots, 15$) for the key K' such that all the bytes of the 16 K'^j are the same as those of K^i for some i except that byte 3 is the same as byte 11 in each K'^j, say t_j, but t_0, $t_1, \cdots$, t_{15} are all pairwise distinct, and bytes 8 and 12 of $K'^j \oplus K^i$ are both a.
- 16 key candidates K'^{*j} for the key K'^* such that bytes 1 and 9 of $K'^{*j} \oplus K'^j$ are both a, and the other bytes of $K'^{*j} \oplus K'^j$ are all 0.

Using these delicately chosen key relationships, we can make 256 key quartets $(K^i, K^{*i}, K'^j, K'^{*j})$ of which one is expected to satisfy the desired key condition, $K^i \oplus K^{*i} = K'^j \oplus K'^{*j} = \Delta K$ and $K^i \oplus K'^j = K^{*i} \oplus K'^{*j} = \Delta K'$ (note that bytes 3 and 11 of $K^i \oplus K'^j$ and $K^{*i} \oplus K'^{*j}$ are both $s_i \oplus t_j$ of which one is expected to be b).

If the above 64 related keys are used in our attack algorithm, the attack works with a data complexity of 2^{124} related-key chosen plaintexts (due to the fact that the attack takes 2^{118} chosen plaintext queries for each key) and with a time complexity of 2^{183} encryptions (due to the fact that Step 4.3 is iterated 2^7 times on average by the 256 key quartets).

5   Related-Key Rectangle Attacks on 8-Round AES-192 and 9-Round AES-256

Similarly, we can construct related-key rectangle attacks on 8-round AES-192 with two related keys ($\Delta K \neq 0$ and $\Delta K' = 0$) and on 9-round AES-256 with four related keys ($\Delta K \neq 0$, $\Delta K' \neq 0$ and $\Delta K \neq \Delta K'$).

The attack on 8-round AES-192 with two related keys recovers bytes 1, 2, 6, 7, 8, 11, 12, 13 of the whitening key pair (K_w, K_w^*) and bytes 3, 6, 9, 12 of the subkey pair (K_7, K_7^*) with a data complexity of about 2^{94} related-key chosen plaintexts, a time complexity of about 2^{120} encryptions and a success rate of 0.9. See Figs. 2 and 4 for a schematic description of our 8-round AES-192 attack (note that the related-key truncated differential in Fig. 2 is used for E^b and E^0 in this attack).

The attack on 9-round AES-256 with four related keys recovers bytes 1, 2, 6, 7, 8, 11, 12, 13 of the whitening key quartet $(K_w, K_w^*, K_w', K_w'^*)$ and bytes 0, 4, 8, 12 of the subkey quartet $(K_8, K_8^*, K_8', K_8'^*)$ with a data complexity of about

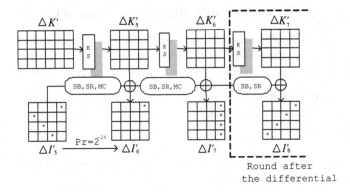

Fig. 4. The second truncated differential for rounds 5-6 (E^1) and the following differential for round 7 (E^f)

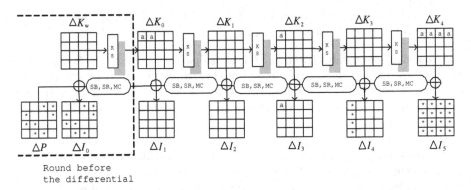

Fig. 5. The first related-key truncated differential for rounds 1-4 including the key addition step of round 0 (E^0), and the preceding differential for round 0 including the whitening key addition step and excluding the key addition step of round 0 (E^b)

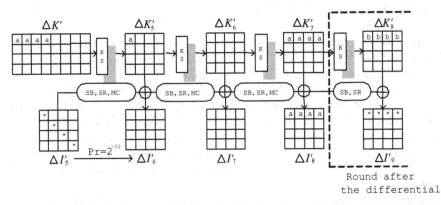

Fig. 6. The second related-key truncated differential for rounds 5-7 (E^1) and the following differential for round 8 (E^f)

2^{99} related-key chosen plaintexts, a time complexity of about 2^{120} encryptions and a success rate of 0.9. See Figs. 5 and 6 for a schematic description of our 9-round AES-256 attack.

6   Comments on the 9-Round AES-192 Attack Presented at Eurocrypt 2005

At Eurocrypt 2005, Biham, Dunkelman and Keller [6] presented a 9-round AES-192 attack which requires 256 related keys, a data complexity of 2^{86} related-key chosen plaintexts and a time complexity of 2^{125} encryptions. However, we have observed that there is some flaw in the key guessing step in their attack. In order to complete their attack, we need to guess in addition 56 bits of the subkey in the last round, hence the attack requires a larger time complexity of $2^{181} = 2^{125} \cdot 2^{56}$ rather than 2^{125} encryptions.

Moreover, their attack can be mounted based on 64 related keys as in our 10-round AES-192 attack. Similarly, it allows the 9-round AES-192 attack to work with a smaller data complexity, but a larger time complexity than the original ones; in our observation their attack works with 64 related keys, a data complexity of 2^{85} related-key chosen plaintexts and a time complexity of 2^{182}. This method (for reducing the number of related keys) can also be used in the 10-round AES-256 attack presented at Eurocrypt 2005. See Table 1 for the attack complexity.

7   Conclusion

In this paper we have presented related-key rectangle attacks on 8-round AES-192 with 2 related keys, 10-round AES-192 with 64 or 256 related keys and 9-round AES-256 with 4 related keys, which are faster than exhaustive key search. All our attacks have been designed based on the key scheduling algorithms of AES-192 and AES-256 which have relatively slow difference propagations.

Our 10-round AES-192 attack leads to the best known attack on AES-192 and our 8-round AES-192, 9-round AES-256 attacks are both better than previously best known attacks on AES-192 with 2 related keys and AES-256 with 4 related keys in terms of the number of attacked rounds and the data or time complexity. It should be clear, however, that none of these attacks presents a realistic threat to the security of the AES.

Acknowledgements. We thank Orr Dunkelman and Nathan Keller for their helpful comments.

References

1. Biham, E., Shamir, A.: Differential Cryptanalysis of DES-like Cryptosystems. In: Menezes, A.J., Vanstone, S.A. (eds.) CRYPTO 1990. LNCS, vol. 537, pp. 2–21. Springer, Heidelberg (1991)
2. Biham, E.: New Types of Cryptanalytic Attacks Using Related Keys. Journal of Cryptology 7(4), 229–246 (1994)

3. Biham, E., Biryukov, A., Shamir, A.: Cryptanalysis of Skipjack Reduced to 31 Rounds Using Impossible Differentials. Journal of Cryptology 18(4), 291–311 (2005)
4. Biham, E., Keller, N.: Cryptanalysis of Reduced Variants of Rijndael, http://csrc.nist.gov/encryption/aes/round2/conf3/aes3papers.html
5. Biham, E., Dunkelman, O., Keller, N.: The Rectangle Attack – Rectangling the Serpent. In: Pfitzmann, B. (ed.) EUROCRYPT 2001. LNCS, vol. 2045, pp. 340–357. Springer, Heidelberg (2001)
6. Biham, E., Dunkelman, O., Keller, N.: Related-Key Boomerang and Rectangle Attacks. In: Cramer, R.J.F. (ed.) EUROCRYPT 2005. LNCS, vol. 3494, pp. 507–525. Springer, Heidelberg (2005)
7. Biham, E., Dunkelman, O., Keller, N.: A Related-Key Rectangle Attack on the Full KASUMI. In: Roy, B. (ed.) ASIACRYPT 2005. LNCS, vol. 3788, pp. 443–461. Springer, Heidelberg (2005)
8. Biham, E., Dunkelman, O., Keller, N.: Related-Key Impossible Differential Attacks on AES-192. In: Pointcheval, D. (ed.) CT-RSA 2006. LNCS, vol. 3860, pp. 21–31. Springer, Heidelberg (2006)
9. Biryukov, A.: The Boomerang Attack on 5 and 6-Round AES. In: Dobbertin, H., Rijmen, V., Sowa, A. (eds.) Advanced Encryption Standard – AES. LNCS, vol. 3373, pp. 11–16. Springer, Heidelberg (2005)
10. Blunden, M., Escott, A.: Related Key Attacks on Reduced Round KASUMI. In: Matsui, M. (ed.) FSE 2001. LNCS, vol. 2355, pp. 277–285. Springer, Heidelberg (2002)
11. Cheon, J., Kim, M., Kim, K., Lee, J., Kang, S.: Improved Impossible Differential Cryptanalysis of Rijndael and Crypton. In: Kim, K.-c. (ed.) ICISC 2001. LNCS, vol. 2288, pp. 39–49. Springer, Heidelberg (2002)
12. Daemen, J., Rijmen, V.: The Design of Rijndael: AES – the Advanced Encryption Standard. Springer, Heidelberg (2002)
13. Dunkelman, O., Keller, N., Kim, J.: Related-Key Rectangle Attack on the Full SHACAL-1. In: Proceedings of Selected Areas in Cryptography 2006, SAC 2006, LNCS, vol. 4356, Springer, Heidelberg (to appear, 2007)
14. Ferguson, N., Kelsey, J., Schneier, B., Stay, M., Wagner, D., Whiting, D.: Improved Cryptanalysis of Rijndael. In: Schneier, B. (ed.) FSE 2000. LNCS, vol. 1978, pp. 213–230. Springer, Heidelberg (2001)
15. Hawkes, P.: Differential-Linear Weak-Key Classes of IDEA. In: Nyberg, K. (ed.) EUROCRYPT 1998. LNCS, vol. 1403, pp. 112–126. Springer, Heidelberg (1998)
16. Hong, S., Kim, J., Lee, S., Preneel, B.: Related-Key Rectangle Attacks on Reduced Versions of SHACAL-1 and AES-192. In: Gilbert, H., Handschuh, H. (eds.) FSE 2005. LNCS, vol. 3557, pp. 368–383. Springer, Heidelberg (2005)
17. Jakimoski, G., Desmedt, Y.: Related-Key Differential Cryptanalysis of 192-bit Key AES Variants. In: Matsui, M., Zuccherato, R.J. (eds.) SAC 2003. LNCS, vol. 3006, pp. 208–221. Springer, Heidelberg (2004)
18. Kelsey, J., Schneier, B., Wagner, D.: Key Schedule Cryptanalysis of IDEA, G-DES, GOST, SAFER, and Triple-DES. In: Koblitz, N. (ed.) CRYPTO 1996. LNCS, vol. 1109, pp. 237–251. Springer, Heidelberg (1996)
19. Kelsey, J., Schneir, B., Wagner, D.: Related-Key Cryptanalysis of 3-WAY, Biham-DES, CAST, DES-X, NewDES, RC2, and TEA. In: Han, Y., Quing, S. (eds.) ICICS 1997. LNCS, vol. 1334, pp. 233–246. Springer, Heidelberg (1997)
20. Kelsey, J., Kohno, T., Schneier, B.: Amplified Boomerang Attacks Against Reduced-Round MARS and Serpent. In: Schneier, B. (ed.) FSE 2000. LNCS, vol. 1978, pp. 75–93. Springer, Heidelberg (2001)

21. Kim, J., Kim, G., Hong, S., Lee, S., Hong, D.: The Related-Key Rectangle Attack – Application to SHACAL-1. In: Wang, H., Pieprzyk, J., Varadharajan, V. (eds.) ACISP 2004. LNCS, vol. 3108, pp. 123–136. Springer, Heidelberg (2004)
22. Kim, J., Kim, G., Lee, S., Lim, J., Song, J.: Related-Key Attacks on Reduced Rounds of SHACAL-2. In: Canteaut, A., Viswanathan, K. (eds.) INDOCRYPT 2004. LNCS, vol. 3348, pp. 175–189. Springer, Heidelberg (2004)
23. Knudsen, L.R.: Cryptanalysis of LOKI91. In: Zheng, Y., Seberry, J. (eds.) AUSCRYPT 1992. LNCS, vol. 718, pp. 196–208. Springer, Heidelberg (1993)
24. Knudsen, L.R.: Truncated and Higher Order Differentials. In: Preneel, B. (ed.) Fast Software Encryption. LNCS, vol. 1008, pp. 196–211. Springer, Heidelberg (1995)
25. Ko, Y., Hong, S., Lee, W., Lee, S., Kang, J.: Related Key Differential Attacks on 26 Rounds of XTEA and Full Rounds of GOST. In: Roy, B., Meier, W. (eds.) FSE 2004. LNCS, vol. 3017, pp. 299–316. Springer, Heidelberg (2004)
26. Langford, S.K., Hellman, M.E.: Differential-Linear Cryptanalysis. In: Desmedt, Y.G. (ed.) CRYPTO 1994. LNCS, vol. 839, pp. 17–25. Springer, Heidelberg (1994)
27. Lu, J., Kim, J., Keller, N., Dunkelman, O.: Related-Key Rectangle Attack on 42-Round SHACAL-2. In: Katsikas, S.K., Lopez, J., Backes, M., Gritzalis, S., Preneel, B. (eds.) ISC 2006. LNCS, vol. 4176, pp. 85–100. Springer, Heidelberg (2006)
28. Lu, J., Lee, C., Kim, J.: Related-Key Attacks on the Full-Round Cobra-F64a and Cobra-F64b. In: De Prisco, R., Yung, M. (eds.) SCN 2006. LNCS, vol. 4116, pp. 95–110. Springer, Heidelberg (2006)
29. Lucks, S.: Attacking Seven Rounds of Rijndael under 192-bit and 256-bit Keys. In: Proceedings of AES 3, NIST (2000)
30. NESSIE — New European Schemes for Signatures, Integrity and Encryption, Performance of Optimized Implementations of the NESSIE Primitives, version 2.0, https://www.cosic.esat.kuleuven.be/nessie/deliverables/D21-v2.pdf
31. Phan, R.C.-W.: Impossible Differential Cryptanalysis of 7-round Advanced Encryption Standard (AES). Information Processing Letters 91(1), 33–38 (2004)
32. Wagner, D.: The Boomerang Attack. In: Knudsen, L.R. (ed.) FSE 1999. LNCS, vol. 1636, pp. 156–170. Springer, Heidelberg (1999)

An Analysis of XSL Applied to BES

Chu-Wee Lim and Khoongming Khoo

DSO National Laboratories, 20 Science Park Drive, S118230, Singapore
lchuwee@dso.org.sg, kkhoongm@dso.org.sg

Abstract. Currently, the only plausible attack on the Advanced Encryption System (AES) is the XSL attack over F_{256} through the Big Encryption System (BES) embedding. In this paper, we give an analysis of the XSL attack when applied to BES and conclude that the complexity estimate is too optimistic. For example, the complexity of XSL on BES-128 should be at least 2^{401} instead of the value of 2^{87} from current literature. Our analysis applies to the eprint version of the XSL attack, which is different from the compact XSL attack studied by Cid and Leurent at Asiacrypt 2005. Moreover, we study the attack on the BES embedding of AES, while Cid and Leurent studies the attack on AES itself. Thus our analysis can be considered as a parallel work, which together with Cid and Leurent's study, disproves the effectiveness of both versions of the XSL attack against AES.

Keywords: XSL algorithm, AES, BES, linearisation.

1 Introduction

In 2001, after a standardisation period of 5 years, NIST finally adopted the Rijndael block cipher as the AES standard. A year later, Courtois and Pieprzyk surprised the cryptographic community by proposing an algebraic attack [3,4] on Rijndael and Serpent, which could obtain the key faster than an exhaustive search. This attack was named XSL for eXtended Sparse Linearisation [3,4], which is a modification of the earlier XL (eXtended Linearisation) attack proposed by Courtois et al [2]. In XSL, the authors exploited the algebraic simplicity of Rijndael's S-box and obtained a number of quadratic equations. Then in order to apply linearisation, equations are multiplied by S-box monomials (as compared to arbitrary monomials in the case of the XL attack). This enables the number of occurring monomials to be kept within a manageable range, and so upon linearisation, the system of linear equations can be solved faster than an exhaustive key search.

At around the same time, Murphy and Robshaw [6] proposed a method of reinterpreting the AES system by writing its equations over the field F_{256} instead of the smaller field F_2. Over F_{256}, the input x and output y of the Rijndael S-box satisfies the simple equation $xy = 1$. In this way, the number of equations of each S-box remains the same while the number of monomials involved is reduced by half. The authors later noted [7] that the BES embedding can lead to a dramatic

A. Biryukov (Ed.): FSE 2007, LNCS 4593, pp. 242–253, 2007.

decrease in the complexity of XSL attack. For example, as Courtois observed in §7 of [4], AES-128 can be broken with complexity 2^{87}.

However, Murphy and Robshaw also expressed skepticism at the practicality of XSL in [7], and concluded that it was unlikely to work. T. T. Moh and D. Coppersmith were similarly skeptical and remarked that XSL was unlikely to work.

In Asiacrypt 2005, Cid and Leurent [1] gave an analysis of the compact XSL attack [4] on AES-128 and proved that it is equivalent to a substitution-then-XL (sXL) method. They concluded that XSL attack is essentially an XL attack on a system of equations larger than that of the original AES. Thus compact XSL is not an effective attack against the AES cipher. This partly answers some uncertainties of the compact XSL attack and suggests that it may not be an effective method against block ciphers.

However, it does not give us the full answer on whether XSL is effective against AES. This is because:

1. In [1], only the compact XSL attack [4] is analysed. But the eprint version of the XSL attack (eprint XSL) [3] is different from the compact XSL attack as it uses a larger set of monomials. Moreover, the bounds derived from the eprint XSL are better than that of the compact XSL.
2. In [1], only the complexity of the compact XSL attack on AES over F_2 is analyzed. However the best attack on AES in the current literature is the eprint XSL attack on AES over the bigger field F_{256}.

In this paper, we provide a more complete answer to the effectiveness of the XSL attacks on AES by analyzing the above two points. We will focus on the embedding of the AES in the Big Encryption System (BES) over F_{256}. Because of the nice algebraic structure of the equations of the Rijndael S-box over F_{256}, we can provide an exact analysis of the linear dependencies that exist between the equations of BES. This allows us to give a more accurate estimate of the number of linearly independent equations in the eprint XSL attack. Based on our estimate, we deduce that the complexity of the eprint XSL attack on BES-128 is *at least* 2^{401} instead of 2^{87} from the known literature. Similar complexity estimates for BES-192 and BES-256 proves that the eprint XSL attack is ineffective against them.

2   The XSL Attack on BES

There are currently two versions of XSL. In this paper, we shall consider the version in [3]. Furthermore, in that paper, the authors described two forms of XSL - the first and the second. We shall look at the second version, which only requires at most two plaintext-ciphertext pairs. This section is essentially a summary of [3], [6] and [7], but we need it to establish notations which will be used in subsequent sections.

Note: we often mention linear equations from the AES cipher, when a technically more accurate name would be *affine*. This is in accordance with the literature.

2.1   A Summary of the XSL Attack

First, recall the S-box in the AES cipher. This comprises of the *inverse* map[1] on the finite field F_{256}, followed by an F_2-affine map on the output. Thus, the only nonlinear component of the cipher is the map $x \mapsto x^{-1}$ on the finite field F_{256}. The authors of [3] believe that this algebraic definition of the S-box presents an opportunity for algebraic attacks on the cipher. They noted that if $y = x^{-1}$ in F_{256}, then we get $xy = 1$, $x^2 y = x$ and $xy^2 = y$. The Frobenius map $x \mapsto x^2$ is F_2-linear on F_{256}, so by taking the components, we get $3 \times 8 = 24$ equations in F_2. There is a slight caveat though: since $xy = 1$ is only true most of the time, it only produces 7 equations instead of 8. The eighth equation only holds $\frac{255}{256}$ of the time. Nevertheless, for the sake of our argument, we shall assume that we get 24 equations. It can be easily verified that up to linear dependence, these are the only equations we can get.

Thus, let $x_0, \ldots, x_7$ and $y_0, \ldots, y_7$ denote the input and output bits of an S-box respectively, We get 24 equations which are linear combinations of the monomials $1, x_i, y_j, x_i y_j$ ($0 \le i, j \le 7$); i.e. , we have 24 equations involving 81 monomials. Such a monomial is said to *belong* to the S-box. Furthermore, if S and S' are distinct S-boxes, then the set of monomials belonging to S and S' are *disjoint*.

On the other hand, there are also F_2-linear equations for the Shiftrows and MixColumns algorithms. Hence, we can write down all the S-box and linear equations and solve them in order to extract the key. We shall use the following notation subsequently:

S = number of S-boxes in the *cipher and key schedule*,

L = number of linear equations in the *cipher and key schedule*,

r = number of equations in each S-box = 24,

t = number of terms used in the equations for each S-box = 81,

P = XSL parameter (to be described later).

The values for S and L for AES-128, -192 and -256 respectively are given in Table 1 below:

Table 1. Parameters for Various AES Ciphers

	AES-128	AES-192	AES-256
S	201	417	501
L	1664	3520	4128

Note that the values of S and L in Table 1 are based on one plaintext-ciphertext pair for the second XSL attack on AES-128 and two plaintext-ciphertext pairs for

---

[1] Strictly speaking, this is not true since the map $x \mapsto x^{-1}$ is not defined at 0. It would be correct mathematically to write this map as $x \mapsto x^{254}$.

the second XSL attack on AES-192, AES-256. They are derived from Section 7 and Appendix B of [3].

The XSL method can be summarised in the following steps, for a parameter P:

1. Pick P distinct S-boxes, and select one of them to be *active*; the others are declared *passive*. From the active S-box, select one equation out of the r quadratic ones; from each of the $P - 1$ passive S-boxes, select a monomial t_i ($i = 2, \ldots, P$). We then multiply the active S-box equation with $t_2 t_3 \ldots t_P$ to obtain a new equation. Let us call such an equation an **extended S-box equation**. Take the collection Σ_S of all extended S-box equations. Also, we call the monomials which occur in the equations of Σ_S the **extended S-box monomials**.

2. Pick a linear equation and select $P - 1$ distinct S-boxes. From each of the selected S-boxes, choose a monomial t_i ($i = 2, 3, \ldots, P$). We then multiply the linear equation by $t_2 t_3 \ldots t_P$ to obtain a new equation. Let us call such an equation an **extended linear equation**. Take the collection Σ_L of all extended linear equations.

3. Take the equations in $\Sigma := \Sigma_S \cup \Sigma_L$ and solve them via linearisation. In other words, replace each occurring monomial by a new variable and solve the set of linear equations by (say) Gaussian elimination. Furthermore, these linear equations are sparse so we can apply advanced techniques like the block Lanczos algorithm [5] to solve them.

4. If there are not enough equations for a complete solution, we can apply the T'-method to produce more equations. This comprises of (i) fixing a variable, say x_i, and (ii) attempting to find equations whose degrees remain the same upon multiplication by x_i.

In [3], it was noted that the equations in Σ_S have a lot of linear dependencies after linearisation. For example, if we pick two active S-boxes and equations eqn_1 and eqn_2 from them, as well as $P - 2$ passive S-boxes with monomials $t_3, t_4, \ldots, t_P$, then by expanding the equation $eqn_1 \cdot eqn_2 \cdot (t_3 \ldots t_P)$ we obtain linear dependencies between extended S-box equations of the form $eqn_1 \cdot (t_2 t_3 \ldots t_P)$ and $t_1 \cdot eqn_2 \cdot (t_3 \ldots t_P)$. After removing some obvious dependencies like these, we get

$$R = \binom{S}{P}(t^P - (t - r)^P) \tag{1}$$

equations. Likewise, by fixing $t - r$ linearly independent terms in each S-box and expressing the remaining terms as linear combinations of them, the number of extended linear equations[2] is (after removing obvious linear dependencies):

$$R' = L \times (t - r)^{P-1} \binom{S}{P - 1}. \tag{2}$$

---

[2] Note that in [3], the author used R' and R'' to denote the extended linear equations from the cipher and the key schedule respectively. Here we denote all extended linear equations by R'.

Hence, the total number of equations obtained is $R + R'$.

On the other hand, it was assumed that the monomials in Σ_L are also extended S-box monomials. Hence, the total number of terms occurring in Σ is the number of extended S-box monomials:

$$T = t^P \binom{S}{P}. \tag{3}$$

Finally, for the T'-method, the authors of [3] remarked that to apply T', we need at least $99.4\% \times T$ equations in the first place. In other words, T'-method can only be applied if the number of equations we have is very very close to being sufficient; in which case, the method helps to increase the number of equations slightly. Thus we shall leave it out in our discussion.

The objective of XSL is to select an appropriate P so that we get enough equations. The following computations apply for AES-128, AES-192 and AES-256.

1. **AES-128:** The smallest P where $R + R' > T$ is $P = 7$. The parameters are $R = 4.95 \times 10^{25}$, $R' = 4.85 \times 10^{24}$, $T = 5.41 \times 10^{25}$. We have $(R + R')/T = 1.004$ and the complexity of XSL attack is $T^{2.376} \approx 2^{203}$.

2. **AES-192:** The smallest P where $R + R' > T$ is $P = 7$. The parameters are $R = 8.65 \times 10^{27}$, $R' = 8.50 \times 10^{26}$, $T = 9.46 \times 10^{27}$. We have $(R + R')/T = 1.004$ and the complexity of XSL attack is $T^{2.376} \approx 2^{221}$.

3. **AES-256:** The smallest P where $R + R' > T$ is $P = 7$. The parameters are $R = 3.15 \times 10^{28}$, $R' = 3.02 \times 10^{27}$, $T = 3.45 \times 10^{28}$. We have $(R + R')/T = 1.002$ and the complexity of XSL attack is $T^{2.376} \approx 2^{225}$.

2.2   A Summary of the BES Cipher

At Crypto 2002, Robshaw and Murphy introduced the Big Encryption System (BES) embedding for the AES cipher [6]. In this embedding, the quadratic equations describing the S-box input-output become much simpler. This results in a substantial reduction in the number of monomials occurring in the system of equations, which can lead to an exponential reduction in the complexity of the XSL attack [7].

While AES performs its operations in the field F_2, BES achieves the same purpose by performing operations in the field F_{256}. The advantage of this rewriting lies in the simplicity of the S-box equation: we have $xy = 1$ immediately instead of 8 quadratic equations in the input and output bits. This, of course, conveniently ignores the case when $x = y = 0$. However this can be countered by other means.

The problem occurs with some of the linear equations, which are F_2-linear but not F_{256}-linear. This is overcome by introducing the conjugates of each variable $x \in F_{256}$, i.e. we have to consider $x_i := x^{2^i}$ for $0 \leq i \leq 7$. Upon introducing these variables, all F_2-linear equations can be expressed as F_{256}-linear equations, by virtue of the following result.

Lemma 1. *Consider the finite field* $K = F_{2^n}$. *Then any* F_2-*affine map* $K \to K$ *can be written in the form:*

$$f(x) = c + a_0 x + a_1 x^2 + \cdots + a_{n-1} x^{2^{n-1}},$$

for some constants $c, a_0, a_1, \ldots, a_{n-1} \in K$.

We will skip the proof, though it suffices to say that the result follows easily from dimension counting (over F_2).

In introducing the conjugates to the S-boxes, we have to express their relationship as $x_{i+1} = x_i^2$, where the subscript is taken from $\mathbb{Z}/8\mathbb{Z}$ (the integers modulo 8). This gives 24 equations for each S-box: indeed, if we denote the input and output variables by $x_0, x_1, \ldots, x_7 \in F_{256}$ and $y_0, y_1, \ldots, y_7 \in F_{256}$ respectively, then

$$x_0 y_0 = 1, \ x_1 y_1 = 1, \ x_2 y_2 = 1, \ \ldots, x_7 y_7 = 1,$$
$$x_0^2 = x_1, \ x_1^2 = x_2, \ x_2^2 = x_3, \ \ldots, x_7^2 = x_0,$$
$$y_0^2 = y_1, \ y_1^2 = y_2, \ y_2^2 = y_3, \ \ldots, y_7^2 = y_0.$$

For convenience, we make the following definition.

Definition 1. *Let* x_i *be an input variable of an S-box and* y_i *be the corresponding output variable such that* $x_i y_i = 1$. *We shall say* x_i *and* y_i *are* **dual** *to each other.*

Although the number of equations per S-box remains $r = 24$, we now only have 41 monomials! These are: $1, x_i, y_i, x_i^2, y_i^2, x_i y_i$ for $0 \le i \le 7$. Hence, we can apply the technique of XSL to this cipher, in which case formulae (1), (2) and (3) in the previous section still hold, with $t = 81$ replaced with $t = 41$. With this new value of t, it turns out that we can pick a smaller P and dramatically reduce the complexity:

1. **BES-128:** The smallest P where $R + R' > T$ is $P = 3$. The parameters are

$$R = 85341866400, \ R' = 9666009600, \ T = 91892369300.$$

So $(R + R')/T = 1.03$ and the complexity of XSL attack is $T^{2.376} \approx 2^{87}$.

2. **BES-192:** The smallest P where $R + R' > T$ is $P = 3$. The parameters are

$$R = 767998707840, \ R' = 88234798080, \ T = 826947240080.$$

So $(R + R')/T = 1.04$ and the complexity of XSL attack is $T^{2.376} \approx 2^{94}$.

3. **BES-256:** The smallest P where $R + R' > T$ is $P = 3$. The parameters are

$$R = 1333494666000, \ R' = 149422248000, \ T = 1435848423250.$$

So $(R + R')/T = 1.03$ and the complexity of XSL attack is $T^{2.376} \approx 2^{96}$.

Finally, the T'-method for BES was not mentioned in [7]. Although the premise of the method should remain the same - find equations whose degree remain the same upon multiplication by some variable - it's not entirely clear if this is effective. This is because the set of monomials involved in the S-box equations forms a very small subset of the set of monomials, thus multiplying an equation by a variable would almost certainly introduce new monomials even if the degree of the equation does not increase.

3   An Analysis of This Attack

In this section, we shall provide an in-depth analysis of the XSL attack when applied to the BES cipher [6]. Due to the nice structure of the BES S-box equations, we can obtain accurate numbers in many cases. Throughout this section, let us fix the XSL parameter P.

3.1   Analysing the Extended S-Box Equations

First, let us consider the S-box equations, each of which is an equality of two monomials. Hence, each extended S-box equation is also of the form $(monomial_1)$ $= (monomial_2)$. Solving them linearly is rather easy: we get a collection of equivalence classes of monomials, where two monomials are considered equivalent if and only if we can obtain one from the other by a finite number of extended S-box equations.

Example 1. Consider three S-boxes, given by S_1, S_2 and S_3. The input and output pairs of these S-boxes are given by (a_i, b_i), (c_i, d_i) and (e_i, f_i) respectively, for $0 \leq i \leq 7$. To clarify the notation further, $a_i b_i = 1$ and $e_i^2 = e_{i+1}$ are examples of their equations. Then the monomials $a_3 b_3 c_2^2 e_5$ and $c_3 e_5$ are considered equivalent:

$$(a_3 b_3) c_2^2 e_5 = (1) c_2^2 e_5 = (1) c_3 e_5,$$

since the first equality follows from an extended S_1-equation, while the second follows from an extended S_2-equation.

Inspired by this example, we make the following definition.

Definition 2. *Let* $\alpha = \alpha_1 \alpha_2 \dots \alpha_Q$ *be an extended S-box monomial, where each* α_i *is a variable belonging to some S-box. Then* α *is said to be reduced if no two variables belong to the same S-box. The set of reduced S-box monomials of degree* Q *is denoted by* Φ_Q.

It follows that a reduced monomial of degree Q is a product of monomials from Q distinct S-boxes, and so $Q \leq P$. Furthermore, we have the following theorem:

Theorem 1. *Every extended S-box monomial* α *is equivalent to a unique reduced monomial* β. *Furthermore, it is possible to obtain the equivalence via:*

$$\alpha = \gamma_1 = \gamma_2 = \gamma_3 = \cdots = \gamma_r = \beta,$$

where each equality is an extended S-box equation, and the degree of each term is strictly less than the previous one.

Proof. Let $\alpha = \alpha_1 \alpha_2 \dots \alpha_Q$ be an extended S-box monomial, where each α_i is an S-box variable. If α were reduced, there is nothing to do. Otherwise, if α_i and α_j belong to the same S-box then they are either identical or *dual*.

    In the first case, we have $\alpha_i = \alpha_j = x_k$ for some input/output variable x_k; hence upon removing α_i and α_j and adding an x_{k+1}, we get an equivalent monomial of degree $Q - 1$. In the second case, we have $\alpha_i \alpha_j = 1$ and we get an equivalent monomial of degree $Q - 2$. This gives us an extended S-box equation

$$\alpha = \gamma_1,$$

where $\deg \gamma_1 < \deg \alpha$. Repeating the above process with α replaced by γ_1, we get the desired result.

Finally, we need to prove uniqueness, i.e. two distinct reduced monomials are not equivalent. This is quite easy: write $\alpha_1 \alpha_2 \ldots \alpha_Q = \beta_1 \beta_2 \ldots \beta_{Q'}$, where the α_i and β_i are S-box variables. Now, the α_i all belong to distinct S-boxes, as do the β_i. For equality to hold, some α_i and β_j must belong to the same S-box. It immediately follows that $\alpha_i = \beta_j$, so we may cancel them from the equation and repeat. $\qquad\square$

Note that each Φ_i has cardinality $\binom{S}{i} 16^i$. Thus, upon solving the extended S-box equations, the number of linearly independent terms is exactly

$$D_0 = \sum_{i=0}^{P} |\Phi_i| = \sum_{i=0}^{P} \binom{S}{i} 16^i.$$

According to theoretical bound in [3], that should have been:

$$T - R = \binom{S}{P}(t - r)^P = \binom{S}{P} 17^P.$$

Hence, we see that the estimate in [3] was rather close.

3.2 Adding the Extended Linear Equations

The second step is to multiply each linear equation by an extended S-box monomial. However, by Theorem 1 it is equivalent to multiply each linear equation by a reduced S-box monomial of degree at most $P - 1$ (note: we cannot multiply by a reduced monomial of degree P since we only multiply $P - 1$ monomials from passive S-boxes). In a nutshell, the XSL method is equivalent to the following:

1. Obtain the set Σ_S of extended S-box equations.
2. For each linear equation, we multiply it by a reduced monomial from $\Phi_0 \cup \Phi_1 \cup \cdots \cup \Phi_{P-1}$ and obtain the set Σ'_L of extended linear equations.
3. Solve $\Sigma_S \cup \Sigma'_L$ together via linearisation.

Consider hypothetically the case where we just do steps 2 and 3 (i.e. step 3 is performed with $\Sigma_S = \emptyset$). In short, we took a bunch of linear equations, multiply them by some monomials and attempt to solve them by linearisation. The question is: how many linearly independent terms will we get from this attempt?

It is not difficult to give a lower bound for this number. First, consider the set of non-reduced S-box equations and linear equations. If we fix the 8 input

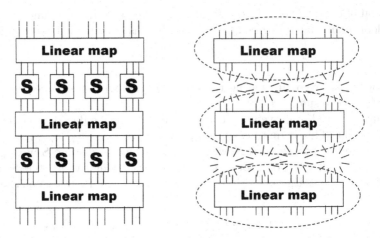

Fig. 1. Removal of the S-boxes gives $8S$ free variables

variables of an S-box, then its output is known. Thus the removal of each S-box contribute 8 free variables. Then the removal of the S S-boxes results in the introduction of $8S$ totally free variables, as the diagram in figure 1 shows.

Without loss of generality, we take the $8S$ input variables of the S-boxes to be the free variables. In other words, the equations in Σ'_L can be satisfied regardless of the values these $8S$ variables take. Hence, the number of linearly independent terms is at least the number of reduced monomials formed by these $8S$ free variables:

$$D_1 = \sum_{i=0}^{P} \binom{S}{i} 8^i.$$

Next, we ask ourselves: *does step 1 provide sufficiently many equations to remove this number of linearly independent terms?*

By theorem 1, we can replace each monomial in Σ'_L by a corresponding reduced one. Furthermore, after such replacements, the extended S-box equations are no longer of any use. This brings us to the next theorem.

Theorem 2. *When solving Σ'_L with the extended S-box equations, we only need to include those extended S-box equations of the form:*

$$(v)(m_1) = (m_2),$$

where:

1. *$m_1, m_2 \in \Phi_0 \cup \Phi_1 \cup \cdots \cup \Phi_{P-1}$, i.e. m_1 and m_2 are reduced monomials of degree at most $P - 1$;*
2. *v is an S-box variable such that it or its dual occurs in m_1;*
3. *the remaining variables in m_1 are among the $8S$ free (input) variables.*

Proof. Recall that each equation of Σ'_L is of the form: $l \cdot m = 0$, where l is a linear equation and m is a *reduced* monomial of degree at most $P-1$. Hence, after linearisation, the only occurring monomials in Σ'_L are those of the form $v \cdot m$, where v is an S-box variable and m is a reduced monomial of degree $\leq P - 1$. If neither v nor its dual occur in m, then this monomial itself is already reduced. Otherwise, we can use an extended S-box equation satisfying conditions 1 and 2 to reduce the monomial further. By Theorem 1, these are the only extended S-box equations we need.

Finally, note that any variable which is not one of the $8S$ free variables, can be expressed as a linear combination of these $8S$ variables. Hence, if $(v)(m_1) = (m_2)$ is an equation satisfying conditions 1 and 2, we can replace each of the remaining variables in m_1 with a linear combination of the $8S$ variables and expand the resulting monomial. Thus, the set of extended S-box equations satisfying conditions 1-3 suffices.   □

We say that an extended S-box equations is "**relevant**" if it satisfies conditions 1-3 in theorem 2.

Now let us compute the number of such equations. Take $(v)(m_1) = (m_2)$, and v' occurs in m_1 where $v' = v$ or the dual of v.

For the case $v = v'$, there are $16S$ choices for the pair $(v, v') = (v, v)$, where v can be one of 16 input or output variables of an S-box. For the case where v' is the dual of v, there are $8S$ choices for the unordered pair (v, v'), which is one of 8 dual pairs of an S-box satisfying $vv' = 1$. Thus there are $16S + 8S = 24S$ choices for the unordered pair (v, v') in the equation $(v)(m_1) = (m_2)$.

For the remaining $P - 2$ variables in m_1, there are $\sum_{i=0}^{P-2} \binom{S-1}{i} 8^i$ choices. Thus, the number of "relevant" equations is

$$D_2 = 24S \times \sum_{i=0}^{P-2} \binom{S-1}{i} 8^i.$$

For the equations in $\Sigma_S \cup \Sigma'_L$ to be solved via linearisation, we must have $D_2 \geq D_1$. However, the values in Table 2 indicate that the stipulated values of P which were supposed to work (see §2.2), actually don't.

To be able to solve for the secret key, we need $D_2 > D_1$. From Table 2, the smallest P where this condition is satisfied is when $P = 23, 33, 36$ for BES-128, BES-192 and BES-256 respectively. In that case, we have the following complexity for the XSL attack:

1. **BES-128:** $S = 201$, $P = 23$.

   Complexity $= D_1^{2.376} = (5.9 \times 10^{50})^{2.376} \approx 2^{401}$.

2. **BES-192:** $S = 417$, $P = 33$.

   Complexity $= D_1^{2.376} = (5.857 \times 10^{78})^{2.376} \approx 2^{622}$.

3. **BES-256:** $S = 501$, $P = 36$.

   Complexity $= D_1^{2.376} = (3.798 \times 10^{87})^{2.376} \approx 2^{691}$.

In comparison, the XL attack against the AES-128 cipher has complexity 2^{330} [3, Section 5.2]. *Thus we see that the XSL attack is not effective against the BES cipher. In fact, it gives worse complexity than the XL attack against AES-128.*

Table 2. Number of linearly independent monomials D_1 from the extended linear equations and number of relevant extended S-box equations D_2 for various P

	BES-128	BES-192	BES-256
$P = 2$	$D_1 = 1.288 \times 10^6$ $D_2 = 4.824 \times 10^3$	$D_1 = 5.554 \times 10^6$ $D_2 = 1.001 \times 10^4$	$D_1 = 8.02 \times 10^6$ $D_2 = 1.202 \times 10^4$
$P = 3$	$D_1 = 6.839 \times 10^8$ $D_2 = 7.723 \times 10^6$	$D_1 = 6.149 \times 10^9$ $D_2 = 3.332 \times 10^7$	$D_1 = 5.093 \times 10^{10}$ $D_2 = 4.811 \times 10^7$
$P = 4$	$D_1 = 2.71 \times 10^{11}$ $D_2 = 6.152 \times 10^9$	$D_1 = 5.093 \times 10^{12}$ $D_2 = 5.532 \times 10^{10}$	$D_1 = 1.063 \times 10^{13}$ $D_2 = 9.605 \times 10^{10}$
$\vdots$	$\vdots$	$\vdots$	$\vdots$
$P = 23$	$D_1 = 5.9 \times 10^{50}$ $D_2 = 6.245 \times 10^{50}$	$\vdots$	$\vdots$
$P = 33$	$\vdots$	$D_1 = 5.857 \times 10^{78}$ $D_2 = 6.02 \times 10^{78}$	$\vdots$
$P = 36$	$\vdots$	$\vdots$	$D_1 = 3.798 \times 10^{87}$ $D_2 = 3.849 \times 10^{87}$

3.3   Further Analysis

The above results show that there are many hidden linear dependencies which were not accounted for in the computations in §2. Here, we attempt to account for some of this discrepancy.

(a) *It is not true that all the monomials which occur are extended S-box monomials.* The problem occurs with the extended linear equations. It can happen that l is a linear equation which involves (say) x_2 from a certain S-box, while the chosen passive S-boxes also include the term (say) y_5 from the same S-box; in which case, the term $x_2 y_5$ is part of an occurring monomial. This monomial is then not an extended S-box monomial. Heuristically, the number of monomials unaccounted for is not very significant.

(b) *The second and worse problem is the presence of inherent linear dependencies among the extended linear equations.* This is essentially identical to the linear dependencies among the extended S-box equations mentioned in §2.1. For example, suppose $linear_1 = 0$ and $linear_2 = 0$ are two different linear equations. By taking terms from $P - 2$ distinct passive S-boxes, we get the monomial $t_3 \ldots t_P$; expanding the equation $(linear_1)(linear_2)(t_3 \ldots t_P) = 0$ results in a linear relation between the equations extended from $linear_1 = 0$ and those extended from $linear_2 = 0$. Notice that this linear dependency is inherent among the extended linear equations. It is completely unrelated to the S-box equations or their extended counterparts. Unfortunately, this has been neglected in the original estimates.

While (a) appears to be a minor oversight, (b) has a much larger effect on the estimates. As mentioned in [3], the removal of obvious linear dependencies on the S-box equations causes the number of such equations to reduce from

$R_{old} = rSt^{P-1}\binom{S-1}{P-1}$ to $R = \binom{S}{P}(t^P - (t-r)^P)$, which is quite a significant difference. It is likely that similar considerations on the extended linear equations would also result in a significant reduction of useful equations.

4   Conclusion

The purpose of our paper is to analyse XSL when applied to BES, and determine if it would work in practice. Due to the nice S-box equations of BES, we can explicitly deduce the linear dependencies between the equations. *Our conclusion is that if XSL works on BES, then it is worse than brute foce.* However, it leaves open the question of whether XSL works for some P at all.

Furthermore, our computations do not carry over to the original Rijndael (with equations over F_2 instead of F_{256}) or to the Serpent cipher. Due to the complexity of the S-box equations, an explicit list of linearly independent terms of the extended S-box equations cannot be easily described. Naturally, we ask if XSL works in those cases.

First, it should be noted that the linear dependencies among the extended linear equations - as mentioned in §3.3 - hold in general, so it is likely that the number of linearly independent equations obtained after linearisation is actually smaller than expected. Second, in the XSL method, the final ratio (number of equations)/(number of terms) after linearisation is about 1.004, which is precariously close to 1. Hence, the presence of a large number of linear dependencies can have an adverse effect on the solvability of the system. Finally, note that we had left the T'-method out of the discussion totally. However we had noted that the T'-method is only effective if the number of linearly independent equations is already extremely close to being sufficient (about 99.4% of what is needed). Thus, even the application of the T'-method is unlikely to help.

References

1. Cid, C., Leurent, G.: An Analysis of the XSL Algorithm. In: Roy, B. (ed.) ASIACRYPT 2005. LNCS, vol. 3788, pp. 333–352. Springer, Heidelberg (2005)
2. Courtois, N., Klimov, A., Patarin, J., Shamir, A.: Efficient Algorithms for Solving Systems of Multivariate Polynomial Equations. In: Preneel, B. (ed.) EUROCRYPT 2000. LNCS, vol. 1807, pp. 392–407. Springer, Heidelberg (2000)
3. Courtois, N., Pieprzyk, J.: Cryptanalysis of Block Ciphers with Overdefined Systems of Equations. IACR eprint server, 2002/044 (March 2002), http://www.iacr.org
4. Courtois, N., Pieprzyk, J.: Cryptanalysis of Block Ciphers with Overdefined Systems of Equations. In: Zheng, Y. (ed.) ASIACRYPT 2002. LNCS, vol. 2501, pp. 267–287. Springer, Heidelberg (2002)
5. Montgomery, P.L.: A Block Lanczos Algorithm for Finding Linear Dependencies over GF(2). In: Guillou, L.C., Quisquater, J.-J. (eds.) EUROCRYPT 1995. LNCS, vol. 921, pp. 106–120. Springer, Heidelberg (1995)
6. Murphy, S., Robshaw, M.: Essential Algebraic Structure Within the AES. In: Yung, M. (ed.) CRYPTO 2002. LNCS, vol. 2442, pp. 1–16. Springer, Heidelberg (2002)
7. Murphy, S., Robshaw, M.: Comments on the Security of the AES and the XSL Technique. Electronic Letters 39, 26–38 (2003)

On the Security
of
IV Dependent Stream Ciphers*

Come Berbain and Henri Gilbert

Côme Berbain and Henri Gilbert

France Télécom R&D
38–40, rue du Général Leclerc
92794 Issy les Moulineaux Cedex 9 — France
{firstname.lastname}@orange-ftgroup.com

Abstract. Almost all the existing stream ciphers are using two inputs: a secret key and an initial value (IV). However recent attacks indicate that designing a secure IV-dependent stream cipher and especially the key and IV setup component of such a cipher remains a difficult task. In this paper we first formally establish the security of a well known generic construction for deriving an IV-dependent stream cipher, namely the composition of a key and IV setup pseudo-random function (PRF) with a keystream generation pseudo-random number generator (PRNG). We then present a tree-based construction allowing to derive a IV-dependent stream cipher from a PRNG for a moderate cost that can be viewed as a subcase of the former generic construction. Finally we show that the recently proposed stream cipher QUAD [3] uses this tree-based construction and that consequently the security proof for QUAD's keystream generation part given in [3] can be extended to incorporate the key and IV setup.

Keywords: stream cipher, PRNG, IV setup, provable security.

1 Introduction

Stream ciphers and block ciphers are the two most popular families of symmetric encryption algorithms. Unlike block ciphers, stream ciphers do not produce a key-dependent permutation over a large blocks space, but a key-dependent sequence of numbers over a small alphabet, typically the binary alphabet $\{0,1\}$. To encrypt a plaintext sequence, each plaintext symbol is combined with the corresponding symbol of the keystream sequence using a group operation, usually the exclusive or operation over $\{0,1\}$.

Nearly all stream ciphers specified recently use two inputs to generate a keystream sequence: a secret key and an additional parameter named the initial value (IV), that is generally not secret. The purpose of the initial value is to allow to derive several "independent" keystream sequences from one single

* The work described in this paper has been supported by the European Commission through the IST Program under Contract IST-2002-507932 ECRYPT.

A. Biryukov (Ed.): FSE 2007, LNCS 4593, pp. 254–273, 2007.
© International Association for Cryptologic Research 2007

key, and thus to provide a convenient method for encrypting several plaintext sequences under the same secret key, by "resynchronizing" the stream cipher each time with a new IV value. This represents an obvious practical advantage over formerly proposed stream ciphers which single input was the secret key. But on the other hand, the use of an IV input has considerable impacts on the cryptanalysis and on the formalization of the security requirements on stream ciphers.

As for cryptanalytic implications, the quite numerous attacks of IV-dependent stream ciphers published in the past years clearly indicate that IVs result in additional attack opportunities, and that the key and IV setup procedure still represents one of the less well understood aspects of stream ciphers design. As a matter of fact an adversary can compare the keystream sequences associated with several known, related or chosen IV values, and potentially derive information upon the corresponding internal state values that could not be derived from one single keystream sequence. This is illustrated by Fluhrer, Mantin, and Shamir's attack on the key and IV loading method of the RC4-based cipher used in certain WiFi systems [10], by Ekdahl and Johansson's cryptanalysis of the GSM cipher A5/1 [9], by Joux and Muller's differential known or chosen IV attacks on various ciphers [16,17], by Daemen, Govaerts, and Vandewalle's and by Armknecht, Lano, and Preneel's resynchronization attacks [8,1] or more recently by attacks against some of the eSTREAM candidates.

As for the implications of IVs on the formalization of security requirements on stream ciphers, they can be outlined as follows:

In the case of a stream cipher without IV, the requirements are conveniently captured by the theory of pseudo-random numbers generators which has been stemming from the seminal work of Shamir [18], Yao [19], Blum and Micali [6] in the early 80's. A stream cipher is considered secure if the associated key to keystream function is a pseudo-random number generator (PRNG), i.e. an input-expanding function allowing to expand a short seed (the key) into a strictly longer output (the keystream) in such a way that if the secret input seed is uniformly distributed, then the probability distribution of the corresponding output is computationally indistinguishable with non negligible probability from the uniform distribution.

In the case of an IV-dependent stream cipher, no as unanimously accepted formalization of the security requirements has emerged so far. However, most cryptologists would probably agree that a sufficient security condition is that the associated function generator which maps the secret key onto the IV to keystream function be a pseudo-random function generator (PRF), i.e. a random function generator indistinguishable with non negligible probability from a perfect random function generator. To quote an example, this is the condition Halevi, Coppersmith, and Jutla are using to express the security requirements on the IV-dependent stream cipher Scream [15]. We will briefly discuss the validity of this PRF-based formalization in Section 3 hereafter, and conclude that it indeed captures the most natural generalization to IV-dependent stream ciphers of the well accepted (PRNG based) formalization of IV less stream ciphers.

One might argue that since constructing a secure PRF can be expected to be more demanding than constructing a secure PRNG and nearly as difficult as constructing a block cipher, introducing IVs in stream ciphers looses all performance advantages of stream ciphers over block ciphers and requires the same kind of techniques than designing a block cipher. This is however not necessarily the case, as will be shown in the sequel.

The purpose of this paper is twofold. Firstly, to clarify the security requirements upon an IV-dependent stream cipher (Section 3) and to identify sufficient conditions on its key and IV setup and key generation parts in order for the whole stream cipher to be secure (Section 4). Secondly, to propose a practical construction allowing to meet these conditions (Section 5), and therefore to derive an IV-dependent stream cipher with a provable security argument. Finally we show that as an application of this construction the security arguments of QUAD can be extended in order to include the key and IV setup (Section 6). An overview of our main results is given in Section 2.

2   Outline of Our Results

For all the stream ciphers we consider in this paper, the keystream derivation is split, as in nearly all existing IV-dependent stream ciphers, into the two following separate phases, according to the generic construction illustrated in Figure 1:

- (1) **Key and IV setup:** an m-bit initial state value is derived from the key and IV value.
- (2) **Keystream generation:** the keystream is derived from the m-bit initial state obtained in the key and IV-setup phase. For that purpose, the m-bit initial state is taken as the seed input of a number generator [1].

We formally establish, in Section 4, the validity of the following "folklore" belief implicitly invoked in the security argumentation of several existing IV-dependent stream ciphers [5,2]: if the family $\{F_K\}$ of IV to initial state functions parametrized by the key K is a PRF and if the number generator g is a PRNG, then the family $\{G_K = g \circ F_K\}$ of IV to keystream sequence functions is a PRF. This provides useful sufficient conditions in order for the IV-dependent stream cipher resulting from the generic construction of Figure 1 to be secure. Our security proof relies upon a simple "composition theorem". A specific construction directly suggested by the former security results might consist of using a trusted block cipher for the IV-setup, as done for instance in the stream cipher candidates LEX [5] and SOSEMANUK [2], both selected as focus ciphers for third

---

[1] Though our constructions are potentially applicable to any number generator with a sufficiently long input size, they are mainly intended for number generators based upon the iterated invocation of a finite state machine (FSM). The keystream sequence generation procedure of nearly all existing stream ciphers has this specific structure, i.e. uses the state transition function of a FSM to update an m-bit internal state and derive at each iteration a t-bit keystream portion by means of a fixed output function.

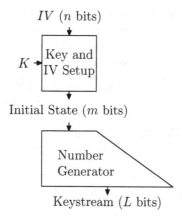

Fig. 1. IV-dependent stream cipher: (generic construction)

evaluation phase of the stream cipher initiative eSTREAM of the European network ECRYPT. One may however argue that this construction results in a lack of design unity when (unlike in LEX) the keystream generation does not reuse the trusted block cipher used for the key and IV setup.

We propose, in Section 5 hereafter, another specific construction also supported by former security results, which has the additional advantage that it better preserves the design unity of stream ciphers. This construction consists of applying the so called tree-based construction proposed by Goldreich, Goldwasser, and Micali in [13] for deriving the PRF needed for the key and IV setup from any n-bit to $2n$-bit PRNG. This PRNG can be essentially the same as the one used in the keystream generation phase. The later option allows to even better preserve the unity of the design, and to achieve substantial savings in the hardware and software implementation complexity of the stream cipher, since the key and IV setup and the keystream generation are then using the same computational ingredients.

Last of all, we focus in Section 6 on a particular stream cipher where the tree-based construction of Section 5 is applied in the key and IV setup, namely the recently proposed stream cipher QUAD [3]. We show that the partial proof of security of [3] (which gives some evidence that the keystream generation part of QUAD is secure) can be extended to incorporate the key and IV setup. This allows to reduce the security of the whole stream cipher to the difficulty of solving a random multivariate quadratic system.

3   Security Model

3.1   Basic Security Notions

We first recall definitions of advantages for distinguishing a number generator from a perfect random generator and a function generator from a perfect random function generator, and the notions of Pseudo-Random Number Generator

(PRNG) and Pseudo-Random Function (PRF). All the security definitions used throughout this paper relate to the concrete (non asymptotic) security model. In the sequel, when we state that a value u is randomly chosen in a set U, we implicitly mean that u is drawn according to the uniform law over U.

Single-query distinguisher for a number generator: let us consider a number generator $g : \{0,1\}^n \longrightarrow \{0,1\}^L$ with input and output lengths $L > n$, used to expand an n-bit secret random seed into an L-bit sequence. A distinguisher in time t for g is a probabilistic testing algorithm A which when input with an L-bit string outputs either 0 or 1 with time complexity at most t. We define the advantage of A for distinguishing g from a perfect random generator as

$$\mathbf{Adv}_g^{prng}(A) = \left| \mathrm{Pr}_{x \in \{0,1\}^n}(A(g(x)) = 1) - \mathrm{Pr}_{y \in \{0,1\}^L}(A(y) = 1) \right|,$$

where the probabilities are not only taken over the value of an unknown randomly chosen $x \in \{0,1\}^n$ (resp. of a randomly chosen $y \in \{0,1\}^L$), as explicitly stated in the above formula, but also over the random choices of the probabilistic algorithm A.

We define the advantage for distinguishing the function g in time t as

$$\mathbf{Adv}_g^{prng}(t) = \max_A \{\mathbf{Adv}_g^{prng}(A)\},$$

where the maximum is taken over all testing algorithms of time complexity at most t.

A function g is said to be a PRNG if $\mathbf{Adv}_g^{prng}(t)$ is negligible (for example less than 2^{-40}) for values of t strictly lower than a fixed threshold (for example 2^{80} or 2^{128}). The definition of a PRNG is therefore dependent upon thresholds reflecting the current perception of an acceptably secure number generator.

Multiple-query distinguisher for a number generator: let us still consider a function g from n bits to L bits. A q-query distinguisher in time t for g is a probabilistic testing algorithm A which when input with a q-tuple of L-bit words outputs either 0 or 1 with time complexity at most t. We define the advantage of A for distinguishing g from a perfect random generator as

$$\mathbf{Adv}_g^{prng}(A) = \left| \mathrm{Pr}(A(g(x_1), \ldots, g(x_q)) = 1) - \mathrm{Pr}(A(y_1, \ldots, y_q) = 1) \right|,$$

where the probabilities are taken over the q-tuples of n-bit values x_i (resp. of L-bit values y_i) and on the random choices of the probabilistic algorithm A. We also define the advantage for distinguishing the function g in time t with q queries as

$$\mathbf{Adv}_g^{prng}(t, q) = \max_A \{\mathbf{Adv}_g^{prng}(A)\},$$

where the maximum is taken over all testing algorithms A having time-complexity at most t and using q inputs.

Distinguisher for a function generator: let us now consider a function generator, i.e. a family $F = \{f_K\}$ of $\{0,1\}^n \longrightarrow \{0,1\}^m$ functions indexed by a key K randomly chosen from $\{0,1\}^k$. A distinguisher in time t with q queries for F is a probabilistic testing algorithm A^f able to query an n-bit to m-bit oracle function f up to q times. Such an algorithm allows to distinguish a randomly chosen function f_K of F from a perfect random function f^* randomly chosen in the set $F_{n,m}^*$ of all $\{0,1\}^n \longrightarrow \{0,1\}^m$ functions with a distinguishing advantage

$$\mathbf{Adv}_F^{prf}(A) = \left| \Pr(A^{f_K} = 1) - \Pr(A^{f^*} = 1) \right|,$$

where the probabilities are taken over $K \in \{0,1\}^k$ (resp $f^* \in F_{n,m}^*$) and over the random choices of A. We define the advantage for distinguishing the family F in time t with q queries as

$$\mathbf{Adv}_F^{prf}(t,q) = \max_A \{\mathbf{Adv}_F^{prf}(A)\},$$

where the maximum is taken over all testing algorithms A working in time at most t and capable to query an n-bit to m-bit oracle function up to q times.

A family of functions $F = \{f_K\}$ is said to be a PRF if $\mathbf{Adv}_F^{prf}(t,q)$ is negligible for values of t and q strictly lower than the respective threshold (for example 2^{80} or 2^{128} for t and 2^{40} for q).

3.2   Security Requirements for an IV-Dependent Stream Ciphers

Let us consider an IV-dependent stream cipher, i.e. a family $G = \{g_K\}_{K \in \{0,1\}^k}$ of IV to keystream functions $g_K : \{0,1\}^n \longmapsto \{0,1\}^L$, where k is the size of the key, n is the size of the IVs and L is the maximum number of keystream bits that can be produced for a given IV.

Such an IV-dependent stream cipher can be viewed as a special number generator, allowing to expand a k-bit secret seed onto an exponentially long sequence of 2^n L-bit keystream words $\{Z_{IV} = g_K(IV)\}_{IV \in \{0,1\}^n}$, with the additional property that while this sequence is too long to be entirely accessed in a sequential manner, it can be directly accessed, i.e. that for any value of $IV \in \{0,1\}^n$ the computational cost for accessing the L-bit subsequence Z_{IV} is constant.

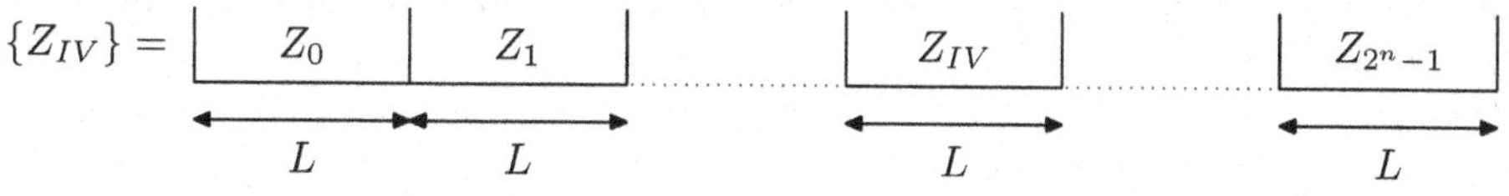

Fig. 2. Exponentially long sequence with direct access associated to an IV-dependent stream cipher g

This observation can be used in order to try to generalize the well accepted formalization of the security requirements on an IV-less stream cipher by means of a PRNG in a natural manner. An IV-less stream cipher is considered secure if and only if no testing algorithm, when given access either to a Λ-bit output of the generator corresponding to a random secret input or to a random Λ-bit sequence, can distinguish both situations in time less than a sufficiently large threshold (say 2^{80}) with a non-negligible advantage. It can be reasonably argued that in the case of an IV-dependent stream cipher, the most natural generalization of the above security definition is to require that no testing algorithm, when given a sufficiently large number q (say for instance $\min(2^{80}, 2^n)$) of direct accesses to L-bit subsequences of a sequence $\{Z_{IV}\}$ associated with a random unknown key K or to a uniformly drawn sequence of 2^n L-bit subsequences, can distinguish both situations in time less than a sufficiently large threshold (say 2^{80}) with a non-negligible advantage.

But it is easy to see that both the sequence $\{Z_{IV}\}$ and the uniformly drawn sequence of $2^n \cdot L$ bits can be viewed as n-bit to L-bit functions, and that direct accesses to these sequences can be viewed as oracle queries to these functions. Therefore the requirements formulated above are strictly equivalent to saying that the function family $G = \{g_K\}_{K \in \{0,1\}^k}$ is a PRF.

In other words, we have given some evidence that an IV-dependent stream cipher $G = \{g_K\}$ for $K \in \{0,1\}^k$ with $g_K : \{0,1\}^n \longrightarrow \{0,1\}^L$ can be viewed as a family of functions, and can be considered secure if and only if it is a PRF , i.e. sufficiently indistinguishable from a perfect random function.

Related key attacks, which relevance, when it comes to security requirements on symmetric ciphers, is a controversial issue [4], are not covered by our security model.

4   Security of the Generic Construction

4.1   A Simple Composition Theorem

In this Section, we define the composition G of a family of function F and a function g, relate the indistinguishability of G to the one of F and g, and show that this composition theorem results in a secure construction allowing to derive a secure IV-dependent stream cipher from a PRF and a PRNG.

Definition 1. *The composition $G = g \circ F$ of an n-bit to m-bit family of functions $F = \{f_K\}$ and of an m-bit to L-bit function g is the n-bit to L-bit family of functions*

$$G = \{g \circ f_K\}.$$

Theorem 1. *Let us consider a PRF $F = \{f_K\}$ where $f_K : \{0,1\}^n \longrightarrow \{0,1\}^m$ and a PRNG $g : \{0,1\}^m \longrightarrow \{0,1\}^L$ that produces L bits in time T_g^L. The advantage in time t with q queries of $G = g \circ F = \{g \circ f_K\}$ can be upper bounded as follows*

$$\mathbf{Adv}_G^{prf}(t, q) \leq \mathbf{Adv}_F^{prf}(t + qT_g^L) + q\mathbf{Adv}_g^{prng}(t + qT_g^L).$$

In order to prove Theorem 1 we first establish a useful lemma which relates the single-query and multiple-queries advantages of any PRNG.

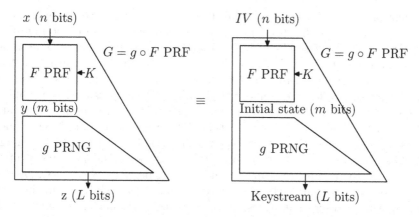

Fig. 3. Composing a PRF and a PRNG gives a PRF

Lemma 1. *Let $g : \{0,1\}^m \longrightarrow \{0,1\}^L$ be a PRNG which can be computed in time T_g^L. The q-query advantage for distinguishing g in time t is related to the single-query advantage for distinguishing g by the inequality*

$$\mathbf{Adv}_g^{prng}(t,q) \le q\mathbf{Adv}_g^{prng}(t + qT_g^L).$$

A proof of the above lemma is given in Appendix 1; this proof is similar to the one of a proposition relating the single-sample indistinguishability and the multiple-sample indistinguishability of polynomial-time constructible ensembles established by Goldreich and Krawczyk [14,12]. Using this lemma, we can prove Theorem 1. Our proof is given in Appendix 2. Theorem 1 is illustrated on the left part of Figure 3.

Application to the Security of IV-Dependent Stream Ciphers. A direct application of Theorem 1 is depicted on the right part of Figure 3. As said in Section 3, an IV-dependent stream cipher can be considered secure if and only if the IV to keystream function g_K parametrized by the key is a PRF. Theorem 1 implies that this is indeed the case, i.e. that the stream cipher is secure if:

1) the n-bit to m-bit IV to initial state function parametrized by the key representing the IV setup of a stream cipher is a PRF;
2) the m-bit to L-bit initial state to keystream function is a PRNG;
3) the upper bounds on the advantage for distinguishing $\{g_K\}$ given by Theorem 1 guarantee a sufficient resistance against attacks.

From now on, we consider the following stream cipher design problem. Let us assume that a trusted number generator g allowing to expand an m-bit initial state into an L-bit sequence is available. We are now faced with the issue of constructing a key and IV setup PRF in order to compose this PRF with g to get a secure IV-dependent stream cipher.

A straightforward construction for such a PRF, directly suggested by the former security results, would consist of using a trusted block cipher with a sufficient block-length to fit the initial state length m of g. As a matter of fact it is usual to conjecture that the family of key dependent encryption permutations associated with a secure block cipher represents both a pseudo-random family of permutations (PRP) and a PRF. The value of m must be typically at least 160 bits in the frequent case where g has a finite state automaton structure, in order to avoid generic time-memory trade-offs. This suggests that one could for instance use the variant of Rijndael with a 256-bit block size, or a truncated instance of this cipher if m is smaller than 256, or an appropriate "one to many blocks" mode of operation of any block cipher [11] for initial state sizes larger than 256 bits. This would allow to accommodate keys and IVs of size up to one block. Such an approach can certainly be considered more conservative than the key and IV setup procedure of many existing stream ciphers. On the other hand, one may argue that it results in a strong performance penalty for the encryption of short messages and an increased implementation complexity, and in a lack of design unity if the trusted number generator g does not reuse the same ingredients as the trusted block cipher.

5   A Tree Based Stream Cipher Construction

Is this Section we present a key and IV setup procedure derived from the Tree Based Construction introduced by Goldreich, Goldwasser, and Micali in [13]. This construction allows to derive a PRF from a PRNG and to relate their securities. Though initially introduced for theoretical purposes (namely show in the asymptotic model that the existence of a PRNG implies the existence of a PRF) it is also of practical interest since it allows, as shown here, to build a stream cipher from two number generators: the Tree Based Construction is used to transform the first number generator into an efficiently computable key and IV setup; the second number generator is initialized with the value given by the key and IV setup, and generates the keystream. These two number generators can advantageously be the same. Thus it becomes possible to build an IV-dependent stream cipher from one single number generator.

5.1   The Tree Based Construction

The Tree Based Construction allows to derive a PRF F^g from a PRNG g. Let us consider a PRNG $g : \{0,1\}^m \longrightarrow \{0,1\}^L$ where $L \geq 2m$ and let us denote the L bit image of $y \in \{0,1\}^m$ by $g(y) = z_0 z_1 \ldots z_{L-1}$. We derive from g two functions $g_0 : y \in \{0,1\}^m \longmapsto z_0, \ldots, z_{m-1}$ and $g_1 : y \in \{0,1\}^m \longmapsto z_m, \ldots, z_{2m-1}$ from m bits to m bits which on input $y \in \{0,1\}^m$ respectively produce the first and the second m-bit string generated by g when input with y.

The PRF F^g is the family of functions $\{f_y\}_{y \in \{0,1\}^m}$ where

$$f_y : \{0,1\}^n \longrightarrow \{0,1\}^m$$
$$(x_1, x_2, \ldots, x_n) \longmapsto f_y(x_1, x_2, \ldots, x_n) = g_{x_n} \circ g_{x_{n-1}} \ldots \circ g_{x_1}(y)$$

This construction is illustrated on Figure 4: the input bits determine a path trough the binary tree leading to the output of the function.

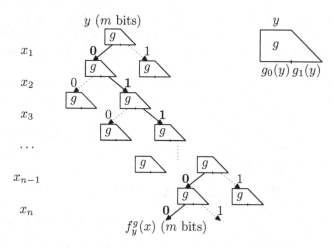

Fig. 4. Tree Based Construction

Theorem 2. *Let* $g : \{0,1\}^m \longrightarrow \{0,1\}^L$ *be a PRNG which generates* $L \geq 2m$ *outputs bits and produces its* $2m$ *first output bits in time* T_g^{2m} *and let* $F^g = \{f_y\}_{y \in \{0,1\}^m}$ *be the family of n-bit to m-bit functions derived from g by the Tree Based Construction. The* (t,q) *advantage of PRF* F^g *is related to the single-query advantage of PRNG g by the following inequality:*

$$\mathbf{Adv}_{F^g}^{prf}(t,q) \leq nq\mathbf{Adv}_g^{prng}(t + q(n+1)T_g^{2m}).$$

A proof of Theorem 2 is given in Appendix 3. This proof is essentially the same as the security proof of the Tree Based Construction due to Goldreich, Goldwasser, and Micali[13], a detailed version of which is given by Goldreich in [12], up to the fact that we consider the concrete security model instead of the asymptotic (polynomial time indistinguishability) security model.

5.2 Resulting Stream Cipher Construction

To build an IV-dependent stream cipher from a m-bit to L-bit PRNG g representing an IV-less stream cipher, we apply the Tree Based Construction to a m to L'-bit PRNG g' ($L' \geq 2m$) typically equal to g itself, in order to derive the key and IV setup function, which produces the initial state of g, following the construction of Figure 4. For a key K and an IV IV, the value $f_K^{g'}(IV)$ is the initial state of g. Thanks to Theorem 1 and since the Tree Based Construction provides a PRF, the resulting stream cipher is also a PRF. The security of the final stream cipher only depends on the security of the PRNG.

In case K is smaller than m bits, we need to extend K to a value randomly chosen in $\{0,1\}^m$. In order to achieve this we can use an additional PRNG $h : \{0,1\}^k \longrightarrow \{0,1\}^m$. The proof for the Tree Based Construction can easily be extended and we have:

$$\mathbf{Adv}_{F^g}^{prf}(t,q) \leq nq\mathbf{Adv}_{g'}^{prng}(t + q(n+1)T_{g'}^{2m}) + q\mathbf{Adv}_h^{prng}(t + q(n+1)T_h^m).$$

Theorem 3. *Let* $g : \{0, 1\}^m \longrightarrow \{0, 1\}^L$ *be a PRNG which generates L outputs bits in time* T_g^L *and* $g' : \{0, 1\}^m \longrightarrow \{0, 1\}^{2m}$ *be a PRNG. We can define a stream cipher* $G = \{G_K\} = g \circ F^{g'}$, *where* $F^{g'}$ *is derived from* g' *using the Tree Based Construction*

$$G_K(IV) = g \circ f_K^{g'}(IV)$$

Moreover G is a PRF and we have:

$$\mathbf{Adv}_G^{prf}(t, q) \leq nq\mathbf{Adv}_{g'}^{prng}(t + qT_g^L + q(n+1)T_{g'}^{2m}) + q\mathbf{Adv}_g^{prng}(t + qT_g^L)$$

If g and g' are equal, then we have

$$\mathbf{Adv}_G^{prf}(t, q) \leq q(n+1)\mathbf{Adv}_g^{prng}(t + q(n+2)T_g^L)$$

Proof. To prove this result we only have to use Theorem 1 and Theorem 2.

The above key and IV setup construction is of practical interest: suppose we have a trusted PRNG, for example the Shrinking Generator [7], we can build an IV-dependent stream cipher based on this PRNG without introducing any additional feature for a moderate computational cost.

5.3 Efficiency Considerations

Let us assume for instance that we want to build from a PRNG g of initial state length 160 bits a stream cipher with a 160-bit key, a 80-bit IV, and a target security of 2^{80}, using the previously described construction. Then the time required to compute the key and IV setup with the above construction is the time required by g to produce 3200 bytes. Considering a very fast PRNG, running at 5 cycles per byte, the key and IV setup requires about 16000 cycles on a standard PC. This is slower than using a block cipher like AES, which would require about 1000 cycles. Therefore for software applications where resynchronization has to be done frequently, this construction is not at all efficient and using a block cipher for the key and IV setup should be considered. However for applications where the keystream generated for a single IV is very long compared to these 3200 bytes our construction can be competitive.

Now in the case of hardware applications, the above construction can be of real interest, in order to minimize the hardware complexity since it uses a single PRNG for the key and IV setup and the keystream generation. Then only a few additional gates are required to implement the key and IV setup. If a block cipher were used instead, then the total number of gates required to implement the stream cipher would be much higher than the number of gates required for a PRNG.

6   Application to the QUAD Stream Cipher

The stream cipher QUAD is a practical stream cipher with some provable security which was introduced [3] by Berbain, Gilbert, and Patarin at Eurocrypt 2006.

The provable security argument relates, in the $GF(2)$ case, the indistinguishability of the keystream generated by QUAD to the conjectured hardness of solving random quadratic systems. QUAD iterates a one way function, namely a quadratic system, upon an internal state and extracts a certain number of bits of this step at each iteration.

The keystream generation makes use of two systems $S_{\text{in}} = (Q_1, \ldots, Q_m)$ and $S_{\text{out}} = (Q_{m+1}, \ldots, Q_{km})$ of multivariate quadratic equations both sharing the same m unknowns over $GF(q)$, typically $GF(2)$ as described on Fig. 5. The first system S_{in} is used to update the internal state and thus contains m equations, whereas the second system S_{out} produces the keystream and contains $(k-1)m$ equations. As explained in [3], the quadratic systems S_{in} and S_{out}, though randomly generated, are both publicly known.

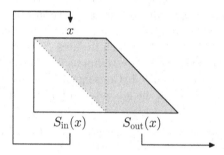

Fig. 5. Stream cipher QUAD

Given an internal state $x = (x_1, \ldots, x_m)$, the keystream generation amounts to iterating the following steps:

- compute $\big(S_{\text{in}}(x), S_{\text{out}}(x)\big) = \big(Q_1(x), \ldots, Q_{km}(x)\big)$, from the internal state x;
- output the sequence $S_{\text{out}}(x) = \big(Q_{m+1}(x), \ldots, Q_{km}(x)\big)$ of $(k-1)m$ keystream elements of $GF(q)$;
- update the internal state x with the sequence $S_{\text{in}}(x) = \big(Q_1(x), \ldots, Q_m(x)\big)$.

Before generating any keystream the internal state x needs to be initialized, with the key K and the initialization vector IV, which are respectively represented by a sequence of $GF(q)$ elements of length $|K|$ and a binary sequence of $\{0, 1\}$ values of length $|IV|$. We will assume in the sequel that $|K|$ is equal to m. The initialization is done as follows: two publicly known chosen multivariate quadratic systems S_0 and S_1 of m equations over m unknowns are used. The initial state is filled with the key. Then for each of the $|IV|$ bits IV_1 to $IV_{|IV|}$ of the IV value the internal state x is updated as follows: if $IV_i = 0$, x is replaced by the $GF(q)^m$ value $S_0(x)$; if $IV_i = 1$, x is replaced by the $GF(q)^m$ value $S_1(x)$. Finally the cipher is clocked m additional times as described before, but without outputting the keystream.

We now extend the partial proof over $GF(2)$ given in [3] to incorporate the key and IV Setup. We denote by $S = (S_{\text{in}} || S_{\text{out}})$ the randomly chosen system of

km equations on m variables and $S' = (S_0||S_1)$ the randomly chosen system of $2m$ equations in m variables. We also denote by $g^S : \{0,1\}^m \longrightarrow \{0,1\}^L$ and $g^{S'} : \{0,1\}^m \longrightarrow \{0,1\}^{2m}$ the corresponding PRNGs.

The key and IV setup proposed in [3] can be divided into two parts: in the first part the Tree Based Construction for $g^{S'}$ is applied and the value $y = f_K^{g^{S'}}(IV)$ is computed. Then in the second part, y is used to initialize g^S which is clocked m times but the corresponding $(k-1)m^2$ bits of keystream are not used. The value of the internal state after these m clocks is used to produce L bits of keystream. The security of the first part is related to the security of $g^{S'}$ thanks to Theorem 2.

$$\mathbf{Adv}^{prf}_{F^{g^{S'}}}(t,q) \leq 2q\mathbf{Adv}^{prng}_{g^{S'}}(t + q(n+2)T^{2m}_{g^{S'}})$$

The security of the PRNG g_{real} which starts by running m clocks like g^S without producing any keystream to reflect the runup of QUAD and then produces L bits of keystream as g^S does, is related to the security of generator $\tilde{g}^S : \{0,1\}^m \longrightarrow \{0,1\}^{L+(k-1)m^2}$ which iterates S to produce $L + (k-1)m^2$ bits, since g_{real} produces the same keystream as $\tilde{g}^S$ up to the fact that the first $(k-1)m^2$ bits of $g^{S'}$ are discarded. Consequently a distinguisher on g_{real} is also a distinguisher for $\tilde{g}^S$. Thus the advantage of g_{real} is upper-bounded by the advantage of $\tilde{g}^S$.

$$\mathbf{Adv}^{prng}_{g_{real}}(t) \leq \mathbf{Adv}^{prng}_{\tilde{g}^S}(t + T^{L+(k-1)m^2}_{\tilde{g}^S})$$

Finally the security of the stream cipher is related to the securities of $\tilde{g}^S$ and of $g^{S'}$ by the composition theorem of Section 4.

$$\mathbf{Adv}^{prf}_{\text{QUAD}}(t,q) \leq 2q\mathbf{Adv}^{prng}_{g^{S'}}(t+q(n+3)T^L_{g^{S'}})+q\mathbf{Adv}^{prng}_{\tilde{g}^S}(t+qT^L_{\tilde{g}^S}+T^{L+(k-1)m^2}_{\tilde{g}^S})$$

Those two generators are based on the iteration of a randomly chosen quadratic system of km equations in m variables (with $k = 2$ for S'). We can use the main result of [3], which relates the security of this kind of PRNG to the difficulty of inverting a randomly chosen multivariate quadratic system to say that the security of QUAD as a stream cipher is related to the difficulty of the MQ Problem, i.e. an adversary able to distinguish QUAD from a PRF in time t with q queries is then able to construct a MQ solver:

$$\mathbf{Adv}^{prf}_{\text{QUAD}}(t,q) \leq 3q\mathbf{Adv}^{prng}_{g^S}(t + qT^{L+(k-1)m^2}_{\tilde{g}^S})$$
$$\leq \mathbf{Adv}^{MQinversion}(t'[t + qT^{L+(k-1)m^2}_{\tilde{g}^S}]),$$

where $t'[t + qT^{L+(k-1)m^2}_{\tilde{g}^S}]$ is the running time of the MQ inverter algorithm given by the main reduction theorem of [3], which is recalled in Appendix.

7 Conclusion

In this paper we investigated security issues arising for IV-dependent stream ciphers. We confirmed the "folklore" belief that the composition of a key and IV setup PRF and a key generation PRNG provides a secure stream cipher, which furnishes a proof that initializing a PRNG with a block cipher is secure provided that the block cipher's block length is sufficiently large. Moreover we described a practical construction that allows to derive an IV-dependent stream cipher from a PRNG (or equivalently an IV-less stream cipher) for a moderate additional cost. This construction is quite simple and does not require additional components. Finally we showed an application of this provably secure construction to the stream cipher QUAD, by incorporating the key and IV setup in the security proof given by the authors of QUAD. The resulting extended proof relates the security of the whole stream cipher (not only the keystream generation part) to the conjectured intractability of the MQ problem.

References

1. Armknecht, F., Lano, J., Preneel, B.: Extending the Resynchronization Attack. In: Handschuh, H., Hasan, M.A. (eds.) SAC 2004. LNCS, vol. 3357, p. 19. Springer, Heidelberg (2004)
2. Berbain, C., Billet, O., Canteaut, A., Courtois, N., Gilbert, H., Goubin, L., Gouget, A., Granboulan, L., Lauradoux, C., Minier, M., Pornin, T., Sibert, H.: Sosemanuk, a Fast Software-Oriented Stream Cipher. eSTREAM, ECRYPT Stream Cipher Project, Report 2005/001 (2005), http://www.ecrypt.eu.org/stream
3. Berbain, C., Gilbert, H., Patarin, J.: QUAD: a Practical Stream Cipher with Provable Security. In: Vaudenay, S. (ed.) EUROCRYPT 2006. LNCS, vol. 4004, Springer, Heidelberg (2006)
4. Bernstein, D.J.: Related-key attacks: Who cares? eSTREAM, ECRYPT Stream Cipher Project (2006), http://www.ecrypt.eu.org/stream/phorum
5. Biryukov, A.: A new 128 bit key Stream Cipher : LEX. eSTREAM, ECRYPT Stream Cipher Project, Report 2005/001 (2005), http://www.ecrypt.eu.org/stream
6. Blum, M., Micali, S.: How to Generate Cryptographically Strong Sequences of Pseudo-Random Bits. SIAM J. Comput. 13(4), 850–864 (1984)
7. Coppersmith, D., Krawczyk, H., Mansour, Y.: The Shrinking Generator. In: Stinson, D.R. (ed.) CRYPTO 1993. LNCS, vol. 773, pp. 22–39. Springer, Heidelberg (1994)
8. Daemen, J., Govaerts, R., Vandewalle, J.: Resynchronization Weaknesses in Synchronous Stream Ciphers. In: Helleseth, T. (ed.) Advances in Cryptology – EUROCRYPT '93. LNCS, vol. 765, pp. 159–167. Springer, Heidelberg (1993)
9. Ekdahl, P., Johansson, T.: Another Attack on A5/1. IEEE Transactions on Information Theory 49(1), 284–289 (2003)
10. Fluhrer, S.R., Mantin, I., Shamir, A.: Weaknesses in the Key Scheduling Algorithm of RC4. In: Selected Areas in Cryptography, pp. 1–24 (2001)
11. Gilbert, H.: The Security of "One-Block-to-Many" Modes of Operation. In: Johansson, T. (ed.) FSE 2003. LNCS, vol. 2887, pp. 376–395. Springer, Heidelberg (2003)
12. Goldreich, O.: The Foundations of Cryptography - Volume 1. Cambridge University Press, Cambridge (2001)
13. Goldreich, O., Goldwasser, S., Micali, S.: How to Construct Random Functions. J. ACM 33(4), 792–807 (1986)

14. Goldreich, O., Krawczyk, H.: Sparse Pseudorandom Distributions. In: Brassard, G. (ed.) CRYPTO 1989. LNCS, vol. 435, pp. 113–127. Springer, Heidelberg (1990)
15. Halevi, S., Coppersmith, D., Jutla, C.S.: Scream: A Software-Efficient Stream Cipher. In: Daemen, J., Rijmen, V. (eds.) FSE 2002. LNCS, vol. 2365, pp. 195–209. Springer, Heidelberg (2002)
16. Joux, A., Muller, F.: A Chosen IV Attack Against Turing. In: Matsui, M., Zuccherato, R.J. (eds.) SAC 2003. LNCS, vol. 3006, pp. 194–207. Springer, Heidelberg (2004)
17. Muller, F.: Differential Attacks against the Helix Stream Cipher. In: Roy, B., Meier, W. (eds.) FSE 2004. LNCS, vol. 3017, Springer, Heidelberg (2004)
18. Shamir, A.: On the Generation of Cryptographically Strong Pseudo-Random Sequences. In: Even, S., Kariv, O. (eds.) Automata, Languages and Programming. LNCS, vol. 115, pp. 544–550. Springer, Heidelberg (1981)
19. Yao, A.: Theory and Applications of Trapdoor Function. In: Foundations of Cryptography FOCS 1982 (1982)

Appendix

Proof of Lemma 1

Lemma 1. Let us consider a PRNG $g : \{0,1\}^m \longrightarrow \{0,1\}^L$ which can be computed in time T_g^L. Then we have

$$\mathbf{Adv}_g^{prng}(t,q) \leq q\mathbf{Adv}_g^{prng}(t + qT_g^L).$$

Proof. Suppose there is an algorithm A that distinguishes q L-bit keystream sequence produced by g from q unknown randomly chosen initial internal state $x_i \in \{0,1\}^m$ from q random L-bit sequences in time t with advantage ϵ. Then we are going to build an algorithm B that distinguishes $g(x)$ corresponding to an unknown random input x, from a random value of size L in time $t' = t + qT_g^L$ with advantage $\frac{\epsilon}{q}$.

We introduce the hybrid probability distributions D^i over $\{0,1\}^L$ for any $0 \leq i \leq q$ respectively associated with the random variables

$$Z^i = (g(x_1), g(x_2), \ldots, g(x_i), r_{i+1}, \ldots, r_q)$$

where the r_j and x_i are random independent uniformly distributed values of $\{0,1\}^L$ and $\{0,1\}^m$ respectively. Consequently D^q is the distribution of the q L-bit keystream produced by g and D^0 is the distribution of q L-bit uniformly distributed values of $\{0,1\}^L$.

We denote by p^i the probability that A accepts a random qL-bit sequence distributed according to D^i.

We have supposed that algorithm A distinguishes between D^0 and D^q with advantage ϵ, in other words that $|p^0 - p^q| \geq \epsilon$.

Algorithm B works as follows : on input $y \in \{0,1\}^L$ it selects randomly an i such that $1 \leq i \leq q$ and constructs the vector

$$Z(y) = (g(x_1), g(x_2), \ldots, g(x_{i-1}), y, r_{i+1}, r_{i+2}, \ldots, r_q)$$

with x_i and r_i randomly chosen values. If y is distributed accordingly to the output distribution of g, i.e. $y = g(x)$ for a uniformly distributed value of x, then

$$Z(y) = (g(x_1), g(x_2), \ldots, g(x_{i-1}), g(x), r_{i+1}, r_{i+2}, \ldots, r_q)$$

is distributed according to D^i. Now if y is distributed according to the uniform distribution, then

$$Z(y) = (g(x_1), g(x_2), \ldots, g(x_{i-1}), y, r_{i+1}, r_{i+2}, \ldots, r_q).$$

Thus $Z(y)$ is distributed according to D^{i-1}. In order to distinguish the output distribution of g from the uniform law, algorithm B calls algorithm A with inputs $Z(y)$ and returns the value returned by A. Thus

$$\left| \Pr_x(B(g(x)) = 1) - \Pr_y(B(y) = 1) \right|$$
$$= \left| \frac{1}{q} \sum_{i=0}^{q-1} p^i - \frac{1}{q} \sum_{i=1}^{q} p^i \right| = \frac{1}{q} |p^0 - p^q| \geq \frac{\epsilon}{q}.$$

Thus B distinguishes the output distribution of g from the uniform distribution with probability at least $\frac{\epsilon}{q}$ in time $t + qT_g^L$. Consequently we have:

$$\mathbf{Adv}_g^{prng}(A) = q\mathbf{Adv}_g^{prng}(B)$$
$$\leq q\mathbf{Adv}_g^{prng}(t + qT_g^L)$$

which gives us the final result:

$$\mathbf{Adv}_g^{prng}(t, q) \leq q\mathbf{Adv}_g^{prng}(t + qT_g^L).$$

Proof of Theorem 1

Theorem 1. Let us consider $F = \{f_K\}$ where $f_K : \{0,1\}^n \longrightarrow \{0,1\}^m$ a PRF and $g : \{0,1\}^m \longrightarrow \{0,1\}^L$ a PRNG that produces L bits in time T_g^L. The advantage in time t with q queries of $G = g \circ F = \{g \circ f_K\}$ can be upper bounded as follows

$$\mathbf{Adv}_G^{prf}(t, q) \leq \mathbf{Adv}_F^{prf}(t + qT_g^L) + q\mathbf{Adv}_g^{prng}(t + qT_g^L).$$

Proof. We want to upper bound the advantage of an algorithm A that in time t with q requests distinguishes a random instance $g_K = g \circ f_K$ of $G = g \circ F$ from a perfect random function $g^* \in F_{n,L}^*$. We can write the advantage of A as

$$\mathbf{Adv}_G^{prf}(A) = \left| \Pr(A^{g_K} = 1) - \Pr(A^{g^*} = 1)) \right|.$$

A is making at most q distinct queries to an oracle function instantiated by g_K, resp. g^*. In order to upper bound $\mathbf{Adv}_G^{prf}(A)$, we consider the intermediate situation where the oracle function is neither g_k nor g^*, but a random instance

$g \circ f^*$ of the composition of a perfect random function $f^* \in F^*_{n,m}$ and g. Due to the triangular inequality, we have

$$\mathbf{Adv}_G^{prf}(A) \le \left| \Pr(A^{g_K} = 1) - \Pr(A^{g \circ f^*} = 1) \right|$$
$$+ \left| \Pr(A^{g \circ f^*} = 1) - \Pr(A^{g^*} = 1) \right|.$$

We denote the first and the second absolute values of the right expression by δ_1 and δ_2.

Let us first upper bound δ_2. It is easy to see that instantiating the oracle function of A with $g \circ f^*$ (resp. g^*) amounts to answering the up to q distinct oracle queries of A with a q-tuple $(g(y_1), \ldots, g(y_q))$ of L-bit values, where the q-tuple $(y_1, \ldots, y_q)$ is a randomly drawn from $\{0,1\}^{mq}$, resp. with a q-tuple $(z_1, \ldots, z_q)$ of answers randomly drawn from $\{0,1\}^{Lq}$. In both case, the q-tuple of oracle answers is independent of the up to q distinct values of the oracle queries. Using this fact we can derive an algorithm B that distinguishes q values $(g(x_1), \ldots, g(x_q))$ from q random values of $\{0,1\}^L$. Algorithm B works as follows: on input $(y_1, \ldots, y_q)$ it runs algorithm A. Consequently it has to answer A's n-bit oracle queries x_i with L-bit responses. On each distinct query x_i, B simply answers y_i. When A halts, B halts also and returns the output of A. We can easily see that $\Pr(B(g(x_1), \ldots g(x_q)) = 1) = \Pr(A^{g \circ f^*} = 1)$ and that $\Pr(B(y_1, \ldots y_q) = 1) = \Pr(A^{g^*} = 1)$. Consequently we have

$$\delta_2 = \mathbf{Adv}_g^{prng}(B) \le \mathbf{Adv}_g^{prng}(t, q).$$

Lemma 1 now provides:

$$\delta_2 \le q\mathbf{Adv}_g^{prng}(t + qT_g^L).$$

In order to upper-bound $\mathbf{Adv}_G^{prf}(A)$ we still have to upper bound δ_1. This can be done by deriving from algorithm A an algorithm C that is able by invoking A one single time to distinguish a random instance of the n-bit to m-bit PRF F from a perfect random function $f^* \in F^*_{n,m}$ with an advantage also equal to δ_1.

Algorithm C has access to an n-bit to m-bit oracle function f. C works as follows: first it invokes algorithm A. Consequently it has to answer A's n-bit oracle queries x_i with L-bit responses. For such a query, C queries its own oracle function f with the same query value x_i, gets an m-bit answer $y_i = f(x_i)$, computes the L-bit value $g(y_i)$, and answers this value to algorithm A. When A halts, C halts as well and outputs the same output as A. Therefore we have $\Pr(C^{f_K} = 1) = \Pr(A^{g \circ f_K} = 1) = \Pr(A^{g_K} = 1)$ and $\Pr(C^{f^*} = 1) = \Pr(A^{g \circ f^*} = 1)$. This implies that $\left| \Pr(C^{f_K} = 1) - \Pr(C^{f^*} = 1) \right|$ is equal to $\left| \Pr(A^{g_K} = 1) - \Pr(A^{g \circ f^*} = 1) \right|$, i.e.

$$\mathbf{Adv}_F^{prf}(C) = \delta_1.$$

Furthermore the time required for algorithm C is equal to the time required for algorithm A plus q times the time of computing g. Therefore $\mathbf{Adv}_F^{prf}(C)$ is upper-bounded by $\mathbf{Adv}_F^{prf}(t + qT_g^L, q)$, i.e. $\delta_1 \leq \mathbf{Adv}_F^{prf}(t + qT_g^L, q)$. Finally we have for any A

$$\mathbf{Adv}_G^{prf}(A) \leq \delta_1 + \delta_2 \leq \mathbf{Adv}_F^{prf}(t + qT_g^L, q) + q\mathbf{Adv}_g^{prng}(t + qT_g^L).$$

Consequently

$$\mathbf{Adv}_G^{prf}(t, q) \leq \mathbf{Adv}_F^{prf}(t + qT_g^L, q) + q\mathbf{Adv}_g^{prng}(t + qT_g^L). \quad \square$$

Proof of Theorem 2

Theorem 2. Let $g : \{0,1\}^m \longrightarrow \{0,1\}^L$ be a PRNG which generates $L \geq 2m$ outputs bits and produces its $2m$ first output bits in time T_g^{2m} and let $F^g = \{f_y\}_{y \in \{0,1\}^m}$ be the family of n-bit to m-bit functions derived from g by the Tree Based Construction. The (t, q) advantage of PRF F^g is related to the single-query advantage of PRNG g by the following inequality:

$$\mathbf{Adv}_{F^g}^{prf}(t, q) \leq nq\mathbf{Adv}_g^{prng}(t + q(n+1)T_g^{2m}).$$

Proof. First we define, for $0 \leq i \leq n$, a family F_i^g of $\{0,1\}^n \longrightarrow \{0,1\}^m$ functions; each F_i^g can be viewed as an intermediate PRF between F^g and the set $F_{n,m}^*$ of perfect random n-bits to m-bit functions.

- $F_0^g = \{f_{y_0}^g\}_{y_0 \in \{0,1\}^n}$ where $f_{y_0}^g : (x_1, \dots, x_n) \longmapsto g_{x_n} \circ \dots \circ g_{x_1}(y_0)$.

- $F_1^g = \{f_{y_0,y_1}^g\}_{(y_0,y_1) \in \{0,1\}^{2n}}$
  where $f_{y_0,y_1}^g : (x_1, \dots, x_n) \longmapsto g_{x_n} \circ \dots \circ g_{x_2}(y_{x_1})$.

- $F_i^g = \{f_{y_0,y_1,\dots,y_{2^i-1}}^g\}_{y_0,y_1,\dots,y_{2^i-1} \in \{0,1\}^{2^i n}}$
  where $f_{y_0,\dots,y_{2^i-1}}^g : (x_1, \dots, x_n) \longmapsto g_{x_n} \circ \dots \circ g_{x_{i+1}}(y_{x_1 \dots x_i})$
  (in the former expression $y_{x_1,\dots x_i}$ represents $y_{\sum_{t=1}^i x_i 2^{i-1}}$)

- $F_n^g = \{f_{y_0,y_1,\dots,y_{2^n-1}}^g\}_{y_0,y_1,\dots,y_{2^n-1} \in \{0,1\}^{2^n}}$
  where $f_{y_0,\dots,y_{2^n-1}}^g : (x_1, \dots, x_n) \longmapsto (y_{x_1 \dots x_n})$.

It is easy to see that F_0^g is equal to F^g, and that F_n^g is the set $F_{n,m}^*$ of all n-bit to m-bit functions.

Let us consider any (t, q) distinguishing algorithm A for F^g, i.e. a testing algorithm capable to query an n-bit to m-bit oracle function up to q times, and let us denote its distinguishing probability by

$$\epsilon = \left| \mathrm{Pr}_{f \in F^g}(A^f = 1) - \mathrm{Pr}_{f \in F_{n,m}^*}(A^f = 1) \right|.$$

We denote $\mathrm{Pr}_{f \in F_i^g}(A^f = 1)$ by p_i. Thus we have

$$\epsilon = |p_0 - p_n|.$$

We now construct a q-query distinguisher B for g, which when input with a q-tuple $(z_1, \ldots, z_q)$ of 2m-bit words is using one invocation of algorithm A to output either 0 or 1. In order to processes an input q-tuple $(z_1, \ldots, z_q)$, B first randomly draws an integer i comprised between 0 and $n-1$, and then inputs (z_1, z_q) to a testing algorithm B_i, and outputs B_i's binary output. Each testing algorithm B_i is defined as follows: B_i invokes algorithm A, and computes the answers to the up to q distinct n-bit oracle queries of A. For that purpose, B_i uses its random generation capability to simulate an auxiliary random function $\alpha : \{0,1\}^i \longrightarrow \{1, q\}$ that is initially undetermined. At each novel n bit oracle query $x^j = (x_1^j, \ldots, x_n^j)$ of A, algorithm B_i uses:

- the bits x_1^j to x_i^j to determine a 2m-bit value z_k as follows: B_i first checks if α is defined on point $(x_1^j, \ldots, x_i^j)$. If not, it selects randomly a value in $\{1, q\}$, affects it to $\alpha(x_1^j, \ldots x_i^j)$ and stores the new point of α. Otherwise B_i simply read the previously stored value. In both case, we denote by k the obtained value of $\alpha(x_1^j, \ldots x_i^j)$; k is used to select the k-th input z_k from B_i's input $(z_1, \ldots, z_q)$;
- the bit x_{i+1}^j to select an m-bit word y equal to the substring of the m left bits of z_k if $x_{i+1}^j = 0$, and of the m right bits of z_k if $x_{i+1}^j = 1$;
- the bits x_{i+2}^j to x_n^j to compute A^s L-bit oracle response $g_{x_n^j} \circ \ldots \circ g_{x_{i+2}^j}(y)$.

Finally when A halts, B_i halts also and returns A's binary output.
It is not too difficult to see that:

- if B_i's input is $(g(a_1), \ldots, g(a_q))$, where $(a_1, \ldots, a_q)$ is a randomly drawn q-tuple of m-bit, then A's oracle queries and response pairs have exactly the same probability distribution as if A were run with an n-bit to m-bit oracle function f randomly drawn from the family F_i^g:

$$\Pr(B_i((g(a_1), \ldots, g(a_q)) = 1) = \Pr_{f \in F_i^g}(A^f = 1) = p_i.$$

- if B_i's input is is a randomly drawn q-tuple $(z_1, \ldots, z_q)$ of 2m-bit values, then A's oracle queries and response pairs have exactly the same probability distribution as if A was run with an n-bit to m-bit oracle function f randomly drawn from the family F_{i+1}^g:

$$\Pr(B_i(z_1, \ldots, z_q) = 1) = \Pr_{f \in F_{i+1}^g} = (A^f = 1) = p_{i+1}.$$

The above equalities imply:

$$\left| \Pr(B((g(a_1), \ldots, g(a_q)) = 1) - \Pr(B(z_1, \ldots, z_q) = 1) \right|$$

$$= \left| \frac{1}{n} \sum_{i=0}^{n-1} p_i - \frac{1}{n} \sum_{i=1}^{n} p_i \right| = \frac{1}{n} |p_0 - p_n| = \frac{\epsilon}{n}.$$

In other words:

$$\mathbf{Adv}_g^{prng}(B) = \frac{1}{n} \mathbf{Adv}_{F_g}^{prf}(A).$$

However, algorithm B requires at most $t+qnT_g^{2m}$, where t is the time required by A and $T_{g'}^{2m}$ is the time required by g' to produce $2m$ bits. Therefore for any A we have

$$\mathbf{Adv}_{F_g}^{prf}(A) \leq n\mathbf{Adv}_g^{prng}(t + qnT_g^{2m}, q).$$

Finally, since $\mathbf{Adv}_g^{prng}(t + qnT_g^{2m}, q) \leq q\mathbf{Adv}_g^{prng}(t + q(n + 1)T_g^{2m})$ due to Lemma 1, we obtain

$$\mathbf{Adv}_{F_g}^{prf}(t,q) \leq qn\mathbf{Adv}_g^{prng}(t+q(n+1)T_g^{2m}). \qquad \square$$

Main Reduction Theorem of [3]

Theorem 4. *Let $L = \lambda(k - 1)n$ be the number of keystream bits produced by in time λT_S using λ iterations of our construction. Suppose there exists an algorithm A that distinguishes the L-bit keystream sequence associated with a known randomly chosen system S and an unknown randomly chosen initial internal state $x \in \{0,1\}^n$ from a random L-bit sequence in time T with advantage ϵ. Then there exists an algorithm C, which given the image $S(x)$ of a randomly chosen (unknown) n-bit value x by a randomly chosen n-bit to m-bit quadratic system S produces a preimage of $S(x)$ with probability at least $\frac{\epsilon}{2^3\lambda}$ over all possible values of x and S in time upper bounded by T'.*

$$T' = \frac{2^7n^2\lambda^2}{\epsilon^2}\left(T + (\lambda + 2)T_S + \log\left(\frac{2^7n\lambda^2}{\epsilon^2}\right) + 2\right) + \frac{2^7n\lambda^2}{\epsilon^2}T_S \ .$$

Two General Attacks on Pomaranch-Like Keystream Generators

Håkan Englund, Martin Hell, and Thomas Johansson

Dept. of Information Technology, Lund University,
P.O. Box 118, 221 00 Lund, Sweden

Abstract. Two general attacks that can be applied to all versions and variants of the Pomaranch stream cipher are presented. The attacks are demonstrated on all versions and succeed with complexity less than exhaustive keysearch. The first attack is a distinguisher which needs keystream from only one or a few IVs to succeed. The attack is not only successful on Pomaranch Version 3 but has also less computational complexity than all previously known distinguishers for the first two versions of the cipher. The second attack is an attack which requires keystream from an amount of IVs exponential in the state size. It can be used as a distinguisher but it can also be used to predict future keystream bits corresponding to an IV if the first few bits are known. The attack will succeed on all versions of Pomaranch with complexities much lower than previously known attacks.

Keywords: Stream ciphers, distinguishing attack, resynchronization attack, eSTREAM, Pomaranch.

1 Introduction

Pomaranch is one of many cipher constructions in the eSTREAM stream cipher project. The Pomaranch family consists of several versions and variants. The first two versions have been cryptanalyzed in [2,10,6]. For each new version, the cipher has been changed such that the attacks on the previous versions would not be successful.

In this paper we present two general attacks that can be applied to all versions and variants of the Pomaranch family of stream cipher. The first attack is a statistical distinguisher, which can be applied to all Pomaranch-like ciphers having one or several types of jump registers and both linear and nonlinear filter function. We improve the computational complexity of all known distinguishers on Version 1 and Version 2. Our attack is also applied to Pomaranch Version 3 and it is shown that the attack will succeed on the 80-bit variant with computational complexity $2^{71.10}$, significantly less than exhaustive key search. For the 128-bit variant, the attack will have computational complexity 2^{126}, almost that of exhaustive key search. The complexity of the attack given in the theorems can be seen as design criteria for subsequent versions of Pomaranch.

The second attack is an IV attack. It first stores many samples from the keystream corresponding to one IV in a table. Given short keystream samples

A. Biryukov (Ed.): FSE 2007, LNCS 4593, pp. 274–289, 2007.

from several other IVs, we can find collisions with the samples in the stored table. When a collision is found, future keystream bits can be predicted. The key will not be recovered but the attack is still more powerful than a distinguisher. Also this attack can be applied to all versions and variants of Pomaranch, e.g., by using a table of size $2^{71.3}$ bytes, 2^{98} different IVs and a computational complexity of 2^{104}, the 128 bit version of Pomaranch Version 3 can be attacked.

The outline of the paper is as follows. Section 2 will describe the Pomaranch stream ciphers and Section 3 will briefly describe the previous attack on Pomaranch. In Section 4 the first attack is given and in Section 5 we give the second attack. Section 6 concludes the paper.

2   Description of Pomaranch

In this section, we will give a brief overview of the design of Pomaranch. There are 3 versions of Pomaranch. First the overall design idea is presented and then we give the specific parameters for the different versions. The attacks described in this paper are independent of the initialization procedure and thus, only the keystream generation will be described here. For more details we refer to the respective design documents, see [7,8,9].

Pomaranch is a synchronous stream cipher designed primarily for being efficient in hardware. It follows the classical design of a filter generator where the contents of an internal state is filtered through a Boolean function to produce the keystream. The design of Pomaranch is illustrated in Figure 1.

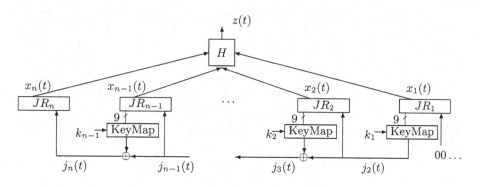

Fig. 1. General overview of Pomaranch

Pomaranch is based on a cascade of n Jump Registers (JR). A jump register can be seen as an irregularly clocked Linear Finite State Machine (LFSM). The clocking of a register can be done in one of two ways, either it jumps c_0 steps or it jumps c_1 steps. The clocking is decided by a binary jump control sequence, denoted by $j_i(t)$ for register i at time t.

$$j_i(t) = \begin{cases} 0, \ JR_i \text{ is clocked } c_0 \text{ times} \\ 1, \ JR_i \text{ is clocked } c_1 \text{ times} \end{cases}$$

The jump registers are implemented using two different kinds of delay shift cells, S-cells and F-cells. The S-cell is a normal D-element and the F-cell is a D-element where the output is fed back and XORed with the input. Half of the cells are implemented as S-cells and half are implemented as F-cells. When the register is clocked it will jump c_0 steps. When the jump control $j_i(t)$ is one, all cells in JR_i are switched to the opposite mode, i.e., all S-cells become F-cells and vice versa. This switch of cells results in a jump through the state space corresponding to c_1. The jump index of a register is the number of c_0 clockings that is equivalent to one c_1 clocking.

Notations and Assumptions. As seen in Figure 1 we denote Jump Register i by JR_i and its period by T_i. The key is denoted by K and the sub-key used for JR_i is denoted k_i. The length of the registers is denoted by L. The filter function is denoted by H (more specifically we will use the notation $H^{(1)}, H_{80}^{(2)}, H_{128}^{(2)}, H_{80}^{(3)}, H_{128}^{(3)}$ for the different filter functions used, where the superscript denotes the version of Pomaranch of interest, and the subscript denotes the key length used in the version). Similarly we will denote the number of registers used by $n^{(1)}, n_{80}^{(2)}, n_{128}^{(2)}, n_{80}^{(3)}, n_{128}^{(3)}$. The bit taken from the register i as input to the filter function at time t is denoted by $x_i(t)$. The keystream bit produced at time t is denoted $z(t)$ and sometimes only z. Both addition modulo 2 and integer addition is denoted by $+$ since there should be no risk of confusion. Only absolute values of biases of approximations will be given.

2.1   Pomaranch Version 1

The first version of Pomaranch was introduced in [7]. In this version, 128 bit keys were used together with an IV in the range of 64 to 112 bits. The cipher is built upon $n^{(1)} = 9$ identical jump registers. Each register uses 14 memory cells together with a characteristic polynomial with the jump index 5945, i.e., $(c_0, c_1) = (1, 5945)$. From register i 9 bits are taken as input to a key dependent function, denoted KeyMap in Figure 1. The output of this function is XORed to the jump sequence from register $i - 1$, $j_i(t)$ to produce the jump sequence for the next register, $j_{i+1}(t)$. The keystream bit z is given as the XOR of the bit in cell 13 from all the registers, i.e.,

$$z = H^{(1)}(x_1, \ldots, x_9) = x_1 + x_2 + \ldots + x_9.$$

2.2   Pomaranch Version 2

The second version of Pomaranch [8], comes in two variants, an 80-bit with $n_{80}^{(2)} = 6$, and a 128-bit key variant using $n_{128}^{(2)} = 9$ registers. The registers are still 14 memory cells long but to prevent the attacks on the first version, different tap positions are used as input to the KeyMap function in Figure 1. The characteristic polynomial is also changed and the new jump index is 13994. A new initialization procedure was introduced to prevent the attacks in [2,5]. The keystream is still taken as the XOR of the bits in cell 13 of all registers and is given by

$$z = \begin{cases} H_{80}^{(2)}(x_1, \ldots, x_6) = x_1 + x_2 + \ldots + x_6 \\ H_{128}^{(2)}(x_1, \ldots, x_9) = x_1 + x_2 + \ldots + x_9 \end{cases}.$$

2.3 Pomaranch Version 3

As in the second version, there are two variants of Pomaranch Version 3 [9], an 80-bit and a 128-bit key variant. The number of registers used are still the same as in Pomaranch Version 2, i.e., $n_{80}^{(3)} = 6$ respectively $n_{128}^{(3)} = 9$. In the case of Pomaranch Version 3, two different jump registers are used, the first using jump index 84074 which is referred to as type I and the second using jump index 27044, referred to as type II. The type I registers are used for odd numbered sections in Figure 1, and type II for the even numbered sections. Both registers are built on 18 memory cells, and have primitive characteristic polynomials, i.e., when only clocked with zeros or ones they have a period of $2^{18} - 1$. From each register one bit is taken from cell 17 and is fed into H. The filter functions used are

$$z = \begin{cases} H_{80}^{(3)}(x_1, \ldots, x_6) = G(x_1, \ldots, x_5) + x_6 \\ H_{128}^{(3)}(x_1, \ldots, x_9) = x_1 + x_2 + \ldots + x_9 \end{cases},$$

where

$$G(x_1, \ldots, x_5) = x_1 + x_2 + x_5 + x_1 x_3 + x_2 x_4 + x_1 x_3 x_4 + x_2 x_3 x_4 + x_3 x_4 x_5$$

is a 1-resilient Boolean function. The keystream length per IV/key pair is limited to 2^{64} bits in Version 3.

3 Previous Attacks on the Pomaranch Stream Ciphers

The first attack on Pomaranch, given in [2], was an attack on the initialization procedure. Soon after, an attack on the keystream generation was presented in [10]. This attack considered the best linear approximation of bits distance $L+1$ apart. Register JR_1 was exhaustively searched since this register is always fed with the all zero jump control sequence and the linear approximation is not valid for this register. When the state of JR_1 was known, JR_2 and 16 bits of the key was guessed. This was iterated until the full key was recovered. A distinguisher was not explicitly mentioned in [10] but it is very easy to see that if the attack is stopped after JR_1 is recovered, it would be equivalent to a distinguishing attack. This distinguisher needed $2^{72.8}$ keystream bits and computational complexity $2^{86.8}$.

Pomaranch Version 2 was designed to resist the attacks in [2,10]. However, it was still possible to find biased linear approximations by looking at keystream bits further apart. This was done in [6] and the new approximation made it possible to mount distinguishing and key recovery attacks on both the 80-bit and the 128-bit variants. The distinguisher for the 80-bit variant needed $2^{44.59}$ keystream bits and a computational complexity of $2^{58.59}$. For the 128-bit variant the complexities were $2^{73.53}$ and $2^{87.53}$ respectively.

4   Distinguishing Attacks on Pomaranch-Like Ciphers

In this section we will present general distinguishers for Pomaranch-like ciphers, in particular we will study the constructions that have been proposed in Pomaranch Version 1-3. We will give general results for these cipher families that can be used as design criteria for future Pomaranch ciphers.

4.1   Period of Registers

The first jump register, denoted JR_1 in Figure 1, is during keystream generation mode fed by the jump sequence only containing zeros. Hence JR_1 is a LFSM. The period of the register is denoted by T_1, hence $x_1(t) = x_1(t + T_1)$ with $T_1 = 2^L - 1$.

From JR_1 a jump control sequence is calculated which controls the jumping of JR_2. Assume that after T_1 clocks of JR_1, register JR_2 has jumped C steps. Then after T_1^2 clocks JR_2 has jumped CT_1 steps, a multiple of T_1 and is thus back to its initial state. If primitive characteristic polynomials are used for the registers, it can be shown that the period for register JR_i is

$$T_i = T_1^i,$$

and hence

$$x_i(t) = x_i(t + T_1^i).$$

Consequently, at time t and $t + T_1^p$, the filter function H has p inputs with exactly the same value, namely the contribution from registers $JR_1, \ldots, JR_p$. This observation will be used in our attack.

4.2   Filter Function

The filter function used in Pomaranch can be a nonlinear Boolean function or just the linear XOR of the output bits of each jump register. Our attack can be applied to both variants. The keystream bit at time t, denoted by $z(t)$, can be described as

$$z(t) = H\big(x_1(t), \ldots, x_n(t)\big).$$

Using the results from Section 4.1 and taking our samples as $z(t) + z(t + T_1^p)$ we can write the expression for the samples as

$$z(t) + z(t + T_1^p) = H\big(x_1(t), \ldots, x_n(t)\big) + H\big(x_1(t + T_1^p), \ldots, x_n(t + T_1^p)\big). \quad (1)$$

4.2.1   Linear Filter Function

When the filter function H is linear, i.e., $H(x_1, \ldots, x_n) = \sum_{i=1}^{n} x_i$, and our samples are taken as $z(t) + z(t + T_1^p)$ we know from Section 4.1 that p inputs to the filter function are the same at time t and at time $t + T_1^p$, hence we can rewrite (1) as

$$z(t) + z(t + T_1^p) = \sum_{i=p+1}^{n} x_i(t) + x_i(t + T_1^p).$$

4.2.2 Nonlinear Filter Function

When the filter function H is nonlinear, $x_i(t)$ and $x_i(t + T_1^p)$ will not cancel out in the keystream with probability one, as in Section 4.2.1. But, the input to H at time t and $t + T_1^p$ has p inputs $x_1, \ldots, x_p$ with the exact same value. This might lead to a biased distribution,

$$\Pr\left(H\big(x_1(t), \ldots, x_n(t)\big) + H\big(x_1(t + T_1^p), \ldots, x_n(t + T_1^p)\big) = 0\right) =$$
$$\Pr\left(z(t) + z(t + T_i^p) = 0\right) = \frac{1}{2}(1 \pm \epsilon),$$

where ϵ denotes the bias and $|\epsilon| \le 1$.

4.3 Linear Approximations of Jump Registers

In our attack, we need to find a linear approximation for the output bits of the jump registers that is biased, i.e., that holds with probability different from one half. We assume that all states of the register are equally probable, except for the all zero state which has probability 0. Further, it is assumed that all jump control sequences have the same probability. Finding the best linear approximation can be done by exhaustive search. Under certain circumstances much faster approaches can be used, see [6]. We search for a set $\mathcal{A}$ of size w such that

$$\Pr(\sum_{i \in \mathcal{A}} x(t + i) = 0) = \frac{1}{2}(1 \pm \varepsilon), \quad |\varepsilon| \le 1,$$

i.e., the weight of the approximation is w and the terms are given by the set $\mathcal{A}$. For our attack to work it is important that the bias of this approximation is sufficiently high.

It is assumed that jump register JR_1 will always have the all zero jump control sequence. Hence, the linear approximation will never apply for this register.

4.3.1 Different Registers

In the case when not all registers use the same kind of jump registers, we are not interested in the most biased linear approximation of single registers. Instead we have to search for a linear approximation that has a good bias for all types of registers at the same time. This is much harder to find than a single approximation for one register. This is the case in Pomaranch Version 3.

4.4 Attacking Different Versions of Pomaranch

A Pomaranch stream cipher can be designed using one or several types of jump registers. It can also use a linear or a nonlinear Boolean filter function. In this section we take a closer look at the different design possibilities that has been used and give an expression for the number of samples needed in a distinguisher for each possibility.

The general expressions for the amount of keystream needed in an attack can be seen as a new design criteria for Pomaranch-like stream ciphers.

4.4.1 One Type of Registers with Linear Filter Function

In this family of Pomaranch stream ciphers we assume that all jump register sections use the same type of register and that the filter function H is linear, i.e., $H(x_1, \ldots, x_n) = \sum_{i=1}^{n} x_i$. Pomaranch Version 1 and Pomaranch Version 2 are both included in this family.

Assume that we have found a linear approximation, as described in Section 4.3, of weight w of the register used. We consider samples at t and $t + T_1^p$ such that p positions into H are the same according to Section 4.1. Our samples will be taken as

$$\sum_{i \in \mathcal{A}} z(t+i) + \sum_{i \in \mathcal{A}} z(t+i+T_1^p) = \sum_{j=1}^{n} \sum_{i \in \mathcal{A}} (x_j(t+i) + x_j(t+i+T_1^p)) \qquad (2)$$

$$= \sum_{j=p+1}^{n} \sum_{i \in \mathcal{A}} (x_j(t+i) + x_j(t+i+T_1^p)).$$

Since the bias of $\sum_{i \in \mathcal{A}} x_i(t+i)$ is ε and we have $2(n-p)$ such relations the total bias of the samples is given by

$$\varepsilon_{tot} = \varepsilon^{2(n-p)}.$$

Using the approximation that $1/\varepsilon_{tot}^2$ samples are needed to reliably distinguish the cipher from a truly random source, we get an estimate of the keystream length needed. (In practice, this number should be multiplied by a small constant.)

Theorem 1. *The computational complexity and the number N of keystream bits needed to reliably distinguish the Pomaranch family of stream ciphers using a linear filter function and n jump registers of the same type is bounded by*

$$N = T_1^p + \frac{1}{\varepsilon^{4(n-p)}}, \quad p > 0,$$

where ε is the bias of the best linear approximation of the jump register.

4.4.2 Different Registers with Linear Filter Function

In this family of generators different types of jump registers are used and the filter function is assumed to be $H(x_1, \ldots, x_n) = \sum_{i=1}^{n} x_i$.

This case is very similar to the case when all registers are of the same type. The difference is that, in this case, we are not looking for the best linear approximation of the registers separately. Instead, we have to find a linear approximation that have a bias for all the registers $JR_{p+1}, \ldots, JR_n$. This can be difficult if there are several types of registers. Approximations with a large bias for one type might have a very small bias for other types. Anyway, assume that we have found such a linear approximation. Our samples will still be taken as in (2). If we denote the bias for the approximation of register i by ε_i, then the total bias will be given as

$$\varepsilon_{tot} = \prod_{i=p+1}^{n} \varepsilon_i^2.$$

Theorem 2. *Assume that there is a linear relation that is biased in all registers. The computational complexity and the number N of keystream bits needed to reliably distinguish the Pomaranch family of stream ciphers using a linear filter function and n jump registers of different types is bounded by*

$$N = T_1^p + \frac{1}{\prod\limits_{i=p+1}^{n} \varepsilon_i^4}, \quad p > 0,$$

where ε_i is the bias of jump register JR_i.

The 128-bit variant of Pomaranch Version 3 belongs to a special subclass of this family, namely all registers in odd positions are of type I and registers in even positions are of type II. In this case we only have to search for a linear approximation that is biased for type I and type II registers at the same time. The bias of $\sum_{i \in \mathcal{A}} x_i(t+i)$ is denoted $\varepsilon_{type\ I}$ and $\varepsilon_{type\ II}$, respectively, for the different registers. In total we have $2\lceil \frac{n-p}{2} \rceil$ type I relations and $2\lfloor \frac{n-p}{2} \rfloor$ type II relations when n is odd. Hence, the total bias of the samples is given by

$$\varepsilon_{tot} = \varepsilon_{type\ I}^{2\lceil \frac{n-p}{2} \rceil} \varepsilon_{type\ II}^{2\lfloor \frac{n-p}{2} \rfloor}.$$

If we apply Theorem 2 to the 128-bit variant of Pomaranch Version 3, the number of samples in the distinguisher is given by

$$N = T_1^p + \frac{1}{\varepsilon_{type\ I}^{4\lceil \frac{n-p}{2} \rceil} \varepsilon_{type\ II}^{4\lfloor \frac{n-p}{2} \rfloor}}. \tag{3}$$

4.4.3   Nonlinear Filter Function

Now we consider the case when the Boolean filter function is a nonlinear function. We only consider the case when the filter function H can be written in the form

$$H(x_1, \ldots, x_n) = G(x_1, \ldots, x_{n-1}) + x_n. \tag{4}$$

The attack can easily be extended to filter functions with more (or less) linear terms but to simplify the presentation, and the fact that the 80-bit variant of Pomaranch Version 3 is in this form, we only consider this special case in this paper.

Attacks on this family use a biased linear approximation of JR_n, see Section 4.3, together with the fact that the input to G at time t and $t + T_1^p$ have p inputs in common and hence in some cases a biased distribution, see Section 4.2.2.

Let ϵ denote the bias of $G(x_1(t), \ldots, x_{n-1}(t)) + G(x_1(t+T_1^p), \ldots, x_{n-1}(t+T_i^p))$, and ε the bias of our linear approximation for JR_n, $\sum_{i \in \mathcal{A}} x_i(t+i)$. The samples are taken as

$$\sum_{i \in \mathcal{A}} z(t+i) + \sum_{i \in \mathcal{A}} z(t+i+T_1^p) = \sum_{i \in \mathcal{A}} x_n(t+i) + \sum_{i \in \mathcal{A}} x_n(t+i+T_1^p)$$

$$+ \sum_{i \in \mathcal{A}} G\big(x_1(t+i), \dots, x_{n-1}(t+i)\big) + G\big(x_1(t+i+T_1^p), \dots, x_{n-1}(t+i+T_1^p)\big),$$

and the bias of the samples is given by

$$\varepsilon_{tot} = \varepsilon^2 \epsilon^w. \tag{5}$$

This relation tells us that we need to keep the weight of the linear approximation of JR_n as low as possible, i.e., there is a trade off between the bias ε of the approximation and the number of terms w in the relation.

Theorem 3. *The computational complexity and the number N of keystream bits needed to reliably distinguish the Pomaranch family of stream ciphers using a filter function of the form (4) is bounded by*

$$N = T_1^p + \frac{1}{(\varepsilon^2 \epsilon^w)^2}.$$

where ε is the bias of the approximation of weight w of register JR_n and ϵ is the bias of $G\big(x_1(t), \dots, x_{n-1}(t)\big) + G\big(x_1(t+T_1^p), \dots, x_{n-1}(t+T_i^p)\big)$.

Note that in this presentation it does not matter if all registers are of the same type or if they are of different types. Since only register JR_n is completely linear in the output function H, we only need to have an approximation of this register.

4.5   Attack Complexities for the Existing Versions of the Pomaranch Family

In this section, we look at the existing versions and variants of Pomaranch that have been proposed so far. These are Pomaranch Version 1, the 80-bit and 128-bit variants of Pomaranch Version 2 and the 80-and 128-bit variants of Pomaranch Version 3. Applying the attack proposed in this paper, we show that we can find distinguishers with better complexity than previously known for *all* 5 ciphers.

4.5.1   Pomaranch Version 1
In Pomaranch Version 1 all registers are the same, so the attack will be according to Section 4.4.1. The best known linear approximation for this register, as given in [10], is

$$\varepsilon = |2\Pr(x(t) + x(t+8) + x(t+14) = 0) - 1| = 2^{-4.286}.$$

Using Theorem 1 for different values of p we get Table 1. We see that the best attack is achieved when $p = 5$. The computational complexity and the amount of keystream needed is then $2^{70.46}$.

Table 1. Number of samples and computational complexity needed to distinguish Pomaranch Version 1 from random

p	1	2	3	4	5	6	7
$N^{(1)}$	$2^{137.15}$	$2^{120.01}$	$2^{102.86}$	$2^{85.72}$	$2^{70.46}$	$2^{83.99}$	$2^{97.99}$

4.5.2 Pomaranch Version 2

Similarly as in Pomaranch Version 1, in Pomaranch Version 2 all registers are the same and the attack will be performed according to Section 4.4.1. The best bias of a linear approximation for the registers used was found in [6] and is given by

$$\varepsilon = |2\Pr(x(t) + x(t+2) + x(t+6) + x(t+18) = 0) - 1| = 2^{-4.788}.$$

Using Theorem 1 for different values of p gives Table 2. For the 80-bit variant the computational complexity and the number of samples is $2^{56.00}$ and for the 128-bit variant it is $2^{76.62}$.

Table 2. Number of samples needed to distinguish Pomaranch Version 2 according to Theorem 1

p	1	2	3	4	5	6
$N^{(2)}_{80}$	$2^{95.76}$	$2^{76.61}$	$2^{57.46}$	$2^{56.00}$	$2^{70.00}$	$2^{84.00}$
$N^{(2)}_{128}$	$2^{153.22}$	$2^{134.06}$	$2^{114.91}$	$2^{95.76}$	$2^{76.62}$	$2^{84.00}$

4.5.3 Pomaranch Version 3

There is a significant difference between the 80-bit and the 128-bit variants of Pomaranch Version 3, so this section will be divided into two parts.

80-bit Variant. The 80-bit variant of Pomaranch Version 3 uses a non-linear filter function, the attack will hence follow the procedure described in Section 4.4.3. We started by estimating the bias of

$$G\big(x_1(t), \ldots, x_5(t)\big) + G\big(x_1(t + T_1^p), \ldots, x_5(t + T_1^p)\big).$$

The results for different p are summarized in Table 3. The keystream per IV/key pair of Pomaranch Version 3 is limited to 2^{64}. Because of this we limit p to $p \in \{1, 2, 3\}$, otherwise $T_1^p > 2^{64}$. We looked for a linear relation of JR_6 that, together with a value of $p \in \{1, 2, 3\}$, minimizes the amount of keystream needed as given by Theorem 3. The best approximation found was

$$\Pr\big(x_6(t) + x_6(t+5) + x_6(t+7) + x_6(t+9) + x_6(t+12) + x_6(t+18) = 0\big) = \frac{1}{2}(1 - 2^{-8.774}),$$

using $p = 3$. The total bias of our samples using this approximation is

$$\varepsilon_{tot} = (2^{-8.774})^2 \cdot (2^{-3})^6 = 2^{-35.548},$$

according to (5). The samples used in the attack are taken according to

$$\sum_{i \in \mathcal{A}} z(t+i) + \sum_{i \in \mathcal{A}} z(t+i+T_1^3),$$

where $\mathcal{A} = \{0, 5, 7, 9, 12, 18\}$. According to Theorem 3, the amount of keystream needed is $2^{54} + 2^{71.096} = 2^{71.096}$. This is also the computational complexity of the attack. In the specification of Pomaranch Version 3 the frame length (keystream per IV/key pair) is limited to 2^{64}. This does not prevent our attack since all samples will have this bias regardless of the key and IV used. We only need to consider 2^{64} keystream bits from $\lceil 2^{7.096} \rceil = 137$ different key/IV pairs.

Table 3. The bias of $G(x_1(t), \ldots, x_5(t)) + G(x_1(t+T_1^p), \ldots, x_5(t+T_1^p))$ in the 80 bit variant of Pomaranch Version 3 for different values of p

p	1	2	3	4	5
ϵ	0	2^{-4}	2^{-3}	2^{-2}	1

128-bit Variant. In Pomaranch Version 3 two different registers are used, so we start by searching for a linear approximation that is good for both types of registers. The best approximation we found was

$$x(t)+x(t+1)+x(t+2)+x(t+5)+x(t+7)+x(t+11)+x(t+12)+x(t+15)+x(t+21),$$

which has the same bias for both types of registers, namely

$$\varepsilon_{even} = \varepsilon_{odd} = 2^{-10.934}.$$

Using (3) for different values of p we get Table 4. Our best distinguishing attack needs $2^{126.00}$ keystream bits. This figure is determined by $T_1^7 = 2^{126.00}$ so it is not possible to look at different key/IV pairs in this case since the distance between the bits in each sample has to be $2^{126.00}$. Since the frame length is limited to 2^{64} it will not be possible to get any biased samples at all with $p = 7$.

Table 4. Number of samples needed to distinguish the 128-bit variant of Pomaranch Version 3 according to (3)

p	1	2	3	4	5	6	7	8
$N_{128}^{(3)}$	$2^{349.89}$	$2^{306.15}$	$2^{262.42}$	$2^{218.68}$	$2^{174.94}$	$2^{131.21}$	$2^{126.00}$	$2^{144.00}$

5   Square Root IV Attack

In this section we will give an attack that works for all families of Pomaranch where the key is longer than half of the total register length. The size of the state in Pomaranch is always larger than twice the key size, e.g., the 128-bit variant of Pomaranch Version 3 has a state size of 290 bits. Thus, the generic time-memory tradeoff attacks will not be applicable in general. Our attack is a variant of the time-memory tradeoff attack and is generic for all stream ciphers. Let us divide the internal state of the cipher into two parts,

$$State = (State_K, State_{K+IV}),$$

where $State_K$ is a part of the state that statically holds the key and $State_{K+IV}$ is a part of the state that is updated, depending on both the key and the IV. If the key size $|K| > |State_{K+IV}|/2$, then the attack will always succeed with complexity below exhaustive key search. In Pomaranch, $State_K$ will consist of the $|K|$ key bits and $State_{K+IV}$ will consist of the register cells.

In the attack scenario, we assume that the key is fixed and that the cipher is initialized with many different IVs. Further, we assume that we have access to one long keystream sequence produced from one of the IVs, denoted IV_0. We intercept the ciphertext corresponding to many other IVs and we know the first l plaintext bits corresponding to every ciphertext. Our goal is to recover the rest of the plaintext for one of the messages.

The key map used to produce the jump control bits is key dependent but independent of the IV. Hence, a fixed key will define a state graph of size $(2^L - 1)^n \approx 2^{nL}$ states, where L is the register length and n the number of registers. We can apply the following attack.

Let a sample of l consecutive keystream bits at time t be denoted $S(t) = \big(z(t), z(t+1), \ldots, z(t+l-1)\big)$. If the sample stems from IV_i we denote it by $S_{IV_i}(t)$. We first store a large amount of samples from IV_0 in a table. We would like to find another IV, denoted IV_c, that results in a sample such that

$$S_{IV_c}(t_c) = S_{IV_0}(t_0).$$

If a collision is found, then with high probability the following keystream of IV_0 and IV_c will also be identical. That means that if we just know the first l keystream bits generated by IV_c, we can predict future keystream bits from IV_c. The attack is visualized in Figure 2.

Assume that $2^{\beta nL}$ ($0 \leq \beta \leq 1$) samples of length l from a keystream sequence of $2^{\beta nL} + l$ bits, originating from IV_0 and key K, is saved in a table. The table is then sorted with complexity $O(\beta nL \cdot 2^{\beta nL})$. This table covers a fraction of $2^{-(1-\beta)nL}$ of the entire cycle. The number of samples (IVs with l known keystream bits) we need to test to find a collision is geometrically distributed with expected value $2^{(1-\beta)nL}$. For each sample, a logarithmic search with complexity $O(\beta nL)$ in the table is performed to see if there is a collision. To be sure that a collision in the table actually means that we have found a collision in the state cycle, l must be $l \approx nL$. The attack complexities are then given by

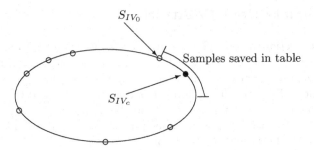

Fig. 2. State graph for a fixed key, a sample is visualized by a small ring

$$Keystream: \quad \begin{array}{l} 2^{\beta nL} + nL, \text{ from one IV and} \\ nL, \text{ from } 2^{(1-\beta)nL} \text{ IVs} \end{array}$$

$$Time: \quad \beta nL2^{\beta nL} + \beta nL2^{(1-\beta)nL}$$

$$Memory: \quad nL2^{\beta nL}$$

where $0 \leq \beta \leq 1$. By decreasing β it is possible to achieve smaller memory complexity at the expense of more IVs and higher time complexity. We can also see that the best time complexity is achieved when $\beta = 0.5$ for large nL. The proposed attack is summarized in Figure 3. In the figure, $S_{IV_i}(t)$ represents a sample from the keystream from IV_i at time t, and T represents the table where samples are stored.

$$
\begin{aligned}
&\textbf{for } i = 1, \ldots, 2^{\frac{nL}{2}} \\
&\quad T[i] = S_{IV_0}(t + i) \\
&\textbf{end for} \\
&\text{sort } T \\
&\textbf{for } i = 1, \ldots, 2^{\frac{nL}{2}} \\
&\quad \textbf{if } S_{IV_i}(t) \in T \\
&\quad\quad \textbf{return } cipher \\
&\textbf{end for} \\
&\textbf{return } random
\end{aligned}
$$

Fig. 3. Summary of square root attack

5.1   Attack Complexities on Pomaranch

In this section we will look at the existing versions of Pomaranch and show that the square root IV attack can be mounted with complexity significantly less than exhaustive key search. We assume $\beta = 0.5$ so the time complexity and the memory complexity in bits are equal. The 128-bit variants of Pomaranch

Version 1 and Version 2 can be attacked using a table of size $2^{67.0}$ bytes together with keystream from $2^{63.0}$ different IVs. The 80-bit variant of Pomaranch Version 2 can be attacked using only a table of $2^{45.4}$ bytes and $2^{42.0}$ different IVs. Pomaranch Version 3 uses larger registers, and the complexity of the attack on the 80-bit variant is a table of size $2^{57.8}$ bytes and $2^{54.0}$ IVs. The 128-bit variant needs a table of $2^{85.3}$ bytes and 2^{81} IVs. However, if we respect the maximum frame length of 2^{64} bits, we need to choose $\beta = 0.395$. Then we need a table of $2^{71.3}$ bytes and 2^{98} IVs. The time complexity is in this case 2^{104}.

The success probability of the attack has been simulated on a reduced version of the 128 bit variant of Pomaranch Version 3, using two registers. Choosing $\beta = 0.5$ implies that we know 2^{18} keystream bits from IV_0, we store all samples of length $nL = 36$ in a table, and that we need samples from 2^{18} different IVs in order to find a collision. The simulation results are summarized in Table 5. We also verified the attack using 3 registers. The attack given in this section suggests a new design criteria for the Pomaranch family of stream ciphers, namely that the total register length must be at least twice the keysize.

Table 5. Simulation results using 2 register Pomaranch version 3 with linear filter funtion, the table summarizes how many times the attack succeeds out of 100 attacks for a specific table size and number of IVs

		Number of IVs				
Success rate		2^{17}	2^{18}	2^{19}	2^{20}	2^{21}
	2^{17}	28	45	64	87	98
Table size	2^{18}	38	67	91	98	100
	2^{19}	71	86	98	100	100

6   Conclusions

We have presented two general attacks on Pomaranch-like keystream generators. The attack complexities are summarized in Table 6.

The first attack is a distinguishing attack that can be applied to all known versions and variants of the stream cipher Pomaranch. For Pomaranch Version 1 and Pomaranch Version 2, the distinguisher will succeed with better computational complexity than any other known distinguisher for these versions. For the 80-bit variant of Pomaranch Version 3, the attack will succeed using $2^{71.1}$ bits and computational complexity $2^{71.1}$. Since the frame length is restricted to 2^{64} bits we can instead collect 2^{64} bits from 137 key/IV pairs. For the 128-bit variant our distinguisher needs about 2^{126} samples and will not succeed if the frame length is restricted.

We have also presented a general IV attack that works for all ciphers where the key size is larger than half of the state size, when the part of the state only affected by the key is not considered. The attack was demonstrated on all versions and variants of Pomaranch with complexity far below exhaustive search. This attack has much lower complexity than the first and will work even if the frame length is restricted. The attack will not recover the key but is different from a distinguishing attack. It will recover the plaintext corresponding to one IV if only the ciphertext together with the first few keystream bits are known. On the other hand, this attack will require keystream from a large amount of IVs. This attack will also be applicable to the stream cipher LEX [1] and to any block cipher used in OFB mode of operation as shown in [3]. The attack scenario here is somewhat similar to the attack scenarios in disk encryption, see e.g., [4], where the adversary has write access to the disk encryptor and read access to the storage medium.

Table 6. Summary of attack complexities, for the two proposed attacks, on all versions and variants of Pomaranch

Attack Complexities	Distinguishing Attack Keystream/Compl.	Square Root IV attack Memory/IVs/Compl.
Pomaranch v1 128 bit	2^{71} / 2^{71}	2^{67} / 2^{63} / 2^{63}
Pomaranch v2 80 bit	2^{56} / 2^{56}	2^{45} / 2^{42} / 2^{42}
Pomaranch v2 128 bit	2^{77} / 2^{77}	2^{67} / 2^{63} / 2^{63}
Pomaranch v3 80 bit	2^{71} / 2^{71}	2^{58} / 2^{54} / 2^{54}
Pomaranch v3 128 bit	* 2^{126} / 2^{126}	2^{71} / 2^{98} / 2^{104}

* Without frame length restriction.

References

1. Biryukov, A.: The design of a stream cipher LEX. Selected Areas in Cryptography—SAC, 2006. Preproceedings (2006)
2. Cid, C., Gilbert, H., Johansson, T.: Cryptanalysis of Pomaranch. IEE Proceedings - Information Security 153(2), 51–53 (2006)
3. Englund, H., Hell, M., Johansson, T.: A note on distinguishing attacks. The State of the Art of Stream Ciphers, Workshop Record, SASC 2007, Bochum, Germany (January 2007)
4. Gjøsteen, K.: Security notions for disk encryption. In: di Vimercati, S.d.C., Syverson, P.F., Gollmann, D. (eds.) ESORICS 2005. LNCS, vol. 3679, pp. 455–474. Springer, Heidelberg (2005)
5. Hasanzadeh, M., Khazaei, S., Kholosha, A.: On IV setup of Pomaranch. eSTREAM, ECRYPT Stream Cipher Project, Report 2005/082 (2005), http://www.ecrypt.eu.org/stream
6. Hell, M., Johansson, T.: On the problem of finding linear approximations and cryptanalysis of Pomaranch version 2. Selected Areas in Cryptography—SAC 2006. Preproceedings (2006)

7. Jansen, C.J.A., Helleseth, T., Kholosha, A.: Cascade jump controlled sequence generator (CJCSG). eSTREAM, ECRYPT Stream Cipher Project, Report 2005/022 (2005), http://www.ecrypt.eu.org/stream
8. Jansen, C.J.A., Helleseth, T., Kholosha, A.: Cascade jump controlled sequence generator and Pomaranch stream cipher (version 2). eSTREAM, ECRYPT Stream Cipher Project, Report 2006/006 (2006), http://www.ecrypt.eu.org/stream
9. Jansen, C.J.A., Helleseth, T., Kholosha, A.: Cascade jump controlled sequence generator and Pomaranch stream cipher (version 3). eSTREAM, ECRYPT Stream Cipher Project, http://www.ecrypt.eu.org/stream
10. Khazaei, S.: Cryptanalysis of Pomaranch (CJCSG). eSTREAM, ECRYPT Stream Cipher Project, Report 2005/065 (2005), http://www.ecrypt.eu.org/stream

Analysis of QUAD

Bo-Yin Yang[1], Owen Chia-Hsin Chen[2],
Daniel J. Bernstein[3], and Jiun-Ming Chen[4]

[1] Academia Sinica and Taiwan Information Security Center
by@moscito.org
[2] National Taiwan University
owenhsin@gmail.com
[3] University of Illinois at Chicago
djb@cr.yp.to
[4] Nat'l Taiwan University and Nat'l Cheng Kung University
jmchen@ntu.edu.tw

Abstract. In a Eurocrypt 2006 article entitled "QUAD: A Practical Stream Cipher with Provable Security," Berbain, Gilbert, and Patarin introduced QUAD, a parametrized family of stream ciphers. The article stated that "the security of the novel stream cipher is provably reducible to the intractability of the MQ problem"; this reduction deduces the infeasibility of attacks on QUAD from the hypothesized infeasibility (with an extra looseness factor) of attacks on the well-known hard problem of solving systems of multivariate quadratic equations over finite fields. The QUAD talk at Eurocrypt 2006 reported speeds for QUAD instances with 160-bit state and output block over the fields GF(2), GF(16), and GF(256).

This paper discusses both theoretical and practical aspects of attacking QUAD and of attacking the underlying hard problem. For example, this paper shows how to use XL-Wiedemann to break the GF(256) instance QUAD(256, 20, 20) in approximately 2^{66} Opteron cycles, and to break the underlying hard problem in approximately 2^{45} cycles. For each of the QUAD parameters presented at Eurocrypt 2006, this analysis shows the implications and limitations of the security proofs, pointing out which QUAD instances are not secure, and which ones will never be proven secure. Empirical data backs up the theoretical conclusions; in particular, the 2^{45}-cycle attack was carried out successfully.

1 Introduction

1.1 Questions

Berbain, Gilbert, and Patarin introduced QUAD at Eurocrypt 2006. Their article [7] is titled "<u>A</u> Practical Stream Cipher," but QUAD is not actually a single stream cipher; it is a parametrized family of stream ciphers. The most important parameters are the QUAD field size, the number of QUAD variables, and the number of QUAD outputs per round. We will write QUAD(q, n, r) to mean any instance of QUAD that has field size q, uses n variables, and produces r outputs per round.

A. Biryukov (Ed.): FSE 2007, LNCS 4593, pp. 290–308, 2007.

The speed and security of QUAD are (obviously) functions of the QUAD parameters. The Performance section [7, Section 6] of the QUAD paper reported a speed of 2915 Pentium-IV cycles/byte for the "recommended version of QUAD," namely QUAD(2, 160, 160). The section closed with the following intriguing comment: "Though QUAD is significantly slower than AES, which runs at 25 cycles/byte, it is much more efficient than other provably secure pseudo random generator. Moreover, implementations of QUAD with quadratic system over larger fields (e.g. GF(16) or GF(256) are much faster and even reach 106 cycles/byte."

Further performance details were revealed in the QUAD talk at Eurocrypt 2006. In that talk, the QUAD authors reported the following implementation results:

- QUAD(2, 160, 160): 2930 cycles/byte for 32-bit architecture, 2081 for 64-bit;
- QUAD(16, 40, 40): 990 cycles/byte for 32-bit architecture, 745 for 64-bit;
- QUAD(256, 20, 20): 530 cycles/byte for 32-bit architecture, 417 (confirmed in [5], correcting the erroneous "106" in [7]) for 64-bit.

We wrote privately to the QUAD authors to ask about parameters. The authors confirmed in response that the Eurocrypt 2006 speed reports were for "160 bits of internal state, and 160 bits produced at each round," i.e., QUAD(q, n, n) with $q^n = 2^{160}$; and that the talk reported the above speeds for $q = 2$, 16, and 256. Speed was never reported for the "proven secure" QUAD(2, 256, 256) or QUAD(2, 350, 350).

Which of these QUAD instances are actually secure? What is the security level under the best known attack? What does the "provable security" of QUAD mean for these parameter choices? What about other parameter choices?

1.2 Conclusions

This paper discusses both theoretical and practical aspects of solving the nonlinear multivariate systems associated with the security of QUAD. Our analysis produces the following conclusions regarding the security of QUAD:

- Section 4: For a surprisingly wide range of parameters (which we call "broken" parameters), a feasible computation distinguishes the QUAD output stream from uniform; in fact, an attacker can compute QUAD's secret internal state from a few output blocks. For example, QUAD(256, 20, 20), one of the three QUAD instances for which timings were reported in the QUAD talks at Eurocrypt 2006 and SAC 2006 [5], is breakable in no more than 2^{66} cycles.
- Section 5: For an even wider range of parameters (which we call "unprovable"), a feasible attack breaks the "hard problem" underlying QUAD. For example, the "hard problems" underlying QUAD(16, 40, 40) and QUAD(256, 20, 20) are breakable in approximately 2^{71} and 2^{45} cycles respectively. An "unprovable" instance can *never* be provably secure. *This does not mean that the QUAD instance is breakable!*
- Section 6: For an extremely wide range of parameters (which we call "unproven"), including QUAD(256, 20, 20), QUAD(16, 40, 40), QUAD(2, 160, 160), and QUAD(2, 256, 256), a "loosely feasible" attack breaks the "hard problem" underlying QUAD. "Loosely feasible" means that the attack time is smaller than

$2^{80}L$ where L is the looseness factor in QUAD's "provable security" theorem. *This does not mean that the proof is in error, and it does not mean that the QUAD instance is breakable!*

Our algorithm analysis—in particular, our analysis of XL-Wiedemann—may also be useful for readers interested in other applications of solving systems of multivariate equations.

We do not claim that *all* QUAD parameters are broken, or unprovable, or unproven. All of our attacks against QUAD(q, n, n) have cost growing exponentially with n when q is fixed. The problem for QUAD is that—especially for large q—the base of the exponential is surprisingly small, so n must be surprisingly large to achieve proven security against these attacks.

We see ample justification for future investigation of QUAD. The QUAD security argument is that a feasible attack against QUAD(q, n, n) for moderately large n would imply an advance in attacks against the $\mathcal{MQ}$ problem, the problem of solving systems of multivariate quadratic equations over finite fields:

Problem $\mathcal{MQ}$: Solve the system $P_1 = P_2 = \cdots = P_m = 0$, where each P_i is a quadratic polynomial in $\mathbf{x} = (x_1, \ldots, x_n)$. All coefficients and variables are in the field $K = \mathrm{GF}(q)$.

This is a well-known difficult problem [20] and the basis for multivariate-quadratic public-key cryptosystems [16,27]. However, we are concerned by the choice of QUAD$(2, 160, 160)$, QUAD$(16, 40, 40)$, and QUAD$(256, 20, 20)$ as the subjects of speed reports in the QUAD talk at Eurocrypt 2006. QUAD$(256, 20, 20)$ is now broken; no extension of the security arguments in [7] will ever prove QUAD$(16, 40, 40)$ secure; and QUAD$(2, 160, 160)$, while apparently unbroken, is currently rather far from being provably secure. There is no single QUAD stream cipher that simultaneously provides the advertised levels of speed and provable security. We recommend that future QUAD evaluations stop describing QUAD as "a stream cipher" and start carefully distinguishing between different choices of the QUAD parameters.

1.3 Previous Work

QUAD is a new proposal, but we are certainly not the first to observe difficulties in the parameter choices for "provably secure" cryptosystems.

Consider, for example, the famous BBS stream generator. Blum, Blum, and Shub [8] proved that an attack against this generator can be converted into an integer-factorization algorithm with a polynomially bounded loss of efficiency and effectiveness. Koblitz and Menezes [22, Section 6.1], after comparing the latest refinements in the BBS security theorem, the speeds of existing factorization algorithms, and the BBS parameter choices in the literature, concluded that common BBS parameter choices destroy the "provable security" of BBS. The speed comparison between 1024-bit BBS and QUAD$(2, 160, 160)$ in [7, Section 6] issues the same warning regarding 1024-bit BBS ("far from the number of bits of the internal state required for proven security"), but does not highlight the comparable lack of proof for "recommended" QUAD(2, 160, 160). See [21] and [22] for more discussions of the limits of "provable security."

Of course, a cryptosystem can be unbroken, and perhaps acceptable to users, without having a security proof. We are not aware of any feasible attacks on any proposed BBS parameter choices. The situation for QUAD is qualitatively different: The QUAD authors reported speeds for QUAD(256, 20, 20). This cipher is not merely unprovable but has now actually been broken.

The QUAD paper contains some analysis of parameter choices: [7, page 121] summarizes Bardet's estimates [2] of equation-solving time for $q = 2$, and concludes that there is no "contradiction"—i.e., the QUAD security proof does not guarantee 80-bit security for 2^{40} output bits against known attacks—for QUAD(2, n, n) for $n < 350$. We analyze QUAD in much more depth. We go beyond $q = 2$, showing that QUAD(16, 40, 40) is unprovable; we consider not just the cost of solving the underlying "hard problem" but also the extra cost of attacking QUAD, showing that QUAD(256, 20, 20) is not merely unprovable but also broken; and we cover both theoretical and practical attack complexity, for example discussing the implications of parallelizing communication costs.

1.4 Future Work

We suggest using the same classification of levels of danger for parameter choices in other "provably secure" systems—systems having security theorems that deduce the infeasibility of attacks on the cryptosystem, assuming the loose infeasibility of attacks on the underlying "hard problem":

- Unproven parameter choices (e.g., 2048-bit BBS or QUAD(2, 160, 160) as "recommended"): Known attacks on the underlying "hard problem" are loosely feasible. They might not be feasible, but the gap is smaller than the looseness of the security proof. It is unjustified and somewhat misleading to label these parameters as "provably secure."
- Unprovable parameter choices (e.g., 512-bit BBS or QUAD(16, 40, 40)): Known attacks on the underlying "hard problem" are feasible. These parameters would not be "provably secure" even if the proof were tightened.
- Broken parameter choices (e.g., QUAD(256, 20, 20)): Feasible attacks are known on the cryptosystem per se. Users must avoid these parameters.

It would be interesting to see a careful comparison of speeds for "provably secure" systems with parameters that avoid all of these dangers. This is a quite different task from comparing systems such as 1024-bit BBS and QUAD(2, 160, 160), both of which are unbroken but have no security proof; the motivation for "provably secure" systems does not apply to unproven parameter choices.

Our security analysis is essentially independent of the choice of polynomials in QUAD. [7, page 113, second paragraph] suggests, but neither quantifies nor justifies, a few "extra precautions" regarding "weak" choices of polynomials. Are some polynomial choices weaker than others? For example, one can save stream-generation time by reducing the density of the quadratic polynomials in

QUAD from $1/2$ to some small ε; what effect does this have on security? How much time can we save in our attacks as ε decreases? Exactly how small would ε have to be before the Raddum-Semaev attack [26] becomes a problem? There are many obvious directions for future security analysis.

2   The QUAD Family of Stream Ciphers

2.1   Definition of QUAD

To specify a particular QUAD stream cipher one must specify a prime power q; positive integers n and r; an "output filter" $\mathbf{P} : \mathrm{GF}(q)^n \to \mathrm{GF}(q)^r$ consisting of r quadratic polynomials $P_1, P_2, \ldots, P_r$ in n variables; and an "update function" $\mathbf{Q} : \mathrm{GF}(q)^n \to \mathrm{GF}(q)^n$ consisting of n quadratic polynomials $Q_1, Q_2, \ldots, Q_n$ in n variables. These quantities $q, n, r, \mathbf{P}, \mathbf{Q}$ are not meant to be secret; they can be published and standardized.

The QUAD cipher expands a secret initial state $\mathbf{x}_0 \in \mathrm{GF}(q)^n$ into a sequence of secret states $\mathbf{x}_0, \mathbf{x}_1, \mathbf{x}_2, \mathbf{x}_3, \ldots \in \mathrm{GF}(q)^n$ and a sequence of output vectors $\mathbf{y}_0, \mathbf{y}_1, \mathbf{y}_2, \mathbf{y}_3, \ldots \in \mathrm{GF}(q)^r$ as follows:

$$\mathbf{x}_0 \longrightarrow \mathbf{x}_1 = \mathbf{Q}(\mathbf{x}_0) \longrightarrow \mathbf{x}_2 = \mathbf{Q}(\mathbf{x}_1) \longrightarrow \mathbf{x}_3 = \mathbf{Q}(\mathbf{x}_2) \longrightarrow \cdots$$

$$\mathbf{y}_0 = \mathbf{P}(\mathbf{x}_0) \qquad \mathbf{y}_1 = \mathbf{P}(\mathbf{x}_1) \qquad \mathbf{y}_2 = \mathbf{P}(\mathbf{x}_2) \qquad \mathbf{y}_3 = \mathbf{P}(\mathbf{x}_3) \qquad \cdots$$

Typically q is a power of 2, allowing each output vector $\mathbf{y}_i \in \mathrm{GF}(q)^r$ to encrypt the next $r \lg q$ bits of plaintext in a straightforward way.

2.2   Parameter Restrictions

Can users expect QUAD to be secure no matter how the parameters are chosen? Certainly not. For example, QUAD$(2, 20, 20, \mathbf{P}, \mathbf{Q})$ has only 20 bits in its initial state $\mathbf{x}_0$, so the initial state can be discovered from some known plaintext by a brute-force search. As another example, QUAD$(2, 512, 512, 0, \mathbf{Q})$ is the "all-zero cipher," a silly stream cipher that always outputs 0. There are other, less obvious, attacks; users considering QUAD need to understand which parameters are broken or potentially so.

The QUAD paper [7, page 114] requires that the public polynomials $P_1, \ldots, P_r$ and $Q_1, \ldots, Q_n$ be "chosen randomly." We see conflicting statements regarding the distribution of these random variables: [7, page 112] defines "chosen randomly" as a uniform random choice (coefficients "uniformly and independently drawn"), but [7, page 113] suggests checking for and discarding some choices. Either way, one can reasonably conjecture that "bad" choices of $\mathbf{P}$ and $\mathbf{Q}$ have a negligible chance of occurring. There is no proof of this conjecture, and as mentioned in the introduction we would like to know exactly how $\mathbf{P}$ and $\mathbf{Q}$ affect security, but that is not the focus of this paper; we consider attacks that work for most choices of $\mathbf{P}$ and $\mathbf{Q}$.

2.3   Example: QUAD(256, 20, 20)

Choose $q = 256$, $n = 20$, and $r = 20$. Also choose 40 quadratic polynomials $P_1, \ldots, P_{20}, Q_1, \ldots, Q_{20}$ in 20 variables. These choices specify a particular QUAD stream cipher. The cipher starts with a secret 20-byte state $\mathbf{x}_0 \in GF(256)^{20}$; computes another secret 20-byte state $\mathbf{x}_1 = (Q_1(\mathbf{x}_0), \ldots, Q_{20}(\mathbf{x}_0))$; computes another secret 20-byte state $\mathbf{x}_2 = (Q_1(\mathbf{x}_1), \ldots, Q_{20}(\mathbf{x}_1))$; and so on. The cipher outputs $\mathbf{y}_0 = (P_1(\mathbf{x}_0), \ldots, P_{20}(\mathbf{x}_0))$; $\mathbf{y}_1 = (P_1(\mathbf{x}_1), \ldots, P_{20}(\mathbf{x}_1))$; $\mathbf{y}_2 = (P_1(\mathbf{x}_2), \ldots, P_{20}(\mathbf{x}_2))$; and so on.

We specifically warn against using QUAD(256, 20, 20). Given the public polynomials $P_1, \ldots, P_{20}, Q_1, \ldots, Q_{20}$ and the first 40 bytes $\mathbf{y}_0, \mathbf{y}_1$ of the output stream, our attacks compute the secret 20-byte initial state $\mathbf{x}_0$ in approximately 2^{66} cycles. See Section 4.2 for details.

2.4   Nonces

Old-fashioned stream ciphers, such as RC4, compute an output stream starting from a secret key. Modern stream ciphers, such as the eSTREAM submissions, compute an output stream starting from a secret key *and a nonce*, allowing the same secret key to be used for many separate output streams.

We have presented QUAD as an old-fashioned stream cipher. The QUAD paper actually presents a modern stream cipher. The modern stream cipher begins with an "initialization," converting a secret key and a nonce into a secret initial state $\mathbf{x}_0$. The modern stream cipher then generates output in exactly the way we have described. Details of the initialization are not relevant to this paper; our attacks recover $\mathbf{x}_0$ no matter how $\mathbf{x}_0$ was generated, breaking both the old-fashioned stream cipher and the modern stream cipher.

Stream ciphers can have "refresh" rules to disrupt the state after some given number of clocks. We do not concern ourselves with this either: initialization and refresh are both considered to be perfectly secure hereafter. For reference, [6] proves one particular setup procedure secure *with an extra loss of efficiency*.

3   How to Solve Multivariate Systems

Modern methods for system all descend spiritually from Buchberger's algorithm of finding Gröbner Bases [9], which is still widespread in symbolic mathematics packages both commercial and free (e.g., Maple and Singular). We present some state-of-the-art improvements below, not nearly as well known in general.

3.1   History of Lazard-Faugère Solvers: F$_4$, F$_5$, XL, XL2, FXL

Macaulay generalized Sylvester's matrix to multivariate polynomials [24]. The idea is to construct a matrix whose lines contain the multiples of the polynomials in the original system, the columns representing a basis of monomials up to a given degree. It was observed by D. Lazard [23] that for a large enough degree, ordering the columns according to a monomial ordering and performing row reduction without column pivoting on the matrix is equivalent to Buchberger's algorithm. In this correspondence, reductions to 0 correspond to lines that are linearly dependent upon the previous ones and the leading term of a polynomial is given by the leftmost nonzero entry in the corresponding line.

Lazard's idea was rediscovered in 1999 by Courtois, Klimov, Patarin, and Shamir [12] as **XL**. Courtois *et al* proposed several adjuncts [11,13,14] to XL. One tweak called XL2 merits a mention as an easy to understand precursor to F_4. Another of these proved to be a real improvement to F_4/F_5 as well as XL. This is FXL, where F means "fixing" (guessing at) variables.

Some time prior to this, J. -C. Faugère had proposed a much improved Gröbner bases algorithm called F_4 [17]. A later version, F_5 [18], made headlines [19] when it was used to solve HFE Challenge 1 in 2002. Commercially, F_4 is only implemented in the computer algebra system MAGMA [10].

3.2 Algorithm XL (eXtended Linearization) at Degree D

For the rest of this paper, we will denote the monomial $x_1^{b_1} x_2^{b_2} \cdots x_n^{b_n}$ by $\mathbf{x}^\mathbf{b}$, and its total degree $|\mathbf{b}| = b_1 + \cdots + b_n$. The set of degree-$D$-or-lower monomials is denoted $\mathcal{T} = \mathcal{T}^{(D)} = \{\mathbf{x}^\mathbf{b} : |\mathbf{b}| \leq D\}$. $|\mathcal{T}|$ is the number of degree $\leq D$ monomials and denoted $T^{(D)} = T$.

Start by multiplying each equation p_i, $i = 1 \cdots m$ by all monomials $\mathbf{x}^\mathbf{b} \in \mathcal{T}^{(D-2)}$. Reduce as a linear system of the equations $\mathcal{R}^{(D)} = \{\mathbf{x}^\mathbf{b} p_j(\mathbf{x}) = 0 : 1 \leq j \leq m, |\mathbf{b}| \leq D - 2\}$, with the monomials $\mathbf{x}^\mathbf{b} \in \mathcal{T}^{(D)}$ as independent variables. Repeat with higher D until we have a solution, a contradiction, or reduce the system to a univariate equation in some variable. The number of equations and independent equations are denoted $R^{(D)} = R = |\mathcal{R}|$ and $I^{(D)} = I = \dim(\mathrm{span}\mathcal{R})$.

If we accept solutions in arbitrary extensions of $K = \mathrm{GF}(q)$, then $T = \binom{n+D}{D}$ regardless of q. However, most crypto applications require solutions in $\mathrm{GF}(q)$ only. The above expression for T then only holds for large q, since we may identify x_i^q with x_i and cut substantially the number of monomials we need to manage. This "Reduced XL" (cf. C. Diem [15]) can lead to extreme savings compared to "Original XL," e.g., if $q = 2$, then $T = \sum_{j=0}^{D} \binom{n}{j}$.

Proposition 3.3 ([3,30]). *The number of monomials is* $T = [t^D] \dfrac{(1 - t^q)^n}{(1 - t)^{n+1}}$ *which reduces to* $\binom{n+D}{D}$ *when q is large. We can then find* $R = R^{(D)} = mT^{(D-2)}$.

We note that the XL of [12,13] terminates more or less reliably when $T - I \leq \min(D, q - 1)$, but sparse matrix computation is only possible when $T - I \leq 1$ [29]. Further, Lazard-Faugère methods work for equations of any degree [4,30]. If $\deg(p_i) = d$, we will only multiply the equation p_i with monomials up to degree $D - d$ in generating $\mathcal{R}^{(D)}$. The principal result is:

Proposition 3.4 ([30, Theorem 7]). *If the equations p_i, with $\deg p_i := d_i$, and the relations $\mathcal{R}^{(D)}$ has no other dependencies than the obvious* *ones generated by* $p_i p_j = p_j p_i$ *and* $p_i^q = p_i$, *then*

$$T - I = [t^D]\, G(t) = [t^D]\, \frac{(1 - t^q)^n}{(1 - t)^{n+1}} \prod_{j=1}^{m} \left(\frac{1 - t^{d_j}}{1 - t^{q\,d_j}} \right). \tag{1}$$

After a certain degree D_{XL}, called the degree of regularity for XL, such that $D_{XL} := \min\{D : [t^D]\, G(t) \leq 0\}$ is the smallest D such that Eq. 1 cannot hold if the system has a solution, because the right hand side of Eq. 1 goes nonpositive. If the boldfaced condition above holds for as long as possible (which means for degrees up to D_{XL}), we say that the system is K-**semi-regular** or q-**semi-regular** (cf. [3,30]). Diem proves [15] for char 0 fields, and conjectures for all K that (i) a generic system (no algebraic relationship betweem the coefficients) is K-semi-regular and (b) if $(p_i)_{i=1\cdots m}$ are *not* K-semi-regular, I can only decrease from the Eq. 1 prediction. Most experts [15] believe the conjecture that a *random* system behaves like a generic system with probability close to 1.

Corollary 3.5. $T - I = [t^D]\left((1 - t)^{-n-1} \prod_{j=1}^{m} \left(1 - t^{d_i} \right) \right)$ *for generic equations if* $D \leq \min(q, D_{XL}^{\infty})$, *where* D_{XL}^{∞} *is the degree of the lowest term with a non-positive coefficient in* $G(t) = \left((1 - t)^{-n-1} \prod_{j=1}^{m} \left(1 - t^{d_i} \right) \right)$.

We would note that (F)XL is a solver and not a true Gröbner basis method as $\mathbf{F_4}$ and $\mathbf{F_5}$ is. However, the analysis much parallels that of $\mathbf{F_4}$-$\mathbf{F_5}$ by Dr. Faugère et al, hence the categorical name "Lazard-Faugère" solvers.

3.6   XL2, $\mathbf{F_4}$ and $\mathbf{F_5}$

XL2 [13] is a tweak of XL as follows: Tag each equation with its maximal degree. Run an elimination on the system with monomials in degree-reverse-lex. In the remaining (row echelon form) system, multiply by each variable $x_1, x_2 \cdots$ *all remaining equations with the maximum tagged degree* and eliminate again. When we cannot eliminate all remaining monomials of the maximum degree, increment the operating degree and reallocate more memory.

XL+XL2 can be considered a primitive or inferior matrix form of $\mathbf{F_4}$ or $\mathbf{F_5}$ [1]. $\mathbf{F_4}$ inserts elimination between expansion stages, which compresses the number of rows that needs to be handled. $\mathbf{F_5}$ is a further refinement of $\mathbf{F_4}$. The set of equations is actually generated one by one (or the matrix row by row). In the process, an algebraic criterion is used to determine, ahead of an elimination process, whether a row will be reduced to zero or not and only the meaningful rows are retained. A complication resulting from the tagging is that the elimination must be done in a strictly ordered way. This corresponds in the matrix form to no row exchanges in a Gaussian. There are two separate degrees in $\mathbf{F_4}/\mathbf{F_5}$, an apparent operating degree D_{F4} and a higher intrinsic degree equal to that of the equivalent XL system. For the full power of $\mathbf{F_4}$ or $\mathbf{F_5}$, auxillary algorithms such as FGLM are needed. See [17,18] for complete details.

Proposition 3.7 ([3]). *If the eqs. p_i are q-semi-regular, at the operating degree*

$$D_{reg} := \min \left\{ D : [t^D] \frac{(1-t^q)^n}{(1-t)^n} \prod_{i=1}^{m} \left(\frac{1-t^{d_i}}{1-t^{qd_i}} \right) < 0 \right\}$$

both $\mathbf{F_4}$-$\mathbf{F_5}$ *will terminate. Note that by specializing to a large field, we find*

$$D_{reg}^{\infty} := \min \left\{ D : [t^D] (1-t)^{-n} \prod_{i=1}^{m} (1-t^{d_i}) < 0 \right\}. \tag{2}$$

If we compare this formula with Cor. 3.5, we see that the only difference is a substitution of n for $n+1$. In other words, we are effectively running with one fewer variable in the large field case. This explains why $\mathbf{F_4}$-$\mathbf{F_5}$ can be much faster than XL. However, the savings is smaller over small fields like GF(2), and even for large fields, removing one variable may not be enough of a savings, because the systems that we aim to solve will spawn millions of monomials (variables). Eliminating in the usual way means that we will run out of memory before time.

According to the description we received from the MAGMA project and Dr. Faugère, even though memory management is very critical, elimination is still relatively straightforward in current implementations of $\mathbf{F_4}$-$\mathbf{F_5}$, and in the process we see reasonably dense matrices, not extremely sparse ones. All said, $\mathbf{F_4}$-$\mathbf{F_5}$ are still the most sophisticated general system-solving algorithms today.

3.8   Practicalities: XL with Wiedemann, and Ramifications

Table 1 lists our tests in solving generic equations with $\mathbf{F_4}$ over GF(256). That we only have 2GB main memory is not critical to our inability to solve equations in the realm of 20 byte-sized variable in as many equations, because we also ran into a SIGMEM (out of memory) error with a 15-variable, 15-equation system on a 16 GB RAM system.

Table 1. System-solving time (sec): MAGMA 2.12, 2GB RAM, Athlon64x2 2.2GHz

$m-n$	D_{XL}	D_{reg}	$n=9$	$n=10$	$n=11$	$n=12$	$n=13$
0	2^m	m	6.090	46.770	350.530	3322.630	sigmem
1	m	$\lceil \frac{m+1}{2} \rceil$	1.240	8.970	53.730	413.780	2538.870
2	$\lceil \frac{m+1}{2} \rceil$	$\lceil \frac{m+2-\sqrt{m+2}}{2} \rceil$	0.320	2.230	12.450	88.180	436.600

Conclusion: We will run out of memory using any Lazard-Faugère solver, including $\mathbf{F_4}$-$\mathbf{F_5}$, if our equation-solving is not tailored to sparse matrices. *Many authors used $T^{2.8}$ (or $T^{2.376}$!) as the cost function. This is why we don't.*

Example 1. If we try to attack QUAD directly using $\mathbf{F_4}$-$\mathbf{F_5}$, we will be solving 20 variables from 20 quadratic equations and 20 quartic equations over (say

GF(256)). We will run into 6906900 monomials at degree 9 with (eventually) a fairly dense matrix. This is not feasible.

We must take advantage of the sparsity of Macaulay matrix. Solving the same system via XL with a Krylov subspace method means that we will run into 30045105 monomials at degree 10, but with a sparse matrix of only 200 or so entries per row with 5 bytes per entry. The whole thing can fit in a single machine with 32GB or (much easier) distributed among a cluster.

Three well-known methods (all utilizing the existence of Krylov subspaces) adapt well to sparse matrices: Conjugate Gradient, Lanczos, and Wiedemann. There are two reasons to prefer Wiedemann. While Lanczos and CG usually takes about N (as opposed to $3N$) multiplications by the matrix A, they are restricted to symmetric matrices. Furthermore, Wiedemann has no worries about the self-orthogonality issue and is easier to program.

Algorithm: XL (with homogenous Wiedemann)

1. Create the extended Macaulay matrix of the system to a certain degree D_{XL}.
2. Randomly delete some rows then add some columns to form a square system, $A\mathbf{x} = 0$ where $\dim A = \beta T + (1 - \beta)R$. Usually we can succeed with $\beta = 1$.
3. Apply the homogeneous version of Wiedemann's method to solve for $\mathbf{x}$:

   (a) Set $k = 0$ and $g_0(z) = 1$, and take a random $\mathbf{b}$.
   (b) Choose a random $\mathbf{u}_{k+1}$ [usually the $(k + 1)$-st unit vector].
   (c) Find the sequence $\mathbf{u}_{k+1}A^i\mathbf{b}$ starting from $i = 0$ and going up to $2N - 1$.
   (d) Apply g_k as a difference operator to this sequence, and run the Berlekamp-Massey over GF(q) on the result to find the minimal polynomial f_{k+1}.
   (e) Set $g_{k+1} := f_{k+1}g_k$ and $k := k + 1$. If $\deg(g_k) < N$ and $k < n$, go to (b).

4. Compute the solution $\mathbf{x}$ using the minpoly $f(z) = g_k(z) = c_m z^m + c_{m-1}z^{m-1} + \cdots + c_\ell z^\ell$: Take another random $\mathbf{b}$. Start from $\mathbf{x} = (c_m A^{m-\ell} + c_{m-1}A^{m-\ell-1} + \cdots + c_\ell 1)\mathbf{b}$, continuing to multiply by A until we find a solution to $Ax = 0$.
5. In the event that the random choice in Step 2 went awry and we dropped an essential equation, or if the system had more than one solution to begin with, the nullity ℓ will be more than 1, and we will have to repeat the check below at every point of a linear subspace (q points).
6. Obtain the solution from the last few elements of $\mathbf{x}$ and check its correctness.

Proposition 3.9 (cf. [4,30]). *The expected running time of XL is roughly $C_{XL} \sim 3\tau T\mathfrak{m}$ where $\tau = kT$ is the total (and k the average) number of terms in an equation, one multiplication $\mathfrak{m} \approx (c_0 + c_1 \lg T)$ cycles, and c_0, c_1 depend on the architecture. E.g. on x86, when $q = 256$, $T < 2^{24}$, each multiplication cost about 14 cycles on a P4 (3 consecutively dependent loads from L1 cache plus change). On a K8 or P-M/Core, it takes 11 cycles on the same serial code, but in x86-64 mode, some loop-unrolling can get it down to about 8.*

4   Broken Parameters for QUAD

4.1   Overview: Using Algebraic Attacks Against QUAD

The QUAD paper [7, Section 5] contains the following statement regarding algebraic attacks: "QUAD was designed to resist algebraic attack techniques. As a matter of fact, the key and IV loading and keystream generation mechanisms of QUAD are based upon the iteration of quadratic systems whose associated equations are conjectured to be computationally impossible to solve."

Can we try to solve $\mathbf{P}(\mathbf{x}_0) = \mathbf{y}_0$ directly? Yes, but even for $q = 256$, $r = 20$, that takes about 2^{80} cycles [29]. We propose instead to solve the equations $\mathbf{P}(\mathbf{x}_0) = \mathbf{y}_0$ and $\mathbf{P}(\mathbf{Q}(\mathbf{x}_0)) = \mathbf{y}_1$ for the initial state $\mathbf{x}_0$ using (F)XL-Wiedemann.

The attacker is given the quadratic polynomials $\mathbf{P}$ and $\mathbf{Q}$, which presumably are public, and the two initial stream-cipher outputs $\mathbf{y}_0, \mathbf{y}_1$, for example from a successful guess regarding the initial bytes of plaintext. The unknown state $\mathbf{x}_0$ consists of n variables from $\mathrm{GF}(q)$. The equation $\mathbf{P}(\mathbf{x}_0) = \mathbf{y}_0$ consists of r quadratic equations in those n variables. The equation $\mathbf{P}(\mathbf{Q}(\mathbf{x}_0)) = \mathbf{y}_1$ consists of r quartic equations in those n variables. The attack solves the combined quadratic-quartic system to find $\mathbf{x}_0$. At this point the attacker can compute the subsequent stream-cipher output $\mathbf{y}_2, \mathbf{y}_3, \ldots$.

Note: we can also frame the attack as *given any consecutive stream-cipher output blocks* $\mathbf{y}_i$, $\mathbf{y}_{i+1}$, *compute the state* $\mathbf{x}_i$ *and all subsequent state and output blocks*. This means that known-plaintext attacks will be particularly effective.

The cost of this attack is, aside from negligible setup costs, the cost of solving r quadratic equations and r quartic equations in n variables over $\mathrm{GF}(q)$, which in turn is dominated by solving a large sparse matrix equation. In the rest of this section we consider the cost of (F)XL-Wiedemann for various choices of (q, n, r). In particular, we show that the attack is feasible against QUAD$(256, 20, 20)$.

4.2   Example: Breaking QUAD(256, 20, 20)

Proposition 4.3. *For $q > 10$, a q-semiregular system of 20 quadratics and 20 quartics can be solved over $\mathrm{GF}(q)$ in no more than 2^{63} $\mathrm{GF}(q)$-multiplications (for $q = 256$, about 2^{66} cycles).*

Proof. The minimum degree to find a nonpositive coefficient in $(1 - t)^{-21}(1 - t^2)^{20}(1 - t^4)^{20}$ is $D_{XL} = 10$, we have $T = \binom{30}{10} = 30045015$, the number of initial equations is $R = 20 \times \binom{28}{8} + 20 \times \binom{26}{6} = 66766700$, and the total number of terms in those equations is $\tau = 20\binom{28}{8}\binom{22}{2} + 20\binom{26}{6}\binom{24}{4} = 63287924700$. Hence the number of multiplications needed is bounded by $3T\tau \lesssim 2^{63}$, or about 2^{66} cycles. **If we can cut down to a $T \times T$ system (which usually succeeds), then it takes $3T(\tau T/R) \sim 2^{60}$ multiplications, or about 2^{63} cycles.**

In contrast, QUAD$(16, 40, 40)$ is not directly breakable by this attack. Here, the first non-positive coefficient of $(1 - t)^{-41}(1 - t^2)^{40}(1 - t^4)^{40}$ happens at $D_{XL} = 14$. So for $q > 14$, $T = 3245372870670$. We find solving a q-semiregular system of 40 quadratics and 40 quartics over $GF(q)$ XL-Wiedemann to take $\lesssim 2^{95}m$ (about 2^{100} cycles). Guessing variables may cut the memory requirement and aid parallelization but does not decrease the number of multiplications. This means that $QUAD(q, 40, 40)$ is considerably below a 128-bit security level, but we know of no attack below the 80-bit security level considered in [7].

Observation: Unless it is possible to run the elimination in $\mathbf{F_4}$-$\mathbf{F_5}$ with a speed that matches the sparse matrix solvers, these more sophisticated methods would be dragged down by the amount of memory and memory operations required. E.g., against $QUAD(256, 20, 20)$, even though we are operating $\mathbf{F_5}$ with a lower $D_{F5} = 9$. $T_{F5} = 6906900$, and there is just too much memory needed.

Conjecture: For generic (i.e., most random) $\mathbf{P}, \mathbf{Q}$ and $\mathbf{y_0}, \mathbf{y_1}$, $\mathbf{P(x)} = \mathbf{y_0}$ and $\mathbf{P(Q(x))} = \mathbf{y_1}$ is 256-semiregular.

We cannot produce a hard proof, but we include a set of our test results (all using i386 code on a P4) as Table 2. The timings for $n = 6$ all the way up to $n = 15$ are consistent with our theoretical predictions. Furthermore, the timings for a $QUAD(256, n, n)$ attack are identical to the timings for n randomly generated quadratics plus n randomly generated quartics in n variables.

Table 2. Direct XL-Wiedemann attack on $QUAD(256, n, n)$, MS C++ 7; P-D 3.0GHz, 2GB DDR2-533

n	6	7	8	9	10	11	12	13	14	15
D	6	6	7	7	7	7	8	8	8	8
C_{XL}	1.22	4.49	$6.08 \cdot 10$	$2.29 \cdot 10^2$	$7.55 \cdot 10^2$	$2.30 \cdot 10^3$	$5.12 \cdot 10^4$	$1.54 \cdot 10^5$	$4.39 \cdot 10^5$	$1.17 \cdot 10^6$
lgC_{XL}	$2.85 \cdot 10^{-1}$	2.17	5.92	7.84	9.56	$1.12 \cdot 10$	$1.56 \cdot 10$	$1.72 \cdot 10$	$1.87 \cdot 10$	$2.02 \cdot 10$
T	$9.24 \cdot 10^2$	$1.72 \cdot 10^3$	$6.44 \cdot 10^3$	$1.14 \cdot 10^4$	$1.94 \cdot 10^4$	$3.28 \cdot 10^4$	$1.26 \cdot 10^5$	$2.03 \cdot 10^5$	$3.20 \cdot 10^5$	$4.90 \cdot 10^5$
aTm	49	65	96	120	147	177	245	288	335	385
clks	28.8	23.5	15.3	14.6	13.6	12.1	13.1	12.9	12.8	12.7

T: #monomials, aTm: average terms in a row, clks: number of clocks per multiplication.

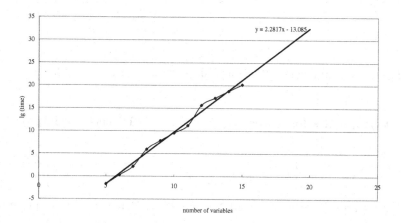

Fig. 1. Logarithmic growth of direct QUAD attack time over GF(256), optimal D

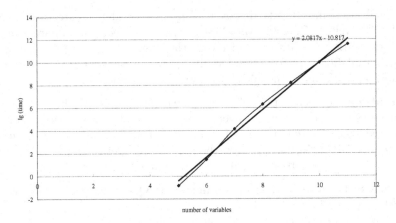

Fig. 2. Logarithmic growth of direct QUAD attack time over GF(256), $D = 7$

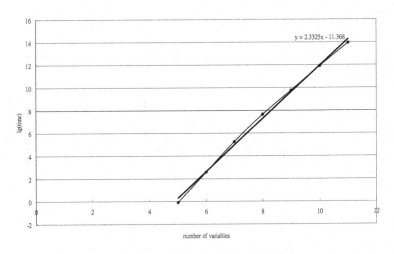

Fig. 3. Logarithmic growth of direct QUAD attack time over GF(256), $D = 8$

4.3   Asymptotics for Attacking QUAD(q, n, n) Directly (q large)

$$D_{XL} = \min \left\{ D : [t^D] \; \frac{1}{(1-t)^{n+1}} \; \left((1-t^2)(1-t^4) \right)^n < 0 \right\},$$

gives the degree D of the eventual XL operation. Here, $D = (w + o(1))n$, and $w \approx$ smallest positive zero w of $\oint \frac{(1-z^2)^n (1-z^4)^n}{(1-z)^{n+1} z^{wn+1}} dz = \oint \frac{dz}{z(1-z)} \left(\frac{(1+z)(1-z^4)}{z^w} \right)^n$ (cf. [28]). As usual in such situations, the expression on the RHS can only vanish when the saddle point equation of this integral has double roots (a "monkey saddle"), and here the saddle point equation is $(w - 5)z^4 + z^3 - z^2 + z - w = 0$

which has a double root when w is very close to 0.2 (actually ≈ 0.200157957). So $\lg T/n \to (1 + w) \lg(1 + w) - w \lg w \approx 0.78$. Assuming Wiedemann solving to continue without a problem, this means that the system will be solvable in $2^{1.56n+o(n)} \times$ (polynomial in n) asymptotically.

Suppose we use, in addition to the degree-2 and degree-4 equations, the degree-8 equations $\mathbf{P}(\mathbf{Q}(\mathbf{Q}(\mathbf{x}_0))) = \mathbf{y}_2$. This reduces the final degree to

$$D_{XL} = \min\left\{D : [t^D] \frac{1}{(1-t)^{n+1}} \left((1-t^2)(1-t^4)(1-t^8)\right)^n < 0\right\},$$

differentiating with respect to z and then finding the root of the discriminant gets us $w \approx 0.1991777370$ which confirms our suspicion that higher-degree equations provide very little improvement in the cost of the attack for large q.

4.4   Asymptotics for Attacking QUAD$(2, n, n)$ Directly

For $q = 2$ there is noticeably more benefit from using higher-degree equations and from guessing variables. We combine n quadratic equations $\mathbf{P}(\mathbf{x}_0) = \mathbf{y}_0$, n quartic equations $\mathbf{P}(\mathbf{Q}(\mathbf{x}_0)) = \mathbf{y}_1$, and so on through n equations $\mathbf{P}(\mathbf{Q}^\ell(\mathbf{x}_0)) = \mathbf{y}_\ell$ of degree 2^ℓ. The degree of operation of XL is

$$D_{XL} \gtrsim \min\left\{D : [t^D] \frac{(1+t)^n}{(1-t)} \left((1+t^2)(1+t^4)\cdots(1+t^{2^\ell})\right)^{-n} < 0\right\},$$

where equality holds for 2-semi-regular systems, and which for $\ell \to \infty$ becomes $D_{XL} = \min\left\{D : [t^D] (1-t)^{n-1}(1+t)^{2n} < 0\right\}$. If we guess at f variables, then we have $D_{FXL} = \min\left\{D : [t^D] (1-t)^{n-1}(1+t)^{2n-f} < 0\right\}$. Let $c = f/n$. We would like to find a c for which the running time is smallest.

Proposition 4.5 ([4]). *We have the following asymptotics*

1. *If $f/m \to \alpha$ as $f, m \to \infty$, then*

$$D_{reg} = \min\left\{D : [t^D] (1-t)^f(1+t)^m < 0\right\} \sim \left(\frac{1}{2} - \sqrt{\alpha} + \frac{\alpha}{2}\right) m + o(m^{1/3}).$$

2. *If $D/n \to w$, then $\lg T/n \to -w \lg w - (1-w) \lg(1-w)$.*

Using the above, we see that $D_{FXL} \sim \left(\frac{1}{2} - \sqrt{2-c} + \frac{2-c}{2}\right) n$, hence for XL-Wiedemann, as $n \to \infty$, the limiting exponential factor $\dfrac{\lg C_{FXL}}{n} \to$

$$2\left((2-c)\lg(2-c) - (\tfrac{3-c}{2} - \sqrt{2-c})\lg(\tfrac{3-c}{2} - \sqrt{2-c}) - (\tfrac{1-c}{2} + \sqrt{2-c})\lg(\tfrac{1-c}{2} + \sqrt{2-c})\right) + c.$$

Due to constraints in asymptotic analysis which are too lengthy to discuss here, this expression is only guaranteed to hold for small c and may not continue to hold as c goes up. But it is clear that asymptotically, FXL is the correct approach here. A rough estimate is that the security under the direct attack is $2^{1.02n}$ with XL, and $2^{0.86n}$ with best FXL.

4.6   Which Field Is Best?

Our analysis suggests that after taking into account all known improvements, to achieve any particular level of security against known attacks, $q = 2$ needs considerably more variables than a large q, the ratio being about $1.56/0.86 \approx 1.81$. The QUAD user can try to save time by increasing q and reducing the number of variables. Our current impression is that $q = 256$ is not a good choice: the extra cost of arithmetic in $GF(q)$ offsets the smaller number of variables and polynomial coefficients to be manipulated. However, $q = 16$ or $q = 4$ may be better. These estimates will have to be updated if faster attacks are discovered.

4.7   The State of System-Solving Cryptanalysis

We have shown that Lazard-Faugère solvers, particularly XL with a sparse matrix solver, can be very useful in attacking generic systems, or at least systems that have generic or randomly created elements. In some cases such simple approaches are more useful than the really sophisticated variations.

There is a problem we ran into when trying to organize a medium-scale parallel cluster to help crack a QUAD instance. It is desirable to partition the matrix because memory capacity is always a problem. Yet in Block Lanczos and Block Wiedemann each computer must hold a copy of the entire matrix. If we do the naive parallelization, then the communications cost goes up squarely as the cluster size but the efficiency gain only linearly. This will take a good tuning.

5   Unprovable Parameters for QUAD

5.1   Overview: Attacking the Underlying Hard Problem

The QUAD paper [7, page 110] states that the security of QUAD "is provably reducible to the intractability of the MQ problem [15], which consists of finding a solution (if any) to a multivariate quadratic system of m quadratic equations in n variables over a finite field $GF(q)$, typically $GF(2)$." Here $m = n + r$. The quoted statement means that there is a proof that *assumes* that $\mathcal{MQ}$ is intractable and *concludes* that QUAD is secure.

We emphasize that the intractability of the $\mathcal{MQ}$ problem is not a theorem; it is a hypothesis. For some parameter choices, the hypothesis is plausible. For other parameter choices, the hypothesis is false. For example, we show in this section how to break 80 quadratic equations in 40 variables over $GF(16)$.

For parameters where the associated $\mathcal{MQ}$ instance is shown to be not "hard," any attempt to show provable security starting from the hardness of the $\mathcal{MQ}$ problem becomes hopeless; hence our terminology "unprovable." Note that unprovable parameters can be analyzed even before a proof has been claimed or written down; unprovability is deduced purely from the parameter choices for the underlying "hard problem."

5.2 Example: QUAD(256, 20, 20)

We return to QUAD(256, 20, 20) as an illustrative example. What the quoted statement says in this case is that the security of QUAD(256, 20, 20) follows from the difficulty of solving 40 quadratic equations in 20 variables over GF(256).

Is solving 40 random quadratic equations in 20 variables over GF(256) actually difficult? No, not at all! We can compute from Prop. 3.5 that $D_{XL} = 5$ and $T = 53130$. The maximum number of terms per equation is $k \lesssim 231$, so on a P4, $C_{XL} \approx 9 \times 10^{12} \lesssim 2^{45}$. To summarize: XL-Wiedemann takes $< 2^{45}$ cycles, which is only a few hours on a decent computer.

To verify our estimates, we carried out this computation the same evening that we began to study the security of QUAD. XL-Wiedemann solved the equations on schedule. The fairly mature $\mathbf{F_4}$ in MAGMA version 2.12 did the same job in about a quarter of the time, presumably using more memory.

Of course, after establishing the breakability of QUAD(256, 20, 20) in Section 4, we could apply the contrapositive of the "provable security" of QUAD to deduce the breakability of the corresponding $\mathcal{MQ}$ system. However, this circuitous approach would be considerably harder to verify than a direct attack on $\mathcal{MQ}$. We also observe a significant gap between the cost of the $\mathcal{MQ}$ attack and the cost of the QUAD attack. Showing unprovability of a set of QUAD parameters is easier than showing breakability.

5.3 Example: QUAD(16, 40, 40)

For QUAD(16, 40, 40), we can verify (cf. Prop. 3.5) $D = 8$, $T = 377348994$, and $k \lesssim 861$. Since the storage will come to about 800 GB among a small cluster, we will assume AMD64 or a similar architecture, and it will take about 8 cycles a multiplication before tuning, and a total of $3 \times 8 \times 377348994^2 \times 861 \approx 3 \times 10^{20} \lesssim 2^{72}$ cycles. With a little work, such as unrolling by hand the critical loop, we can expect to cut this down to 2^{71} on an Opteron.

These attacks are considerably below the specified 2^{80} security level. We conclude that QUAD(16, 40, 40) is unprovable.

5.4 Examples: QUAD(2, 160, 160), (2, 256, 256), (2, 350, 350)

For QUAD(2, 160, 160): $D = 13$ (this time cf. Prop. 3.4), but this time $T = 4801989157032669149$, and $k \lesssim 12881$ which makes it rather difficult to fit everything into memory. An optimistic 6-cycles-per-multiplication hypothesis, ignoring space problems, leads to an estimate of approximately 2^{140} cycles.

For QUAD(2, 256, 256): $D = 19$, $T = 257079680477996666061215057489$ or $\lesssim 2^{94}$. $C_{XL} \approx 2^{205}$m. For QUAD (2,350,350), $D = 24$, $T \lesssim 2^{123}$. $C_{XL} \lesssim 2^{263}$m. [In the astronomical realm, a multiplication may take rather more than 6 cycles.]

6 Proven and Unproven Parameters for QUAD

There are several limitations on what was actually proven in [7]. We focus on one limitation, namely the lack of tightness in [7, Theorem 4].

The theorem does not claim that a QUAD attack implies an $\mathcal{MQ}$ attack *with the same efficiency*. It says that a QUAD attack implies an $\mathcal{MQ}$ attack *with a bounded loss of efficiency*: specifically, if λn bits of output from QUAD$(2, n, r)$ can be distinguished from uniform with advantage ϵ in time T, then a random $\mathcal{MQ}$ system of $n + r$ equations in n variables over GF(2) can be solved with probability $2^{-3}\epsilon/\lambda$ in time

$$T' \leq \frac{2^7 n^2 \lambda^2}{\epsilon^2} \left(T + (\lambda + 2)T_S + \log\left(\frac{2^7 n \lambda^2}{\epsilon^2}\right) + 2 \right) + \frac{2^7 n \lambda^2}{\epsilon^2} T_S,$$

where $T_S :=$ time to run one block of QUAD$(2, n, r)$.

How could we conclude the security of QUAD—a lower bound on T, such as $T \geq 2^{80}$—from this theorem? A minor point is that the theorem would need to be extended to all q, not just $q = 2$, and proven; we will make the (questionable) assumption that this is done, and that it does not produce worse time bounds. The major point is that we would need to assume a much larger lower bound on T', compensating for (among other things) a factor of $2^{10} n^2 \lambda^3 / \epsilon^3$. For example, as in [7] let's accept ϵ as large as 0.01, and let's put $L = \lambda n = 2^{40}$. The extra factor is then $2^{150}/n$. The theorem cannot conclude $T \geq 2^{80}$ without assuming that $T' \geq 2^{230}/n$. Let's check this against cases studied in Section 5:

- QUAD$(256, 20, 20)$: Unproven. Our (computer-verified) estimate is $T' \leq 2^{45}$.
- QUAD$(16, 40, 40)$: Unproven, we estimate $T' \leq 2^{71}$.
- QUAD$(2, 350, 350)$: *Proven*, if there are no better $\mathcal{MQ}$ attacks. We estimate $T' \approx 2^{263}$m. A 2^{80} distinguishing attack would lead to an $\approx 2^{221}$m expected-time solution. Note $T' \approx 2^{222}$m for QUAD$(2, 320, 320)$, which should suffice.
- QUAD$(2, 160, 160)$: Unproven ($T' \leq 2^{140}$, [7] specified this as unproven).
- QUAD$(2, 256, 256)$: *Proven* for the parameters [7] $L = 2^{22}$, $\epsilon = 0.01$, if there are no better $\mathcal{MQ}$ attacks. We estimate $T' \approx 2^{205}$m where we need 2^{168}m.

Why did [7] overestimate n? The $N^{2.376}$ *formula discounts an expected big coefficient, compensating for our improvement to* $N^{2+\epsilon}$ *and then some.*

The bottom line is that *all three* QUAD parameter choices used for speed reports in the QUAD talk at Eurocrypt 2006 are unproven, although the gap in the QUAD$(2, 160, 160)$ case might be closed by a tighter security proof.

Acknowledgement

The authors would all like to thank the TWISC project (Taiwan Information Security Center, NSC95-2218-E-001-001) for sponsoring a series of lectures on cryptology at its Nat'l Taiwan U. of Sci. and Tech. location (NSC95-2218-E-011-015) in 2006, the discussions following which led to this work. BY would like to thank the Taiwan's National Science Council for support under project 95-2115-M-001-021 and indirectly via TWISC. JMC was partially supported by TWISC at NCKU (NSC94-3114-P-006-001-Y). Date of this document: 2007.04.18.

References

1. Ars, G., Faugère, J.-C., Imai, H., Kawazoe, M., Sugita, M.: Comparison between XL and Gröbner Basis algorithms. In: AsiaCrypt [25], pp. 338–353.
2. Bardet, M.: Étude des systèmes algébriques surdétermininés. Applications aux codes correcteurs et á la cryptographie. PhD thesis, Université Paris VI (2004)
3. Bardet, M., Faugère, J.-C., Salvy, B.: On the complexity of gröbner basis computation of semi-regular overdetermined algebraic equations. In: Proceedings of the International Conference on Polynomial System Solving, pp. 71–74 (2004)
4. Bardet, M., Faugère, J.-C., Salvy, B., Yang, B.-Y.: Asymptotic expansion of the degree of regularity for semi-regular systems of equations. In: Gianni, P. (ed.) MEGA 2005 Sardinia, Italy (2005)
5. Berbain, C., Billet, O., Gilbert, H.: Efficient implementations of multivariate quadratic systems. In: Workshop record distributed at 13th Annual Workshop on Selected Areas in Cryptography (August 2006)
6. Berbain, C., Gilbert, H.: On the security of IV dependent stream ciphers. In: Workshop record distributed at 14th Annual Workshop on Fast Software Encryption (March 2007)
7. Berbain, C., Gilbert, H., Patarin, J.: QUAD: A practical stream cipher with provable security. In: Vaudenay, S. (ed.) EUROCRYPT 2006. LNCS, vol. 4004, pp. 109–128. Springer, Heidelberg (2006)
8. Blum, L., Blum, M., Shub, M.: Comparison of two pseudo-random number generators, pp. 61–78. Plenum Press, New York (1983)
9. Buchberger, B.: Ein Algorithmus zum Auffinden der Basiselemente des Restklassenringes nach einem nulldimensionalen Polynomideal. PhD thesis, Innsbruck (1965)
10. Computational Algebra Group, University of Sydney. The MAGMA Computational Algebra System for Algebra, Number Theory and Geometry, http://magma.maths.usyd.edu.au/magma/
11. Courtois, N.: Algebraic attacks over $GF(2^k)$, application to HFE challenge 2 and Sflash-v2. In: Bao, F., Deng, R.H., Zhou, J. (eds.) PKC 2004. LNCS, vol. 2947, pp. 201–217. Springer, Heidelberg (2004)
12. Courtois, N.T., Klimov, A., Patarin, J., Shamir, A.: Efficient algorithms for solving overdefined systems of multivariate polynomial equations. In: Preneel, B. (ed.) EUROCRYPT 2000. LNCS, vol. 1807, pp. 392–407. Springer, Heidelberg (2000), http://www.minrank.org/xlfull.pdf
13. Courtois, N.T., Patarin, J.: About the xl algorithm over gf(2). In: Joye, M. (ed.) CT-RSA 2003. LNCS, vol. 2612, pp. 141–157. Springer, Heidelberg (2003)
14. Courtois, N.T., Pieprzyk, J.: Cryptanalysis of block ciphers with overdefined systems of equations. In: Zheng, Y. (ed.) ASIACRYPT 2002. LNCS, vol. 2501, pp. 267–287. Springer, Heidelberg (2002)
15. Diem, C.: The xl-algorithm and a conjecture from commutative algebra. In AsiaCrypt [25], pp. 323–337
16. Ding, J., Gower, J., Schmidt, D.: Multivariate Public-Key Cryptosystems. Advances in Information Security. Springer, Heidelberg (2006)
17. Faugère, J.-C.: A new efficient algorithm for computing Gröbner bases (F_4). Journal of Pure and Applied Algebra 139, 61–88 (1999)
18. Faugère, J.-C.: A new efficient algorithm for computing Gröbner bases without reduction to zero (F_5). In: International Symposium on Symbolic and Algebraic Computation — ISSAC 2002, pp. 75–83. ACM Press, New York (2002)

19. Faugère, J.-C., Joux, A.: Algebraic cryptanalysis of Hidden Field Equations (HFE) using Gröbner bases. In: Boneh, D. (ed.) CRYPTO 2003. LNCS, vol. 2729, pp. 44–60. Springer, Heidelberg (2003)
20. Garey, M.R., Johnson, D.S.: Computers and Intractability — A Guide to the Theory of NP-Completeness. In: W.H. Freeman and Company (1979) ISBN 0-7167-1044-7 or 0-7167-1045-5.
21. Koblitz, N., Menezes, A.: Another look at "provable security". Cryptology ePrint Archive, Report 2004/152 (2004), http://eprint.iacr.org/
22. Koblitz, N., Menezes, A.: Another look at "provable security". ii. In: Barua, R., Lange, T. (eds.) INDOCRYPT 2006. LNCS, vol. 4329, pp. 148–175. Springer, Heidelberg (2006)
23. Lazard, D.: Gröbner-bases, Gaussian elimination and resolution of systems of algebraic equations. In: van Hulzen, J.A. (ed.) ISSAC 1983 and EUROCAL 1983. LNCS, vol. 162, pp. 146–156. Springer, Heidelberg (1983)
24. Macaulay, F.S.: The algebraic theory of modular systems. Cambridge Mathematical Library, vol. xxxi. Cambridge University Press, Cambridge (1916)
25. Lee, P.J. (ed.): ASIACRYPT 2004. LNCS, vol. 3329, pp. 3–540. Springer, Heidelberg (2004)
26. Raddum, H., Semaev, I.: New technique for solving sparse equation systems. Cryptology ePrint Archive, Report 2006/475 (2006), http://eprint.iacr.org/
27. Wolf, C., Preneel, B.: Taxonomy of public key schemes based on the problem of multivariate quadratic equations. Cryptology ePrint Archive, Report 2005/077, 64 pages (May 12 [th] 2005), http://eprint.iacr.org/2005/077/
28. Wong, R.: Asymptotic approximations of integrals, vol. 34 of Classics in Applied Mathematics. Society for Industrial and Applied Mathematics (SIAM), Philadelphia, PA, (2001) Corrected reprint of the 1989 original
29. Yang, B.-Y., Chen, J.-M.: All in the XL family: Theory and practice. In: Park, C.-s., Chee, S. (eds.) ICISC 2004. LNCS, vol. 3506, pp. 67–86. Springer, Heidelberg (2005)
30. Yang, B.-Y., Chen, J.-M.: Theoretical analysis of XL over small fields. In: Wang, H., Pieprzyk, J., Varadharajan, V. (eds.) ACISP 2004. LNCS, vol. 3108, pp. 277–288. Springer, Heidelberg (2004)

Message Freedom in MD4 and MD5 Collisions: Application to APOP

Gaëtan Leurent

Laboratoire d'Informatique de l'École Normale Supérieure,
Département d'Informatique,
45 rue d'Ulm, 75230 Paris Cedex 05, France
gaetan.leurent@ens.fr

Abstract. In Wang's attack, *message modifications* allow to deterministically satisfy certain sufficient conditions to find collisions efficiently. Unfortunately, message modifications significantly change the messages and one has little control over the colliding blocks. In this paper, we show how to choose small parts of the colliding messages. Consequently, we break a security countermeasure proposed by Szydlo and Yin at CT-RSA '06, where a fixed padding is added at the end of each block.

Furthermore, we also apply this technique to recover part of the passwords in the Authentication Protocol of the Post Office Protocol (POP). This shows that collision attacks can be used to attack real protocols, which means that finding collisions is a real threat.

Keywords: Hash function, MD4, MD5, message modification, meaningful collisions, APOP security.

1 Introduction

At EUROCRYPT '05 and CRYPTO '05, Wang *et al.* described a new class of attacks on most of the hash functions of the MD4 family, MD4, MD5, HAVAL, RIPEMD, SHA-0 and SHA-1 in [21,23,24,22], which allows to find collisions for these hash functions very efficiently. However, the practical impact is unclear as many real-life applications of hash functions do not just rely on collision resistance.

One drawback with Wang's attacks when used against practical schemes is that due to the message modification technique, the colliding blocks cannot be chosen and look random. However, these attacks work with any IV, so one can choose a common prefix for the two colliding messages, and the Merkle-Damgård construction allows to add a common suffix to the colliding messages. Therefore, an attacker can choose a prefix and a suffix, but he must somehow hide the colliding blocks (1 block in MD4 and SHA-0, and 2 blocks in MD5 and SHA-1). The poisoned message attack [6] exploits this property to create two different PostScript files that display two different chosen texts but whose digests are equal. In this construction, the two different texts are in both PS files and the collision blocks are used by an `if-then-else` instruction to choose which part

A. Biryukov (Ed.): FSE 2007, LNCS 4593, pp. 309–328, 2007.

to display. This attack was extended to other file formats in [8]. Lenstra and de Weger also used the free prefix and free suffix property to create different X.509 certificates for the same *Distinguished Name* but with different secure RSA moduli in [12]. Here, the colliding blocks are hidden in the second part of the RSA moduli.

Recently, more concrete attacks have appeared: Stevens, Lenstra and de Weger in [19] found colliding X.509 certificates for two different *Distinguished Name*. In this work, the technique used is far more complex and allows to find messages colliding under MD5 with two *different* chosen prefixes. They used an approach suggested by Wang to find a near-collision for different IVs and used different differential paths to absorb the remaining differences. However, the messages B, B' are not controlled, and this randomness must still be hidden in the moduli.

Other applications of Wang collisions have been proposed to attack HMAC with several hash functions in [3,9]. The techniques use Wang's differential path as a black box but with particular messages to recover some keys in the related-key model or to construct advanced distinguishers.

Our Results. In the paper we try to extend Wang's attack to break more hash function uses. More precisely, we address the question of message freedom *inside* the colliding blocks, and we show that this can be used to attack APOP, a challenge-response authentication protocol.

The first contribution of this paper is a technique to gain partial control over the colliding blocks; this can be combined with previous work to make the colliding blocks easier to hide. More concretely, we show that we can select some part at the end of the messages which will collide. Our attack can use any differential path, and only requires a set a sufficient conditions. We are able to choose the last three message words in a one-block MD4 collision, and three specific message words in a two-block MD5 collision with almost no overhead. We are also able to choose the 11 last words of a one-block MD4 collision with a work factor of about 2^{31} MD4 computations.

An important point is that the technique used is nearly as efficient as the best message modifications on MD4 or MD5, even when we choose some parts of the messages. This contradicts the usual assumption that Wang's collisions are mostly random. As a first application of this new message modification technique, we show that a countermeasure recently proposed by Szydlo and Yin at CT-RSA '06 in [20] is almost useless for MD4 and can be partially broken for MD5. This can also be used to handle the padding *inside* the colliding blocks.

The second contribution is a partial password-recovery attack against the APOP authentication protocol, based on the message freedom in MD5 collisions. We are able to recover 3 characters of the password, therefore greatly reducing its entropy. Even though we do not achieve the full recovery of the password, we reduce the complexity of the exhaustive search and it is sufficient in practice to reduce this search to a reasonable time for small passwords, *i.e.* less than 9 characters.

Related Work. The first MD4 collision was found by Dobbertin [7], and his attack has a time complexity of about 2^{20} MD4; it combines algebraic techniques and optimization techniques such as genetic algorithms. His attack also allows to choose a large part of the colliding blocks at some extra cost: one can choose 11 words out of 16 with a complexity of about 2^{30} MD4 computations (little details are given and only an experimental time complexity).

Our work is based on Wang's collision attack [21,23], which have the following advantages over Dobbertin's:

- it can be adapted to other hash functions (Dobbertin's method can give collisions on the MD5 compression function, but has not been able to provide MD5 collisions);
- it is somewhat more efficient.

More recently, Yu *et al.*[25] proposed a differential path for MD4 collisions with a small number of sufficient conditions. This allows to build a collision (M, M') which is close to a given message $\bar{M}$ (about 44 different bits). This is quite different from what we are trying to do since the changed bits will be spread all over the message. We are trying to choose many consecutive bits, which is useful for different applications. However we studied their work and propose some improvements in Appendix C.

De Cannière and Rechberger announced at the rump session of CRYPTO '06 that they can find reduced-SHA-1 collisions and chose up to 25% of the message. However, they gave few details on their technique, and the conference version does not talk about this aspect of their work. Their idea seems to be to compute a differential path with the chosen message as conditions.

Organization of the Paper. This paper is divided in three sections: in the first part we describe APOP, how the attack works, and give background on MD4, MD5 and Wang's attack; then we describe our new collision finding algorithm and how to choose a part of the message; and eventually we describe some applications of these results, including the practical attack against APOP.

2 Background and Notation

2.1 APOP

APOP is a command of the Post Office Protocol Version 3 [13] implementing a simple challenge-response authentication protocol; it was introduced in the POP protocol to avoid sending the password in clear over the network. Servers implementing the APOP command send a challenge (formatted as a message identifier, or msg-id) in their greeting message, and the client authenticates itself by sending the username, and the MD5 of the challenge concatenated with the password: $MD5(msg\text{-}id\|passwd)$. The server performs the same computation on his side, and checks if the digests match. Thanks to this trick, an eavesdropper will not learn the password, provided MD5 is a partial one-way function. As there is no integrity protection and no authentication of the server, this protocol is subject to a man-in-the-middle attack, but the man-in-the-middle should not be able to learn the password or to re-authenticate later. Quoting RFC 1939 [13]:

It is conjectured that use of the APOP command provides origin identification and replay protection for a POP3 session.

This challenge-response can be seen as a MAC algorithm, known as the suffix method: $\text{MAC}_k(M) = \text{MD5}(M\|k)$. This construction is weak for at least two reasons: first, it allows off-line collision search so there is a generic forgery attack with $2^{n/2}$ computations and one chosen-text MAC; second, the key-recovery attack against the envelope method of Preneel and van Oorschot [15] can be used on the suffix method with 2^{67} offline computations and 2^{13} chosen text MACs.

This is a hint that APOP is weak, but these attacks require more computations than the birthday paradox, and the first one is mostly useless in a challenge-response protocol. However, we can combine these weaknesses with the collision attack against MD5 to build a practical attack against APOP.

The APOP Attack. In this attack, we will act as the server, and we will send some specially crafted challenges to the client. We will use pairs of challenges, such that the corresponding digest will collide only if some part of the password was correctly guessed.

Let us assume we can generate a MD5 collision with some specific format: $M = \text{"<???...???>x"}$ and $M' = \text{"<ịịị...ịịị>x"}$, where M and M' have both size 128 bytes (2 MD5 blocks). The '?' and 'ị' represent any character chosen by the collision finding algorithm. We will send "<???...???>" and "<ịịị...ịịị>" as a challenge, and the client returns $\text{MD5}(\text{"<???...???>}p_0p_1p_2...p_{n-1}\text{"})$ and $\text{MD5}(\text{"<ịịị...ịịị>}p_0p_1p_2...p_{n-1}\text{"})$, where "$p_0p_1p_2...p_{n-1}$" is the user password (the p_i's are the characters of the password).

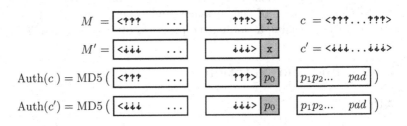

Fig. 1. APOP password recovery through MD5 collisions

As we can see in Figure 1, if $p_0 = \text{'x'}$, the two hashes will collide after the second block, and since the end of the password and the padding are the same, we will see this collision in the full hashes (and it is very unlikely that the two hashes collide for $p_0 \neq \text{'x'}$). Therefore we are able to test the first password character without knowing the others. We will construct pairs of challenge to test the 256 possible values, and stop once we have found the first password character.

Then we generate a new collision pair to recover the second character of the password: $M =$ "`<???...???>`p_0y" and $M' =$ "`<↓↓↓...↓↓↓>`p_0y", so as to test if $p_1 =$ 'y'. Thus, we can learn the password characters one by one in linear time.

This motivates the need for message freedom in MD5 collisions: we need to generate collisions with a specific format, and some chosen characters. We will see in section 4.1 that this attack can be used in practice to recover the first 3 characters of the password. We believe most passwords in real use are short enough to be found by exhaustive search once we know these first characters, and this attack will allow us to re-authenticate later. We stress that this attack *is* practical: it needs less than one hour of computation, and a few hundreds authentications.

Since people often read their mail from different places (including insecure wireless networks and Internet cafés), we believe that this man-in-the-middle setting is rather realistic for an attack against APOP. Moreover most mail clients automatically check the mailbox on a regular basis, so it seems reasonable to ask for a few hundred authentications.

2.2   MD4 and MD5

MD4 and MD5 follow the Merkle-Damgård construction. Their compression functions are designed to be very efficient, using 32-bit words and operations implemented in hardware in most processors:

- rotation ⋘;
- addition mod 2^{32} ⊞;
- bitwise boolean functions Φ_i.

The message block M is first split into 16 words $\langle M_i \rangle_{i=0}^{15}$, then expanded to provide one word m_i for each step of the compression function. In MD4 and MD5 this message expansion is simple, it just reuses many times the words M_i. More precisely, the full message is read in a different order at each round: we have $m_i = M_{\pi(i)}$ (π is given below).

The compression function uses an internal state of four words, and updates them one by one in 48 steps for MD4, and 64 steps for MD5. Here, we will assign a name to every different value of these registers, so the description is different from the standard one: the value changed on step i is called Q_i (this follows the notations of Daum [5]).

Then MD4 is given by:

Step update: $Q_i = (Q_{i-4} \boxplus \Phi_i(Q_{i-1}, Q_{i-2}, Q_{i-3}) \boxplus m_i \boxplus k_i) \lll s_i$
Input: $Q_{-4} \| Q_{-1} \| Q_{-2} \| Q_{-3}$
Output: $Q_{-4} \boxplus Q_{44} \| Q_{-1} \boxplus Q_{47} \| Q_{-2} \boxplus Q_{46} \| Q_{-3} \boxplus Q_{45}$

$\pi(\ 0..15)$:	0	1	2	3	4	5	6	7	8	9	10	11	12	13	14	15
$\pi(16..31)$:	0	4	8	12	1	5	9	13	2	6	10	14	3	7	11	15
$\pi(32..47)$:	0	8	4	12	2	10	6	14	1	9	5	13	3	11	7	15

And for MD5, we have:

Step update: $Q_i = Q_{i-1} \boxplus (Q_{i-4} \boxplus \Phi_i(Q_{i-1}, Q_{i-2}, Q_{i-3}) \boxplus m_i \boxplus k_i) \lll s_i$
Input: $Q_{-4} \| Q_{-1} \| Q_{-2} \| Q_{-3}$
Output: $Q_{-4} \boxplus Q_{60} \| Q_{-1} \boxplus Q_{63} \| Q_{-2} \boxplus Q_{62} \| Q_{-3} \boxplus Q_{61}$

$\pi(\ 0..15)$:	0	1	2	3	4	5	6	7	8	9	10	11	12	13	14	15
$\pi(16..31)$:	1	6	11	0	5	10	15	4	9	14	3	8	13	2	7	12
$\pi(32..47)$:	5	8	11	14	1	4	7	10	13	0	3	6	9	12	15	2
$\pi(48..64)$:	0	7	14	5	12	3	10	1	8	15	6	13	4	11	2	9

The security of the compression function was based on the fact that the operations are not "compatible" and mix the properties of the input.

We will use $x^{[k]}$ to represent the $k+1$-th bit of x, that is $x^{[k]} = (x \ggg k) \bmod 2$ (note that we count bits and steps starting from 0).

2.3 Wang's Attack Against MD4 and MD5

Wang *et al.* attacks against the hash functions of the MD4 family are differential attacks, and have the same structure with two main parts:

1. A precomputation phase:
   - choose a message difference Δ
   - find a differential path
   - compute a set of sufficient conditions
2. Search for a message M that satisfies the conditions;
   then $H(M) = H(M + \Delta)$.

The differential path specifies how the computations of $H(M)$ and $H(M+\Delta)$ are related: it tells how the differences introduced in the message will evolve in the internal state Q_i. If we choose Δ with a low hamming weight, and some extra properties, we can find some differences in the Q_i's that are very likely. Then we look at each step of the compression function, and we can express a set of sufficient conditions that will make the Q_i's follow the path. These conditions are on the bits of Q_i, so we can not directly find a message satisfying them; however some of them can be fulfilled deterministically through message modifications, and the rest will be statistical by trial and error.

3 A New Approach to Collision Finding

In this paper we assume that we are given a set of sufficient conditions on the internal state variables Q_i that produces collisions. We will try to find a message M such that when one computes a hash of this message, the conditions on the Q_i's hold. We will first describe the general idea that applies to both MD4 and MD5, and we will then study in more details those two hash functions.

In contrast to previous works [14,11,2], we will not focus on a particular path and give message modification techniques for every single condition, but we will give a generic algorithm that can take any path as input, like Klima in [10] and Stevens in [18]. Our method is based on two important facts:

1. We can search a suitable internal state rather than searching a suitable message, because the step update function is invertible: if Q_{i+1}, Q_{i+2} and Q_{i+3} are known, then we can compute any one of Q_i, Q_{i+4} or m_i from the two others (see Algorithm 2 in Appendix B for explicit formulas).

2. We do not need to search for the internal state from the beginning to the end. If we start from the middle we can satisfy the conditions in the first round and some conditions in the second round at the time; additionally we can choose the end of the message before running the search with little extra cost.

3.1   Previous Works

Wang's method to find a message satisfying a set of conditions is roughly described in Algorithm 3 in Appendix B: one basically picks many messages at random, modifies them to fulfill some of the conditions, and checks if the other conditions are fulfilled.

The best message modifications known allow to satisfy every condition up to round 22 in MD4 (which gives a collision probability of 2^{-2}) and up to round 24 in MD5 (which gives a collision probability of 2^{-29}). Basically, message modifications for the conditions in the first round are very easy, but in the second round it becomes much more difficult because we cannot freely change message words without breaking the Q_i in the first round (and therefore also in the second round). At the beginning of the second round it is still possible to find message modifications, but it becomes increasingly difficult as we go forward. Wang's differential paths are chosen with this constraint in mind, and most of their conditions are in the first round and at the beginning of the second.

The algorithm can be rewritten more efficiently: instead of choosing a random message and modify it, we can choose the Q_i in the first round and compute the corresponding message. Since all the conditions in MD4 and MD5 are on the Q_i's, this will avoid the need for message modifications in the first round.

To further enhance this algorithm, Klima introduced the idea of *tunnels* in [11], which is closely related to Biham and Chen's *neutral bits* used in the cryptanalysis of SHA-0 [1]. A tunnel is a message modification that does not affect the conditions up to some step $p_v - 1$ (*point of verification*). Therefore, if we have one message that fulfill the conditions up to $p_v - 1$ and τ tunnels, we can generate 2^τ messages that fulfills conditions up to step $p_v - 1$. This does not change the number of messages we have to try, but it greatly reduces the cost of a single try, and therefore speeds up collision search a lot. This is described in Algorithm 4 in Appendix B.

In MD4 and MD5, the point of verification will be in the second round, and we put it after the last condition in the second round (step 22 in MD4, 24 in MD5). We have message modifications for almost every condition before the point of verification, and it seems impossible to find message modifications for round 3 and later.

3.2 Our Method

Our method is somewhat different: we will *not* fix the Q_i from the beginning to the end, but we will start from the middle, and this will allow us to deal with the first round and the beginning of the second round at the same time.

First we choose a *point of choice* p_c of and a *point of verification* p_v. The point of verification is the step where we will start using tunnels, and the point of choice is the first step whose conditions will not be satisfied deterministically. The value of p_c depends on the message expansion used in the second round: we must have $\pi(16) < \pi(17) < ... < \pi(p_c - 1) < 12$, so we will choose $p_c = 19$ in MD4 ($\pi(18) = 8$), and $p_c = 19$ in MD5 ($\pi(18) = 11$).

The key idea of our collision search is to first choose the end of the first round, *i.e.* Q_{12} to Q_{15}. Then we can follow the computations in the first round and in the second round *at the same time*, and choose m_i's that satisfy the conditions in both rounds. There is no difficulty when the first round meets the values we fixed in the beginning: since we only fixed the Q_i's, we just have to compute the corresponding m_i's. More precisely, we will choose the Q_i from step 0 to $\pi(p_c - 1)$, and when we hit a message m_i that is also used in the second round with $i = \pi(j)$, we can modify it to generate a good Q_j since we have already fixed Q_{j-4}, Q_{j-3}, Q_{j-2} and Q_{j-1}. Thus, we can fulfill conditions up to round $p_c - 1$ almost for free.

In the end, we will make random choices for the remaining steps ($Q_{\pi(p_c-1)+1}$ to Q_{11}), until we have a message that follows the path up to step $p_v - 1$, and finally we use the tunnels. For a more detailed description of the algorithm, see Algorithm 1 in Appendix B, and take $t = 0$ (this algorithm is more generic and will be described in the next section).

Since we do not choose the Q_i's in the natural order, we have to modify a little bit the set of sufficient conditions: if we have a condition $Q_{12}^{[k]} = Q_{11}^{[k]}$, we will instead use $Q_{11}^{[k]} = Q_{12}^{[k]}$ because we choose Q_{12} before Q_{11}.

Compared to standard message modifications, our algorithm has an extra cost when we try to satisfy conditions in steps p_c to $p_v - 1$. However, this is not so important for two reasons:

- Testing one message only requires to compute a few steps, and we will typically have less than 10 conditions to satisfy, so this step will only cost about a hundred hash computations.
- This cost will be shared between all the messages which we find with the tunnels. In the case of MD5, we have to use a lot of tunnels to satisfy the 29 conditions in round 3 and 4 anyway, and in MD4 we will use a lot of tunnels if we look for many collisions (if one needs a single MD4 collision, the collision search cost should not be a problem).

3.3 Choosing a Part of the Message

This method can be extended to allow some message words to be fixed in the collision search. This will make the search for a first message following the path up to the point of verification harder, and it will forbid the use of some tunnels. Actually, we are buying some message freedom with computation time, but we can still find collisions very efficiently. We will show which message words can be chosen, and how to adapt the algorithm to find a first message following the path up to the step $p_v - 1$ with some message words chosen.

Choosing the Beginning. If the first steps in the first round are such that in the second round $m_0...m_i$ are only used after the step p_c or in a step j with no condition on Q_j, then we can choose $m_0...m_i$ before running the algorithm. The choice of $m_0...m_i$ will only fix $Q_0...Q_i$ (if there are some conditions on $Q_0...Q_i$, we must make sure they are satisfied by the message chosen), and the algorithm will work without needing any modification.

On MD5, this allows to choose m_0. Using Wang's path, there are no conditions on Q_0 for the first block, so m_0 is really free, but in the second block there are many conditions. On MD4, we have $\pi(16) = 0$, so this can only be used if we use $p_c = 16$, which will significantly increase the cost of the collision search.

Choosing the End. The main advantage of our algorithm is that it allows to choose the end of the message. This is an unsuspected property of Wang's attack, and it will be the core of our attack against APOP. Our idea is to split the search in two: first deal with fixed message words, then choose the other internal state variables. This is made possible because our algorithm starts at the end of the first round; the conditions in those steps do not directly fix bits of the message, they also depend on the beginning of the message.

More precisely, if we are looking for collisions where the last t words are chosen, we begin by fixing Q_{12-t}, Q_{13-t}, Q_{14-t} and Q_{15-t}, and we compute Q_{16-t} to Q_{15} using the chosen message words. We can modify Q_{12-t} if the conditions on the first state Q_{16-t} are not satisfied, but for the remaining $t - 1$ steps this is impossible because it would break the previous steps. So, these conditions will be fulfilled only statistically, and we might have to try many choices of Q_{12-t}, Q_{13-t}, Q_{14-t}, Q_{15-t} (note that each choice does not cost a full MD4 computation, but only a few steps).

Once we have a partial state that matches the chosen message, we run the same algorithm as in the previous section, but we will be able to deal with less steps of the second round due to the extra fixed states Q_{12-t} to Q_{11}. The full algorithm is given in Algorithm 1 in Appendix B.

3.4   MD4 Message Freedom

Using Wang's EUROCRYPT path [21]. If we use Wang's EUROCRYPT path, we will choose $p_v = 23$ so as to use tunnels only for the third round. Therefore, the tunnels will have to preserve the values of m_0, m_4, m_8, m_{12}, m_1, m_5 and m_9 when they modify the Q_i's in the first round.

There are two easy tunnels we can use, in Q_2 and Q_6. If we change the value of Q_2, we will have to recompute m_2 to m_6 as we do not want to change any other Q_i, but if we look at step 4, we see that Q_2 is only used trough $IF(Q_3, Q_2, Q_1)$.

So some bits of Q_2 can be changed without changing Q_4: if $Q_3^{[k]} = 0$ then we can modify $Q_2^{[k]}$. The same thing happens is step 5: Q_2 is only used in $\mathrm{IF}(Q_4, Q_3, Q_2)$, and we can switch $Q_2^{[k]}$ if $Q_4^{[k]} = 1$. So on the average, we have 8 bits of Q_2 that can be used as a tunnel. The same thing occurs in Q_6: if $Q_7^{[k]} = 0$ and $Q_8^{[k]} = 1$, then we can change $Q_6^{[k]}$ without altering m_8 and m_9. If we add some extra conditions on the path we can enlarge these tunnels, but we believe it's not necessary for MD4.

We can use our collision finding algorithm with up to 5 fixed words; Table 1 gives the number of conditions we will have to satisfy probabilistically. Of course, the cost of the search increases with t, but with $t = 3$ it should be about one 2^9 MD4 computations, which is still very low. Note that this cost is only for the first collision; if one is looking for a bunch of collisions, this cost will be shared between all the collisions found through the tunnels, and we expect 2^{14} of them. Another important remark is that this path has a non-zero difference in m_{12}; therefore when choosing more than 3 words, the chosen part in M and M' will have a one bit difference.

Table 1. Complexity estimates for MD4 collision search with t fixed words, and comparison with Dobbertin's technique [7]. We assume that a single trial costs 2^{-3} MD4 on the average due to early abort techniques.

message words chosen: t	0	1	2	3	4	5	0	11	0	11
path used			[21]				[25]		[7]	
point of choice: p_c	19	19	19	18	18	18	19	17		
point of verification: p_v	23	23	23	23	23	23	23	17		
conditions in steps $17 - t$ to 15	0	0	6	12	18	24	0	17		
conditions in steps p_c to $p_v - 1$	8	8	8	11	11	11	11	0		
conditions in steps p_v to N	2	2	2	2	2	2	17	34		
complexity (MD4 computations $\log_2$)	5	5	5	9	15	21	14	31	20	30

Using Yu _et al._'s CANS path [25]. To push this technique to the limit, we will try to use $t = 11$: this leaves only m_0 to m_4 free, which is the minimum freedom to keep a tunnel. In this setting, the conditions in steps 6 to 15 can only be satisfied statistically, which will be very expensive with Wang's path [21] (its goal was to concentrate the conditions in the first round). Therefore we will use the path from [25], which has only 17 conditions in steps 6 to 15.

Since we fix almost the full message, the second phase of the search where we satisfy conditions in the first and second round at the same time will be very limited, and we have $p_c = 17$. Then we use the tunnel in Q_0, which is equivalent to iterating over the possible Q_{16}'s, computing m_0 from Q_{16}, and then recomputing Q_0 and m_1, m_2, m_3, m_4. There are 34 remaining conditions, so we will have to use the tunnel about 2^{34} times. Roughly, we break the message search in two: first find $m_0..m_4$ such that the message follows steps 0 to 16, then modify it to follow up to the end by changing Q_0. This path is well suited for this approach, with few conditions well spread over the first two rounds.

This gives us a lot of freedom in MD4 collisions, but the collision search becomes more expensive. Another interesting property of this path is that it only introduces a difference in m_4, so the 11 chosen words will be the same in M and M'.

3.5   MD5 Message Freedom

Using Wang's MD5 path [23], we will choose $p_v = 24$ so as to use tunnels only for the third round. Therefore, when we modify the Q_i's in the first round to use the tunnels, we have to keep the values of m_1, m_6, m_{11}, m_0, m_5, m_{10}, m_{15} and m_4. We will not describe the available tunnels here, since they are extensively described in Klima's paper [11].

We use the set of conditions from Stevens [18], which adds the conditions on the rotations that were missing in Wang's paper [23]. We had to remove the condition because it is incompatible with some choices of m_{15}, so we check instead the less restrictive condition on Φ_{15} ($\Phi_{15}^{[31]} = 0$). We also found out that some conditions mark as optimization conditions were actually needed for the set of conditions to be sufficient.

As already stated, we can choose m_0 in the first block, and we will see how many words we can choose in the end of the message. With $t = 0$, we set $p_c = 19$, so we have 7 conditions in steps p_c to $p_v - 1$, and after p_v, there are 29 conditions for the first block, and 22 for the second block. With $t = 1$, we use $p_c = 18$, which increase the number of conditions between steps p_c and $p_v - 1$ to 9, but since we will use a lot of tunnels, this has very little impact on the computing time. We can also try to set $t = 2$, but this adds a lot of conditions when we search the states in the end of the first round. According to our experiments, the conditions on the Q_i's also imply some conditions on m_{14}, so m_{14} could not be chosen freely anyway.

As a summary, in a two-block MD5 collision $(M||N, M'||N')$, we can choose M_0, M_{15} and N_{15}, and the complexity of the collision search is roughly the same as a collision search without extra constraints.

We only implemented a little number of tunnels, but we find the first block in a few minutes with m_0 and m_{15} chosen, and the second block in a few seconds with m_{15} chosen. This is close to Klima's results in [11].

4   Applications

Freedom in colliding blocks can be used to break some protocols and to create collisions of special shape. The applications we show here requires that the chosen part is identical in M and M', *i.e.* the differential path must not use a difference there.

Fixing the Padding. We can use this technique to find messages with the padding included in the colliding block. For instance, this can be useful to build pseudo-collision of the hash function: if there is a padding block after the pseudo-colliding messages, the pseudo-collision will be completely broken.

We can also find collisions on messages shorter than one block, if we fix the rest of the block to the correct padding. An example of a 160 bit MD4 collision is given in Table 2 in Appendix A.

Zeroed Collisions. Szydlo and Yin proposed some message preprocessing techniques to avoid collisions attacks against MD5 and SHA-1 in [20]. Their idea is to impose some restrictions on the message, so that collision attacks become harder. One of their schemes is called *message whitening*: the message is broken into blocks smaller than 16 words, and the last t words are filled with zeroes. Using our technique we can break this strengthening for MD4 and MD5: in Appendix A we show a 11-whitened MD4 collision in Table 3 and a 1-whitened MD5 collision in Table 4.

4.1   The APOP Attack in Practice

To implement this attack, we need to efficiently generate MD5 collisions with some chosen parts; we mainly have to fix the last word, which is precisely what we can do cheaply on MD5.

Additionally, the POP3 RFC [13] requires the challenge to be a msg-id, which means that:

- It has to begin with '<' and end with '>'
- It must contain exactly one '@', and the remaining characters are very restricted. In particular, they should be ASCII characters, but we can not find two colliding ASCII messages with Wang's path.

In practice most mail clients do not check these requirements, leaving us a lot of freedom. According to our experiments with Thunderbird[1] and Evolution[2] there are only four characters which they reject in the challenge:

- 0x00 Null: used as end-of-string in the C language
- 0x3e Greater-Than Sign ('>'): used to mark the end of the msg-id
- 0x0a Line-Feed: used for end-of-line (POP is a text-based protocol)
- 0x0d Carriage-Return: also used for end-of-line

Thunderbird also needs at least one '@' in the msg-id.

We will use Wang's path [23], which gives two-block collisions. The first block is more expensive to find, so we will use the same for every msg-id, and we will fix the first character as a '<', and the last character as a '@'. Then we have to generate a second block for every password character test; each time the last word is chosen, and we must avoid 4 characters in the message. Using the ideas from Section 3.5 we can do this in less than 5 seconds per collision on a standard desktop computer.

Unfortunately, this path uses a difference $\delta M_{14} = 2^{32}$ and this makes a difference in character 60. In order to learn the i-th password character p_{i-1}, we need to generate a collision where we fix the last $i + 1$ characters (i password characters, plus a '>' to form a correct msg-id). Therefore, we will only be able to

---

[1] Available at http://www.mozilla.com/en-US/thunderbird/

[2] Available at http://www.gnome.org/projects/evolution/

retrieve 3 characters of the password with Wang's path. This points out a need for new paths following Wang's ideas, but adapted to other specific attacks; here a path less efficient for collision finding but with better placed differences could be used to learn more characters of the password.

Complexity. To estimate the complexity of this attack, we will assume the user's password is 8-characters long, and each character has 6 bits of entropy. This seems to be the kind of password most people use (when they don't use a dictionary word...). Under these assumptions, we will have to generate 3×2^5 collisions, and wait for about 3×2^6 identifications. Each collision takes about 5 seconds to generate; if we assume that the client identifies once per minute this will be the limiting factor, and our attack will take about 3 hours. In a second phase we can do an offline exhaustive search over the missing password characters. We expect to find them after 2^{30} MD5 computations and this will take about half an hour according to typical OpenSSL benchmarks.

Recent Developments. The attack against APOP had been independently found by Sasaki *et al.* almost simultaneously; they put a summary of their work on the eprint [16]. Moreover, Sasaki announced at the rump session of FSE '07 that he had improved the attack; he can recover 31 password characters using a new differential path.

Recommendations. We believe APOP is to be considered broken, and we suggest users to switch to an other authentication protocol if possible. Since the current attack needs non-RFC-compliant challenges, an easy countermeasure in the mail user agent is to strictly check if the challenge follows the RFC, but maybe this could be defeated by an improved attack.

Acknowledgement

We thank the anonymous reviewers of FSE, for their helpful comments and for pointing out related work. Thanks are also due to Phong Nguyen and Pierre-Alain Fouque for their precious help and proofreading. Finally, we would also like to thank Yu Sasaki for the explanations of his improvements over the APOP attack.

Part of this work is supported by the Commission of the European Communities through the IST program under contract IST-2002-507932 ECRYPT, and by the French government through the Saphir RNRT project.

References

1. Biham, E., Chen, R.: Near-Collisions of SHA-0. In: Franklin, M. (ed.) CRYPTO 2004. LNCS, vol. 3152, pp. 290–305. Springer, Heidelberg (2004)
2. Black, J., Cochran, M., Highland, T.: A Study of the MD5 Attacks: Insights and Improvements. In: Robshaw, M. (ed.) FSE 2006. LNCS, vol. 4047, pp. 262–277. Springer, Heidelberg (2006)

3. Contini, S., Yin, Y.L.: Forgery and Partial Key-Recovery Attacks on HMAC and NMAC Using Hash Collisions. In: Lai, X., Chen, K. (eds.) ASIACRYPT 2006. LNCS, vol. 4284, Springer, Heidelberg (2006)
4. Cramer, R.J.F. (ed.): EUROCRYPT 2005. LNCS, vol. 3494. Springer, Heidelberg (2005)
5. Daum, M.: Cryptanalysis of Hash Functions of the MD4-Family. PhD thesis, Ruhr-University of Bochum (2005)
6. Daum, M., Lucks, S.: Hash Collisions (The Poisoned Message Attack) "The Story of Alice and her Boss". Presented at the rump session of Eurocrypt '05. http://th.informatik.uni-mannheim.de/people/lucks/HashCollisions/
7. Dobbertin, H.: Cryptanalysis of MD4. J. Cryptology 11(4), 253–271 (1998)
8. Gebhardt, M., Illies, G., Schindler, W.: A Note on the Practical Value of Single Hash Collisions for Special File Formats.In: Dittmann, J. (ed.) Sicherheit, vol. 77 of LNI, pp. 333–344. GI (2006)
9. Kim, J., Biryukov, A., Preneel, B., Hong, S.: On the Security of HMAC and NMAC Based on HAVAL, MD4, MD5, SHA-0 and SHA-1. In: De Prisco, R., Yung, M. (eds.) SCN 2006. LNCS, vol. 4116, pp. 242–256. Springer, Heidelberg (2006)
10. Klima, V.: Finding MD5 Collisions on a Notebook PC Using Multi-message Modifications. Cryptology ePrint Archive, Report 2005/102 (2005), http://eprint.iacr.org/
11. Klima, V.: Tunnels in Hash Functions: MD5 Collisions Within a Minute. Cryptology ePrint Archive, Report 2006/105 (2006), http://eprint.iacr.org/
12. Lenstra, A.K., Weger, B.d.: On the Possibility of Constructing Meaningful Hash Collisions for Public Keys.. In: Boyd, C., González Nieto, J.M. (eds.) ACISP 2005. LNCS, vol. 3574, pp. 267–279. Springer, Heidelberg (2005)
13. Myers, J., Rose, M.: Post Office Protocol - Version 3. RFC 1939 (Standard) (May 1996) Updated by RFCs 1957, 2449.
14. Naito, Y., Sasaki, Y., Kunihiro, N., Ohta, K.: Improved Collision Attack on MD4 with Probability Almost 1. In: Won, D.H., Kim, S. (eds.) ICISC 2005. LNCS, vol. 3935, pp. 129–145. Springer, Heidelberg (2006)
15. Preneel, B., van Oorschot, P.C.: On the Security of Two MAC Algorithms. In: EUROCRYPT, pp. 19–32 (1996)
16. Sasaki, Y., Yamamoto, G., Aoki, K.: Practical password recovery on an md5 challenge and response. Cryptology ePrint Archive, Report 2007/101(2007), http://eprint.iacr.org/
17. Shoup, V. (ed.): CRYPTO 2005. LNCS, vol. 3621, pp. 14–18. Springer, Heidelberg (2005)
18. Stevens, M.: Fast Collision Attack on MD5. Cryptology ePrint Archive, Report 2006/104 (2006), http://eprint.iacr.org/
19. Stevens, M., Lenstra, A., de Weger, B.: Target Collisions for MD5 and Colliding X.509 Certificates for Different Identities. Cryptology ePrint Archive, Report 2006/360 (2006), http://eprint.iacr.org/
20. Szydlo, M., Yin, Y.L.: Collision-Resistant Usage of MD5 and SHA-1 Via Message Preprocessing. In: Pointcheval, D. (ed.) CT-RSA 2006. LNCS, vol. 3860, pp. 99–114. Springer, Heidelberg (2006)
21. Wang, X., Lai, X., Feng, D., Chen, H., Yu, X.: Cryptanalysis of the Hash Functions MD4 and RIPEMD. In: Cramer [4], pp. 1–18.
22. Wang, X., Yin, Y.L., Yu, H.: Finding Collisions in the Full SHA-1. In: Shoup [17], pp. 17–36 (2005)
23. Wang, X., Yu, H.: How to Break MD5 and Other Hash Functions. In: Cramer [4], pp. 19–35 (2005)

24. Wang, X., Yu, H., Yin, Y.L.: Efficient Collision Search Attacks on SHA-0. In: Shoup [17], pp. 1–16 (2005)
25. Yu, H., Wang, G., Zhang, G., Wang, X.: The Second-Preimage Attack on MD4. In: Desmedt, Y.G., Wang, H., Mu, Y., Li, Y. (eds.) CANS 2005. LNCS, vol. 3810, pp. 1–12. Springer, Heidelberg (2005)

A   Collision Examples

Table 2. A 160-bit MD4 collision

Message M			
42 79 2d 65	f0 f8 4f d8	d5 7d 86 bf	78 54 9d 67
3f b3 8c aa			
Message M'			
42 79 2d 65	f0 f8 4f d8	d5 7d 86 bf	78 54 9d 67
3f b3 8c ac			
Message M padded to one block			
42 79 2d 65	f0 f8 4f d8	d5 7d 86 bf	78 54 9d 67
3f b3 8c aa	80 00 00 00	00 00 00 00	00 00 00 00
00 00 00 00	00 00 00 00	00 00 00 00	00 00 00 00
00 00 00 00	00 00 00 00	a0 00 00 00	00 00 00 00
Message M' padded to one block			
42 79 2d 65	f0 f8 4f d8	d5 7d 86 bf	78 54 9d 67
3f b3 8c ac	80 00 00 00	00 00 00 00	00 00 00 00
00 00 00 00	00 00 00 00	00 00 00 00	00 00 00 00
00 00 00 00	00 00 00 00	a0 00 00 00	00 00 00 00
MD4			
46 6d cb bd	04 66 2c 43	75 12 18 f6	f4 e5 68 71

Table 3. A 11-whitened MD4 collision

Whitened Message M			
b9 39 4f 51	3b 43 68 dd	d6 1d 6f 1c	5d b6 a0 b2
44 d4 69 18	00 00 00 00	00 00 00 00	00 00 00 00
00 00 00 00	00 00 00 00	00 00 00 00	00 00 00 00
00 00 00 00	00 00 00 00	00 00 00 00	00 00 00 00
Whitened Message M'			
b9 39 4f 51	3b 43 68 dd	d6 1d 6f 1c	5d b6 a0 b2
44 d4 69 1a	00 00 00 00	00 00 00 00	00 00 00 00
00 00 00 00	00 00 00 00	00 00 00 00	00 00 00 00
00 00 00 00	00 00 00 00	00 00 00 00	00 00 00 00
MD4 without padding			
1d ba e9 89	02 22 9f a6	a9 bb 88 f8	30 c1 38 ab
MD4 with padding			
e1 54 1e 65	46 d8 4b 79	db b3 5b b2	13 00 06 9b

Table 4. A 1-whitened MD5 collision

Whitened Message M											
00	00	00	00	23	f9	5a	1c	c8	4f	18	59	1b	ef	74	a9
02	7a	b6	bf	ff	47	53	be	c3	29	a9	dd	b1	1e	62	94
d1	2c	24	05	07	5e	b4	42	1b	e2	58	72	25	83	b2	52
12	97	d8	24	ca	8c	ae	13	e1	e9	34	77	00	00	00	00
54	01	42	6f	b4	b5	4a	77	d2	15	90	5a	7a	42	cf	dd
9f	76	5b	37	90	dd	7e	3d	0a	fd	77	d7	d1	4c	55	de
49	ff	3e	f2	f5	52	b8	86	72	c0	49	7e	80	ac	d1	1f
c7	38	b4	96	a8	d3	73	f0	4b	5c	d8	f2	00	00	00	00

Whitened Message M'											
00	00	00	00	23	f9	5a	1c	c8	4f	18	59	1b	ef	74	a9
02	7a	b6	3f	ff	47	53	be	c3	29	a9	dd	b1	1e	62	94
d1	2c	24	05	07	5e	b4	42	1b	e2	58	72	25	03	b3	52
12	97	d8	24	ca	8c	ae	13	e1	e9	34	f7	00	00	00	00
54	01	42	6f	b4	b5	4a	77	d2	15	90	5a	7a	42	cf	dd
9f	76	5b	b7	90	dd	7e	3d	0a	fd	77	d7	d1	4c	55	de
49	ff	3e	f2	f5	52	b8	86	72	c0	49	7e	80	2c	d1	1f
c7	38	b4	96	a8	d3	73	f0	4b	5c	d8	72	00	00	00	00

MD5 without padding											
98	28	90	7a	75	75	ae	7a	25	f9	80	94	62	ea	52	76

MD5 with padding											
7c	c4	3c	db	a7	a6	f6	9e	c5	10	8e	46	95	00	fd	82

Table 5. APOP MD5 collision. These two msg-id's collide if padded with "`bar`".

Message M											
`<xxxÑÕç\ᴴs∅ºä4ᴾᴅ<ᴴ₀`	3c 78 78 78	d1 d5 e7 5c	88 d8 ba e4	34 8b 3c 81
`mXÀᴬ¢n]ᴱʙ∅\4UᴮsµiQᴾ₁`	6d 58 c0 9f	6e 5d 17 d8	5c 34 55 08	b5 69 51 91
`Åì"ᴬ§xxõ!¿ïFsc'áᴰ`	c5 ec 22 06	02 78 f4 21	bf ef 46 73	63 27 e1 d2
`Ýì*sôGᴱcû49¾₄Vᴴxxx@`	dd ec 8a f4	47 1b fb 34	39 be 56 89	78 78 78 40
`ÚcCòßPꝋᴱqÙÖᴴᴿᴹᴿᴵRéý`	da 63 43 f2	df 50 ba 05	d9 d6 09 83	8d 52 e9 fd
`Äꝯc4$¾₄MH-:y6ÇÊ@Üᶠs`	c4 94 34 24	bd 4d 48 2d	3a 79 36 c7	ca a9 dc 1c
`-0:Þñᴹʷ│enᴾʋë5{ᴰcꝓ¾₄`	2d 30 3a de	f1 95 7c 65	6e 8c eb 35	7b 90 fe be
`ïL´ï²∅#»ì])ü>`	ef 4c b4 ef	aa d8 23 bb	ec 5d 29 fc	3e

Message M'											
`<xxxÑÕç\ᴴs∅ºä4ᴾᴅ<ᴴ₀`	3c 78 78 78	d1 d5 e7 5c	88 d8 ba e4	34 8b 3c 81
`mXÀᵁsn]ᴱʙ∅\4UᴮsµiQᴾ₁`	6d 58 c0 1f	6e 5d 17 d8	5c 34 55 08	b5 69 51 91
`Åì"ᴬ§xxõ!¿ïFsc§áᴰ`	c5 ec 22 06	02 78 f4 21	bf ef 46 73	63 a7 e1 d2
`Ýì*sôGᴱcû49¾₄Vᴴxxx@`	dd ec 8a f4	47 1b fb 34	39 be 56 09	78 78 78 40
`ÚcCòßPꝋᴱqÙÖᴴᴿᴹᴿᴵRéý`	da 63 43 f2	df 50 ba 05	d9 d6 09 83	8d 52 e9 fd
`Äꝯc4∅¾₄MH-:y6ÇÊ@Üᶠs`	c4 94 34 a4	bd 4d 48 2d	3a 79 36 c7	ca a9 dc 1c
`-0:Þñᴹʷ│enᴾʋë5{ᴰ↓ꝓ¾₄`	2d 30 3a de	f1 95 7c 65	6e 8c eb 35	7b 10 fe be
`ïL´ï²∅#»ì])│>`	ef 4c b4 ef	aa d8 23 bb	ec 5d 29 7c	3e

MD5(M‖"bar")	b8 98 53 57	f8 06 8c 23	72 cf f8 c2	4c 22 c3 81
MD5(M'‖"bar")	b8 98 53 57	f8 06 8c 23	72 cf f8 c2	4c 22 c3 81
MD5(M‖"ban")	40 c6 ee cc	6f e1 e5 2b	53 74 a0 e8	3e f7 4f 54
MD5(M'‖"ban")	02 c1 8c 29	49 91 04 99	8f 88 33 77	a1 eb 81 be

B   Algorithms

All the algorithms take the set of condition as an implicit input, and will modify
a shared state containing the m_i's and the Q_i's.

Algorithm 1. Our collision finding algorithm

1: **procedure** FINDMESSAGE(p_v, p_c, t)
2:    **repeat**
3:      choose Q_{12-t}, Q_{13-t}, Q_{14-t}, Q_{15-t}
4:      **if** $t \neq 0$ **then**
5:        STEPFORWARD($16 - t$)
6:        FIXSTATE($16 - t$)
7:        STEPBACKWARD($16 - t$)
8:        **if not** CHECKCONDITIONS($12 - t$) **then**
9:          **goto** 3
10:      **for** $17 - t \leq i < 16$ **do**
11:        STEPFORWARD(i)
12:        **if not** CHECKCONDITIONS(i) **then**
13:          **goto** 3
14:      $i \leftarrow 0$
15:      **for** $16 \leq j < p_c$ **do**
16:        **while** $i < \pi(j)$ **do**
17:          choose Q_i
18:          STEPMESSAGE(i)
19:          $i \leftarrow i + 1$
20:        STEPFORWARD(j)
21:        FIXSTATE(j)
22:        STEPMESSAGE(j)
23:        STEPFORWARD(i)
24:        **if not** CHECKCONDITIONS(i) **then**
25:          **goto** 16
26:      **for** $\pi(p_c - 1) + 1 \leq i < 12 - t$ **do**
27:        choose Q_i
28:        STEPMESSAGE(i)
29:      STEPMESSAGE(12-t ... 15-t)
30:      **for** $p_c \leq i < p_v$ **do**
31:        STEPFORWARD(i)
32:        **if not** CHECKCONDITIONS(i) **then**
33:          **goto** 26
34:      **for all** tunneled message **do**
35:        **for** $p_v \leq i < N$ **do**
36:          STEPFORWARD(i)
37:          **if not** CHECKCONDITIONS(i) **then**
38:            use the next message
39:    **until** all conditions are fulfilled

Algorithm 2. Step functions

1: **function** MD4STEPFORWARD(i)
2:     $Q_i \leftarrow (Q_{i-4} \boxplus \Phi_i(Q_{i-1}, Q_{i-2}, Q_{i-3}) \boxplus m_i \boxplus k_i) \lll s_i$
3: **function** MD4STEPBACKWARD(i)
4:     $Q_{i-4} \leftarrow (Q_i \ggg s_i) \boxminus \Phi_i(Q_{i-1}, Q_{i-2}, Q_{i-3}) \boxminus m_i \boxminus k_i$
5: **function** MD4STEPMESSAGE(i)
6:     $m_i \leftarrow (Q_i \ggg s_i) \boxminus Q_{i-4} \boxminus \Phi_i(Q_{i-1}, Q_{i-2}, Q_{i-3}) \boxminus k_i$
7: **function** MD5STEPFORWARD(i)
8:     $Q_i \leftarrow Q_{i-1} \boxplus (Q_{i-4} \boxplus \Phi_i(Q_{i-1}, Q_{i-2}, Q_{i-3}) \boxplus m_i \boxplus k_i) \lll s_i$
9: **function** MD5STEPBACKWARD(i)
10:     $Q_{i-4} \leftarrow (Q_i \boxminus Q_{i-1}) \ggg s_i \boxminus \Phi_i(Q_{i-1}, Q_{i-2}, Q_{i-3}) \boxminus m_i \boxminus k_i$
11: **function** MD5STEPMESSAGE(i)
12:     $m_i \leftarrow (Q_i \boxminus Q_{i-1}) \ggg s_i \boxminus Q_{i-4} \boxminus \Phi_i(Q_{i-1}, Q_{i-2}, Q_{i-3}) \boxminus k_i$

Algorithm 3. Wang's message finding algorithm

1: **procedure** FINDMESSAGEWANG
2:     **repeat**
3:         choose a random message
4:         **for** $0 \le i < N$ **do**
5:             STEPFORWARD(i)
6:             **if** not CHECKCONDITIONS(i) **then**
7:                 try to modify the message
8:     **until** all conditions are fulfilled

Algorithm 4. Klima's message finding algorithm

1: **procedure** FINDMESSAGEKLIMA
2:     **repeat**
3:         **for** $0 \le i < 16$ **do**
4:         choose Q_i
5:         STEPMESSAGE(i)
6:         **for** $16 \le i < p_v$ **do**
7:             STEPFORWARD(i)
8:             **if** not CHECKCONDITIONS(i) **then**
9:                 modify the message
10:         **for all** tunneled message **do**
11:             **for** $p_v \le i < N$ **do**
12:                 STEPFORWARD(i)
13:                 **if** not CHECKCONDITIONS(i) **then**
14:                     use the next message
15:     **until** all conditions are fulfilled

C Collisions with a High Number of Chosen Bits

In this paper, we considered the problem of finding collisions with some chosen *words* but some other works addressed the problem of choosing *bits* (mainly [25]). We believe it is more useful to choose consecutive bits and the applications we give in Section 4 all need this property. Specifically, the APOP attack requires to choose consecutive bits in the end of the block; it will fail if one of these bits is uncontrolled. However let us say a few words about collisions with many chosen bits.

Following the ideas of Yu *et al.*[25], we will use a collision path with very few conditions. Such paths are only known in the case of MD4, and we found out that the path from [25] can be slightly enhanced: if we put the difference in the bit 25 instead of the bit 22, we get only 58 conditions (instead of 62). Now the basic idea is to take a message M, and apply message modifications in the first round: this will give a message M^* that has about 10 bit difference from M (there are 20 conditions in the first round) and it gives a collision $(M^*, M^* + \Delta)$ with probability 2^{-38}. Therefore we will generate about 2^{38} messages M_i close to M and the corresponding M_i^*, and one of them will give a collision.

Little detail is given in Yu *et al.* paper, but we can guess from their collision example that they generated the M_i's by changing m_{14} and m_{15}. This makes the attack more efficient since the M_i^* will all have the same first 14 words, but it will modify about 32 extra bits. Actually, one only needs to iterate over 38 bits, which gives on the average 19 modified bits, but Yu *et al.* used the whole 64 bits.

In fact, if the goal is to have a high number of chosen bits, it is better to choose the M_i in another way: instead of iterating over some bits, we will switch a few bits in the whole message, and iterate over the positions of the differences. We have $\binom{512}{5} \approx 2^{38}$, so it should be enough to select 5 positions, but we will have to run the full message modifications in the first round for every message, which is quite expensive (about 2^{37} MD4 computations). Instead, one can choose 4 positions in the first 480 bits, and two in the last 32 bits: we have $\binom{480}{4}\binom{32}{2} \approx 2^{40}$,

Table 6. A MD4 collision close to 1^{512}

Message M															
ff	ff	ff	ff	bf	ff	ff	ff	ff	f7	ff	ff	ff	ff	df	ff
ff	ff	ff	fd	ff	ff	df	ff	ff	ff	fd	ff	ff	ef	ff	ff
ff	ff	ff	ef	ff	ff	ff	fe	ff	ff	ef	7f	ff	7f	ff	ff
7f	ff	fd	7f	ff	bf	ff	ff	ff	ff	ff	ff	ff	bf	ff	fd

Message M'															
ff	ff	ff	ff	bf	ff	ff	ff	ff	f7	ff	ff	ff	ff	df	ff
ff	ff	ff	ff	ff	ff	df	ff	ff	ff	fd	ff	ff	ef	ff	ff
ff	ff	ff	ef	ff	ff	ff	fe	ff	ff	ef	7f	ff	7f	ff	ff
7f	ff	fd	7f	ff	bf	ff	ff	ff	ff	ff	ff	ff	bf	ff	fd

MD4 without padding											
ff	a3	b5	2d	51	63	59	36	11	e5	9a	d0	a6	cf	8b	33

MD4 with padding											
59	93	19	84	d0	6f	55	9f	f3	d0	87	4b	c6	24	f4	8d

and the message modification on the first 15 words will only be run every 2^9 messages; the main cost will be that of testing 2^{38} messages for the second and third rounds: using early abort techniques, this cost will be about 2^{33} MD4[3].

On the average, we expect to have 6 bit differences coming from this iteration, plus 10 coming from the message modification in the first round. An example of such message is given in Table 6, it has 18 bit differences from the target (a block consisting only of 1's), which is much better than achieved by Yu *et al.* (43 bit differences).

---

[3] There are two conditions on step 15, three on step 16 and one on step 17: so $3 \cdot 2^{36}$ messages will stop after 1 step, $7 \cdot 2^{33}$ after 2 steps, 2^{32} after 3 steps, and the remaining 2^{32} messages will need at most 16 steps. This gives less than $95 \cdot 2^{32}$ MD4 steps, that is less than 2^{33} full MD4.

New Message Difference for MD4

Yu Sasaki, Lei Wang, Kazuo Ohta, and Noboru Kunihiro

The University of Electro-Communications
Chofugaoka 1-5-1, Chofu-shi, Tokyo, 182-8585, Japan
{yu339,wanglei,ota,kunihiro}@ice.uec.ac.jp

Abstract. This paper proposes several approaches to improve the collision attack on MD4 proposed by Wang et al. First, we propose a new local collision that is the best for the MD4 collision attack. Selection of a good message difference is the most important step in achieving effective collision attacks. This is the first paper to introduce an improvement to the message difference approach of Wang et al., where we propose a new local collision. Second, we propose a new algorithm for constructing differential paths. While similar algorithms have been proposed, they do not support the new local collision technique. Finally, we complete a collision attack, and show that the complexity is smaller than the previous best work.

Keywords: Hash Function, Collision Attack, MD4, Local Collision, Message Difference, Differential Path.

1 Introduction

Hash functions play an important role in modern cryptology. Hash functions must hold the property of *collision resistance*. This means that it must be computationally hard to find a pair of messages M and M' such that $H(M) = H(M')$ and $M \neq M'$.

MD4 is a hash function that was proposed by Rivest in 1990 [5]. MD4 has the Merkle-Damgård structure, and is designed for fast calculation. MD4 has been used to design other hash functions such as MD5 or SHA-1, which are widely used today. Therefore, cryptanalysis on MD4 is important since it affects the cryptanalysis of other hash functions based on MD4.

Several papers have found and reported the weaknesses of MD4. In 1996, the first collision attack was proposed by Dobbertin [1]. This attack finds a collision with probability of 2^{-22}. In 1998, Dobbertin pointed out that the first two rounds of MD4 did not have an property of one way. In 2004, at the CRYPTO rump session, Wang presented that collisions of MD4 could be generated very rapidly [7]. In 2005, Wang et al. reported the details of this attack which finds a collision with probability of 2^{-2} to 2^{-6}, and its complexity is less than 2^8 MD4 operations. This attack is highly efficient and many papers have suggested further improvements to this attack. In 2005, Naito et al. presented an improved attack [4]. They pointed out the mistakes of the sufficient conditions detailed in [8], and made improvements to the message modification techniques. This

A. Biryukov (Ed.): FSE 2007, LNCS 4593, pp. 329–348, 2007.
© International Association for Cryptologic Research 2007

attack finds a collision with complexity less than 3 MD4 operations, and it is the fastest approach up to this time. In 2006, Schläffer and Oswald [6] analyzed how message differences given by [8] worked, and proposed an automated differential path search algorithm. As a result, they found a differential path that needed fewer sufficient conditions than Wang et al.'s.

An important characteristic of previous attacks is that they all used the same local collision, i.e. the message differences and differential path as proposed by Wang et al. In [8], the authors claim that their message differences, which are derived from their local collision, are optimized for collision attack. However, our research shows that local collision of [8] is not optimized.

For collision attacks, selecting good message differences is very important. Message differences are derived from a local collision, so selecting a good local collision is an important aspect of this work. If better local collisions or message differences are found, attack complexity can be drastically reduced. Moreover, an effective way to find a good local collision or message difference for MD4 may be applicable to other hash functions. This fact motivated our research on MD4.

This paper makes two contributions.

1. We propose a new local collision and new message differences that appear to be the best for collision attack on MD4. We use less than one half of the non-negligible sufficient conditions for new differences needed by Wang et al.'s approach.
2. We show that the differential path construction algorithm proposed by [6] does not work for the new local collision technique. Therefore, we analyze the problems of [6], and develop a new algorithm.

The improved collision attack is realized by combining the above two contributions and the message modification technique described by [3,4]. Since local collision and message differences are improved, we can find a collision with complexity less than 2 MD4 operations, which is the fastest of all existing attacks. We show the new local collision in the right side of Table 1 and Figure 2.2, new message differences and differential path in Table 7, new sufficient conditions in Table 8, and all message modification procedures in Table 9 to Table 20; an example of a collision generated by our attack is shown in Table 4.

The organization of this paper is as follows. In section 2, we explain the specification of MD4 and related works. In section 3, we explain why the local collision of [8] is not best, and propose a new local collision for MD4. In section 4, we explain why the algorithm of [6] does not support the new local collision, and propose a new algorithm that constructs differential paths for the first round of MD4. Section 5 completes the improved attack, and compares its efficiency to previous works. Finally, we conclude this paper in section 6.

2   Preliminaries

2.1   Specification of MD4 [5]

MD4 input is an arbitrary length message M, and MD4 output is 128-bit data $H(M)$. MD4 has a Merkle-Damgård structure. First, the input message is padded to be a multiple of 512 bits. Since padding does not affect collision attack, we omit its explanation. The padded message M^* is divided into 512-bit messages $M^* = (M_0, \ldots, M_{n-1})$. The initial value (IV) for the hash value is set to be $H_0 = (0x67452301, 0xefcdab89, 0x98badcfe, 0x10325476)$. The output of the compression function H_1 is calculated using M_0 and H_0. In this paper, we call the calculation performed in a single run of the compression function 1 block. Next, H_2 is calculated using M_1 and H_1. Similarly, the compression function is calculated until the last message M_{n-1} is used. Let H_n be the hash value of M.

Compression Function of MD4
All calculations in the compression function are 32-bit. We omit the notation of "mod 2^{32}". The input to the compression function is a 512-bit message M_j and a 128-bit value H_j. First, M_j is divided into $(m_0, \ldots m_{15})$, where each m_i is a 32-bit message. The compression function consists of 48 steps. Steps 1-16 are called the first round (1R). Steps 17-32 and 33-48 are the second and third rounds (2R and 3R), respectively. In step i, chaining variables $a_i, b_i, c_i, d_i (1 \leq i \leq 48)$ are updated by the following expression (a_0, b_0, c_0, d_0 are the IV).

$a_i = d_{i-1},$
$b_i = (a_{i-1} + f(b_{i-1}, c_{i-1}, d_{i-1}) + m_k + t) \lll s_i,$
$c_i = b_{i-1},$
$d_i = c_{i-1},$

where f is a Boolean function defined in each round, m_k is one of $(m_0, \ldots m_{15})$, and index k is defined in each step. If $m_0, \ldots m_{15}$ are fixed, all other $m_i(16 \leq i \leq 47)$ are also fixed. In this paper we call this "Message Expansion". t is a constant number defined in each round, $\lll s_i$ denotes left rotation by s_i bits, and s_i is defined in each step. Details of f and t are as follows.

$$1R : t = 0x00000000, f(X, Y, Z) = (X \wedge Y) \vee (\neg X \wedge Z),$$
$$2R : t = 0x5a827999, f(X, Y, Z) = (X \wedge Y) \vee (Y \wedge Z) \vee (X \wedge Z),$$
$$3R : t = 0x6ed9eba1, f(X, Y, Z) = X \oplus Y \oplus Z.$$

After 48 steps are calculated, H_{j+1} is calculated as follows.
$aa_0 \leftarrow a_{48} + a_0, \qquad bb_0 \leftarrow b_{48} + b_0,$
$cc_0 \leftarrow c_{48} + c_0, \qquad dd_0 \leftarrow d_{48} + d_0,$
$H_{j+1} \leftarrow (aa_0 | bb_0 | cc_0 | dd_0).$

2.2 Related Work 1: Collision Attack by Wang et al. and Its Improvement by Previous Works

The first collision attack on MD4 was proposed by Wang et al. in 2005. Many researchers have improved this attack since its publication. Our paper also improves this attack. In this section, we explain the attack procedure and which part of the attack has been improved so far.

This attack is a differential attack, and generates a collision with complexity less than 2^8 MD4 operations. Let m and m' be a pair of messages that yield a collision. Difference Δ is defined to be the value yielded by subtracting value for

m from value for m'. For example, $\Delta m = m' - m$. The attack procedure is as follows.

1. Find the "Message Difference (ΔM)" that yields a collision with high probability.
2. Determine how the impact of ΔM propagates. The propagation of this difference is called the "Differential Path (DP)."
3. Generate "Sufficient Conditions (SC)" for realizing the differential path on the value of chaining variables for calculating the hash value of m.
4. Determine the procedures of "Message Modification (MM)" that satisfy sufficient conditions.
5. Locate a message that satisfies all sufficient conditions by randomly generating messages and applying Message Modification. Let such a message be M_*.
6. Calculate $M'_* = M_* + \Delta M$. Finally, M_* and M'_* become a collision pair.

ΔM of Wang et al.'s attack is as follows.

$$\Delta M = M' - M = (\Delta m_0, \Delta m_1, \ldots, \Delta m_{15}),$$
$$\Delta m_1 = 2^{31}, \Delta m_2 = 2^{31} - 2^{28}, \Delta m_{12} = -2^{16}, \Delta m_i = 0, 0 \le i \le 15, i \ne 1, 2, 12.$$

[8] introduces two kinds of message modification: Single-Step Modification and Multi-Step Modification. Later, these names were changed to Basic Message Modification (BMM) and Advanced Message Modification (AMM). They are related as follows.

- Basic message modification is the technique that can satisfy all sufficient conditions in the first round with probability of 1.
- Advanced message modification can satisfy several sufficient conditions in the second round. (If the sufficient conditions exist in an early step of the second round, they may be satisfied by advanced message modification. On the other hand, if they exist in a later step of the second round, it's almost impossible to satisfy them.)
- No technique is known to satisfy any sufficient conditions in the third round.

For details about the differential path, sufficient condition and message modification, please refer to [8].

Strategy of Selecting ΔM

The strategy of selecting ΔM is explained by Schläffer and Oswald [6]. They showed that ΔM was determined by the Local Collision (LC) in the third round, although the reason was not explained. They then showed how Wang et al.'s local collision worked. We show the local collision of Wang et al. in the left side of Table 1, and its diagram in Figure 2.2.

Improvement of Collision Attack

So far, several papers [4,6] have improved the collision attack proposed by Wang et al. Naito et al. [4] pointed out the sufficient condition mistakes in [8], and made improvements to message modification. They used the same ΔM, differential

Table 1. Comparison of Wang et al.'s Local Collision and Ours

Wang et al.'s Local Collision				
Step	Δa_k	Δb_k Δc_k Δd_k		Δm_{k-1}
i	2^{31}			2^{31-s_i}
$i+1$		2^{31} 2^{31}		$2^{31-s_{i+1}}, 2^{31}$
$i+2$		2^{31} 2^{31}		
$i+3$	2^{31}	2^{31}		
$i+4$	2^{31}			
$i+5$				2^{31}

New Local Collision				
Step	Δa_k	Δb_k Δc_k Δd_k		Δm_{k-1}
i	2^{31}			2^{31-s_i}
$i+1$		2^{31}		2^{31}
$i+2$		2^{31}		2^{31}
$i+3$	2^{31}			2^{31}
$i+4$				2^{31}

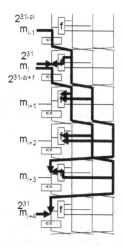 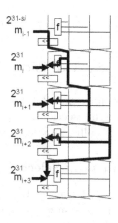

Fig. 1. Diagram of Wang et al.'s Local Collision and Ours

path and sufficient condition as given by [8], but proposed advanced message modification that could satisfy all sufficient conditions in the second round and quickly satisfy sufficient conditions in the third round[1]. Their attack finds a collision of MD4 with complexity less than 3 MD4 operations. Schläffer and Oswald [6] proposed an algorithm that could construct differential paths when ΔM was given. They used the same ΔM as given by [8], and showed a differential path that needed fewer sufficient conditions than Wang et al.'s.

For Wang et al.'s attack, selection of ΔM gives the most significant impact to the attack complexity. However, no paper has attempted to improve the ΔM of Wang et al. In this paper, we propose new ΔM in section 3.

2.3 Related Work 2: Differential Path Search by Schläffer and Oswald [6]

At FSE06, Schläffer and Oswald [6] proposed an algorithm on how to construct differential paths for MD4 collision attack, when message differences are given.

---

[1] This technique is restarting collision search from an intermediate step, which is different from message modification for the third round.

They used Wang et al.'s message differences. We show this algorithm can not be applied to the new ΔM (see section 4.2 and 4.3). In this section, we explain their algorithm as follows.

1. Calculate "Target Differences." In step i, target differences mean differences that are used to cancel Δm in steps $i + 4k(k = 0, \pm 1, \pm 2 \cdots)$.
2. Determine the actual output differences of f function. f function can not produce differences that are not elements of target differences. Resulted differential path usually has some contradictions in sufficient conditions.
3. Resolve contradictions in the sufficient conditions to construct a differential path.

For more details, please refer to [6].

3   New Local Collision

As mentioned in section 2.2, no paper has proposed a better ΔM than that of Wang et al. In this section, we propose a superior local collision and ΔM that are the best for collision attack on MD4.

3.1   Importance of Selecting ΔM

[6] determined ΔM was set by local collision in the third round, however, they didn't mention a reason for this. This motivated us to start with an explanation.

Thanks to the various message modification techniques which are available to us, following the differential path during the first two rounds (32 steps) of MD4 has very low cost. The computational cost of the attack comes from the probability of following the path during the third round. As a consequence, in order to get a more efficient collision attack on MD4, minimizing the number of sufficient conditions in the third round is the most important. Therefore, we start by looking for a very efficient local collision in the third round.

3.2   Problem of Wang et al.'s Local Collision and Constructing New Local Collision

In order to minimize the number of sufficient conditions in the third round, the method of Wang et al. applies local collision there. Their local collision is shown in left side of Table 1. They insert single message difference to step i in the third round, and insert other message differences to minimize the impact of the propagation of the inserted difference. As a result, the inserted difference is cancelled in following 6 steps. In Wang et al.'s local collision, two differences are made in the MSB of b_i and b_{i+1} for ensuring that differences don't propagate in step $i+2$, $i+3$ and $i+4$ (obviates the need to insert other message differences). Here, due to the addition and left rotation, the following fact is achieved. (We show the proof of this in Appendix A.)

Fact: To make $\Delta b_i = \pm 2^{31}$ from $\Delta m_k = \pm 2^{31-s_i}$, we need the following sufficient condition:

$$\begin{cases} b_{i,31} = 0 & (\text{if } \Delta m_k = +2^{31-s_i}) \\ b_{i,31} = 1 & (\text{if } \Delta m_k = -2^{31-s_i}) \end{cases}$$

Wang et al.'s local collision makes two differences. Therefore, it needs two sufficient conditions to realize the local collision.

In a collision attack on MD4, at least one message difference in the third round is necessary since collision messages must be different. Considering the Fact, it is obvious that if a local collision yielding one difference in the MSB of b_i can be constructed, that local collision is the best.

3.3   Construction of New Local Collision

We construct a new local collision by making only one difference in the MSB of b_i. The new local collision is summarized in right side of Table 1 and Figure 2.2. In step i, we make $\Delta b_i = \pm 2^{31}$ by inserting $\pm 2^{31-s_i}$ on m_{i-1}. From the Fact, we need one sufficient condition to make $\Delta b_i = \pm 2^{31}$. In step $i+1$, $\pm 2^{31}$ of b_i propagates through function $f = (b_i \oplus c_i \oplus d_i)$. Therefore, we insert $\pm 2^{31}$ on m_i in order to cancel $\pm 2^{31}$ from f. A similar situation occurs in step $i+2$ and $i+3$, so we insert $\pm 2^{31}$ on m, to cancel the difference that propagates from f. In step 4, we cancel $\Delta a_{i+3} = \pm 2^{31}$ by $\Delta m_{i+3} = \pm 2^{31}$, and finally all differences are cancelled. (Since all Δm are in MSB, any sign for Δm is acceptable.)

3.4   New Message Difference

As described in section 3.3, the new local collision makes only one difference, therefore, the new local collision needs only one sufficient condition in the third round, and this is the best local collision for MD4. In order to complete the entire collision attack, the impact of the message expansion on the previous rounds needs to be considered. The local collision constructed in section 3.3 does not fix step i that is the initial step of the local collision. Therefore, we analyze the impact of message expansion, and select the best i that minimizes the impact on the previous rounds.

We show details of this analysis in Appendix B. As a result, we found that $i = 33$ is the best, and obtained following message differences.

$$\Delta m_0 = 2^{28}, \Delta m_2 = 2^{31}, \Delta m_4 = 2^{31}, \Delta m_8 = 2^{31}, \Delta m_{12} = 2^{31},$$
$$\Delta m_i = 0 \text{ for } i = 1, 3, 5, 6, 7, 9, 10, 11, 13, 14, 15.$$

The differential path for the third round is also determined by local collision. It is shown in Table 7 in Appendix C.

Remaining work is to construct a differential path for the first and second round. In the first round, all sufficient conditions are satisfied by basic message modification, whereas, some of sufficient conditions in the second round may not be satisfied by advanced message modification. Therefore, we should minimize the number of sufficient conditions in the second round even if the differential path for the first round becomes more complicated. Since constructing the differential path for the second round is simple, we show only the result in Table 7 in Appendix C.

4   Differential Path Search Algorithm

In this section, we describe the DP construction algorithm, denoted as DPC.

4.1   Definition of Good DPC

Input of DPC is ΔM. A good DPC should not just work for one specified ΔM. It should be able to work for all ΔM. Moreover, it should be finished in reasonable time. If a DPC is able to work for all ΔM in reasonable time, we will call it "Good DPC."

Crucial technique to Realize Good DPC

For DPC, how to control difference propagation through f function is important. Therefore, in design of good DPC, it is crucial to control difference propagation through f function to search DP in a large space efficiently.

4.2   Overview of Problems of Previous Work [6] and Our Improvement

Schläffer and Oswald proposed an algorithm to construct differential paths, based on Wang et al.'s message differences. Unfortunately the algorithm can not work for our new message differences. Therefore, we will propose a new DPC, which is able to work for our new ΔM. We will give an overview of comparison of these two algorithms.

Problem of Schläffer and Oswald's Algorithm

By Schläffer and Oswald's algorithm, the differences produced by f function in step i can only be used in step $i + 4k$, and no others. Although their rule reduced the search space considerably, they found, fortunately, a better differential path than Wang et al's algorithm. However, reduced search space is too small, and consequently, their algorithm can not work for our new message differences.

Advantage of Our Algorithm

In our algorithm, we found a way to control difference propagation through f function efficiently in larger search space.

Important improvement: for step i, Schläffer and Oswald's algorithm only produces difference of f to cancel ΔM in step $i + 4k$. By our algorithm, besides this kind of difference, f can also produce difference to guarantee that ΔM in other steps can be cancelled.

The search space of our algorithm is larger, which makes our algorithm work for the new ΔM proposed in section 3.

4.3   Details of Our New Proposed Algorithm for the First Round

The goal of our algorithm is to construct a differential path for the first round with given chaining variable differences after step 16 and given message differences. Our algorithm consists of three major parts: Forward search, Backward search and Joint algorithm.

i. Forward search
Forward search is done from step 1 to step 4. As mentioned in section 4.1, the crucial technique for making good DPC is controlling difference propagation through f function in a large search space. In forward search, we exhaustively search the possibility of difference through f. After forward search is finished, there are possible differences for chaining variables after step 4. Possible differences for a_4, b_4, c_4, d_4 are called "Potential Differences."

ii. Backward search
First of all, we will redefine the concept "Target Difference" for backward search. Backward Search is done from step 16 to step 8.

Target Differences: for step i, target differences Δt_i are calculated by differences of chaining variable b_i and message differences Δm_{i-1} as follows:

$$\Delta t_i = -\Delta m_{i-1} + (\Delta b_i \ggg s_i)(8 \le i \le 16)$$

Chaining variable differences after step 16 are already fixed, in other words, differences of $b_i(12 \le i \le 16)$ are fixed, it is easy to calculate target differences from step 16 to step 12. In step i, a target difference can be achieved by differences of a_{i-1} or f function. We need to determine each target difference which should be produced by a_{i-1} or f function. During backward search, we assume that differences of a_{i-1} and input differences of f function can always be produced in previous steps, so any target difference can be produced by a_{i-1}. This is main idea of Schläffer and Oswald's algorithm. We enlarge the search space by considering how to guarantee target differences produced by f function.

If we determine that a target difference will be achieved by f, input chaining variables b_{i-1}, c_{i-1}, or d_{i-1} should have a difference at the same position. If b_{i-1}, c_{i-1}, or d_{i-1} has a difference, it should be cancelled in the following step. If it can not be cancelled, it can not be used.

Considering that the chaining variable differences after step 16 are already fixed, from step 16 to step 12, it is very easy to check whether a difference of b_{i-1}, c_{i-1} or d_{i-1} can be cancelled in the following steps.

After backward search from step 16 to step 12 is done, there are some candidates for chaining variable differences in step 12. For every candidate, backward search from step 12 to step 8 is done by applying the same method.

After backward search from step 16 to 8 is done, there are some candidates for chaining variable differences in step 8, which are called "Aimed Differences."

iii. Joint algorithm
From step 4 to step 8, aimed differences will select potential differences. In the joint algorithm, only potential differences that match aimed differences can be

produced by f function. Selected potential differences need to be cancelled in the following steps, To cancel these differences, carry can be used to expand input differences of f function. More differences can also be introduced.

After all the selected differences are canceled, that is, after the joint algorithm is finished, a differential path is constructed.

5   Attack Implementation

5.1   Message Modification

The results of Section 3 and 4 yield the new ΔM, differential path and sufficient conditions. To complete the collision attack, what remains is to propose advanced message modification procedures for satisfying sufficient conditions in the second and third round. Therefore, we propose advanced message modification, and evaluate the total attack complexity of proposed attack. Papers [3,4] give some good ideas of advanced message modification. Since proposing advanced message modification is outside the scope of this paper, we show only the procedures of advanced message modification in Table 9 to Table 20 in Appendix D. As a result of applying advanced message modification, all sufficient conditions in the second round are satisfied with very high probability. Furthermore, by applying the technique called "Shortcut Modification" by [4] or "Tunnel" by [3], the sufficient condition in the third round is also quickly satisfied.

5.2   Attack Procedure and Complexity Estimation

The attack procedure is shown in Table 2. (Note extra conditions are set for advanced message modification. These are introduced in Appendix D.)

For line 1 to 5: Computation for m_i needs almost 1 MD4 step. Therefore, for line 4, we need 16 MD4 steps. Line 2 and 3 can be done very quickly compared to line 4.

For line 6 to 13: For line 7, computation for b_i is 1 MD4 step. Therefore, for line 7, we need 13 MD4 steps. For line 10, there are 11 message modification procedures. According to our evaluation, each procedure requires less than 3 MD4 steps (See Table 9 to 19 for details). Since each condition is satisfied with probability $1/2$, we expect half of the message modification procedures are executed. Therefore, we need $3 \times 11 \div 2 = 16.5$ MD4 steps.

For line 14 to 18: Line 14 needs 4 steps of MD4 operations. Then, if "$b_{33,31} = 0$" is not satisfied, we apply message modification. According to our evaluation, modifying message for "$b_{33,31} = 0$" takes less than 1 MD4 step, and the probability that this condition is satisfied after repeating line 14 to 16 is $1/2$. Therefore, this condition is expected to be satisfied at most 2 trial. Complexity of satisfying in the first trial is 4 MD4 steps and by the second trial is $4 + 4 + 1 = 9$ MD4 steps. Therefore, average complexity is $(4 + 9) \div 2 = 6.5$ MD4 steps. Complexity for line 18 is negligible.

Table 2. Attack Procedure

```
 1. for (1 ≤ i ≤ 16) {
 2.      Randomly take the value of bᵢ.
 3.      Change bᵢ to satisfy all sufficient conds and extra conds for bᵢ.
 4.      Compute mᵢ₋₁ = (bᵢ ⋙ sᵢ) − aᵢ₋₁ − f(bᵢ₋₁, cᵢ₋₁, dᵢ₋₁) − t.
 5. }
 6. for (17 ≤ i ≤ 29) {
 7.      Compute bᵢ in standard way.
 8.      for (0 ≤ j ≤ 31) {
 9.           if (sufficient conds or extra conds on bᵢ,ⱼ is not satisfied) {
10.                Do message modification described in Appendix D.
11.           }
12.      }
13. }
14. Compute b₃₀ to b₃₃ in standard way.
15. if ("b₃₃,₃₁ = 0" is not satisfied) {
16.      Do message modification for "b₃₃,₃₁ = 0", and goto line 14.
17. }
18. Let the present message be M. Finally, M and M + ΔM become a collision pair.
```

Total complexity: Total complexity is addition of above three parts, that is, $16+13+16.5+7.5=53$ MD4 steps. Considering other trivial complexity, we evaluate that the complexity of our attack is less than 2 MD4 operations($=96$ steps).

5.3   Comparison with Previous Attacks

The efficiencies of the proposed attack and previous attacks are compared in Table 3.

Table 3. Comparison of Previous Attacks and Our Result

Method	ΔM	#SCs in 2R	#SCs in 3R	Total complexity (Unit MD4 operations)
Wang et al.	Wang's ΔM	25	2	Less than 2^8
Naito et al.	Wang's ΔM	25	2	Less than 3
Schläffer and Oswald	Wang's ΔM	22	2	-
Proposed attack	ΔM for new LC	9	1	Less than 2

From Table 3, it can be said that improvement of local collision is very important. With the new local collision, number of sufficient conditions in the second and third round becomes less than half those of previous attacks, and this enables us to further reduce the complexity of generating a collision.

6   Conclusion

This paper proposed a new local collision that is the best for collision attack on MD4. This was achieved by the first improvement to the message difference approach of Wang et al.

For successful collision attacks, ΔM must be carefully selected to reduce the number of sufficient conditions that cannot be satisfied by message modification. In this paper, we showed how efficient the proposed improvement to ΔM is. With the new ΔM, number of sufficient conditions in the second and third round is less than half those of previous studies; attack complexity is also improved.

Improving the attack on MD4 is not the final goal. As future works, we will apply our analysis to other hash functions, and aim to find more efficient ΔM.

We realized that differential path construction algorithm proposed by [6] did not work with the new ΔM. Therefore, we analyzed the problems of [6], and proposed a new algorithm for constructing differential path for the first round.

Finally, we show a collision generated by using the new local collision in Table 4. We note that ΔM of this example is different from all previous works.

Table 4. An Example of MD4 Collision Generated with New Local Collision

M	m_0=0x$bcdd2674$ m_1=0x53$fce1ed$ m_2=0x25$d202ce$ m_3=0x$e87d102e$
	m_4=0x$f45be728$ m_5=0x$acc992cc$ m_6=0x6$acfb3ea$ m_7=0x7$dbb29d4$
	m_8=0x$ed03bf75$ m_9=0x$c6aedc45$ m_{10}=0x$d442b710$ m_{11}=0x$fca27d99$
	m_{12}=0x$a5f5eff1$ m_{13}=0x$fb2ee79b$ m_{14}=0x0$f590d68$ m_{15}=0x4989$f380$
M'	m_0'=0x$ccdd2674$ m_1'=0x53$fce1ed$ m_2'=0x$a5d202ce$ m_3'=0x$e87d102e$
	m_4'=0x745$be728$ m_5'=0x$acc992cc$ m_6'=0x6$acfb3ea$ m_7'=0x7$dbb29d4$
	m_8'=0x6$d03bf75$ m_9'=0x$c6aedc45$ m_{10}'=0x$d442b710$ m_{11}'=0x$fca27d99$
	m_{12}'=0x25$f5eff1$ m_{13}'=0x$fb2ee79b$ m_{14}'=0x0$f590d68$ m_{15}'=0x4989$f380$
Hash	0x $c257b7be$ 324$f26ef$ 69$d3d290$ $b01be001$

Acknowledgements

We would like to thank Antoine Joux for improving our paper and anonymous reviewers for helpful comments.

References

1. Dobbertin, H.: Cryptanalysis of MD4. In: Gollmann, D. (ed.) Fast Software Encryption. LNCS, vol. 1039, pp. 53–69. Springer, Heidelberg (1996)
2. Dobbertin, H.: The First Two Rounds of MD4 are Not One-Way. In: Vaudenay, S. (ed.) FSE 1998. LNCS, vol. 1372, pp. 284–292. Springer, Heidelberg (1998)
3. Klima, V.: Tunnels in Hash Functions: MD5 Collisions Within a Minute. Cryptology ePrint Archive, Report (2006)/105
4. Naito, Y., Sasaki, Y., Kunihiro, N., Ohta, K.: Improved Collision Attack on MD4 with Probability Almost 1. In: Won, D.H., Kim, S. (eds.) ICISC 2005. LNCS, vol. 3935, pp. 129–145. Springer, Heidelberg (2006)

5. Rivest, R.: The MD4 Message Digest Algorithm. In: Menezes, A.J., Vanstone, S.A. (eds.) CRYPTO 1990. LNCS, vol. 537, pp. 303–311. Springer, Heidelberg (1991), http://www.ietf.org/rfc/rfc1320.txt
6. Schläffer, M., Oswald, E.: Searching for Differential Paths in MD4. In: Robshaw, M. (ed.) FSE 2006. LNCS, vol. 4047, pp. 242–261. Springer, Heidelberg (2006)
7. Wang, X., Feng, D., Chen, H., Lai, X., Yu, X.: Collision for Hash Functions MD4, MD5, HAVAL-128 and RIPEMD. In: Franklin, M. (ed.) CRYPTO 2004. LNCS, vol. 3152, Springer, Heidelberg (2004)
8. Wang, X., Lai, X., Feng, D., Chen, H., Yu, X.: Cryptanalysis of the Hash Functions MD4 and RIPEMD. In: Cramer, R.J.F. (ed.) EUROCRYPT 2005. LNCS, vol. 3494, pp. 1–18. Springer, Heidelberg (2005)

A   Proof of the Fact in Section 3.2

Proof. Remember that the calculation of b_i in step i is as follows.

$b_i = (a_{i-1} + f(b_{i-1}, c_{i-1}, d_{i-1}) + m_k + t) \lll s_i$

Here, we make $\Delta b_i = 2^{31}$ or -2^{31} by adding the difference $\Delta m_k = +2^{31-s_i}$ or -2^{31-s_i}. Let the value of $a_{i-1} + f(b_{i-1}, c_{i-1}, d_{i-1}) + m_k + t$ be Σ. We focus on the value of Σ in bit position $31 - s_i$ (In short, Σ_{31-s_i}). Σ_{31-s_i} is 0 or 1. First, we consider the case that Σ_{31-s_i} is 0. In this case, if Δm_k is positive, Σ_{31-s_i} changes from 0 to 1, and other bits are kept unchanged. Therefore, it becomes $\Delta b_i = +2^{31}$ after the rotation. However, if Δm_k is negative, Σ_{31-s_i} changes from 0 to 1, and bit position $31 - s_i + 1$ also changes because of carry. Here, assume that carry is stopped in bit position $31 - s_i + 1$, therefore, the value in bit position $31 - s_i + 1$ changes from 1 to 0, and other bits are unchanged. In this case, after the rotation, it becomes $\Delta b_i = +2^{31} - 2^0$, and thus, we cannot obtain the desired difference. The same analysis is applied in the case of Σ_{31-s_i} is 1. The result of the analysis is summarized below.

Table 5. Summary of Proof

	$\Delta m_k = +2^{31-s_i}$	$\Delta m_k = -2^{31-s_i}$
$\Sigma_{31-s_i} = 0$	Success	Failure
$\Sigma_{31-s_i} = 1$	Failure	Success

Consider the fact that $\Sigma_{31-s_i} = b_{i,31}$, we get following conclusion; To make success, $b_{i,31} = 0$ if $\Delta m_k = +2^{31-s_i}$, or $b_{i,31} = 1$ if $\Delta m_k = -2^{31-s_i}$. (Q.E.D)

B   Selecting ΔM by Considering Message Expansion

As described in section 3.3, the new local collision makes one difference in the third round. This does not fix step i that is the initial step of the local collision.

Therefore, we analyze the impact of message expansion, and select the best i that minimizes the impact on the previous rounds.

However, thanks to the basic message modification, even if the differential path becomes complicated and many sufficient conditions are required, all sufficient conditions in the first round can be satisfied with probability 1. Therefore, we only have to care about the impact on the second round.

The third round consists of step 33-48. Since the new local collision takes 5 steps to realize cancellation, we have 12 choices for the initial step of local collision ($IniLC$), ($33 \leq IniLC \leq 44$). To determine a good $IniLC$, we consider following characteristics of the second round.

- We need to cancel all differences by the final step of the second round since local collision in the third round must start from no-difference state.
- Let the last step with some message difference in the second round be cp, which stands for Cancellation Point. If all differences are cancelled at step cp, remaining steps of the second round are guaranteed to be no-difference without sufficient conditions.

Therefore, in order to reduce number of sufficient conditions in the second round, cp should be as early a step as possible. Finally, we select $IniLC$ to minimize cp. We show the analysis of message expansion in Table 6. In Table 6, gray highlight shows an example of the analysis for $IniLC = 33$. Messages that have differences on the second round and third round are colored. From the table, we see that cp for $IniLC = 33$ is 25. The same analysis is applied for all $IniLC$. cp for all $IniLC$ are listed in the bottom of Table 6.

Table 6. Analysis of Impact of Message Expansion

Message index for each step in 2R and 3R

2R	step number	17	18	19	20	21	22	23	24	25	26	27	28	29	30	31	32
	message index	0	4	8	12	1	5	9	13	2	6	10	14	3	7	11	15

3R	step number	33	34	35	36	37	38	39	40	41	42	43	44	45	46	47	48
	message index	0	8	4	12	2	10	6	14	1	9	5	13	3	11	7	15

The value of cp for all $IniLC$

$IniLC$	33	34	35	36	37	38	39	40	41	42	43	44
cp	25	27	27	28	28	28	28	28	29	31	31	32

From Table 6, finally, we find that $IniLC = 33$ is the best since its cp is the smallest. The obtained ΔM and differential path for the third round is shown in Table 7 in Appendix C[2].

---

[2] We also calculate the good ΔM for Wang et al.'s local collision by using this analysis. Then, we found that the resulting ΔM is the same as proposed by Wang et al.'s.

C   Differential Path and Sufficient Conditions

Table 7. Differential Path of MD4 for New Local Collision

Step	Shift		Δb_i	
i	s_i	Δm_{i-1}	Numerical difference	Difference in each bit
1	3	2^{28}	-2^{31}	$\Delta[-31]$
			-2^0	$\Delta[0,-1]$
2	7		-2^7	$\Delta[7]$
			-2^8	$\Delta[-9]$
3	11	2^{31}	-2^{11}	$\Delta[-11]$
4	19		-2^{19}	$\Delta[19,20,-21]$
5	3	2^{31}	2^3	$\Delta[3]$
			2^{22}	$\Delta[-22,-23,24]$
6	7		-2^{18}	$\Delta[18,19,-20]$
			2^{31}	$\Delta[31]$
7	11		2^1	$\Delta[-1,\cdots,9]$
			-2^{22}	$\Delta[22,23,-24]$
			2^{31}	$\Delta[31]$
8	19			
9	3	2^{31}	2^{22}	$\Delta[-22,23]$
			2^{25}	$\Delta[25]$
10	7		-2^{12}	$\Delta[-12]$
			-2^{25}	$\Delta[25,26,-27]$
			2^{29}	$\Delta[-29,-30,31]$
11	11		2^{12}	$\Delta[12]$
12	19		-2^{12}	$\Delta[-12]$
13	3	2^{31}	2^{25}	$\Delta[25]$
			-2^{28}	$\Delta[28,-29]$
14	7			
15	11			
16	19		-2^{31}	$\Delta[-31]$
17	3	2^{28}	2^{28}	$\Delta[28]$
18	5	2^{31}		
19	9	2^{31}		
20	13	2^{31}		
21	3		2^{31}	$\Delta[31]$
22	5			
23	9			
24	13			
25	3	2^{31}		
$\cdots$				
33	3	2^{28}	2^{31}	$\Delta[31]$
34	9	2^{31}		
35	11	2^{31}		
36	15	2^{31}		
37	3	2^{31}		

The symbol '$\Delta[i]$' means the value of chaining variable in bit position i changes from 0 to 1 by difference. The symbol '$\Delta[-i]$' means it changes from 1 to 0 instead.

Table 8. Sufficient Conditions and Extra Conditions for New Local Collision

Chaining variables	Conditions on bits			
	31 - 24	23 - 16	15 - 8	7 - 0
b_1	1 0 - - - - - -	ā - - - - - - -	- - - - - a -	a - - - ā - 0 1
b_2	1 0 - 0 - ā - -	- - - - 0 - - -	- - - - a - 1 -	0 - - - - - 0 1
b_3	1 1 - 1 ā - ā 0	0 - a a 1 - ā ā	ā ā ā ā 1 ā 0 -	0 - - - 1̄ - 1 0
b_4	1 - ā - 1 0 1 1	a a 1 0 0 - 0̄ 1	1̄ 1̄ 1̄ 1̄ 0 1̄ 1 a	1 a a a a - - -
b_5	a - - - 0̄ 0̄ 0̄ 0	1 1 0 0 0 a 1̄ 0	0̄ 0̄ 0̄ 0̄ 0 0̄ 0 1	1 1 1 1 1 - - -
b_6	0 - 0̄ - 1̄ - 1 1	1 1 1 1 0 0 1̄ 1̄	1̄ 1̄ 1̄ 1̄ - 1̄ 0 0	0 0 0 0 0 a a -
b_7	0 - 1̄ - - - 1 1	0 0 - 0 1 0 - -	- - - - - - 0 1	1 1 1 1 1 1 1 -
b_8	1 - - - - - a 0	0 0 - 1 0 1 - -	- - - 0 - - 0 0	0 0 0 0 1 0 0 -
b_9	0 a a - a a 0 1	0 1 - - - - - -	- - - a - - 1 1	1 1 0 1 1 1 1 -
b_{10}	0 1̄ 1̄ - 1 0 0 -	1 1 - - - ā - -	- - - 1 - 0̄ 0̄ -	- - - - - - - -
b_{11}	0 0 0 - 1 1 0 -	1 1 - - - - - -	- - - 0 - 1̄ 1̄ -	- - - - - - - -
b_{12}	0 1 1 a 0 0 1 -	- - - - - 1̄ - -	- - - 1 - 1̄ 0 -	- - - - - - - -
b_{13}	- - 1 0 - - 0 -	- - - - - 1̄ - -	- - - 0 - 0̄ 0̄ -	- - - - - - - -
b_{14}	- - 0 0 - - 0 -	- - - - ā - - -	- - - 0 - 1̄ 1̄ -	- - - - - - - -
b_{15}	a - 1 1 b - 1 -	- - - - - - b	- - - - - - - -	- - - - - - - -
b_{16}	1 - - a 0̄ - - -	- - - - c̄ - - -	- - - - - - - -	- - - - - - - -
b_{17}	b - - 0 - - - -	- - - - ā - - -	- - - - - - - -	- - - - - - - -
b_{18}	b - - c - - - -	- - - - - - - -	- - - - - - - -	- - - - - - - -
b_{19}	- - - a - - - -	- - - - - - - -	- - - - - - - -	- - - - - - - -
b_{20}	a - - - - - - -	- - - - - - - -	- - - - - - - -	- - - - - - - -
b_{21}	0 - - - - - - -	- b - - - - - -	- - - - - - - -	- - - - - - - -
b_{22}	c - - - - - - -	- - - - - - - -	- - - - - - - -	- - - - - - - -
b_{23}	a - - - - - - -	- - - - - - - -	- - - - - - - -	- - - - - - - -
b_{24}	- - - - - - - -	- - - - - - - -	- - - - - - - -	- - - - - - - -
$\cdots$				
b_{33}	0 - - - - - - -	- - - - - - - -	- - - - - - - -	- - - - - - - -
$\cdots$				

The notation '0' stands for the conditions $b_{i,j} = 0$, the notation '1' stands for the conditions $b_{i,j} = 1$, the notation 'a' stands for the conditions $b_{i,j} = b_{i-1,j}$, 'b' stands for the condition $b_{i,j} \neq b_{i-1,j}$ and 'c' stands for the condition $b_{i,j} = b_{i-2,j}$. Conditions without upper bar are sufficient conditions for the new differential path. Conditions with upper bar are extra conditions that are introduced in Appendix D.

D   List of Message Modification Procedures

Table 9 to Table 20 show the procedure for advanced message modification for each condition. These procedures are executed only if the target condition is not satisfied. Several procedures need to satisfy "extra conditions (EC)" in advance. ECs guarantee that each modification procedure is executed without breaking other sufficient conditions. ECs in the first round are set together with sufficient conditions in the first round by basic message modification.

We explain how these procedures work, by explaining Table 10 as an example. Before we execute Table 10, we check whether the target sufficient condition "$b_{17,28} = 0$" is satisfied or not. If it's satisfied, do nothing. Otherwise, execute procedures shown in Table 10. In the table, the first and second column show the rounds and steps where we are looking on. The third column shows the procedure to be executed. The last column shows how differences are used in each calculation. In case of Table 10, we first modify m_0. Then, by considering the expression in step 1, we modify b_1. We make sure that a carry doesn't occur in b_1. Therefore, if $b_{1,28} = 0$ we choose "+," otherwise, we choose "-." After that, to make sure that the result of step 2 keeps unchanged, we modify m_1. In 3rd step, from the property of f, if $b_{2,28}$ is fixed to be 0, the result of step 3 keeps unchanged. Therefore, we set the extra condition "$b_{2,28} = 0$." Step 4 and 5 are the almost same. After step 5 is calculated, all differences in the first round are cancelled. Since m_0 is modified, step 17 in the second round is also modified. From the expression of b_{17}, b_{17} has difference 2^{28}, and this difference flips the value of $b_{17,28}$. Therefore, the sufficient condition is satisfied.

In this section, "$b_i \leftarrow$ standard computation" means computing $b_i \leftarrow (a_{i-1} + f(b_{i-1}, c_{i-1}, d_{i-1}) + m_k + t) \lll s_i$. "$m_i \leftarrow$ inverse computation" means computing $m_i \leftarrow (b_i \ggg s_i) - a_{i-1} - f(b_{i-1}, c_{i-1}, d_{i-1}) - t$.

Table 9. Modification Procedure for Extra Condition "$b_{17,19} = c_{17,19}$"

Round	Step	Actual procedure	Expression with difference in each step
1R	16	$m_{15} \leftarrow m_{15} \pm 2^{29}$ $b_{16} \leftarrow b_{16} \pm 2^{16}$ (Take sign to avoid b_{16} carry.)	$b_{16} \leftarrow (a_{15} + f(b_{15}, c_{15}, d_{15}) + m_{15} \pm 2^{29} + t) \lll 19$
2R	17	EC: "$c_{16,16} \neq d_{16,16}$" $b_{17} \leftarrow$ standard computation	$b_{17} \leftarrow (a_{16} + f(b_{16}[\pm 16], c_{16}, d_{16}) + m_0 + t) \lll 3$

Table 10. Modification Procedure for "$b_{17,28} = 0$"

Round	Step	Actual procedure	Expression with difference in each step
1R	1	$m_0 \leftarrow m_0 \pm 2^{25}$ $b_1 \leftarrow b_1 \pm 2^{28}$ (Take sign to avoid b_1 carry.)	$b_1 \leftarrow (a_0 + f(b_0, c_0, d_0) + m_0 \pm 2^{25} + t) \lll 3$
	2	$m_1 \leftarrow$ inverse computation	$b_2 \leftarrow (a_1 + f(b_1[\pm 28], c_1, d_1) + m_1 + t) \lll 7$
	3	EC: "$b_{2,28} = 0$"	$b_3 \leftarrow (a_2 + f(b_2, c_2[\pm 28], d_2) + m_2 + t) \lll 11$
	4	EC: "$b_{3,28} = 1$"	$b_4 \leftarrow (a_3 + f(b_3, c_3, d_3[\pm 28]) + m_3 + t) \lll 19$
	5	$m_4 \leftarrow m_4 \mp 2^{28}$	$b_5 \leftarrow (a_4 \pm 2^{28} + f(b_4, c_4, d_4) + m_4 \mp 2^{28} + t) \lll 3$
2R	17	$b_{17} \leftarrow$ standard computation	$b_{17} \leftarrow (a_{16} + f(b_{16}, c_{16}, d_{16}) + m_0 \pm 2^{25} + t) \lll 3$

Table 11. Modification Procedure for "$b_{17,31} \neq b_{16,31}$"

Round	Step	Actual procedure	Expression with difference in each step
1R	1	$m_0 \leftarrow m_0 + 2^{27}$ $b_1 \leftarrow b_1 + 2^{30}$ EC: "$b_{1,30} = 0$"	$b_1 \leftarrow (a_0 + f(b_0, c_0, d_0) + m_0 + 2^{27} + t) \lll 3$
	2	$m_1 \leftarrow$ inverse operation	$b_2 \leftarrow (a_1 + f(b_1[+30], c_1, d_1) + m_1 + t) \lll 7$
	3	EC: "$b_{2,30} = 0$"	$b_3 \leftarrow (a_2 + f(b_2, c_2[+30], d_2) + m_2 + t) \lll 11$
	4	EC: "$b_{3,30} = 1$"	$b_4 \leftarrow (a_3 + f(b_3, c_3, d_3[+30]) + m_3 + t) \lll 19$
	5	$m_4 \leftarrow m_4 - 2^{30}$	$b_5 \leftarrow (a_4 + 2^{30} + f(b_4, c_4, d_4) + m_4 - 2^{30} + t) \lll 3$
1-2R	16	$m_{15} \leftarrow m_{15} + 2^8$ EC: "$b_{16,27} = 0$"	$b_{16} \leftarrow (a_{15} + f(b_{15}, c_{15}, d_{15}) + m_{15} + 2^8 + t) \lll 19$
	17	$b_{17} \leftarrow$ standard operation EC: "$c_{16,27} \neq d_{16,27}$"	$b_{17} \leftarrow (a_{16} + f(b_{16}[+27], c_{16}, d_{16}) + m_0 + 2^{27} + t) \lll 3$

Table 12. Modification Procedure for "$b_{18,28} = 0$"

Round	Step	Actual procedure	Expression with difference in each step
1R	2	$m_1 \leftarrow m_1 \pm 2^{16}$ $b_2 \leftarrow b_2 \pm 2^{23}$ (Take sign to avoid b_2 carry.)	$b_2 \leftarrow (a_1 + f(b_1, c_1, d_1) + m_1 \pm 2^{16} + t) \lll 7$
	3	EC: "$c_{2,23} = d_{2,23}$"	$b_3 \leftarrow (a_2 + f(b_2[\pm 23], c_2, d_2) + m_2 + t) \lll 11$
	4	EC: "$b_{3,23} = 0$"	$b_4 \leftarrow (a_3 + f(b_3, c_3[\pm 23], d_3) + m_3 + t) \lll 19$
	5	$m_4 \leftarrow m_4 \mp 2^{23}$ EC: "$b_{4,23} = 0$"	$b_5 \leftarrow (a_4 + f(b_4, c_4, d_4[\pm 23]) + m_4 \mp 2^{23} + t) \lll 3$
	6	$m_5 \leftarrow m_5 \mp 2^{23}$	$b_6 \leftarrow (a_5 \pm 2^{23} + f(b_5, c_5, d_5) + m_5 \mp 2^{23} + t) \lll 7$
2R	17	$b_{18} \leftarrow$ standard computation	$b_{18} \leftarrow (a_{17} + f(b_{17}, c_{17}, d_{17}) + m_4 \mp 2^{23} + t) \lll 5$

Table 13. Modification Procedure for "$b_{18,31} \neq b_{17,31}$"

Round	Step	Actual procedure	Expression with difference in each step
1R	5	$m_4 \leftarrow m_4 \pm 2^{26}$ $b_5 \leftarrow b_5 \pm 2^{29}$ (Take sign to avoid b_5 carry.)	$b_5 \leftarrow (a_4 + f(b_4, c_4, d_4) + m_4 \pm 2^{26} + t) \lll 3$
	6	EC: "$c_{5,29} = d_{5,29}$"	$b_6 \leftarrow (a_5 + f(b_5[\pm 29], c_5, d_5) + m_5 + t) \lll 7$
	7	EC: "$b_{6,29} = 0$"	$b_7 \leftarrow (a_6 + f(b_6, c_6[\pm 29], d_6) + m_6 + t) \lll 11$
	8	EC: "$b_{7,29} = 1$"	$b_8 \leftarrow (a_7 + f(b_7, c_7, d_7[\pm 29]) + m_7 + t) \lll 19$
	9	$m_8 \leftarrow m_8 \mp 2^{29}$	$b_9 \leftarrow (a_8 \pm 2^{29} + (f(b_8, c_8, d_8) + m_8 \mp 2^{29} + t) \lll 3$
2R	17	$b_{18} \leftarrow$ standard computation	$b_{18} \leftarrow (a_{17} + f(b_{17}, c_{17}, d_{17}) + m_4 \pm 2^{26} + t) \lll 5$

Table 14. Modification Procedure for "$b_{19,28} = b_{18,28}$"

Round	Step	Actual procedure	Expression with difference in each step
1-2R	15	$m_{14} \leftarrow m_{14} \pm 2^8$ $b_{15} \leftarrow b_{15} \pm 2^{19}$ (Take sign to avoid b_{15} carry.)	$b_{15} \leftarrow (a_{14} + f(b_{14}, c_{14}, d_{14}) + m_{14} \pm 2^8 + t) \lll 11$
	16	EC: "$c_{15,19} = d_{15,19}$"	$b_{16} \leftarrow (a_{15} + f(b_{15}[\pm 19], c_{15}, d_{15}) + m_{15} + t) \lll 19$
	17	EC: "$b_{16,19} = d_{16,19}$"	$b_{17} \leftarrow (a_{16} + f(b_{16}, c_{16}[\pm 19], d_{16}) + m_0 + t) \lll 3$
	18	EC: "$b_{17,19} = c_{17,19}$"	$b_{18} \leftarrow (a_{17} + f(b_{17}, c_{17}, d_{17}[\pm 19]) + m_4 + t) \lll 5$
	19	$b_{19} \leftarrow$ standard computation	$b_{19} \leftarrow (a_{18} \pm 2^{19} + f(b_{18}, c_{18}, d_{18}) + m_8 + t) \lll 9$

Table 15. Modification Procedure for "$b_{20,31} = b_{19,31}$"

round	step	Actual procedure	Expression with difference in each step
1R	11	$m_{10} \leftarrow m_{10} \pm 2^7$ $b_{11} \leftarrow b_{11} \pm 2^{18}$ take sign to avoid b_{11} carry	$b_{11} \leftarrow (a_{10} + f(b_{10}, c_{10}, d_{10}) + m_{10} \pm 2^7 + t) \lll 11$
	12	EC: "$c_{11,18} = d_{11,18}$"	$b_{12} \leftarrow (a_{11} + f(b_{11}[\pm 18], c_{11}, d_{11}) + m_{11} + t) \lll 19$
	13	$m_{12} \leftarrow m_{12} \mp 2^{18}$	$b_{13} \leftarrow (a_{12} + f(b_{12}, c_{12}[\pm 18], d_{12}) + m_{12} \mp 2^{18} + t) \lll 3$
	14	EC: "$b_{12,18} = 1$" EC: "$b_{13,18} = 1$"	$b_{14} \leftarrow (a_{13} + f(b_{13}, c_{13}, d_{13}[\pm 18]) + m_{13} + t) \lll 7$
	15	$m_{14} \leftarrow m_{14} \mp 2^{18}$	$b_{15} \leftarrow (a_{14} \pm 2^{18} + f(b_{14}, c_{14}, d_{14}) + m_{14} \mp 2^{18} + t) \lll 11$
2R	20	$b_{20} \leftarrow$ standard operation	$b_{20} \leftarrow (a_{19} + f(b_{19}, c_{19}, d_{19}) + m_{12} \mp 2^{18} + t) \lll 13$

Table 16. Modification Procedure for Extra Condition "$b_{21,22} \neq b_{20,22}$"

Round	Step	Actual procedure	Expression with difference in each step
1R	12	$m_{11} \leftarrow m_{11} \pm 2^{22}$ $b_{12} \leftarrow b_{12} \pm 2^9$ EC: "$b_{12,10} = 1$", "$b_{12,9} = 0$" (if sign is +, $b_3[9]$, else $b_3[-10,9]$.)	$b_{12} \leftarrow (a_{11} + f(b_{11}, c_{11}, d_{11}) + $ $+ m_{11} \pm 2^{22} + t) \lll 19$
	13	$m_{12} \leftarrow m_{12} \mp 2^9$ (Take sign to avoid b_{20} carry.) EC: "$c_{12,10} = 1$", "$d_{12,10} = 0$", "$c_{12,9} = 1$", "$d_{12,9} = 0$"	$b_{13} \leftarrow (a_{12} + f(b_{12}[9]/[-10,9], c_{12}, d_{12}) $ $+ m_{12} \mp 2^9 + t) \lll 3$
	14	EC: "$b_{13,10} = 0$", "$b_{13,9} = 1$"	$b_{14} \leftarrow (a_{13} + f(b_{13}, c_{13}[9]/[-10,9], d_{13}) $ $+ m_{13} + t) \lll 7$
	15	EC: "$b_{14,10} = 1$", "$b_{14,9} = 1$"	$b_{15} \leftarrow (a_{14} + f(b_{14}, c_{14}, d_{14}[9]/[-10,9]) $ $+ m_{14} + t) \lll 11$
	16	$m_{15} \leftarrow m_{15} \mp 2^9$	$b_{16} \leftarrow (a_{15} \pm 2^9 + f(b_{15}, c_{54}, d_{15}) $ $+ m_{15} \mp 2^9 + t) \lll 19$
2R	20	$b_{20} \leftarrow$ standard computation	$b_{20} \leftarrow (a_{19} + f(b_{19}, c_{19}, d_{19}) $ $+ m_{12} \mp 2^9 + t) \lll 13$
	21	$b_{21} \leftarrow$ standard computation ($b_{21,22}$ is never changed by this calculation).	$b_{21} \leftarrow (a_{20} + f(b_{20}[\mp 22], c_{20}, d_{20}) $ $+ m_1 + t) \lll 3$

Table 17. Modification Procedure for "$b_{21,31} = 0$"

Round	Step	Actual procedure	Expression with difference in each step
1R	2	$m_1 \leftarrow m_1 \pm 2^{28}$ $b_2 \leftarrow b_2 \pm 2^3$ (Take sign to avoid b_2 carry.)	$b_2 \leftarrow (a_1 + f(b_1, c_1, d_1) + m_1 \pm 2^{28} + t) \lll 7$
	3	EC: "$c_{2,3} = d_{2,3}$"	$b_3 \leftarrow (a_2 + f(b_2[\pm 3], c_2, d_2) + m_2 + t) \lll 11$
	4	$m_3 \leftarrow$ inverse computation	$b_4 \leftarrow (a_3 + f(b_3, c_3[\pm 3], d_3) + m_3 + t) \lll 19$
	5	EC: "$b_{4,3} = 1$"	$b_5 \leftarrow (a_4 + f(b_4, c_4, d_4[\pm 3]) + m_4 + t) \lll 3$
	6	$m_5 \leftarrow m_5 \mp 2^3$	$b_6 \leftarrow (a_5 \pm 2^3 + f(b_5, c_5, d_5) + m_5 \mp 2^3 + t) \lll 7$
2R	21	$b_{21} \leftarrow$ standard computation	$b_{21} \leftarrow (a_{20} + f(b_{20}, c_{20}, d_{20}) + m_1 \pm 2^{28} + t) \lll 3$

Table 18. Modification Procedure for "$b_{22,31} = b_{21,31}$"

Round	Step	Actual procedure	Expression with difference in each step
1R	3	$m_2 \leftarrow m_2 \pm 2^{15}$ $b_3 \leftarrow b_3 \pm 2^{26}$ (Take sign to avoid b_3 carry.)	$b_3 \leftarrow (a_2 + f(b_2, c_2, d_2) + m_2 \pm 2^{15} + t) \lll 11$
	4	EC: "$c_{3,26} = d_{3,26}$"	$b_4 \leftarrow (a_3 + f(b_3[\pm 26], c_3, d_3) + m_3 + t) \lll 19$
	5	EC: "$b_{4,26} = 0$"	$b_5 \leftarrow (a_4 + f(b_4, c_4[\pm 26], d_4) + m_4 + t) \lll 3$
	6	$m_5 \leftarrow m_5 \mp 2^{26}$ EC: "$b_{5,26} = 0$"	$b_6 \leftarrow (a_5 + f(b_5, c_5, d_5[\pm 26]) + m_5 + t) \lll 7$
	7	$m_6 \leftarrow m_6 \mp 2^{26}$	$b_7 \leftarrow (a_6 \pm 2^{26} + f(b_6, c_6, d_6) + m_6 \mp 2^{26} + t) \lll 11$
2R	21	$b_{22} \leftarrow$ standard computation	$b_{22} \leftarrow (a_{21} + f(b_{21}, c_{21}, d_{21}) + m_5 \mp 2^{26} + t) \lll 5$

Table 19. Modification Procedure for "$b_{23,31} = b_{22,31}$"

Round	Step	Actual procedure	Expression with difference in each step
1R	4	$m_3 \leftarrow m_3 \pm 2^{30}$ $b_4 \leftarrow b_4 \pm 2^{17}$ (Take sign to avoid b_4 carry.)	$b_4 \leftarrow (a_3 + f(b_3, c_3, d_3) + m_3 \pm 2^{30} + t) \lll 19$
	5	EC: "$c_{4,17} = d_{4,17}$"	$b_5 \leftarrow (a_4 + f(b_4[\pm 17], c_4, d_4) + m_4 + t) \lll 3$
	6	$m_5 \leftarrow m_5 \mp 2^{17}$ EC: "$b_{5,17} = 1$"	$b_6 \leftarrow (a_5 + f(b_5, c_5[\pm 17], d_5) + m_5 \mp 2^{17} + t) \lll 7$
	7	EC: "$b_{6,17} = 1$"	$b_7 \leftarrow (a_6 + f(b_6, c_6, d_6[\pm 17]) + m_6 + t) \lll 11$
	8	$m_7 \leftarrow m_7 \mp 2^{17}$	$b_8 \leftarrow (a_7 \pm 2^{17} + f(b_7, c_7, d_7) + m_7 \mp 2^{17} + t) \lll 19$
2R	22	$b_{22} \leftarrow$ standard computation (This breaks SC on $b_{22,31}$ with probability 2^{-9}.)	$b_{22} \leftarrow (a_{21} + f(b_{21}, c_{21}, d_{21}) + m_5 \mp 2^{17} + t) \lll 5$
	23	$b_{23} \leftarrow$ standard computation EC: "$c_{22,22} \neq d_{22,22}$"	$b_{23} \leftarrow (a_{22} + f(b_{22}[\mp 22], c_{22}, d_{22}) + m_9 + t) \lll 9$

Table 20. Modification Procedure for Starting Collision Search from Middle of the Second Round

In order to satisfy a condition "$b_{33,31} = 0$" in the third round, we modify the value of chaining variable in step 29, which is a late step of the second round, This difference will propagate until step 33, and can flip the value of $b_{33,31}$. In this modification, we need to set several extra conditions. We can set all extra conditions when $i = 0, 9, 11, 26, 28, 29, 30, 31$. Therefore, we can make 2^8 values for b_{29} that satisfy all sufficient conditions in previous steps. Since, we don't control how the difference is propagated from step 29 to 33, the probability of satisfying "$b_{33,31} = 0$" is not 1. However, 256 times of trials enable us to quickly find a message that also satisfies "$b_{33,31} = 0$."

Round	Step	Actual procedure	Expression with difference in each step
1R	4	$m_3 \leftarrow m_3 - 2^{i-3}$ $b_4 \leftarrow b_4 - 2^{i+16}$ EC: "$b_{4,i+16} = 0$"	$b_4 \leftarrow (a_3 + f(b_3, c_3, d_3) +$ $m_3 - 2^{i-3} + t) \lll 19$
	5	EC: "$c_{4,i+16} = d_{4,i+16}$"	$b_5 \leftarrow (a_4 + f(b_4[-(i+16)], c_4, d_4) +$ $m_4 + t) \lll 3$
	6	EC: "$b_{5,i+16} = 0$"	$b_6 \leftarrow (a_5 + f(b_5, c_5[-(i+16)], d_5) +$ $m_5 + t) \lll 7$
	7	EC: "$b_{6,i+16} = 1$"	$b_7 \leftarrow (a_6 + f(b_6, c_6, d_6[-(i+16)]) +$ $m_6 + t) \lll 11$
	8	$m_7 \leftarrow m_7 + 2^{i+16}$	$b_8 \leftarrow (a_7 - 2^{i+16} + (f(b_7, c_7, d_7) +$ $m_7 + 2^{i+16} + t) \lll 19$
2R	29	$b_{29} \leftarrow$ standard computation	$b_{29} \leftarrow (a_{28} + f(b_{28}, c_{28}, d_{28}) +$ $m_3 - 2^{i-3} + t) \lll 3$
	30	$b_{30} \leftarrow$ standard computation	
	31	$b_{31} \leftarrow$ standard computation	Difference propagation is out of control.
	32	$b_{32} \leftarrow$ standard computation	Probabilistically flip the value of $b_{33,31}$.
3R	33	$b_{33} \leftarrow$ standard computation	

Algebraic Cryptanalysis of 58-Round SHA-1

Makoto Sugita[1], Mitsuru Kawazoe[2], Ludovic Perret[3], and Hideki Imai[4]

[1] IT Security Center, Information-technology Promotion Agency, Japan
2-28-8 Honkomagome, Bunkyo-ku Tokyo, 113-6591, Japan
m-sugita@ipa.go.jp
[2] Faculty of Liberal Arts and Sciences
Osaka Prefecture University
1-1 Gakuen-cho Naka-ku Sakai Osaka 599-8531 Japan
kawazoe@las.osakafu-u.ac.jp
[3] SPIRAL/SALSA
Site Passy-Kennedy
LIP6 – Paris 6 University
104 avenue du Président Kennedy
75016 Paris France
ludovic.perret@lip6.fr
[4] National Institute of Advanced Industrial Science and Technology (AIST)
Akihabara Dai Bldg., 1-18-13 Sotokanda, Chiyoda-ku, Tokyo 101-0021, Japan
Department of Electrical, Electronic and Communication Engineering
Faculty of Science and Engineering, Chuo University
1-13-27 Kasuga Bunkyo-ku, Tokyo 112-8551 Japan
h-imai@aist.go.jp

Abstract. In 2004, a new attack against SHA-1 has been proposed by a team leaded by Wang [15]. The aim of this article[1] is to sophisticate and improve Wang's attack by using algebraic techniques. We introduce new notions, namely semi-neutral bit and adjuster and propose then an improved message modification technique based on algebraic techniques. In the case of the 58-round SHA-1, the experimental complexity of our improved attack is 2^{31} SHA-1 computations, whereas Wang's method needs 2^{34} SHA-1 computations. We have found many new collisions for the 58-round SHA-1. We also study the complexity of our attack for the full SHA-1.

Keywords: SHA-1, Gröbner basis, differential attack.

1 Introduction

Conceptually, we can split Wang's attack in four different steps.

Step 1. Choose a suitable 80×32-bit vector Γ.
Step 2. Choose a differential characteristic.

---

[1] A part of this work has been done when the third author was invited at Research Center for Information Security (RCIS) at Tokyo (Japan).

A. Biryukov (Ed.): FSE 2007, LNCS 4593, pp. 349–365, 2007.
© International Association for Cryptologic Research 2007

Step 3. Find a set of sufficient conditions on a message m and the chaining variables which guarantees with high probability that the message pair $(m, m + \Gamma)$ follows the differential characteristic. This implies that the two messages do collide.

Step 4. Choose a message m randomly and modify it until all sufficient conditions hold.

Using this method, Wang's team succeeded in finding collisions on the most popular hash functions, namely MD4, MD5, RIPEMD, SHA-0 and 58-round SHA-1 [11,17,15]. The attack is conceptually simple, but its implementation turns out to be very laborious in practice. To fill this gap between theory and practice, several teams decided to compensate their lack of intuition by the power of a computer, that is to say they tried to automatize the different steps of the attack.

For instance, coding theory can be used for finding a suitable difference in Step 1[8] of the attack. Recently, De Cannière and Rechberger [3] presented an algorithm allowing to find optimal differential characteristics in Step 2. The third step is tightly coupled with the previous one; most sufficient conditions follow from the choice of the differential characteristic.

We use algebraic techniques for actually finding collisions on 58-round SHA-1. In this case, the complexity of our method for finding a collision is equivalent to 2^{31} SHA-1 computations (experimentally), whereas Wang's method needs 2^{34} SHA-1 computations. As a proof of concept, we have found many new collisions for 58-round SHA-1, which have never been reported so far. We also apply our method for the case of the full SHA-1, and study the complexity of our approach.

The key idea is to describe the message modification technique into an algebraic framework. This is done by viewing the set of sufficient conditions as a non-linear system of Boolean equations. We hope that this will be a first step towards the use of algebraic tools (such as Gröbner bases) in the cryptanalysis of hash functions.

We will focus our attention on the last step of Wang's et al attack [15]. Namely, find a message satisfying a set of sufficient conditions depending on a disturbance vector and a differential path. This message can be then use to produce a collision. We shall call *conventional message modification* the process [15] permitting to construct such a suitable message. Here, we will present an improved message modification technique. To do so, we introduce the concepts *semi-neutral bits* and *adjusters*.

This paper is organized as follows. In Section 2, we give a description of SHA-1. Along the way, we introduce the notations and definitions that will be used throughout this paper. In Section 3, we describe our improved message modification technique. We explain how to use Gaussian elimination to construct a set controlled relations from the sufficient conditions. We also introduce a new notion, that we called *semi-neutral bit*, and describe then our improved message modification. We also give an algebraic descriptions of our improved message modification. This permits to give an interesting connection between the cryptanalysis of hash functions and the use of Gröbner bases. In Section 4, we present the details of our method on 58-round SHA-1. In the appendix, we provide the details for the full SHA-1.

2   Preliminaries

2.1   Description of SHA-1

The hash function SHA-1 generates a 160-bit hash value (or digest) from a message of length less than 2^{64} bits. The input message is padded and then processed in 512-bit message blocks through the Merkle/Damgard iterative structure.

A 80-step compression function is then applied to each of these 512-bit message blocks. It has two types of inputs: a chaining input of 160 bits and a message input of 512 bits. The initial chaining value (called IV) is a set of fixed constants, and the result of the last call to the compression function is the hash of the message.

In SHA-1, the message expansion is defined as follows: each 512-bit block of the padded message is divided into a 16×32-bit word $(m_0, m_1, \ldots, m_{15})$, and then expanded according to the following linear relation :

$$m_i \leftarrow (m_{i-3} \oplus m_{i-8} \oplus m_{i-14} \oplus m_{i-16}) \lll 1, \text{ for all } i, 16 \leq i \leq 79,$$

$x \lll n$, denoting the n-bit left rotation of a 32-bit word x. The compression function is defined for all $i, 1, \leq i \leq 80$ as follows:

$$a_i \leftarrow (a_{i-1} \lll 5) + f_i(b_{i-1}, c_{i-1}, d_{i-1}) + e_{i-1} + m_{i-1} + k_i$$
$$b_i \leftarrow a_{i-1}$$
$$c_i \leftarrow b_{i-1} \lll 30$$
$$d_i \leftarrow c_{i-1}$$
$$e_i \leftarrow d_{i-1}$$

The initial chaining value IV=$(a_0, b_0, c_0, d_0, e_0)$ being equal to:

$$(0x67452301, 0xefcdab89, 0x98badcfe, 0x10325476, 0xc3d2e1f0).$$

Note that we express as usual 32-bit words as hexadecimal numbers.

For $n, 58 \leq n \leq 80$, we call n-round SHA-1, the restriction of SHA-1 to the first n rounds. The Boolean function f_i and constant k_i employed at each step are defined as in Table 1.

Table 1. Definition of f_i and k_i w.r.t. the step

Step	Boolean function f_i	Constant k_i
$1-20$	IF: $(x \land y) \lor (\neg x \land z)$	$0x5a827999$
$21-40$	XOR: $x \oplus y \oplus z$	$0x6ed6eba1$
$41-60$	MAJ: $(x \land y) \lor (x \land z) \lor (y \land z)$	$0x8fabbcdc$
$61-80$	XOR: $x \oplus y \oplus z$	$0xca62c1d6$

2.2 Definition and Notation

We will identify the ring $\mathbb{Z}/2^{32}\mathbb{Z}$ with $\{0, 1, 2, \ldots, 2^{32} - 1\}$. If we ignore carry effects in the arithmetic of $\mathbb{Z}/2^{32}\mathbb{Z}$, we can identify the ring $\mathbb{Z}/2^{32}\mathbb{Z}$ with the vector space $\mathbb{F}_2^{32}$ by using the canonical bijective mapping ϱ:

$$\varrho : x_{31} 2^{31} + x_{30} 2^{30} + \cdots + x_1 2^1 + x_0 2^0 \in \mathbb{Z}/2^{32}\mathbb{Z} \longmapsto (x_{31}, x_{30}, \ldots, x_0) \in \mathbb{F}_2^{32}.$$

Here and in the rest of the paper, we try to find a collision between two messages $m = (m_0, m_1, \ldots, m_{79})$ and $m' = (m'_0, m'_1, \ldots, m'_{79})$ of $\left(\mathbb{F}_2^{32}\right)^{80}$. The corresponding chaining variables will be denoted by a_i, b_i, c_i, d_i, e_i and $a'_i, b'_i, c'_i, d'_i, e'_i$ respectively. For each $m_i = (m_{i,31}, m_{i,30}, \ldots, m_{i,0})$ and $m'_i = (m'_{i,31}, m'_{i,30}, \ldots, m'_{i,0})$, we define:

$$\Delta m_{i,j} = m_{i,j} \oplus m'_{i,j} \in \mathbb{F}_2,$$
$$\Delta m_i = m_i \oplus m'_i \in \mathbb{F}_2^{32},$$
$$\delta m_i = \varrho^{-1}(m_i) - \varrho^{-1}(m'_i) \in \mathbb{Z}/2^{32}\mathbb{Z}.$$

Moreover, we set :

$$\Delta^+ m_{i,j} = \begin{cases} 1 & \text{if } (m_{i,j}, m'_{i,j}) = (0, 1) \\ 0 & \text{otherwise,} \end{cases} \qquad \Delta^- m_{i,j} = \begin{cases} 1 & \text{if } (m_{i,j}, m'_{i,j}) = (1, 0) \\ 0 & \text{otherwise.} \end{cases}$$

and

$$\Delta^+ m_i = (\Delta^+ m_{i,31}, \ldots, \Delta^+ m_{i,1}, \Delta^+ m_{i,0}), \quad \Delta^- m_i = (\Delta^- m_{i,31}, \ldots, \Delta^- m_{i,1}, \Delta^- m_{i,0}).$$

Note that $\Delta m_i = \Delta^+ m_i \oplus \Delta^- m'_i \in \mathbb{F}_2^{32}$. Similarly, we define $\Delta, \Delta^+, \Delta^-, \delta$ for the chaining variables b_i, c_i, d_i and e_i (resp. b'_i, c'_i, d'_i and e'_i). Using the above definition, a *differential characteristic* and a *differential* are defined as follows.

Definition 1. *We call* **differential characteristic** *the sequence:*

$$\left(\Delta m_i, \Delta a_i, \Delta b_i, \Delta c_i, \Delta d_i, \Delta e_i\right)_{0 \leq i \leq 79},$$

and **differential**:

$$\left(\Delta^+ m_i, \Delta^- m_i, \Delta^+ a_i, \Delta^- a_i, \ldots, \Delta^+ e_i, \Delta^- e_i\right)_{0 \leq i \leq 79}.$$

3 An Improved Message Modification Technique

Here, we will consider a n-round SHA-1, with $n, 58 \leq n \leq 80$. We will focus our attention on the last step of Wang's attack. Thus, we will suppose that a disturbance vector is fixed, as well as a suitable differential. We can then determine sufficient conditions on the messages permitting to produce collisions. Remark that sufficient conditions depend on the choice of a disturbance vector and its differential.

3.1 How to Calculate Sufficient Conditions on the a_i ?

In this step, we only consider expanded messages by ignoring relations arising from the message expansion. We compute the sufficient conditions on chaining variables by adjusting b_i, c_i and d_i such that for all $i, 0 \leq i \leq n - 1$:

$$\delta f_i(b_i, c_i, d_i) = \delta a_{i+1} - (\delta a_i \lll 5) - \delta e_i - \delta m_i.$$

In this calculation, we must adjust carry effects by "hand". It is indeed difficult to calculate this full-automatically. In Table 2 and Table 6, we present the sufficient conditions that we have obtained on the chaining variables for 58-round and the full SHA-1, respectively. Note that sufficient conditions on the messages are also quoted in this table.

3.2 Gaussian Elimination and Controlled Relations

To calculate sufficient conditions on the $\{m_{i,j}\}_{0 \leq i \leq n-1}^{0 \leq j \leq 31}$, we take into account that $\Delta^+ m_{i,j} = 1$ implies $m_{i,j} = 0$ and $\Delta^- m_{i,j} = 0$ implies $m_{i,j} = 1$. We also consider the relations derived from the key expansion:

$$m_{i,j} \leftarrow (m_{i-3,j} \oplus m_{i-8,j} \oplus m_{i-14,j} \oplus m_{i-16,j}) \lll 1, \text{ for all } i, 16 \leq i \leq n - 1.$$

We shall call *controlled relations* a particular set of $\mathbb{F}_2$-linear equations on the $m_{i,j}$ on one hand, and on the chaining variables $a_{i,j}$ on the other. For the $m_{i,j}$, we consider the relations obtained by performing a Gaussian elimination on the linear equations defined by the key expansion, and the equations derived from the sufficient conditions. To perform this Gaussian elimination, we have considered the following order on the $m_{i,j}$:

$$m_{i',j'} < m_{i,j}, \text{ if } i' < i \text{ or } i' = i \text{ and } j' < j.$$

For $n = 58$, we obtain for instance the following controlled relations :

$m_{15,31} = 1$, $m_{15,30} = 1$, $m_{15,29} = 0$, $m_{15,28} + m_{10,28} + m_{8,29} + m_{7,29} + m_{4,28} + m_{2,28} = 1$,
$m_{15,27} + m_{14,25} + m_{12,28} + m_{12,26} + m_{10,28} + m_{9,27} + m_{9,25} + m_{8,29} + m_{8,28} + m_{7,28} +$
$m_{7,27} + m_{6,26} + m_{5,28} + m_{4,26} + m_{3,25} + m_{2,28} + m_{1,25} + m_{0,28} = 1$, $m_{15,26} + m_{10,28} +$
$m_{10,26} + m_{8,28} + m_{8,27} + m_{7,27} + m_{6,29} + m_{5,27} + m_{4,26} + m_{2,27} + m_{2,26} + m_{0,27} = 1$,
$m_{15,25} + m_{11,28} + m_{10,27} + m_{10,25} + m_{9,28} + m_{8,27} + m_{8,26} + m_{7,26} + m_{6,29} + m_{6,28} +$
$m_{5,26} + m_{4,25} + m_{3,28} + m_{2,28} + m_{2,26} + m_{2,25} + m_{1,28} + m_{0,28} + m_{0,26} = 0$, $m_{15,24} +$
$m_{12,28} + m_{11,27} + m_{10,26} + m_{10,24} + m_{9,28} + m_{9,27} + m_{8,29} + m_{8,26} + m_{8,25} + m_{7,25} +$
$m_{6,29} + m_{6,28} + m_{6,27} + m_{5,25} + m_{4,28} + m_{4,24} + m_{3,28} + m_{3,27} + m_{2,27} + m_{2,25} + m_{2,24} +$
$m_{1,28} + m_{1,27} + m_{0,27} + m_{0,25} = 1$, $m_{15,23} + m_{12,28} + m_{12,27} + m_{11,26} + m_{10,25} + m_{10,23} +$
$m_{9,27} + m_{9,26} + m_{8,28} + m_{8,25} + m_{8,24} + m_{7,29} + m_{7,24} + m_{6,28} + m_{6,27} + m_{6,26} + m_{5,24} +$
$m_{4,27} + m_{4,23} + m_{3,27} + m_{3,26} + m_{2,26} + m_{2,24} + m_{2,23} + m_{1,27} + m_{1,26} + m_{0,26} + m_{0,24} = 1$,
$m_{15,22} + m_{14,25} + m_{12,28} + m_{12,27} + m_{11,25} + m_{10,27} + m_{10,24} + m_{10,22} + m_{9,28} + m_{9,27} +$
$m_{9,26} + m_{8,27} + m_{8,24} + m_{8,23} + m_{7,28} + m_{7,27} + m_{7,23} + m_{6,27} + m_{6,25} + m_{5,23} + m_{4,28} +$
$m_{4,27} + m_{4,22} + m_{3,26} + m_{2,28} + m_{2,27} + m_{2,25} + m_{2,23} + m_{2,22} + m_{1,26} + m_{0,25} + m_{0,23} = 0$,
$\ldots$, $m_{5,0} + m_{3,0} + m_{1,31} = 1$, $m_{4,31} = 0$, $m_{4,30} = 0$, $m_{4,29} = 0$, $m_{4,6} = 0$, $m_{4,1} = 1$,

$m_{3,30} = 1$, $m_{3,29} = 0$, $m_{3,6} = 1$, $m_{2,31} = 0$, $m_{2,30} = 1$, $m_{2,29} = 0$, $m_{2,6} = 1$, $m_{2,1} = 1$, $m_{2,0} = 1$, $m_{1,30} = 0$, $m_{1,29} = 1$, $m_{1,5} = 0$, $m_{1,4} = 1$, $m_{1,1} = 1$, $m_{0,31} = 0$, $m_{0,30} = 0$, $m_{0,29} = 0$.

The controlled relations also include a subset of the sufficient conditions on the chaining variables. Precisely, we will only consider the conditions involving $a_{i,j}$, with $i \leq R$. The bound R is a positive integer that will be defined later. We will call *uncontrolled relations*, the sufficient conditions which are not a controlled relation.

We define now the notions of *semi-neutral bit*, *control bit* and *adjuster*. The concept of semi-neutral bit is closely related to Biham and Chen's "neutral bit" [2] and Klima's "tunnels" [22]. Namely, if the effect of flipping a bit corresponding to a chaining variable can be "easily" eliminated (i.e. such that all conditions previously satisfied can be satisfied by modifying few bits), then we shall call this bit a *semi-neutral bit*. Thus, the effect of changing a semi-neutral bit can be eliminated by controlling a little number of bits. We shall call these particular bits *adjusters*. Note that the choice of semi-neutral bits and adjusters is not unique. Thus, we have to choose it heuristically.

We emphasize that each $m_{i,j}$ can be viewed as a polynomial on the $a_{k,\ell}$'s, with $k \leq i+1$. Indeed, each $m_{i,j}$ can be viewed as a Boolean function on the $a_{k,\ell}$'s, with $k \leq i+1$, by the definition of SHA-1. Note that when we view $m_{i,j}$ as a Boolean function, we do not approximate (based on approximating MAJ by XOR, ignoring carry effect, etc.), but consider it as exact polynomial on the $a_{k,\ell}$. *Control bits* are determined for each controlled relation. Control bits are chosen among the $a_{k,\ell}$ which appear as a leading term or a term 'near' leading term in $m_{i,j}$, where $m_{i,j}$ is considered as a Boolean function on the $a_{k,\ell}$. The notion of leading term being related to a term ordering, we mention that we have considered here the following order on the $a_{k,\ell}$:

$$a_{k,\ell} < a_{k',\ell'} \text{ if } k' < k \text{ or } k' = k \text{ and } \ell' < \ell.$$

3.3   Conventional/Advanced Message Modification Techniques

The last step of Wang's attack consists of randomly choosing a message and modify some of its bits until all sufficient conditions are satisfied. To our knowledge, this technique has been described for the first time in [18,19]. We shall call this method *conventional message modification* technique.

Here, we introduce an *improved message modification*. The conventional message modification will be used to obtain a "pre-collision", i.e. a collision from the first round to a given round R. This bound R will depend on the number n of rounds considered. We take $R = 23$ in the case of 58-round SHA-1 and $R = 26$ for the full SHA-1. The improved message modification will then allow to extend the pre-collision into a real collision on n-round SHA-1.

We would like to emphasize that our procedure will modify the chaining variable and not the message. Since IV$=(a_0, b_0, c_0, d_0, e_0)$ is fixed, it is clear that

SHA-1 induces a bijection between $(m_0, m_1, \ldots, m_{15})$ and $(a_1, a_2, \ldots, a_{16})$. This implies that a modification on the $a_{i,j}$ can be mapped into a modification on the $m_{i,j}$.

Using our new terminology, we describe the conventional message modification. For this, we use a list of controlled relations C_R, and a list of control bits C_B.

Algorithm 1. Conventional Message Modification
CMM
Input : *A positive integer R, a list C_R of controlled relations, and a list C_B of control bits*
Output : $\mathbf{a} = (a_1, a_2, \ldots, a_{16}) \in \left(\mathbb{F}_2^{32}\right)^{16}$ *satisfying all the controlled relations*
Randomly choose $(a_1, a_2, \ldots, a_{16}) \in \left(\mathbb{F}_2^{32}\right)^{16}$
$\mathbf{a} \leftarrow (a_1, a_2, \ldots, a_{16})$
While *all the controlled relations C_B are not satisfied* **do**
Perform an exhaustive search on the bits $a_{i,j} \in C_B$
If *all the relations C_B are satisfied* **then** *return the updaded $\mathbf{a}$*
Else $\mathbf{a} \leftarrow$ *Randomly choose $(a_1, a_2, \ldots, a_{16}) \in \left(\mathbb{F}_2^{32}\right)^{16}$*

The CMM algorithm permits then to find a collision on R-round SHA-1. Using semi-neutral bits and adjusters, we present an improved algorithm permitting to find a collision on a n-round SHA-1 (with $n > R$). The new procedure is as follows.

Algorithm 2. Improved Message Modification
IMM
Input : *Positive integers n, R, two lists (SNB, Ad) of semi-neutral bits and adjusters, a list C_R of controlled relations, a list C_B of control bits, and a list S_C of sufficient conditions*
Output : $\mathbf{a} = (a_1, a_2, \ldots, a_{16}) \in \left(\mathbb{F}_2^{32}\right)^{16}$ *satisfying all the sufficient conditions*
$\mathbf{a} = (a_1, a_2, \ldots, a_{16}) \leftarrow CMM(R, C_R, C_B)$
While *all the sufficient conditions of S_C are not satisfied* **do**
Adjust the $a_{i,j}$ of $\mathbf{a}$ corresponding to a semi-neutral bit of SNB or an adjuster of Ad
EndWhile
Return $\mathbf{a}$

Remark 1. We mention that a different version of the CMM and IMM algorithms can be found in [9,10]. These versions could be more suitable for those wishing to actually implement these two algorithms.

We now analyze the complexity of the IMM algorithm for finding a collision on 58-round SHA-1. In this case, we choose $R = 23$, i.e. the CMM algorithm will be used to find a collision on 23-round SHA-1. It will remain 5 uncontrolled relations in rounds 17–23. Therefore, the CMM algorithm needs at most 2^5

iterations for returning a collision on 23-round SHA-1. There are 29 remaining conditions from rounds 23–58. To adjust these 29 conditions, we use 21 semi-neutral bits and 16 adjusters. Experimentally, the total complexity is improved to 2^{31} SHA-1 computation – with our latest implementation – whereas Wang's method needs theoretically 2^{34} SHA-1 computations. Note that the cost of the IMM algorithm is dominated by the exhaustive search among 21 semi-neutral bits, which means that we could neglect the cost of the CMM algorithm.

As a proof of concept, we give here a new collision on 58-round SHA-1.

$$m = 0x1ead6636319fe59e4ea7ddcbc79616420ad9523af98f28db0ad135d0e4d62aec$$
$$6c2da52c3c7160b606ec74b2b02d545ebdd9e4663f1563194f497592dd1506f9$$
$$m' = 0x3ead6636519fe5ac2ea7dd88e7961602ead95278998f28d98ad135d1e4d62acc$$
$$6c2da52f7c7160e446ec74f2502d540c1dd9e466bf1563596f497593fd150699$$

3.4   An Algebraic Description of the Improved Message Modification

We present here an algebraic description of the IMM and CMM algorithms which could be useful for further improvements. For this, we remark that the CMM algorithm is equivalent to the solving of a polynomial system of equations via controlled relations with control bits as unknown variables. Similarly, the while-loop of the IMM algorithm is equivalent to the solving of an algebraic system of equations via sufficient conditions with semi-neutral bits and adjusters as unknown variables.

In other words, let $\mathbf{X} = \{X_{i,j}\}_{1 \leq i \leq n}^{0 \leq j \leq 31}$ and let $\mathbb{F}_2[\mathbf{X}]$ be the polynomial ring over $\mathbb{F}_2$ whose variables are $\mathbf{X}$. Remark that sufficient conditions can be considered as polynomial equations via Boolean functions. Thus, they can be expressed as algebraic polynomials on the $a_{i,j}$. Therefore – by replacing each $a_{i,j}$ by the variable $X_{i,j}$ – we can associate a set of polynomials on $\mathbb{F}_2[\mathbf{X}]$ to the set of sufficient conditions. With an obvious notation, we shall call *controlled polynomial* (resp. *uncontrolled polynomial*) the polynomial associated to a controlled relation (resp. uncontrolled relation).

Let J be an ideal in $\mathbb{F}_2[\mathbf{X}]$ generated by $\{X_{i,j}^2 + X_{i,j}\}_{1 \leq i \leq n}^{0 \leq j \leq 31}$, i.e. $J = \langle X_{i,j}^2 + X_{i,j} \rangle_{1 \leq i \leq n}^{0 \leq j \leq 31}$. Let then B_n be a quotient ring $\mathbb{F}_2[\mathbf{X}]/J$. Note that B_n represents the set of all Boolean functions on the variables $X_{i,j}$.

Let $\mathbf{f} = (f_1, f_2, \dots)$ be the set of controlled polynomials. Note that all controlled polynomials of $\mathbf{f}$ are in the subring $\mathbb{F}_2[\{X_{i,j}\}_{1 \leq i \leq R}^{0 \leq j \leq 31}]$, where R is determined by n (for instance, $R = 23$ when $n = 58$ and $R = 26$ when $n = 80$).

For a randomly taken $\mathbf{a} = (a_1, a_2, \dots, a_{16}) \in (\mathbb{F}_2^{32})^{16}$, let $C_R(\mathbf{a})$ be the system obtain from the sufficient conditions by replacing each variables $X_{i,j}$ – not corresponding to a control bit – by $a_{i,j}$. Similarly, let $S_C(\mathbf{a})$ be the system obtain from the sufficient conditions by replacing each variables $X_{i,j}$ – not corresponding to a semi-neutral bit or adjuster – by $a_{i,j}$. In this setting, the CMM and IMM algorithms can roughly be described as follows:

Algorithm 3. *Algebraic Message Modification*
Randomly choose $(a_1, a_2, \ldots, a_{16}) \in \left(\mathbb{F}_2^{32}\right)^{16}$
$\mathbf{a} \leftarrow (a_1, a_2, \ldots, a_{16})$
Solve the algebraic system of equations $C_R(\mathbf{a})$: The solutions correspond to affectations of control bits verifying all controlled polynomials
Solve the algebraic system of equations $S_C(\mathbf{a})$: The solutions correspond to affectations of semi-neutral bits and adjusters verifying all uncontrolled polynomials
Update $\mathbf{a}$ *according to the solutions of the two previous systems*
Return $\mathbf{a}$

Relation Between Message Modification and Decoding of Error- Correcting Codes. Let S be the set of all points in $F = \left(\mathbb{F}_2^{32}\right)^{16}$ satisfying advanced sufficient conditions on $\{a_{i,j}\}$. Note that S is a non-linear subset of F because there are non-linear conditions. Then, for a given $\mathbf{a} \in F$ which is not necessarily contained in S, to find an element in S by modifying $\mathbf{a}$ is analogous to a decoding problem in error-correcting codes. Hence, a conventional message modification and a proposed improved message modification including changing semi-neutral bits can be viewed as an error-correcting process for a non-linear code S in F. More precisely, for a non-linear code S in F, an error-correction can be achieved by manipulating control bits and semi-neutral bits.

4   Analysis of the 58-Round SHA-1 Using the Improved Message Modification

In this part, we detail the different steps of our technique for finding a collision on 58-round SHA-1. In Table 2, we give the sufficient conditions. Note that we have only quoted the sufficient conditions for the first 20 rounds due to space limitation. For the complete list of the conditions, see [9,10]. The control bits and controlled relations are given in Table 3. Semi-neutral bits and adjusters are given in Table 4.

- 'a' means $a_{i,j} = a_{i-1,j}$
- 'A' means $a_{i,j} = a_{i-1,j} + 1$
- 'b' means $a_{i,j} = a_{i-1,(j+2 \mod 32)}$
- 'B' means $a_{i,j} = a_{i-1,(j+2 \mod 32)} + 1$
- 'c' means $a_{i,j} = a_{i-2,(j+2 \mod 32)}$
- 'C' means $a_{i,j} = a_{i-2,(j+2 \mod 32)} + 1$
- 'L' means the leading term of controlled relation of Table 3
- 'w', 'W': adjust $a_{i,j}$ so that $m_{i+1,j} = 0, 1$, respectively
- 'v', 'V': adjust $a_{i,j}$ so that $m_{i,(j+27 \mod 32)} = 0, 1$, respectively
- 'h': adjust $a_{i,j}$ so that corresponding controlled relation including $m_{i+1,j}$ as leading term holds
- 'r' means to adjust $a_{i,j}$ so that corresponding controlled relation including $m_{i,(j+27 \mod 32)}$ as leading term holds

Table 2. Sufficient condition on the $m_{i,j}$ (resp. $a_{i,j}$)

message variable	31 - 24	23 - 16	15 - 8	8 - 0
m_0	--0-----	--------	--------	--------
m_1	-01-----	--------	--------	--01-1-
m_2	-10-----	--------	--------	-1----11
m_3	--0-----	--------	--------	-1-----
m_4	000-----	--------	--------	-0---1--
m_5	-11-----	--------	--------	------1-
m_6	0-------	--------	--------	-------0
m_7	--------	--------	--------	--1-----
m_8	--------	--------	--------	-----00
m_9	-0------	--------	--------	--0-1-1-
m_{10}	-0------	--------	--------	-0------
m_{11}	101-----	--------	--------	-1-1-1-
m_{12}	1-1-----	--------	--------	--------
m_{13}	0-------	--------	--------	--0-----
m_{14}	--0-----	--------	--------	-------0
m_{15}	--0-----	--------	--------	--11----
m_{16}	0-------	--------	--------	-------0
m_{17}	-0------	--------	--------	-1----0-
m_{18}	00------	--------	--------	-1----01
m_{19}	-0------	--------	--------	--1--1-
m_{20}	--------	--------	--------	-----11

chaining variable	31 - 24	23 - 16	15 - 8	8 - 0
a_0	01100111	01000101	00100011	00000001
a_1	101-----	--------	--------	-1-a10aa
a_2	01100--	------0-	----a---	1--00010
a_3	0010----	-10---1a	------0-	0a-1a0-0
a_4	11010---	-01-----	01aaa---	0-10-100
a_5	10-01a--	-1-01-aa	--00100-	0---01-1
a_6	11--0110	-a-1001-	01100010	1-a111-1
a_7	-1--1110	a1a1111-	-101-001	1---0-10
a_8	-0----10	0000000a	a001a1--	100-0-1-
a_9	00------	11000100	00000000	101-1-1-
a_{10}	0-1-----	11111011	11100000	00--0-1-
a_{11}	1-0-----	-------1	01111110	11----0-
a_{12}	0-1-----	--------	--------	-1-a---
a_{13}	1-0-----	--------	--------	-1---01-
a_{14}	1-------	--------	--------	-1---1--
a_{15}	0-------	--------	--------	---0-0
a_{16}	-1------	--------	--------	----a---
a_{17}	-0------	--------	--------	----100-
a_{18}	1-1-----	--------	--------	-----00-
a_{19}	--------	--------	--------	------0
a_{20}	-C------	--------	--------	----A---

Table 3. Control bits and controlled relations

Control sequence s_i	Control bit b_i	Controlled relation r_i
s_{124}	$a_{16,7}, a_{15,9}, a_{14,9}$	$a_{23,0} = 0$
s_{123}	$a_{16,9}$	$a_{22,2} + a_{21,2} = 1$
s_{122}	$a_{16,13}, a_{15,15}, a_{15,12}, a_{15,11}$	$a_{22,1} = 1$
s_{121}	$a_{16,10}$	$a_{21,3} + m_{20,3} = 0$
s_{120}	$a_{16,8}$	$a_{21,1} = 1$
s_{119}	$a_{16,15}, a_{16,20}$	$a_{20,3} + m_{19,3} = 1$
s_{118}	$a_{16,17}$	$a_{19,0} = 0$
s_{117}	$a_{16,21}$	$a_{18,31} = 1$
s_{116}	$a_{16,19}$	$a_{18,29} = 1$
s_{115}	$a_{13,4}$	$a_{18,2} = 0$
s_{114}	$a_{13,3}$	$a_{18,1} = 0$
s_{113}	$a_{14,15}$	$a_{17,30} = 0$
s_{112}	$a_{16,31}$	$m_{15,31} = 1$
s_{111}	$a_{16,29}$	$m_{15,29} = 0$
s_{110}	$a_{16,28}$	$m_{15,28} + m_{10,28} + m_{8,29} + m_{7,29} + m_{4,28} + m_{2,28} = 1$
s_{109}	$a_{16,27}, a_{13,28}$	$m_{15,27} + m_{14,25} + m_{12,28} + m_{12,26} + m_{10,28} + m_{9,27} + m_{9,25} + m_{8,29} + m_{8,28} + m_{7,28} + m_{7,27} + m_{6,26} + m_{5,28} + m_{4,26} + m_{3,25} + m_{2,28} + m_{1,25} + m_{0,28} = 1$
s_{108}	$a_{16,26}$	$m_{15,26} + m_{10,28} + m_{10,26} + m_{8,28} + m_{8,27} + m_{7,27} + m_{6,29} + m_{5,27} + m_{4,26} + m_{2,27} + m_{2,26} + m_{0,27} = 1$
s_{107}	$a_{16,25}$	$m_{15,25} + m_{11,28} + m_{10,27} + m_{10,25} + m_{9,28} + m_{8,27} + m_{8,26} + m_{7,26} + m_{6,29} + m_{6,28} + m_{5,26} + m_{4,25} + m_{3,28} + m_{2,28} + m_{2,26} + m_{2,25} + m_{1,28} + m_{0,28} + m_{0,26} = 0$
s_{106}	$a_{16,24}$	$m_{15,24} + m_{12,28} + m_{11,27} + m_{10,26} + m_{10,24} + m_{9,28} + m_{9,27} + m_{8,29} + m_{8,26} + m_{8,25} + m_{7,25} + m_{6,29} + m_{6,28} + m_{6,27} + m_{5,25} + m_{4,28} + m_{4,24} + m_{3,28} + m_{3,27} + m_{2,27} + m_{2,25} + m_{2,24} + m_{1,28} + m_{1,27} + m_{0,27} + m_{0,25} = 1$
s_{105}	$a_{16,23}$	$m_{15,23} + m_{12,28} + m_{12,27} + m_{11,26} + m_{10,25} + m_{10,23} + m_{9,27} + m_{9,26} + m_{8,28} + m_{8,25} + m_{7,29} + m_{7,24} + m_{6,28} + m_{6,27} + m_{6,26} + m_{5,24} + m_{4,27} + m_{4,23} + m_{3,27} + m_{3,26} + m_{2,26} + m_{2,24} + m_{2,23} + m_{1,27} + m_{1,26} + m_{0,26} + m_{0,24} = 1$
s_{104}	$a_{16,22}$	$m_{15,22} + m_{14,25} + m_{12,28} + m_{12,27} + m_{11,25} + m_{10,27} + m_{10,24} + m_{10,22} + m_{9,28} + m_{9,27} + m_{9,26} + m_{8,27} + m_{8,24} + m_{8,23} + m_{7,28} + m_{7,27} + m_{7,23} + m_{6,27} + m_{6,25} + m_{5,23} + m_{4,28} + m_{4,27} + m_{4,22} + m_{3,26} + m_{2,28} + m_{2,27} + m_{2,25} + m_{2,23} + m_{2,22} + m_{1,26} + m_{0,25} + m_{0,23} = 0$
s_{103}	$a_{16,6}$	$m_{15,6} = 1$
s_{102}	$a_{16,5}$	$m_{15,5} = 1$
s_{101}	$a_{16,4}$	$m_{15,4} + m_{12,5} + m_{10,4} + m_{4,5} + m_{4,4} + m_{2,5} + m_{2,4} = 1$

$\vdots$

Control sequence s_i	Control bit b_i	Controlled relation r_i
s_3	$a_{1,28}$	$m_{1,1} = 1$
s_2	$a_{1,25}$	$m_{1,30} = 0$
s_1	$a_{1,24}$	$m_{1,29} = 1$
s_0	$a_{1,23}$	$m_{1,29} = 1$

Table 4. Semi-neutral bits and adjusters

message variable	31 - 24	23 - 16	15 - 8	8 - 0
m_0	--0-----	--------	--------	--------
m_1	-01-----	--------	--------	--01--1-
m_2	L10-----	--------	--------	-1----11
m_3	-L0-----	--------	--------	-1------
m_4	000-----	--------	--------	-0----1-
m_5	L11-----	--------	--------	------1L
m_6	0L------	--------	--------	-------0
m_7	LL------	--------	--------	--1----L
m_8	LL------	--------	--------	------00
m_9	LOL-----	--------	--------	--0L1--1L
m_{10}	LOL-----	--------	--------	--0L----L
m_{11}	101-----	--------	--------	-1-1--1L
m_{12}	1L1-----	--------	--------	-------L
m_{13}	OLLLLL-L	LL------	--------	--OLLLLL
m_{14}	LLOLLL-L	LLLL----	--------	--LLLLLO
m_{15}	LLOLLLLL	LL------	--------	--11LLLLO
m_{16}	0-------	--------	--------	-------0
m_{17}	-0------	--------	--------	--1----0-
m_{18}	00------	--------	--------	--1---01
m_{19}	-0------	--------	--------	---1--1-
m_{20}	--------	--------	--------	------11
m_{21}	-0------	--------	--------	--0----1-
m_{22}	01------	--------	--------	-0----10
m_{23}	11------	--------	--------	--1---0-
m_{24}	--------	--------	--------	-------0
m_{25}	-1------	--------	--------	-------1-
m_{26}	10------	--------	--------	-0----10
m_{27}	-1------	--------	--------	--01--0-
m_{28}	1-------	--------	--------	-------0
m_{29}	-1------	--------	--------	--1---0-
m_{30}	-0------	--------	--------	--1---0-
m_{31}	-1------	--------	--------	-------0-
m_{32}	--------	--------	--------	-------1-
m_{33}	--------	--------	--------	---0----
m_{34}	0-------	--------	--------	-------1-
m_{35}	0-------	--------	--------	--------
m_{36}	1-------	--------	--------	-------1-
m_{37}	1-------	--------	--------	---0----
m_{38}	--------	--------	--------	--------
m_{39}	0-------	--------	--------	---1----
m_{40}	1-------	--------	--------	--------
m_{41}	--------	--------	--------	---1----
m_{42}	1-------	--------	--------	--------
m_{43}	--------	--------	--------	---1----
m_{44}	1-------	--------	--------	-------1-
m_{45}	--------	--------	--------	--------
m_{46}	1-------	--------	--------	--------
m_{47}	0-------	--------	--------	--------
$m_i \ (i \geq 48)$	--------	--------	--------	--------

chaining variable	31 - 24	23 - 16	15 - 8	8 - 0
a_0	01100111	01000101	00100011	00000001
a_1	101V--vV	Y-------	--------	-1-a10aa
a_2	01100vVv	------0-	----a---	1-w00010
a_3	0010--Vv	-10---1a	------0-	0aX1a0W0
a_4	11010vv-	-01-----	01aaa---	.0W10-100
a_5	10w01aV-	-1-01-aa	--00100-	0w--01W1
a_6	11W-0110	-a-1001-	01100010	1-a111W1
a_7	w1x-1110	a1a1111-	-101-001	1---0-10
a_8	h0Xvvv10	0000000a	a001a1--	100X0-1h
a_9	00XVrr-V	11000100	00000000	101-1-1y
a_{10}	0w1-rv-v	11111011	11100000	00hW0-1h
a_{11}	1w0--V-V	-------1	01111110	11x---0Y
a_{12}	0w1-rV-V	--------	--------	-1XWa-Wh
a_{13}	1w0--vv-	-rr-----	--------	-1-qq01y
a_{14}	1rhhvvVh	hh------	qNNNNNqN	N1hhh1hh
a_{15}	0rwhhhVh	hhhh---N	qNNqqNqN	NNhhOhhO
a_{16}	W1whhhhh	hhqNqNqN	NNqNNqqq	qWWhahhh
a_{17}	--------	--------	--------	----100-
a_{18}	1-1-----	--------	--------	-----00-
a_{19}	--------	--------	--------	-------0
a_{20}	-C------	--------	--------	----A---
a_{21}	-b------	--------	--------	----a-i-
a_{22}	--------	--------	--------	-----A1-
a_{23}	--------	--------	--------	-------0
a_{24}	-c------	--------	--------	----A---
a_{25}	-B------	--------	--------	-----a--
a_{26}	--------	--------	--------	-----A1-
a_{27}	--------	--------	--------	-------1
a_{28}	-c------	--------	--------	----A---
a_{29}	-B------	--------	--------	----A-0-
a_{30}	--------	--------	--------	------0-
a_{31}	--------	--------	--------	--------
a_{32}	--------	--------	--------	----A---
a_{33}	--------	--------	--------	------1-
a_{34}	--------	--------	--------	--------
a_{35}	--------	--------	--------	--------
a_{36}	--------	--------	--------	----A---
a_{37}	--------	--------	--------	-----1-
a_{38}	--------	--------	--------	----A---
a_{39}	B-------	--------	--------	------0-
a_{40}	C-------	--------	--------	----A---
a_{41}	B-------	--------	--------	------0-
a_{42}	C-------	--------	--------	----A---
a_{43}	B-------	--------	--------	------0-
a_{44}	C-------	--------	--------	--------
a_{45}	B-------	--------	--------	--------
$a_i \ (i \geq 46)$	--------	--------	--------	--------

- 'x', 'y': adjust $a_{i+1,j-1}, a_{i,j-1}$ so that $m_{i,j} = 0$, respectively
- 'X', 'Y': adjust $a_{i+1,j-1}, a_{i,j-1}$ so that $m_{i,j} = 1$, respectively
- 'N': semi-neutral bit
- 'q' : adjust $a_{i,j}$ so that relations after 17-round hold

In this case, the set of bits corresponding to 'q' is exactly same to the set of *adjusters*.

The conditions remaining after the conventional message modification are listed below: $a_{17,3} = 1, a_{17,2} = 0, a_{17,1} = 0, a_{26,1} = 1, a_{27,0} = 1, a_{29,1} = 0, a_{30,1} = 0, a_{33,1} = 1, a_{37,1} = 1, a_{39,1} = 0, a_{41,1} = 0, a_{43,1} = 0, a_{20,30} + a_{18,0} = 1, a_{21,30} + a_{20,0} = 0, a_{24,30} + a_{22,0} = 0, a_{25,30} + a_{24,0} = 1, a_{25,3} + a_{24,3} = 0, a_{26,2} + a_{25,2} = 1, a_{28,30} + a_{26,0} = 0, a_{28,3} + a_{27,3} = 1, a_{29,30} + a_{28,0} = 1, a_{29,3} + a_{28,3} = 1, a_{32,3} + a_{31,3} = 1, a_{36,3} + a_{35,3} = 1, a_{38,3} + a_{37,3} = 1, a_{39,31} + a_{38,1} = 1, a_{40,3} + a_{39,3} = 1, a_{40,31} + a_{38,1} = 1, a_{41,31} + a_{40,1} = 1, a_{42,31} + a_{40,1} = 1, a_{43,31} + a_{42,1} = 1, a_{42,3} + a_{41,3} = 1, a_{44,31} + a_{42,1} = 1, a_{45,31} + a_{44,1} = 1.$

There are five conditions $a_{17,3} = 1, a_{17,2} = 0, a_{17,1=0}, a_{20,30} + a_{18,0} = 1, a_{21,30} + a_{20,0} = 0$ which are only related to first 23 rounds. For other 29 conditions, we adjust by using 21 semi-neutral bits and 11 adjusters as explained in the improved message modification algorithm (see Algorithm 2).

5 A Concluding Note

This paper present an improved method for finding collision on SHA-1. To do so, we use algebraic techniques for describing the message modification technique and propose an improvement. The details of our attack can be found in the appendices. The proposed method improves the complexity of an attack against 58-round SHA-1 and we found many new collisions.

References

1. Biham, E., Chen, R., Joux, A., Carribault, P., Lemuet, C., Jalby, W.: Collisions of SHA-0 and Reduced SHA-1. In: Cramer, R.J.F. (ed.) EUROCRYPT 2005. LNCS, vol. 3494, pp. 36–57. Springer, Heidelberg (2005)
2. Biham, E., Chen, R.: Near-Collisions of SHA-0. In: Franklin, M. (ed.) CRYPTO 2004. LNCS, vol. 3152, pp. 290–305. Springer, Heidelberg (2004)
3. De Cannière, C., Rechberger, C.: Finding SHA-1 Characteristics. In: Lai, X., Chen, K. (eds.) ASIACRYPT 2006. LNCS, vol. 4284, pp. 1–20. Springer, Heidelberg (2006)
4. Chabaud, F., Joux, A.: Differential Collisions in SHA-0. In: Krawczyk, H. (ed.) CRYPTO 1998. LNCS, vol. 1462, pp. 56–71. Springer, Heidelberg (1998)
5. Dobbertin, H.: Cryptanalysis of MD4. In: Gollmann, D. (ed.) Fast Software Encryption. LNCS, vol. 1039, pp. 53–69. Springer, Heidelberg (1996)
6. Hui, L.C.K., Wang, X., Chow, K.P., Tsang, W.W., Chong, C.F., Chan, H.W.: The Differential Analysis of Skipjack Variants from the first Round. In: Advance in Cryptography – CHINACRYPT 2002 Science Publishing House (2002)
7. Joux, A.: Multicollisions in Iterated Hash Functions. Application to Cascaded Constructions. In: Franklin, M. (ed.) CRYPTO 2004. LNCS, vol. 3152, pp. 306–316. Springer, Heidelberg (2004)
8. Pramstaller, N., Rechberger, C., Rijmen, V.: Exploiting Coding Theory for Collision Attacks on SHA-1. In: Smart, N.P. (ed.) Cryptography and Coding. LNCS, vol. 3796, pp. 78–95. Springer, Heidelberg (2005)
9. Sugita, M., Kawazoe, M., Imai, H.: Gröbner Basis Based Cryptanalysis of SHA-1. IACR Cryptology ePrint Archive 2006/098 (2006), http://eprint.iacr.org/2006/098.pdf
10. Sugita, M., Kawazoe, M., Imai, H.: Gröbner Basis Based Cryptanalysis of SHA-1. In: Proc. of SECOND NIST CRYPTOGRAPHIC HASH WORKSHOP (2006)
11. Wang, X., Lai, X., Feng, D., Chen, H., Yu, X.: Cryptanalysis of the Hash Functions MD4 and RIPEMD. In: Cramer, R.J.F. (ed.) EUROCRYPT 2005. LNCS, vol. 3494, pp. 1–18. Springer, Heidelberg (2005)
12. Wang, X., Feng, D., Yu, X.: An Attack on Hash Function HAVAL-128. Science in China Series 48, 545–556 (2005)
13. Wang, X., Yao, A.C., Yao, F.: Cryptanalysis on SHA-1. In: Proc. of NIST Cryptographic Hash Workshop (2005)

14. Wang, X., Yin, Y.L., Yu, H.: New Collision Search for SHA-1. In: Rump Session of CRYPTO (2005)
15. Wang, X., Yin, Y.L., Yu, H.: Finding Collisions in the Full SHA-1. In: Shoup, V. (ed.) CRYPTO 2005. LNCS, vol. 3621, pp. 17–36. Springer, Heidelberg (2005)
16. Wang, X., Yu, H.: How to Break MD5 and Other Hash Functions. In: Cramer, R.J.F. (ed.) EUROCRYPT 2005. LNCS, vol. 3494, pp. 19–35. Springer, Heidelberg (2005)
17. Wang, X., Yu, H., Yin, Y.L.: Efficient Collision Search Attacks on SHA-0. In: Shoup, V. (ed.) CRYPTO 2005. LNCS, vol. 3621, pp. 1–16. Springer, Heidelberg (2005)
18. Wang, X.: The Collision attack on SHA-0 (1997)
19. Wang, X.: The Improved Collision attack on SHA-0 (1998)
20. Wang, X.: Collisions for Some Hash Functions MD4, MD5, HAVAL-128, RIPEMD. In: Rump Session of Crypto'04
21. Wang, X.: Cryptanalysis of Hash Functions and Potential Dangers. In: RSA Conference 2006, San Jose, USA (2006)
22. Klima, V.: Tunnels in Hash Functions: MD5 Collisions Within a Minute. IACR Cryptology ePrint Archive 2006/105 (2006), http://eprint.iacr.org/2006/098.pdf

Appendix

A   Analysis of the Full SHA-1

We present here the details of our method on the full SHA-1.

A.1   Disturbance Vector and Differential

We start from the disturbance vector and differential given by Wang et al [13]. From these, we construct a new differential. We modify Wang's differential by

Table 5. A differential for the full SHA-1

	Δm	$\Delta^+ m$	$\Delta^- m$
$i = 0$	a0000003	00000001	a0000002
$i = 1$	20000030	20000020	00000010
$i = 2$	60000000	60000000	00000000
$i = 3$	e000002a	40000000	a000002a
$i = 4$	20000043	20000042	00000001
$i = 5$	b0000040	a0000000	10000040
$i = 6$	d0000053	d0000042	00000011
$i = 7$	d0000022	d0000000	00000022
$i = 8$	20000000	00000000	20000000
$i = 9$	60000032	20000030	40000002
$i = 10$	60000043	60000041	00000002
$i = 11$	20000040	00000000	20000040
$i = 12$	e0000042	c0000000	20000042
$i = 13$	60000002	00000000	60000000
$i = 14$	80000001	00000001	80000000
$i = 15$	00000020	00000020	00000000
$i = 16$	00000003	00000002	00000001
$i = 17$	40000052	00000002	40000050
$i = 18$	40000040	00000000	40000040
$i = 19$	e0000052	00000002	e0000050
$i = 20$	a0000000	00000000	a0000000
$\vdots$			
$i = 78$	0000004a	00000002	00000048
$i = 79$	0000080a	00000808	00000002
$i = 80$	00000000	00000000	00000000

	Δa	$\Delta^+ a$	$\Delta^- a$
$i = 0$	00000000	00000000	00000000
$i = 1$	e0000001	a0000000	40000001
$i = 2$	20000004	20000000	00000004
$i = 3$	c07fff84	803fff84	40400000
$i = 4$	800030e2	800010a0	00002042
$i = 5$	08408060	08008020	00400090
$i = 6$	80003a00	00001a00	80002000
$i = 7$	0fff8001	08000001	07ff8000
$i = 8$	00000008	00000008	00000000
$i = 9$	80000101	80000100	00000001
$i = 10$	00000002	00000002	00000000
$i = 11$	00000100	00000000	00000100
$i = 12$	00000002	00000002	00000000
$i = 13$	00000000	00000000	00000000
$i = 14$	00000000	00000000	00000000
$i = 15$	00000001	00000001	00000000
$i = 16$	00000000	00000000	00000000
$i = 17$	80000002	80000002	00000000
$i = 18$	80000002	00000002	00000000
$i = 19$	80000002	80000002	00000000
$i = 20$	00000000	00000000	00000000
$\vdots$			
$i = 78$	00000000	00000000	00000000
$i = 79$	00000040	00000000	00000040
$i = 80$	00000000	00000000	00000000

changing $\Delta a_{14,31}^{+}$ and $\Delta a_{14,31}^{-}$. For our method, it is important to choose $\Delta^{+}m$, $\Delta^{-}m$, $\Delta^{+}a$ and $\Delta^{-}a$ carefully. A bad choice of plus/minus could lead to an underdefined system of equations (as result of the Gaussian elimination). For this reason, we have to correct Wang's differential. We present in Table 5 the new differential constructed.

A.2   Sufficient Conditions

For the disturbance vector, and the differential given in the previous step, we give in Table 6 the sufficient conditions for the full SHA-1. In Table 6: 'a' means $a_{i,j} = a_{i-1,j}$, 'A' means $a_{i,j} = a_{i-1,j} + 1$, 'b' means $a_{i,j} = a_{i-1,(j+2 \bmod 32)}$, 'B' means $a_{i,j} = a_{i-1,(j+2 \bmod 32)} + 1$, 'c' means $a_{i,j} = a_{i-2,(j+2 \bmod 32)}$ and 'C' means $a_{i,j} = a_{i-2,(j+2 \bmod 32)} + 1$.

Table 6. Sufficient conditions for the full SHA-1

message variable	31 - 24	23 - 16	15 - 8	8 - 0		chaining variable	31 - 24	23 - 16	15 - 8	8 - 0
m_0	1-1-----	--------	--------	------10		a_0	01100111	01000101	00100011	00000001
m_1	--0-----	--------	--------	--01----		a_1	010----0	-0-01-0-	10-0-10-	---a0101
m_2	-00-----	--------	--------	--------		a_2	-100---1	0aa10a1a	01a1a011	1--a11a1
m_3	101-----	--------	--------	--1-1-1-		a_3	01011---	-1000000	00000000	01--a0a1
m_4	--0-----	--------	--------	-0----01		a_4	0-101--a	---10000	00101000	010---10
m_5	0-01----	--------	--------	-1------		a_5	0-0101-1	-1-11110	00111-00	10010100
m_6	00-0----	--------	--------	-0-1--01		a_6	1-0a1a0a	a0a1aaa-	--10010-	--01-0--
m_7	00-0----	--------	--------	--1---1-		a_7	--0-0111	11111111	111-010-	0-0-0110
m_8	--1-----	--------	--------	--------		a_8	-10---01	11110000	010-111-	1---000-
m_9	-10-----	--------	--------	--00--1-		a_9	00----11	11111111	111----0	----1-01
m_{10}	-00-----	--------	--------	-0---10		a_{10}	-11-----	--------	-----a--	-1-1-0-
m_{11}	--1-----	--------	--------	-1------		a_{11}	100-----	--------	-------1	-1-0---
m_{12}	001-----	--------	--------	-1---1-		a_{12}	--------	--------	--------	-1----0-
m_{13}	-11-----	--------	--------	------0-		a_{13}	0-------	--------	--------	-1--0--
m_{14}	1-------	--------	--------	-------0		a_{14}	1-------	--------	--------	----1--
m_{15}	--------	--------	--------	--0-----		a_{15}	--------	--------	--------	---0--0
m_{16}	--------	--------	--------	-----01		a_{16}	-1------	--------	--------	----1-A-
m_{17}	-1------	--------	--------	-1-1--0-		a_{17}	00------	--------	--------	----0-0-
m_{18}	-1------	--------	--------	--1-----		a_{18}	1-1-----	--------	--------	----a-0-
m_{19}	111-----	--------	--------	-1-1--0-		a_{19}	0-b-----	--------	--------	------0-
m_{20}	1-1-----	--------	--------	--------		a_{20}	--0-----	--------	--------	----a---
⋮						⋮				
m_{78}	--------	--------	--------	-1--1-0-		a_{78}	--------	--------	--------	----b---
m_{79}	--------	--------	---0---	---0-1-		a_{79}	--------	--------	--------	----1-----
m_{80}	--------	--------	--------	--------		a_{80}	--------	--------	--------	--------

By a Gaussian elimination, we reduce all conditions on the m_i. The result of Gaussian elimination is listed below. There are 167 conditions on $m_{i,j}$:

$m_{15,31} = 0$, $m_{15,30} = 1$, $m_{15,29} = 1$, $m_{15,28} + m_{10,28} + m_{4,28} + m_{2,28} = 0$, $m_{15,27} + m_{10,27} + m_{8,28} + m_{4,27} + m_{2,28} + m_{2,27} + m_{0,28} = 1$, $m_{15,26} + m_{10,28} + m_{10,26} + m_{8,28} + m_{8,27} + m_{7,27} + m_{5,27} + m_{4,26} + m_{2,27} + m_{2,26} + m_{0,27} = 0$, $m_{15,25} + m_{11,28} + m_{10,27} + m_{10,25} + m_{9,28} + m_{8,27} + m_{8,26} + m_{7,26} + m_{5,26} + m_{4,25} + m_{3,28} + m_{2,28} + m_{2,26} + m_{2,25} + m_{1,28} + m_{0,28} + m_{0,26} = 0$, $m_{15,24} + m_{12,28} + m_{11,27} + m_{10,26} + m_{10,24} + m_{9,28} + m_{9,27} + m_{8,26} + m_{8,25} + m_{7,25} + m_{6,27} + m_{5,25} + m_{4,28} + m_{4,24} + m_{3,28} + m_{3,27} + m_{2,27} + m_{2,25} + m_{2,24} + m_{1,28} + m_{1,27} + m_{0,27} + m_{0,25} = 1$, $m_{15,23} + m_{12,28} + m_{12,27} + m_{11,26} + m_{10,25} + m_{10,23} + m_{9,27} + m_{9,26} + m_{8,28} + m_{8,25} + m_{8,24} + m_{7,24} + m_{7,0} + m_{6,27} + m_{6,26} + m_{5,24} + m_{4,27} + m_{4,23} + m_{3,27} + m_{3,26} + m_{2,26} + m_{2,24} + m_{2,23} + m_{1,30} + m_{1,27} + m_{1,26} + m_{1,0} + m_{0,26} + m_{0,24} = 0$,

$\cdots$,

$m_{5,0} + m_{1,30} + m_{1,0} = 0$, $m_{4,31} = 0$, $m_{4,30} = 0$, $m_{4,29} = 0$, $m_{4,6} = 0$, $m_{4,1} = 0$, $m_{4,0} = 1$, $m_{3,31} = 1$, $m_{3,30} = 0$, $m_{3,29} = 1$, $m_{3,5} = 1$, $m_{3,3} = 1$, $m_{3,1} = 1$, $m_{3,0} + m_{1,0} = 0$, $m_{2,31} = 1$, $m_{2,30} = 0$, $m_{2,29} = 0$, $m_{2,0} + m_{1,30} = 1$, $m_{1,31} + m_{1,30} = 1$, $m_{1,29} = 0$, $m_{1,5} = 0$, $m_{1,4} = 1$, $m_{0,31} = 1$, $m_{0,30} = 0$, $m_{0,29} = 1$, $m_{0,1} = 1$, $m_{0,0} = 0$.

A.3 Control Bits and Controlled Relations

The control bits and controlled relations are presented in Table 7.

Now we give the semi-neutral bits and adjuster in Table 8. In this table :

- 'a', 'A', 'b', 'B', 'c', 'C': as in Section A.2.
- 'L' means the leading term of controlled relation of Table 7.
- 'w', 'W': adjust $a_{i,j}$ so that $m_{i+1,j} = 0, 1$, respectively.
- 'v', 'V': adjust $a_{i,j}$ so that $m_{i,(j+27 \mod 32)} = 0, 1$, respectively.
- 'h': adjust $a_{i,j}$ so that corresponding controlled relation including $m_{i+1,j}$ as leading term holds.
- 'r' means to adjust $a_{i,j}$ so that corresponding controlled relation including $m_{i,(j+27 \mod 32)}$ as leading term holds.
- 'x', 'y': adjust $a_{i+1,j-1}$, $a_{i,j-1}$ so that $m_{i,j} = 0$, respectively.
- 'X', 'Y': adjust $a_{i+1,j-1}$, $a_{i,j-1}$ so that $m_{i,j} = 1$, respectively.
- 'N': semi-neutral bit.
- 'q' : adjust $a_{i,j}$ so that relations after 17-round hold.
- 'F' : etc.

Table 7. Control bits and controlled relations for the full SHA-1

ctrl. seq.	control bits	controlled relation
$s168$	$a_{15,8}$	$a_{30,2} + a_{29,2} = 1$
$s167$	$a_{16,6}$	$a_{26,2} + a_{25,2} = 1$
$s166$	$a_{15,7}$	$a_{25,3} + a_{24,3} = 0$
$s165$	$a_{13,7}$	$a_{24,3} + a_{23,3} = 0$
$s164$	$a_{13,9}$	$a_{23,0} = 0$
$s163$	$a_{16,10}$	$a_{22,3} + a_{21,3} = 0$
$s162$	$a_{16,11}$	$a_{21,29} + a_{20,31} = 0$
$s161$	$a_{16,8}$	$a_{21,1} = 0$
$s160$	$a_{16,9}$	$a_{20,29} = 0$
$s159$	$a_{15,10}$	$a_{20,3} + a_{19,3} = 0$
$s158$	$a_{15,11}$	$a_{19,31} = 0$
$s157$	$a_{15,9}$	$a_{19,29} + a_{18,31} = 0$
$s156$	$a_{14,8}$	$a_{19,1} = 0$
$s155$	$a_{14,11}$	$a_{18,31} = 1$
$s154$	$a_{15,14}$	$a_{18,29} = 1$
$s153$	$a_{13,8}$	$a_{18,1} = 0$
$s152$	$a_{13,11}$	$a_{17,31} = 0$
$s151$	$a_{13,10}$	$a_{17,30} = 0$
$s150$	$a_{13,13}$	$a_{17,1} = 0$
$s149$	$a_{16,31}$	$m_{15,31} = 0$
$s148$	$a_{16,29}$	$m_{15,29} = 1$
$s147$	$a_{16,28}$	$m_{15,28} + m_{10,28} + m_{4,28} + m_{2,28} = 0$
$s146$	$a_{16,27}$	$m_{15,27} + m_{10,27} + m_{8,28} + m_{4,27} + m_{2,28} + m_{2,27} + m_{0,28} = 1$
$s145$	$a_{16,26}$	$m_{15,26} + m_{10,28} + m_{10,26} + m_{8,28} + m_{8,27} + m_{7,27} + m_{5,27} + m_{4,26} + m_{2,27} + m_{2,26} + m_{0,27} = 0$
$s144$	$a_{16,25}$	$m_{15,25} + m_{11,28} + m_{10,27} + m_{10,25} + m_{9,28} + m_{8,27} + m_{8,26} + m_{7,26} + m_{5,26} + m_{4,25} + m_{3,28} + m_{2,28} + m_{2,26} + m_{2,25} + m_{1,28} + m_{0,28} + m_{0,26} = 0$
$s143$	$a_{16,24}$	$m_{15,24} + m_{12,28} + m_{11,27} + m_{10,26} + m_{10,24} + m_{9,28} + m_{9,27} + m_{8,26} + m_{8,25} + m_{7,25} + m_{6,27} + m_{5,25} + m_{4,24} + m_{3,28} + m_{3,27} + m_{2,27} + m_{2,25} + m_{2,24} + m_{1,28} + m_{1,27} + m_{0,27} + m_{0,25} = 1$
$s142$	$a_{16,23}$	$m_{15,23} + m_{12,28} + m_{12,27} + m_{11,26} + m_{10,25} + m_{10,23} + m_{9,27} + m_{9,26} + m_{8,28} + m_{8,25} + m_{8,24} + m_{7,24} + m_{7,0} + m_{6,27} + m_{6,26} + m_{5,24} + m_{4,27} + m_{4,23} + m_{3,27} + m_{3,26} + m_{2,26} + m_{2,24} + m_{2,23} + m_{1,30} + m_{1,27} + m_{1,26} + m_{1,0} + m_{0,26} + m_{0,24} = 0$

⋮

$s3$	$a_{2,5}$	$m_{1,5} = 0$
$s2$	$a_{1,26}$	$m_{1,31} + m_{1,30} = 1$
$s1$	$a_{1,23}$	$m_{1,29} = 0$

Table 8. Semi-neutral bits and Adjusters

message variable	31 - 24	23 - 16	15 - 8	8 - 0
m_0	1-1-----	--------	--------	------10
m_1	L-0-----	--------	--------	--01----
m_2	L00-----	--------	--------	-------L
m_3	101-----	--------	--------	--1-1-1L
m_4	LL0-----	--------	--------	-0----01
m_5	OL01----	--------	--------	-1-----L
m_6	OOL0----	--------	--------	-0-1--01
m_7	00-0----	--------	--------	--1L-1--
m_8	L-1-----	--------	--------	---L--L-
m_9	L10-----	--------	--------	--00-L1L
m_{10}	L00-----	--------	--------	-OLLLL10
m_{11}	LL1-----	--------	--------	-1LLLLLL
m_{12}	001-----	--------	--------	-1LLL-1L
m_{13}	L11LLLLL	LLLLLLL	L-L-----	--LLLLOL
m_{14}	1LLLLLLL	LLLLLLL	L-LL----	--LLLLLO
m_{15}	LLLLLLLL	LLLLLLL	LL-L----	L-OLLLLL
m_{16}	--------	--------	--------	------01
m_{17}	-1------	--------	--------	-1-1--0-
m_{18}	-1------	--------	--------	-1------
m_{19}	111-----	--------	--------	-1-1--0-
m_{20}	1-1-----	--------	--------	--------
m_{21}	0-------	--------	--------	-1------
m_{22}	--1-----	--------	--------	-------0
m_{23}	--1-----	--------	--------	--11----
m_{24}	1-------	--------	--------	-------1
m_{25}	-1------	--------	--------	-1----1-
m_{26}	10------	--------	--------	--0----10
m_{27}	-1------	--------	--------	--1--0-
m_{28}	--------	--------	--------	------10
m_{29}	-0------	--------	--------	-0----0-
m_{30}	11------	--------	--------	-1----01
m_{31}	11------	--------	--------	--0--1-
m_{32}	--------	--------	--------	-------1
m_{33}	-0------	--------	--------	-----0-
m_{34}	00------	--------	--------	-1----00
m_{35}	-0------	--------	--------	-11--1-
m_{36}	1-------	--------	--------	-------0
m_{37}	-1------	--------	--------	-0----0-
m_{38}	-0------	--------	--------	--1--1-
m_{39}	-1------	--------	--------	-----0-
m_{40}	--------	--------	--------	------1-
m_{41}	--------	--------	--------	-0------
m_{42}	0-------	--------	--------	-----0-
m_{43}	1-------	--------	--------	--------
m_{44}	0-------	--------	--------	-----0-
m_{45}	0-------	--------	--------	-1------
m_{46}	--------	--------	--------	--------
m_{47}	0-------	--------	--------	-1------
m_{48}	0-------	--------	--------	--------
m_{49}	--------	--------	--------	-1------
m_{50}	0-------	--------	--------	--------
m_{51}	--------	--------	--------	-1------
m_{52}	1-------	--------	--------	-----1-
m_{53}	--------	--------	--------	--------
m_{54}	1-------	--------	--------	--------
m_{55}	0-------	--------	--------	--------
m_{56}	--------	--------	--------	--------
m_{57}	--------	--------	--------	--------
m_{58}	--------	--------	--------	--------
m_{59}	--------	--------	--------	--------
m_{60}	--------	--------	--------	--------
m_{61}	--------	--------	--------	--------
m_{62}	--------	--------	--------	--------
m_{63}	--------	--------	--------	--------
m_{64}	--------	--------	--------	--------
m_{65}	--------	--------	--------	--------
m_{66}	--------	--------	--------	----0-
m_{67}	--------	--------	1-------	--------
m_{68}	--------	--------	--------	----0-
m_{69}	--------	--------	--------	--0-0-
m_{70}	--------	--------	------1	-------0
m_{71}	--------	--------	--------	---0-1
m_{72}	--------	--------	--------	--0--0-
m_{73}	--------	--------	-----1-	-----0-
m_{74}	--------	--------	--------	--00-1-
m_{75}	--------	--------	------1	-1-1-0-
m_{76}	--------	--------	----0--	---00--
m_{77}	--------	--------	--------	--1-10-
m_{78}	--------	--------	--------	-1--1-0-
m_{79}	--------	--------	----0---	---0-1-
m_{80}	--------	--------	--------	--------

chaining variable	31 - 24	23 - 16	15 - 8	8 - 0
a_0	01100111	01000101	00100011	00000001
a_1	010-FrF0	y0-01-0-	10-0-10-	F-Fa0101
a_2	F100-Vv1	0aa10a1a	01a1a011	1-wa11a1
a_3	01011VFV	-1000000	00000000	01FFa0a1
a_4	0w101v-a	y--10000	00101000	010XWF10
a_5	0w0101y1	V1-11110	00111-00	10010100
a_6	1w0a1a0a	a0a1aaa-	--10010F	-W01F0Fh
a_7	ww0w0111	11111111	111-010F	0w0W0110
a_8	w10vvv01	11110000	010-111F	1-Wh000F
a_9	00WV--11	11111111	111----0	--F1F01
a_{10}	W11x-Vvv	--------	-----a--	-1ww1h0w
a_{11}	100V----	--------	-------1	-1hh0hWw
a_{12}	wwWF-v--	--------	--------	-1hhhh0h
a_{13}	0wW--V--	-F-F-F--	FNqNqqqq	q1hhh0WW
a_{14}	1WWhhhhh	hhhhhhhh	hNhNqNNq	NNhhh1wh
a_{15}	WWwhhhhh	hhhhhhhh	hqhhqqqq	qNwh0hh0
a_{16}	w1Whhhhh	hhhhhhhh	hhNhqqqq	hqwh1hAh
a_{17}	00------	--------	--------	----0-0-
a_{18}	1-1-----	--------	--------	----a-0-
a_{19}	0-b-----	--------	--------	------0-
a_{20}	--0-----	--------	--------	----a---
a_{21}	--b-----	--------	--------	------0-
a_{22}	--------	--------	--------	----aa--
a_{23}	--------	--------	--------	------00
a_{24}	--c-----	--------	--------	----a---
a_{25}	-B------	--------	--------	----a-0-
a_{26}	--------	--------	--------	-----A1-
a_{27}	--------	--------	--------	------0
a_{28}	-C------	--------	--------	----a---
a_{29}	-b------	--------	--------	----a-1-
a_{30}	--------	--------	--------	-----A0-
a_{31}	--------	--------	--------	-------1
a_{32}	-c------	--------	--------	----a---
a_{33}	-b------	--------	--------	----a---
a_{34}	--------	--------	--------	---AA0-
a_{35}	--------	--------	--------	-----00
a_{36}	-c------	--------	--------	----A---
a_{37}	-B------	--------	--------	----A-1-
a_{38}	--------	--------	--------	------0-
a_{39}	--------	--------	--------	--------
a_{40}	B-------	--------	--------	----A---
a_{41}	--------	--------	--------	------1-
a_{42}	C-------	--------	--------	--------
a_{43}	B-------	--------	--------	--------
a_{44}	--------	--------	--------	----A---
a_{45}	--------	--------	--------	------0-
a_{46}	C-------	--------	--------	----A---
a_{47}	B-------	--------	--------	------0-
a_{48}	C-------	--------	--------	----A---
a_{49}	B-------	--------	--------	------0-
a_{50}	C-------	--------	--------	----A---
a_{51}	--------	--------	--------	------0-
a_{52}	C-------	--------	--------	--------
a_{53}	B-------	--------	--------	--------
a_{54}	--------	--------	--------	--------
a_{55}	--------	--------	--------	--------
a_{56}	--------	--------	--------	--------
a_{57}	--------	--------	--------	--------
a_{58}	--------	--------	--------	--------
a_{59}	--------	--------	--------	--------
a_{60}	--------	--------	--------	--------
a_{61}	--------	--------	--------	--------
a_{62}	--------	--------	--------	--------
a_{63}	--------	--------	--------	--------
a_{64}	--------	--------	--------	--------
a_{65}	--------	--------	--------	--------
a_{66}	--------	--------	--------	---A---
a_{67}	--------	--------	--------	-----0--
a_{68}	--------	--------	--------	-------C
a_{69}	--------	--------	--------	--A---B
a_{70}	--------	--------	--------	----0--
a_{71}	--------	--------	--------	------C-
a_{72}	--------	--------	--------	-A----B-
a_{73}	--------	--------	--------	----0---
a_{74}	--------	--------	--------	--A--C--
a_{75}	--------	--------	--------	A---0B--
a_{76}	--------	--------	--------	--1---C-
a_{77}	--------	--------	--------	----C-B-
a_{78}	--------	--------	--------	-----b--
a_{79}	--------	--------	--------	-1------
a_{80}	--------	--------	--------	--------

For the full SHA-1, the CMM algorithm will be used to find a collision on 26-round SHA-1. After the conventional message modification, it will remain the 9 following conditions; which are only related to rounds 17-26:

$a_{17,3} = 0$, $a_{18,3}+a_{17,3} = 0$, $a_{22,2}+a_{21,2} = 0$, $a_{23,1} = 0$, $a_{24,30}+a_{22,0} = 0$, $a_{24,3}+a_{23,3} = 0$, $a_{25,30} + a_{24,0} = 1$, $a_{25,1} = 0$, $a_{26,1} = 1$.

Uncontrolled Relations. There is 64 uncontrolled relations on the $a_{i,j}$:

$a_{27,0} = 0$, $a_{28,30} + a_{26,0} = 1$, $a_{28,3} + a_{27,3} = 0$, $a_{29,30} + a_{28,0} = 0$, $a_{29,3} + a_{28,3} = 0$, $a_{29,1} = 1$, $a_{30,1} = 0$, $a_{31,0} = 1$, $a_{32,30} + a_{30,0} = 0$, $a_{33,30} + a_{32,0} = 0$, $a_{33,3} + a_{32,3} = 0$, $a_{34,3} + a_{33,3} = 1$, $a_{34,2} + a_{33,2} = 1$, $a_{34,1} = 0$, $a_{35,1} = 0$, $a_{35,0} = 0$, $a_{36,30} + a_{34,0} = 0$, $a_{36,3}+a_{35,3} = 1$, $a_{37,30}+a_{36,0} = 1$, $a_{37,3}+a_{36,3} = 1$, $a_{37,1} = 1$, $a_{38,1} = 0$, $a_{40,31}+a_{39,1} = 1$, $a_{40,3} + a_{39,3} = 1$, $a_{41,1} = 1$, $a_{42,31} + a_{40,1} = 1$, $a_{43,31} + a_{42,1} = 1$, $a_{44,3} + a_{43,3} = 1$, $a_{45,1} = 0$, $a_{46,31} + a_{44,1} = 1$, $a_{46,3} + a_{45,3} = 1$, $a_{47,31} + a_{46,1} = 1$, $a_{47,1} = 0$, $a_{48,31} + a_{46,1} = 1$, $a_{48,3} + a_{47,3} = 1$, $a_{49,31} + a_{48,1} = 1$, $a_{49,1} = 0$, $a_{50,31} + a_{48,1} = 1$, $a_{50,3} + a_{49,3} = 1$, $a_{51,31} + a_{50,1} = 1$, $a_{51,1} = 0$, $a_{52,31} + a_{50,1} = 1$, $a_{53,31} + a_{52,1} = 1$, $a_{66,4} + a_{65,4} = 1$, $a_{67,2} = 0$, $a_{68,0} + a_{66,2} = 1$, $a_{69,5} + a_{68,5} = 1$, $a_{69,0} + a_{68,2} = 1$, $a_{70,3} = 0$, $a_{71,1}+a_{69,3} = 1$, $a_{72,6}+a_{71,6} = 1$, $a_{72,1}+a_{71,3} = 1$, $a_{73,4} = 0$, $a_{74,5}+a_{73,5} = 1$, $a_{74,2}+a_{72,4} = 1$, $a_{75,7}+a_{74,7} = 1$, $a_{75,3} = 0$, $a_{75,2}+a_{74,4} = 1$, $a_{76,5} = 1$, $a_{76,1}+a_{74,3} = 1$, $a_{77,3} + a_{75,5} = 1$, $a_{77,1} + a_{76,3} = 1$, $a_{78,3} + a_{77,5} = 0$, $a_{79,6} = 1$.

To adjust these 64 conditions, we have tried to use semi-neutral bits and adjusters as explained in the IMM algorithm. We use 10 semi-neutral bits (corresponding to 'N' in Table 8) and 8 *adjusters* which are the bits 1-bit-left to 'N'. We present an example of message $m = (m_0, m_1, \ldots, m_{15})$ obtained by Algorithm 2.

$$m = b8550bb2\ 5b2e1a15\ 88a0e568\ b0d7cbaf\ 0b430105\ 1e7f1b5e\ 0637da31\ 0dc9d562$$
$$7d857448\ defac00e\ 9d06ba9e\ 2dd8235a\ 324e9acb\ f7c56578\ c69dfd0e\ 71bf1d08$$

The above m satisfies all message conditions of 0-80 rounds and all chaining variable conditions of 0-28 rounds.

Algebraic Immunity of S-Boxes
and Augmented Functions

Simon Fischer and Willi Meier

FHNW, CH-5210 Windisch, Switzerland
{simon.fischer,willi.meier}@fhnw.ch

Abstract. In this paper, the algebraic immunity of S-boxes and augmented functions of stream ciphers is investigated. Augmented functions are shown to have some algebraic properties that are not covered by previous measures of immunity. As a result, efficient algebraic attacks with very low data complexity on certain filter generators become possible. In a similar line, the algebraic immunity of the augmented function of the eSTREAM candidate Trivium is experimentally tested. These tests suggest that Trivium has some immunity against algebraic attacks on augmented functions.

Keywords: S-box, Stream Cipher, Augmented Function, Algebraic Attack, Filter Generator, Trivium.

1   Introduction

Algebraic attacks can be efficient against stream ciphers based on LFSR's [12], and potentially against block ciphers based on S-boxes [13]. In the case of stream ciphers, the algebraic immunity $\mathcal{AI}$ of the filter function is a measure for the complexity of conventional algebraic attacks. However, it turned out in some cases that large $\mathcal{AI}$ did not help to prevent fast algebraic attacks (FAA's). It is an open question if immunity against FAA's is a sufficient criterion for any kind of algebraic attacks on stream ciphers. In the case of block ciphers, the algebraic immunity of S-boxes is a measure for the complexity of a very general type of algebraic attacks, considering implicit or conditional equations [13, 2]. Present methods for computation of $\mathcal{AI}$ of S-boxes are not very efficient, only about $n = 20$ variables are computationally feasible (except for power mappings, see [20, 11]).

In this paper, we integrate the general approach for S-boxes in the context of stream ciphers and generalise the concept of algebraic immunity of stream ciphers (see Open Problem 7 in [7]). More precisely, we investigate conditional equations for augmented functions of stream ciphers and observe some algebraic properties (to be used in an attack), which are not covered by the previous definitions of $\mathcal{AI}$. As a consequence, immunity against FAA's is not sufficient to prevent any kind of algebraic attack: Depending on the Boolean functions used in a stream cipher, we demonstrate that algebraic properties of the augmented function allow for attacks which need much less known output than

A. Biryukov (Ed.): FSE 2007, LNCS 4593, pp. 366–381, 2007.

established algebraic attacks. This induces some new design criteria for stream ciphers. Time complexity of our attacks is derived by intrinsic properties of the augmented function. Our framework can be applied to a large variety of situations. We present two applications (which both have been implemented). First, we describe efficient attacks on some filter generators. For example, we can efficiently recover the state of a filter generator based on certain Boolean functions when an amount of output data is available which is only linear in the length of the driving LFSR. This should be compared to the data complexity of conventional algebraic attacks, which is about $\binom{n}{e}$, where n is the length of the LFSR and e equals the algebraic immunity of the filter function. Our investigation of the augmented function allows to contribute to open problems posed in [15], and explains why algebraic attacks using Gröbner bases against filter generators are in certain cases successful even for a known output segment only slightly larger than the LFSR length. In a second direction, a large scale experiment carried out with the eSTREAM focus candidate Trivium suggests some immunity of this cipher against algebraic attacks on augmented functions. This experiment becomes feasible as for Trivium with its 288-bit state one can find preimages of 144-bit outputs in polynomial time.

Augmented functions of LFSR-based stream ciphers have previously been studied, e.g. in [1], [16] and [19], where it had been noticed that the augmented function can be weaker than a single output function, with regard to (conditional) correlation attacks as well as to inversion attacks. However, for the first time, we analyse the $\mathcal{AI}$ of (sometimes quite large) augmented functions. Surprisingly, augmented functions did not receive much attention in this context yet.

The paper is organised as follows: In Sect. 2, we investigate some algebraic properties of S-boxes. Our general ideas of algebraic attacks on augmented functions (which are some special S-boxes) are presented in Sec. 3. In Sect. 4, this framework is discussed for filter generators. Sect. 5 and Sect. 6 contain applications of our method, namely for some specific filter generators and for eSTREAM candidate Trivium.

2 Algebraic Properties of S-Boxes

Let $\mathbb{F}$ denote the finite field GF(2), and consider the vectorial Boolean function (or S-box) $S : \mathbb{F}^n \to \mathbb{F}^m$ with $S(x) = y$, where $x := (x_1, \ldots, x_n)$ and $y := (y_1, \ldots, y_m)$. In the case of $m = 1$, the S-box reduces to a Boolean function, and in general, the S-box consists of m Boolean functions $S_i(x)$. These functions give rise to the explicit equations $S_i(x) = y_i$. Here, we assume that y is known and x is unknown.

2.1 Implicit Equations

The S-box can hide implicit equations, namely $F(x, y) = 0$ for each $x \in \mathbb{F}^n$ and with $y = S(x)$. The algebraic normal form of such an equation is denoted

$F(x, y) = \sum c_{\alpha,\beta} x^\alpha y^\beta = 0 \mod 2$, with coefficients $c_{\alpha,\beta} \in \mathbb{F}$, multi-indices $\alpha, \beta \in \mathbb{F}^n$ (which can likewise be identified by their integers) and the notation $x^\alpha := (x_1^{\alpha_1} \cdots x_n^{\alpha_n})$. In the context of algebraic attacks, it is of interest to focus on implicit equations with special structure, *e.g.* on sparse equations or equations of small degree. Let the degree in x be $d := \max\{|\alpha|, c_{\alpha,\beta} = 1\} \leq n$ with the weight $|\alpha|$ of α, and consider an unrestricted degree for the known y, hence $\max\{|\beta|, c_{\alpha,\beta} = 1\} \leq m$. The maximum number of monomials (or coefficients) in an equation of degree d corresponds to $2^m D$, where $D := \sum_{i=0}^{d} \binom{n}{i}$. In order to determine the existence of an implicit equation of degree d, consider a matrix M in $\mathbb{F}$ of size $2^n \times 2^m D$. Each row corresponds to an input x, and each column corresponds to an evaluated monomial (with some fixed order). If the number of columns in M is larger than the number of rows, then linearly dependent columns (*i.e.* monomials) exist, see [9,13]. The rank of M determines the number of solutions, and the solutions correspond to the kernel of M^T. Any non-zero implicit equation (which holds for each input x) may then depend on x and y, or on y only. If it depends on x and y, then the equation may degenerate for some values of y. For example, $x_1 y_1 + x_2 y_1 = 0$ degenerates for $y_1 = 0$.

2.2   Conditional Equations

As the output is assumed to be known, one could investigate equations which are conditioned by the output y, hence $F_y(x) = 0$ for each preimage $x \in S^{-1}(y)$ and of degree d in x. The number of preimages is denoted $U_y := |S^{-1}(y)|$, where $U_y = 2^{n-m}$ for balanced S and $m \leq n$. Notice that conditional equations for different outputs y need not be connected in a common implicit equation, and one can find an optimum equation (*i.e.* an equation of minimum degree) for each output y. Degenerated equations are not existing in this situation, and the corresponding matrix M_y has a reduced size of $U_y \times D$. Similar to the case of implicit equations, one obtains:

Proposition 1. *Consider an S-box $S : \mathbb{F}^n \to \mathbb{F}^m$ and let $S(x) = y$. Then, the number of (independent) conditional equations of degree at most d for some y is $R_y = D - \mathrm{rank}(M_y)$. A sufficient criterion for the existence of a non-zero conditional equation is $0 < U_y < D$.*

The condition $R_y > 0$ requires some minimum value of d, which can depend on y. As already proposed in [2], this motivates the following definition of algebraic immunity for S-boxes:

Definition 1. *Consider an S-box $S : \mathbb{F}^n \to \mathbb{F}^m$. Given some fixed output y, let d be the minimum degree of a non-zero conditional equation $F_y(x) = 0$ which holds for all $x \in S^{-1}(y)$. Then the algebraic immunity $\mathcal{AI}$ of S is defined by the minimum of d over all $y \in \mathbb{F}^m$.*

The $\mathcal{AI}$ can be bounded, using the sufficient condition of Prop. 1. Let d_0 be the minimum degree such that $D > 2^{n-m}$. If the S-box is surjective, then there exists at least one y with a non-zero conditional equation of degree at most d_0, hence $\mathcal{AI} \leq d_0$. In addition, the block size m of the output could be considered as a parameter (by investigating truncated S-boxes S_m, corresponding to partial conditioned equations for S). Let $m_0 := \lfloor n - \log_2 D + 1 \rfloor$ for some degree d. Then, the minimum block size m to find non-zero conditional equations of degree at most d is bounded by m_0. See Tab. 1 for some numerical values of m_0.

Table 1. Theoretical block size m_0 for different parameters n and d

d $\diagdown$ n	16	18	20	32	64	128
1	12	14	16	27	58	121
2	9	11	13	23	53	115
3	7	9	10	20	49	110

A single output y is called *weak*, if non-zero conditional equations of degree d exist for $U_y \gg D$ (or if the output is strongly imbalanced). This roughly corresponds to the condition $d \ll d_0$, or $m \ll m_0$.

2.3 Algorithmic Methods

As already mentioned in [7], memory requirements to determine the rank of M are impractical for about $n > 20$. In the case of conditional equations, the matrix M_y can be much smaller, but the bottleneck is to compute an exhaustive list of preimages, which requires a time complexity of 2^n. However, one could use a probabilistic variant of this basic method: Instead of determining the rank of M_y which contains all U_y inputs x, one may solve for a smaller matrix M'_y with $V < U_y$ random inputs. Then, one can determine the *non-existence* of a solution: If no solution exists for M'_y, then no solution exists for M_y either. On the other hand, if one or more solutions exist for M'_y, then they hold true for the subsystem of V inputs, but possibly not for all U_y inputs. Let the probability p be the fraction of preimages that satisfy the equation corresponding to such a solution. With the heuristical argument $(1 - p)V < 1$, we expect that $p > 1 - 1/V$. However, this argument holds only for $V > D$, because otherwise, there are always at least $D - V$ solutions (which could be balanced). Consequently, if V is a small multiple of D, the probability can be quite close to one. For this reason, all solutions of the smaller system can be useful in later attacks. Determining only a few random preimages can be very efficient: In a naive approach, time complexity to find a random preimage of an output y is about $2^n/U_y$ (which is 2^m for balanced S), and complexity to find D preimages is about $2^n D/U_y$. This is an improvement compared to the exact method if $U_y \gg D$, *i.e.* equations can be found efficiently for weak outputs. Memory requirements of the probabilistic algorithm are about $C_M = D^2$, and time complexity is about $C_T = D2^m + D^3$. Computation of $\mathcal{AI}$ requires about $C_T = D2^m + D^3 2^m = \mathcal{O}(D^3 2^m)$.

3   Algebraic Attacks Based on the Augmented Function

In this section, we focus on algebraic cryptanalysis of S-boxes in the context of stream ciphers. Given a stream cipher, one may construct an S-box as follows:

Definition 2. *Consider a stream cipher with internal state x of n bits, an update function L, and an output function f which outputs one bit of keystream in a single iteration. Then, the augmented function S_m is defined by*

$$S_m : \mathbb{F}^n \to \mathbb{F}^m$$
$$x \mapsto (f(x), f(L(x)), \ldots, f(L^{m-1}(x))) . \tag{1}$$

The update L can be linear (*e.g.* for filter generators), or nonlinear (*e.g.* for Trivium). The input x correspond to the internal state at some time t, and the output y corresponds to an m-bit block of the known keystream. Notice that m is a very natural parameter here. The goal is to recover the initial state x by algebraic attacks, using (potentially probabilistic) conditional equations $F_y(x) = 0$ of degree d for outputs y of the augmented function S_m. This way, one can set up equations for state variables of different time steps t. In the case of a linear update function L, each equation can be transformed into an equation of degree d in the initial state variables x. In the case of a nonlinear update function L, the degree of the equations is increasing with time. However, the nonlinear part of the update is sometimes very simple, such that equations for different time steps can be efficiently combined. Finally, the system of equations in the initial state variables x is solved.

If the augmented function has some weak outputs, then conditional equations can be found with the probabilistic algorithm of Sect. 2.3, which requires about D preimages of a single m-bit output. One may ask if there is a dedicated way to compute random preimages of m-bit outputs in the context of augmented functions. Any stream cipher as in Def. 2 can be described by a system of equations. Nonlinear systems of equations with roughly the same number of equations as unknowns are in general NP-hard to solve. However, due to the special (simple) structure of some stream ciphers, it may be easy to partially invert the nonlinear system. For example, given a single bit of output of a filter generator, it is easy to find a state which gives out this bit. Efficient computation of random preimages for m-bit outputs is called *sampling*. The maximal value of m for which states can be sampled without trial and error is called sampling resistance of the stream cipher. Some constructions have very low sampling resistance, see [5, 4].

The parameters of our framework are the degree d of equations, and the block-size m of the output. An optimal tradeoff between these parameters depends on the algebraic properties of the augmented function. The attack is expected to be efficient, if:

1. There are many low-degree conditional equations for S_m.
2. Efficient sampling is possible for this block size m.

This measure is well adapted to the situation of augmented functions, and can be applied to sometimes quite large augmented functions, see Sect. 5 and 6. This way, we intend to prove some immunity of a stream cipher, or present attacks with reduced complexity.

4   Generic Scenarios for Filter Generators

Our framework is investigated in-depth in the context of LFSR-based stream ciphers (and notably for filter generators), which are the main target of conventional and fast algebraic attacks (see also Appendix A). We describe some elementary conditional equations induced by annihilators. Then, we investigate different methods for sampling, which are necessary to efficiently set up conditional equations. We suggest a basic scenario and estimate data complexity of an attack, the scenario is refined and improved.

4.1   Equations Induced by Annihilators

Let us first discuss the existence of conditional equations of degree $d = \mathcal{AI}$, where $\mathcal{AI}$ is the ordinary algebraic immunity of f here. With $m = 1$, the number of conditional equations for $y = 0$ (resp. $y = 1$) corresponds to the number of annihilators of $f + 1$ (resp. f) of degree d. If one increases m, then all equations originating from annihilators are involved: For example, if there is 1 annihilator of degree d for both f and $f + 1$, then the number of equations is expected to be at least m for any m-bit output y. Notice that equations of fast algebraic attacks are not involved if m is small compared to n.

4.2   Sampling

Given an augmented function S_m of a filter generator, the goal of sampling is to efficiently determine preimages x for fixed output $y = S_m(x)$ of m bits. Due to the special structure of the augmented function, there are some efficient methods for sampling:

Filter Inversion. One could choose a fixed value for the k input bits of the filter, such that the observed output bit is correct (using a table of the filter function). This can be done for about n/k successive output bits, until the state is unique. This way, preimages of an output y of n/k bits can be found in polynomial time, and by partial search, preimages of larger outputs can be computed. Time complexity to find a preimage of $m > n/k$ bits is about $2^{m-n/k}$, *i.e.* the method is efficient if there are only few inputs k.

Linear Sampling. In each time step, a number of l linear conditions are imposed on the input variables of f, such that the filter becomes linear. The linearised filter gives one additional linear equation for each keystream bit. Notice that all variables can be expressed by a linear function of the n variables of the initial state. Consequently, for an output y of m bits, one obtains $(l+1)m$ (inhomogeneous) linear equations for n unknowns, *i.e.* we expect that preimages can be found in polynomial time if $m \leq n/(l+1)$. To find many different preimages, one should have several independent conditions (which can be combined in a different way for each clock cycle).

In practice, sampling should be implemented carefully in order to avoid contradictions (*e.g.* with appropriate conditions depending on the keystream), see [5].

4.3   Basic Scenario

We describe a basic scenario for algebraic attacks on filter generators based on the augmented function: With C_D bits of keystream, one has $C'_D = C_D - m + 1$ (overlapping) windows of m bits. Assume that there are $R := \sum_y R_y$ equations of degree d for m-bit outputs y. For each window, we have about $r := R/2^m$ equations, which gives a total of $N = rC'_D$ equations.[1] Each equation has at most D monomials in the initial state variables, so we need about the same number of equations to solve the system by linearisation. Consequently, data complexity is $C_D = D/r + m - 1$ bits. The initial state can then be recovered in $C_T = D^3$. This should be compared with the complexity of conventional algebraic attacks $C_D = 2E/R_A$ and $C_T = E^3$, where $e := \mathcal{AI}$, $E := \sum_{i=0}^{e} \binom{n}{i}$, and R_A the number of annihilators of degree e. Notice that the augmented function may give low-degree equations, which are not visible for single-bit outputs; this increases information density and may reduce data complexity. Our approach has reduced time complexity if $d < e$, provided that sampling (and solving the matrix) is efficient.

4.4   Refined Basic Scenario

The basic scenario for filter generators should be refined in two aspects, concerning the existence of dependent and probabilistic equations: First, with overlapping windows of m bits, it may well happen that the same equation is counted several times, namely if the equation already exists for a substring of $m' < m$ bits (*e.g.* in the case of equations produced by annihilators). In addition, equations may be linearly dependent by chance. If this is not considered in the computation of R, one may have to enlarge data complexity a little bit. Second, one can expect to obtain probabilistic solutions. However, depending on the number of computed preimages, the probability p may be large and the corresponding equations can still be used in our framework, as they increase R and reduce data complexity, but potentially with some more cost in time. As we need about D (correct and linearly independent) equations to recover the initial state, the probability p should be at least $1 - 1/D$ (together with our estimation for p, this justifies that the number of preimages should be at least D). In the case of a contradiction, one could complement a few equations in a partial search and solve again, until the keystream can be verified. Depending on the actual situation, one may find an optimal tradeoff in the number of computed preimages. Notice that our probabilistic attack deduced from an algebraic attack with

---

[1] From a heuristical point of view, the parameter r is only meaningful if the conditional equations are approximately uniformly distributed over all outputs y.

equations of degree 1 is a powerful variant of a conditional correlation attack, see [19]. A probabilistic attack with nonlinear equations is a kind of higher order correlation attack, see [8].

4.5 Substitution of Equations

It is possible to further reduce data complexity in some cases. Consider the scenario where one has $N = rC'_D$ linear equations. On the other hand, given an annihilator of degree $e := \mathcal{AI}$, one can set up a system of degree e as in conventional algebraic attacks. The N linear equations can be substituted into this system in order to eliminate N variables. This results in a system of $D' := \sum_{i=0}^{e} \binom{n-N}{i}$ monomials, requiring a data complexity of $C_D = D'$ and time complexity $C_T = D'^3$. Notice that data can be reused in this case, which gives the implicit equation in C_D. Obviously, a necessary condition for the success of this method is $rE > 1$. A similar improvement of data complexity is possible for nonlinear equations of degree d. One can multiply the equations by all monomials of degree $e - d$ in order to obtain additional equations of degree e, along the lines of XL [14] and Gröbner bases algorithms.

5 First Application: Some Specific Filter Generators

Many conventional algebraic attacks on filter generators require about $\binom{n}{e}$ output bits where e equals the algebraic immunity of the filter function. On the other hand, in [15], algebraic attacks based on Gröbner bases are presented, which in a few cases require only $n+\varepsilon$ data. It is an open issue to understand such a behavior from the Boolean function and the tapping sequence. We present attacks on the corresponding augmented functions, requiring very low data complexity. This means, we can identify the source of the above behavior, and in addition, we can use our method also for other functions.

5.1 Existence of Equations

In this subsection, we give extensive experimental results for different filter generators. Our setup is chosen as follows: The filter functions are instances of the CanFil family (see [15]) or the Majority functions. These instances all have five inputs and algebraic immunity 2 or 3. Feedback taps correspond to a random primitive feedback polynomial, and filter taps are chosen randomly in the class of full positive difference sets, see Tab. 4 in Appendix B for an enumerated specification of our setups. Given a specified filter generator and parameters d and m, we compute the number R_y of independent conditional equations $F_y(x) = 0$ of degree d for each output $y \in \mathbb{F}^m$. The overall number of equations $R := \sum_y R_y$ for $n = 20$ is recorded in Tab. 2. Thereby, preimages are computed by exhaustive search in order to exclude probabilistic solutions.

In the case of CanFil1 and CanFil2, linear equations exist only for $m \geq m_0 - 1$, independent of the setup. On the other hand, for CanFil5 and Majority5, there

Table 2. Counting the number of linear equations R for the augmented function of different filter generators, with $n = 20$ bit input and m bit output

Filter	m	R for setups $6 - 10$				
CanFil1	14	0	0	0	0	0
	15	3139	4211	3071	4601	3844
CanFil2	14	0	0	0	0	0
	15	2136	2901	2717	2702	2456
CanFil5	6	0	0	0	2	0
	7	0	0	0	8	0
	8	0	0	0	24	0
	9	0	0	0	64	0
	10	6	0	0	163	0
	11	113	0	2	476	0
	12	960	16	215	1678	29
Majority5	9	0	0	0	2	0
	10	1	10	1	18	1
	11	22	437	40	148	56

exist many setups where a large number of linear equations already exists for $m \approx n/2$, see Ex. 1. We conclude that the number of equations weakly depends on the setup, but is mainly a property of the filter function. The situation is very similar for other values of n, see Appendix C. This suggests that our results can be scaled to larger values of n. Let us also investigate existence of equations of higher degree: CanFil1 and CanFil2 have $\mathcal{AI} = 2$ and there is 1 annihilator for both f and $f + 1$, which means that at least m quadratic equations can be expected for an m-bit output. For each setup and $m < m_0 - 1$, we observed only few additional equations, whereas the number of additional equations is exploding for larger values of m. This was observed for many different setups and different values of n.

Example 1. Consider CanFil5 with $n = 20$ and setup 9. For the output $y = 000000$ of $m = 6$ bits, there are exactly 2^{14} preimages, hence the matrix M_y has 2^{14} rows and $D = 21$ columns for $d = 1$. Evaluation of M_y yields a rank of 20, *i.e.* a nontrivial solution exists. The explicit solution is $F_y(x) = x_2 + x_4 + x_5 + x_6 + x_{10} + x_{11} + x_{12} + x_{13} + x_{14} + x_{15} + x_{17} = 0$. □

5.2   Probabilistic Equations

In the previous subsection, the size n of the state was small enough to compute a complete set of preimages for some m-bit output y. However, in any practical situation where n is larger, the number of available preimages is only a small multiple of D, which may introduce probabilistic solutions. Here is an example with $n = 20$, where the probability can be computed exactly:

Example 2. Consider again CanFil5 with $n = 20$ and setup 9. For the output $y = 000000$ of $m = 6$ bits, there are 2^{14} preimages and one exact conditional equation of degree $d = 1$. We picked 80 random preimages and determined all (correct or probabilistic) linear conditional equations. This experiment was repeated 20 times with different preimages. In each run, we obtained between 2 and 4 independent equations with probabilities $p = 0.98, \ldots, 1$. For example, the (probabilistic) conditional equation $F_y(x) = x_2 + x_3 + x_4 + x_7 + x_{10} + x_{16} + x_{17} + x_{18} = 0$ holds with probability $p = 1 - 2^{-9}$. □

In the above example, there are only few probabilistic solutions and they have impressively large probability, which makes the equations very useful in an attack. Notice that experimental probability is in good agreement with our estimation $p > 1 - 1/80 = 0.9875$. The situation is very similar for other parameters. With the above setup and $m = 10$, not only $y = 000\ldots0$ but a majority of outputs y give rise to linear probabilistic equations. In the case of CanFil1 and CanFil2, we did not observe linear equations of large probability for $m < m_0 - 1$. It is interesting to investigate the situation for larger values of n:

Example 3. Consider CanFil5 with $n = 40$ and setup 11. For the output $y = 000\ldots0$ of $m = 20$ bits, we determine 200 random preimages. With $d = 1$, evaluation of M_y yields a rank of 30, *i.e.* 11 (independent) solutions exist. With 2000 random preimages, we observed a rank of 33, *i.e.* only 3 solutions of the first system were detected to be merely probabilistic. An example of an equation is $F_y(x) = x_1 + x_8 + x_{10} + x_{14} + x_{15} + x_{18} + x_{19} + x_{26} + x_{31} + x_{34} = 0$. □

The remaining 8 solutions of the above example may be exact, or probabilistic with very high probability. By sampling, one could find (probabilistic) conditional equations for much larger values of n. For example, with CanFil5, $n = 80$, $m = 40$ and filter inversion, time complexity to find a linear equation for a weak output is around 2^{32}.

5.3   Discussion of Attacks

Our experimental results reveal that some filter functions are very vulnerable to algebraic attacks based on the corresponding augmented function. For CanFil5 with $n = 20$ and setup 9, we observed $R = 163$ exact equations using the parameters $m = 10$ and $d = 1$, which gives a ratio of $r = 0.16$. Including probabilistic equations, this ratio may be even larger. Here, preimages of any y can be found efficiently by sampling: using filter inversion, a single preimage can be found in $2^{m-n/k} = 2^6$ steps, and a single equation in around 2^{13} steps. Provided that equations are independent and the probability is large, data complexity is about $C_D = (n + 1)/r + m - 1 = 140$. The linear equations could also be substituted into the system of degree $\mathcal{AI} = 2$, which results in a data complexity of about $C_D = 66$. Notice that conventional algebraic attacks would require $C_D = E = 211$ bits (and time complexity E^3). As we expect that our observation can be scaled, (*i.e.* that r remains constant for larger values of n and $m = n/2$), data complexity is a linear function in n. Considering time complexity for variable n, the matrix M and the final system of equations can be solved in polynomial time, whereas sampling is subexponential (and polynomial in some cases, where linear sampling is possible).

In [15], CanFil5 has been attacked experimentally with $n + \varepsilon$ data, where $n = 40, \ldots, 70$ and $\varepsilon < 10$. Our analysis gives a conclusive justification for their observation. Other functions such as Majority5 could be attacked in a similar way, whereas CanFil1 and CanFil2 are shown to be much more resistant against this general attack: No linear equations have been found for $m < m_0 - 1$, and only few quadratic equations.

6   Second Application: Trivium

Trivium [6] is a stream cipher with a state of 288 bits, a nonlinear update and a linear output. It has a simple algebraic structure, which makes it an interesting candidate for our framework. We consider the S-box $S_m(x) = y$, where S is the augmented function of Trivium, x the state of $n = 288$ bits, and y the output of m bits. We will first analyse the sampling of S_m, which is very similar to linear sampling of filter generators.

6.1   Sampling

The state consists of the 3 registers $R_1 = (x_1, \ldots, x_{93})$, $R_2 = (x_{94}, \ldots, x_{177})$ and $R_3 = (x_{178}, \ldots, x_{288})$. In each clock cycle, a linear combination of 6 bits of the state (2 bits of each register) is output. Then, the registers are shifted to the right by one position, with a nonlinear feedback to the first position of each register. In the first 66 clocks, each keystream bit is a linear function of the input, whereas the subsequent keystream bit involves a nonlinear expression. Consequently, given any output of $m = 66$ bits, one can efficiently determine some preimages by solving a linear system. It is possible to find preimages of even larger output size. Observe that the nonlinear function is quadratic, where the two factors of the product have subsequent indices. Consequently, one could fix some alternating bits of the state, which results in additional linear equations for the remaining variables. Let c, l, q denote constant, linear, and quadratic dependence on the initial state. Let all the even bits of the initial state be c, see Tab. 3. After update 83, bits 82 and 83 of R_2 are both l. Variable t_2 takes bits 82 and 83 of R_2 to compute the nonlinear term. So after update 84, $t_2 = x_{178}$ is q (where nonlinear terms in t_1 and t_3 appear somewhat later).

Table 3. Evolution of states with partially fixed input

Initial state	After 1 update	After 84 updates
$R_1 = $ lclcl$\ldots$	$R_1 = $ llclcl$\ldots$	$R_1 = $ 11111$\ldots$
$R_2 = $ clclc$\ldots$	$R_2 = $ lclclc$\ldots$	$R_2 = $ 11111$\ldots$
$R_3 = $ clclc$\ldots$	$R_3 = $ lclclc$\ldots$	$R_3 = $ q1111$\ldots$

After 65 more updates, x_{243} is quadratic, where x_{243} is filtered out from R_3 in the next update (after 84 updates, other bits are also q and are filtered out from registers R_1 and R_2, but on a later point in time). Consequently, keystream bit number $66 + 84 = 150$ (counting from 1) is q, and the first 149 keystream bits are linear in the remaining variables.

The number of remaining variables in the state (the degree of freedom) is 144. Consequently, for an output of size $m = 144$ bits, we can expect to find one solution for the remaining variables; this was verified experimentally. The solution (combined with the fixed bits) yields a preimage of y. Notice that we do not exclude any preimages this way. In addition, m can be somewhat larger with partial search for the additional bits.

Example 4. Consider the special output $y = 000 \ldots 0$ of $m = 160$ bits. By sampling and partial exhaustive search, we find the nontrivial preimage

$$x = \begin{array}{l} 1000101111000101110011000101010011010000010010010 \\ 0001001001001100111110110110110110000100110010101000 \\ 110000000010101100111000011111101100110000110101010 \\ 0111000001010100110011011110101010101111111101000001 \\ 0000010000011010001000011110011010101000101010111 \\ 1010000011101001010100110001001110010100010101101 \end{array} \qquad \square$$

6.2   Potential Attacks

The nonlinear update of Trivium results in equations $S_m(x) = y$ of increasing degree for increasing values of m. However, for any output y, there are at least 66 linear equations in the input variables. It is an important and security related question, if there are additional linear equations for some fixed output y. A linear equation is determined by $D = 289$ coefficients, thus we have to compute somewhat more than 289 preimages for this output. By sampling, this can be done in polynomial time. Here is an experiment:

Example 5. Consider a prescribed output y of 144 bits, and compute 400 preimages x such that $S_m(x) = y$ (where the preimages are computed by a uniform random choice of 144 fixed bits of x). Given these preimages, set up and solve the matrix M of linear monomials in x. For 30 uniform random choices of y, we always observed 66 linearly independent solutions. $\qquad \square$

Consequently, Trivium seems to be immune against additional linear equations, that might help in an attack. Because of the lack of probabilistic solutions, Trivium is also supposed to be immune against equations of large probability (compare with CanFil1 and CanFil2). As pointed out in [17], there are some states resulting in a weak output: If R_1, R_2 and R_3 are initialised by some period-3 states, then the whole state (and hence the output) repeats itself every 3 iterations. Each of these states results in $y = 000 \ldots 0$. Here is an extended experiment (with partial exhaustive search) for this special output:

Example 6. Consider the output $y = 000\ldots0$ of 150 bits, and compute 400 random preimages x such that $S_m(x) = y$. By solving the matrix M of linear monomials in x, we still observed 66 linearly independent solutions. $\square$

7   Conclusions

Intrinsic properties of augmented functions of stream ciphers have been investigated with regard to algebraic attacks. Certain properties of the augmented function enable efficient algebraic attacks with lower data complexity than established algebraic attacks. In order to assess resistance of augmented functions against such improved algebraic attacks, a prespecified number of preimages of outputs of various size of these functions have to be found. For a random function, the difficulty of finding preimages increases exponentially with the output size. However, due to a special structure of the augmented function of a stream cipher, this can be much simpler than in the random case. For any such stream cipher, our results show the necessity of checking the augmented function for algebraic relations of low degree for output sizes for which finding preimages is feasible. In this paper, this has been successfully carried out for various filter generators as well as for the eSTREAM candidate Trivium.

Acknowledgments

This work is supported in part by the National Competence Center in Research on Mobile Information and Communication Systems (NCCR-MICS), a center of the Swiss National Science Foundation under grant number 5005-67322. The second author is supported by Hasler Foundation www.haslerfoundation.ch under project number 2005. We would like to thank Steve Babbage for encouraging us to study algebraic immunity of large S-boxes.

References

1. Anderson, R.J.: Searching for the Optimum Correlation Attack. In: Preneel, B. (ed.) Fast Software Encryption - FSE 1994. LNCS, vol. 1008, Springer, Heidelberg (1995)
2. Armknecht, F., Krause, M.: Constructing Single- and Multi-Output Boolean Functions with Maximal Algebraic Immunity. In: Bugliesi, M., Preneel, B., Sassone, V., Wegener, I. (eds.) ICALP 2006. LNCS, vol. 4052, Springer, Heidelberg (2006)
3. Armknecht, F., Carlet, C., Gaborit, P., Künzli, S., Meier, W., Ruatta, O.: Efficient Computation of Algebraic Immunity for Algebraic and Fast Algebraic Attacks. In: Vaudenay, S. (ed.) EUROCRYPT 2006. LNCS, vol. 4004, Springer, Heidelberg (2006)
4. Babbage, S.: A Space/Time Tradeoff in Exhaustive Search Attacks on Stream Ciphers. In: European Convention on Security and Detection. IEE Conference Publication No. 408 (1995)

5. Biryukov, A., Shamir, A.: Cryptanalytic Time/Memory/Data Tradeoffs for Stream Ciphers. In: Okamoto, T. (ed.) ASIACRYPT 2000. LNCS, vol. 1976, Springer, Heidelberg (2000)
6. de Cannière, C., Preneel, B.: Trivium - A Stream Cipher Construction Inspired by Block Cipher Design Principles. In: eSTREAM, ECRYPT Stream Cipher Project, Report 2005/030
7. Canteaut, A.: Open Problems Related to Algebraic Attacks on Stream Ciphers. In: Ytrehus, Ø. (ed.) WCC 2005. LNCS, vol. 3969, Springer, Heidelberg (2006)
8. Courtois, N.: Higher Order Correlation Attacks, XL algorithm and Cryptanalysis of Toyocrypt. In: Cryptology ePrint Archive, Report 2002/087
9. Courtois, N.: Algebraic Attacks on Combiners with Memory and Several Outputs. In: Cryptology ePrint Archive, Report 2003/125
10. Courtois, N.: How Fast can be Algebraic Attacks on Block Ciphers. In: Cryptology ePrint Archive, Report 2006/168
11. Courtois, N., Debraize, B., Garrido, E.: On Exact Algebraic (Non-)Immunity of S-boxes Based on Power Functions. In: Cryptology ePrint Archive, Report 2005/203
12. Courtois, N., Meier, W.: Algebraic Attacks on Stream Ciphers with Linear Feedback. In: Biham, E. (ed.) Advances in Cryptology – EUROCRPYT 2003. LNCS, vol. 2656, Springer, Heidelberg (2003)
13. Courtois, N., Pieprzyk, J.: Cryptanalysis of Block Ciphers with Overdefined Systems of Equations. In: Zheng, Y. (ed.) ASIACRYPT 2002. LNCS, vol. 2501, Springer, Heidelberg (2002)
14. Courtois, N., Shamir, A., Patarin, J., Klimov, A.: Efficient Algorithms for solving Overdefined Systems of Multivariate Polynomial Equations. In: Preneel, B. (ed.) EUROCRYPT 2000. LNCS, vol. 1807, Springer, Heidelberg (2000)
15. Faugère, J.-C., Ars, G.: An Algebraic Cryptanalysis of Nonlinear Filter Generators using Gröbner Bases. In: Rapport de Recherche de l'INRIA (2003)
16. Golić, J.Dj.: On the Security of Nonlinear Filter Generators. In: Gollmann, D. (ed.) Fast Software Encryption. LNCS, vol. 1039, Springer, Heidelberg (1996)
17. Hong, J.: Some Trivial States of Trivium. In: eSTREAM Discussion Forum (2005)
18. Krause, M.: BDD-Based Cryptanalysis of Keystream Generators. In: Knudsen, L.R. (ed.) EUROCRYPT 2002. LNCS, vol. 2332, Springer, Heidelberg (2002)
19. Löhlein, B.: Attacks based on Conditional Correlations against the Nonlinear Filter Generator. In: Cryptology ePrint Archive, Report 2003/020
20. Nawaz, Y., Gupta, K.C., Gong, G.: Algebraic Immunity of S-boxes Based on Power Mappings: Analysis and Construction. In: Cryptology ePrint Archive, Report 2006/322

A   Algebraic and Fast Algebraic Attacks

The algebraic immunity $\mathcal{AI}$ of a Boolean function f is defined by the minimum degree d of a function g, such that $fg = 0$ or $(f + 1)g = 0$. In the case $fg = 0$, one can multiply $y^t = f(L^t(x))$ by g and obtains $g(L^t(x)) \cdot y^t = 0$. For $y^t = 1$, this is an equation of degree d. Similarly, for $(f + 1)g = 0$, one obtains $g(L^t(x)) \cdot (y^t + 1) = 0$. With R_A linearly independent annihilators of degree d for f and $f + 1$, a single output bit can be used to set up (in average) $R_A/2$ equations in x at time t. The number of monomials in these equations is at most $D := \sum_{i=0}^{d} \binom{n}{i}$, hence by linearisation, data complexity of conventional algebraic attacks becomes about $2D/R_A$, and time complexity $C_T = D^3$.

In fast algebraic attacks, one considers equations of type $fg = h$ for $\deg h \geq \mathcal{AI}$ and $\deg g < \mathcal{AI}$. The equation $y^t = f(L^t(x))$ is multiplied by g such that $g(L^t(x)) \cdot y^t = h(L^t(x))$. One can precompute then a linear combination $\sum_i c_i \cdot h(L^{t+i}(x)) = 0$ for all t, such that $\sum_i c_i \cdot g(L^{t+i}(x)) \cdot y^{t+i} = 0$ of lower degree $\deg g$. The linear combination utilises the structure of the LFSR, and helps to cancel out all monomials of degree larger than $\deg g$. However, the equation depends on several output bits y^t. It is a special case of implicit equation, where the degree in y is 1. Depending on the degrees of g and h, time complexity can be smaller than in algebraic attacks, and data complexity is about $C_D = D + E$, where $E := \sum_{i=0}^{e} \binom{n}{i}$. This is not much larger than in algebraic attacks (with the same asymptotic complexity). See [3] for an efficient computation of annihilators and low-degree multiples.

B   Experimental Setup for Filter Generators

In Tab. 4, we collect the setups of our experiments with filter generators, where n is the size of the LFSR, and k the number of inputs to the filter function. The feedback taps are chosen such that the LFSR has maximum period (*i.e.*, the corresponding polynomial is primitive), and filter taps are chosen according to a full positive difference set (*i.e.*, all the positive pairwise differences are distinct). Tap positions are counted from the left (starting by 1), and the LFSR is shifted to the right.

Table 4. Different setups for our experiments with filter generators

Setup	n	k	feedback taps	filter taps
1	18	5	$[2, 3, 5, 15, 17, 18]$	$[1, 2, 7, 11, 18]$
2	18	5	$[1, 2, 5, 7, 9, 14, 15, 16, 17, 18]$	$[1, 3, 7, 17, 18]$
3	18	5	$[3, 5, 7, 15, 17, 18]$	$[1, 5, 8, 16, 18]$
4	18	5	$[4, 5, 6, 10, 13, 15, 16, 18]$	$[1, 6, 7, 15, 18]$
5	18	5	$[2, 3, 5, 7, 11, 15, 17, 18]$	$[1, 3, 6, 10, 18]$
6	20	5	$[7, 10, 13, 17, 18, 20]$	$[1, 3, 9, 16, 20]$
7	20	5	$[1, 2, 4, 7, 8, 10, 11, 12, 13, 15, 19, 20]$	$[1, 5, 15, 18, 20]$
8	20	5	$[2, 3, 4, 5, 6, 7, 8, 11, 13, 14, 19, 20]$	$[1, 4, 9, 16, 20]$
9	20	5	$[1, 2, 3, 4, 5, 8, 10, 11, 12, 13, 15, 17, 19, 20]$	$[1, 2, 15, 17, 20]$
10	20	5	$[1, 2, 6, 7, 9, 11, 15, 20]$	$[1, 5, 13, 18, 20]$
11	40	5	$[3, 8, 9, 10, 11, 13, 14, 15, 18, 19,$ $23, 24, 25, 26, 27, 30, 33, 34, 36, 40]$	$[1, 3, 10, 27, 40]$

C   Additional Experimental Results

In Tab. 5, we present the number of conditional equations for different filters and different parameters, where the size of the LFSR is $n = 18$.

Table 5. Counting the number of linear equations R for the augmented function of different filter generators, with $n = 18$ bit input and m bit output

Filter	m	R for setups 1-5				
CanFil1	12	0	0	0	0	0
	13	625	288	908	335	493
CanFil2	12	0	0	0	0	0
	13	144	346	514	207	418
CanFil3	12	0	0	4	0	0
	13	1272	1759	2173	2097	983
CanFil4	7	0	0	0	0	0
	8	19	4	0	0	0
	9	102	17	1	0	12
	10	533	69	9	20	167
CanFil5	6	1	0	0	0	0
	7	4	0	0	0	0
	8	15	0	0	0	1
	9	55	1	0	0	39
	10	411	61	3	0	360
	11	2142	1017	166	10	1958
CanFil6	8	0	0	0	0	0
	9	0	10	64	0	0
	10	0	97	256	0	0
	11	0	517	1024	0	0
	12	0	2841	3533	1068	0
	13	152	19531	17626	12627	9828
CanFil7	11	0	2	0	0	6
	12	68	191	36	26	178
Majority5	8	1	0	0	0	0
	9	8	3	42	27	14
	10	97	94	401	282	158

Generalized Correlation Analysis of Vectorial Boolean Functions

Claude Carlet[1], Khoongming Khoo[2], Chu-Wee Lim[2], and Chuan-Wen Loe[2]

[1] University of Paris 8 (MAATICAH)
also with INRIA, Projet CODES, BP 105, 78153, Le Chesney Cedex, France
[2] DSO National Laboratories, 20 Science Park Drive, S118230, Singapore
claude.carlet@inria.fr, kkhoongm@dso.org.sg, lchuwee@dso.org.sg,
lchuanwe@dso.org.sg

Abstract. We investigate the security of n-bit to m-bit vectorial Boolean functions in stream ciphers. Such stream ciphers have higher throughput than those using single-bit output Boolean functions. However, as shown by Zhang and Chan at Crypto 2000, linear approximations based on composing the vector output with any Boolean functions have higher bias than those based on the usual correlation attack. In this paper, we introduce a new approach for analyzing vector Boolean functions called generalized correlation analysis. It is based on approximate equations which are linear in the input x but of free degree in the output $z = F(x)$. Based on experimental results, we observe that the new generalized correlation attack gives linear approximation with much higher bias than the Zhang-Chan and usual correlation attacks. Thus it can be more effective than previous methods.

First, the complexity for computing the generalized nonlinearity for this new attack is reduced from $2^{2^m \times n+n}$ to 2^{2n}. Second, we prove a theoretical upper bound for generalized nonlinearity which is much lower than the unrestricted nonlinearity (for Zhang-Chan's attack) or usual nonlinearity. This again proves that generalized correlation attack performs better than previous correlation attacks. Third, we introduce a generalized divide-and-conquer correlation attack and prove that the usual notion of resiliency is enough to protect against it. Finally, we deduce the generalized nonlinearity of some known secondary constructions for secure vector Boolean functions.

Keywords: Vectorial Boolean Functions, Unrestricted Nonlinearity, Resiliency.

1 Introduction

In this paper, we consider n-bit to m-bit vectorial Boolean functions when they are used in stream ciphers. There are two basic designs for such stream ciphers based on linear feedback shift registers (LFSR). One is a combinatorial generator [12] which consists of n LFSR's and a vector function $F(x)$. At each clock, one bit is tapped from the secret state of each LFSR as an input bit of $F(x)$ to

A. Biryukov (Ed.): FSE 2007, LNCS 4593, pp. 382–398, 2007.

produce m bits of output keystream. This keystream is then XORed with the plaintext to form the ciphertext. The other model is the filter function generator [12] where n bits are tapped from one LFSR as input to $F(x)$ to produce the keystream output. The advantage of using vector Boolean functions is that the stream ciphers have higher throughput, i.e. the encryption and decryption speed is m times faster than single output Boolean functions. However, we need to study its security when compared to the single-bit output case.

Basic attacks on these stream ciphers are the correlation attack by Siegenthaler [14] and its improvements (see e.g. [2]). In the case of the filter function model, a linear approximation is formed between the LFSR state bits and output keystream. If the approximation has probability $p \neq 1/2$, then we can recover the secret LFSR bits when enough keystream bits are known. In the case of the combinatorial model, an approximation of the output is made by the combination of t out of the n input bits produced by the LFSRs and it is shown in [2] that the attack is optimal with the linear combination of these t bits. Siegenthaler's attack was described for single-output Boolean functions but it can be generalized naturally to the vector output case where we take any linear combination of output bits [3].

This attack can be improved as shown by Zhang and Chan at Crypto 2000 [15] where they consider linear approximation based on any Boolean function of the output vector (instead of just linear combination of the output vector). Since there are 2^{2^m+n} linear approximations to choose from in the Zhang-Chan approach compared to just 2^{n+m} linear approximations in the usual approach, it is easier to choose one with higher bias, i.e. where probability p is further away from $1/2$.

In Section 2, we introduce the generalized correlation attack by considering linear approximations which are linear in the input x and of free degree in the output $z = F(x)$. Now there are $2^{2^m \times (n+1)}$ linear approximations from which we can choose one with even higher bias than the Zhang-Chan and usual correlation attack. However, choosing the best linear approximation out of that many choices is infeasible. Therefore in Section 3, we reduce the complexity of choosing the best linear approximation for generalized correlation attack from $2^{2^m \times (n+1)+n}$ to 2^{2n}, which is much more manageable.

The generalized nonlinearity is an analogue of the usual nonlinearity, which measures the effectiveness of a function against generalized correlation attack. Based on efficient computation for finding the best generalized linear approximation, we computed the generalized nonlinearity of highly nonlinear vector functions and randomly generated vector functions in Section 3.2. We observe that the generalized nonlinearity is much lower than the usual nonlinearity and unrestricted nonlinearity (corresponding to Zhang-Chan's attack) for these functions. For example, when the inverse function on $GF(2^8)$ is restricted to $5, 6, 7$ output bits, the usual and unrestricted nonlinearities are non-zero while the generalized nonlinearity is already zero. That means the stream cipher can be attacked as a deterministic linear system while the Zhang-Chan and usual correlation attack are still probabilistic. Theoretical results on the generalized nonlinearity are also studied. In Section 4, we derive an upper bound for generalized nonlinearity which is much lower than the upper bound for usual correlation attack (covering radius bound [3]) and that for Zhang-Chan's attack (unrestricted nonlinearity bound [4]). Thus it gives further evidence that generalized correlation attack is more effective than the other correlation attacks on vector Boolean functions.

The Siegenthaler divide-and-conquer attack on combinatorial stream ciphers [14] resulted in the notion of t-th order correlation immune function (called t-resilient when the function is balanced, which is necessary). To protect against this attack, we require the combining function $F(x)$ to be t-resilient for large t. In Section 5.1, we introduce the concept of generalized divide-and-conquer correlation attack and generalized resiliency to protect against it. We observe that usual resiliency is equivalent to generalized resiliency and thus is sufficient to protect the cipher against the generalized divide-and-conquer attack.

In Section 6, we investigate the generalized nonlinearity of two secondary constructions for vector Boolean functions that are resilient and/or possess high nonlinearity. We conclude output composition (e.g. dropping output bits) of balanced vector functions may increase generalized nonlinearity. For a concatenated function to possess high generalized nonlinearity, we require all component functions to possess high generalized nonlinearity.

2 Generalized Correlation Analysis of Vector Output Stream Ciphers

We consider a stream cipher where the state bits of one or more linear feedback shift registers are filtered by a vector Boolean function $F : GF(2)^n \to GF(2)^m$ to form keystream bits. The keystream bits will be XORed with the plaintext to form the ciphertext. Traditionally, an adversary who wants to perform correlation attack on this stream cipher tries to find an approximation of a linear combination of output bits by a linear combination of input bits $u \cdot F(x) \approx w \cdot x$. For correlation attack to be successful, we require that the bias defined by:

$$Bias = |Pr(u \cdot F(x) = w \cdot x) - 1/2|, \ u \in GF(2)^m, \ w \in GF(2)^n,$$

is large. Conversely, if all linear approximations of $u \cdot F(x)$ have small bias, then it is secure against correlation attack. A concept related to the correlation attack is the *Hadamard transform* $\hat{f} : GF(2)^n \to \mathbb{R}$ of a Boolean function $f : GF(2)^n \to GF(2)$ which is defined as $\hat{f}(w) = \sum_{x \in GF(2)^n} (-1)^{f(x)+w \cdot x}$. Based on the Hadamard transform, we can define the *nonlinearity* [3,10] of $F(x)$ as:

$$N_F = 2^{n-1} - 1/2 \max_{0 \neq u \in GF(2)^m, w \in GF(2)^n} |\widehat{u \cdot F}(w)|. \tag{1}$$

From the above equation, we deduce that a high nonlinearity ensures protection against correlation attack. It is well known that $0 \leq N_F \leq 2^{n-1} - 2^{n/2-1}$ [3].

At Crypto 2000, Zhang and Chan [15] observed that instead of taking linear combination of the output bit functions $u \cdot F(x)$, we can compose $F(x)$ with any

Boolean function $g : GF(2)^m \to GF(2)$ and consider the probability:

$$Pr(g(z) = w \cdot x) \text{ where } z = F(x). \tag{2}$$

Because $z = F(x)$ corresponds to the output keystream which is known, then $g(z)$ is also known. Therefore $g(z) \approx w \cdot x$ is a linear approximation which can be used in correlation attacks. Since we are choosing from a larger set of equations now, we can find linear approximations with larger bias $|Pr(g(z) = w \cdot x) - 1/2|$. Let us define the unrestricted nonlinearity [4] which measures the effectiveness of the Zhang-Chan attack. Denote by $wt(f)$ the number of ones among the output of $f : GF(2)^n \to GF(2)$.

Definition 1. *Let* $F : GF(2)^n \to GF(2)^m$ *and let* $\mathcal{G} = $ *Set of m-bit Boolean functions* $g : GF(2)^m \to GF(2)$. *We define the* unrestricted nonlinearity *as:*

$$UN_F = \min\{\min_{0 \neq u \in GF(2)^m} (wt(u \cdot F), 2^n - wt(u \cdot F)), nonlin_{UN}F\}$$

where

$$nonlin_{UN}F = 2^{n-1} - \frac{1}{2} \max_{w \neq 0, g \in \mathcal{G}} \widehat{g \circ F}(w). \tag{3}$$

Remark 1. If $w = 0$ in equation (2), then it does not involve the input x and it is not useful for correlation attack. Thus we let $w \neq 0$ when computing $nonlin_{UN}F$ which gauge the effectiveness of equation (2) for correlation attack. The other part $\min_{u \neq 0}(wt(u \cdot F), 2^n - wt(u \cdot F))$ ensures that $F(x)$ is close to balanced when UN_F is high.

From equation (3), we deduce that a high unrestricted nonlinearity is required for protection against correlation attack on $g \circ F(x)$.

In this paper, we introduce a linear approximation for performing correlation attack, which is more effective than the Zhang-Chan attack [15]. The idea is to consider implicit equations which are linear in the input variable x and of any degree in the output variable $z = F(x)$. In the pre-processing stage, we compute

$$Pr(g(z) + w_1(z)x_1 + w_2(z)x_2 + \cdots + w_n(z)x_n = 0), \tag{4}$$

where $z = F(x)$ and $w_i : GF(2)^m \to GF(2)$. In other words, we consider the probability $Pr(g(F(x)) + w_1(F(x))x_1 + w_2(F(x))x_2 + \cdots + w_n(F(x))x_n = 0)$, where x uniformly ranges over F_2^n. Then in the attack, because $z = F(x)$ corresponds to the output keystream which is known, $g(z)$ and $w_i(z)$ are known for all $i = 1, \ldots, n$. This means that we can substitute the known values $z = F(x)$ and treat equation (4) as a linear approximation.

We call the attack based on this linear approximation the *generalized correlation attack*. This attack can be considered as a generalization of Zhang-Chan's correlation attack because if we let $w_i(z)$ constant for $i = 1 \ldots n$, equation (4) becomes equation (2). Since we are choosing from a larger set than that of Zhang and Chan, it is easier to find a linear approximation with larger bias $|Pr(g(z) + w_1(z)x_1 + w_2(z)x_2 + \cdots w_n(z)x_n = 0) - 1/2|$.

In relation to the approximation of equation (4), we make the following definition:

Definition 2. *Let $F : GF(2)^n \to GF(2)^m$. The generalized Hadamard transform $\hat{F} : (GF(2)^{2^m})^{n+1} \to \mathbb{R}$ is defined as:*

$$\hat{F}(g(\cdot), w_1(\cdot), \ldots, w_n(\cdot)) = \sum_{x \in GF(2)^n} (-1)^{g(F(x)) + w_1(F(x))x_1 + \cdots w_n(F(x))x_n},$$

where the input is an $(n+1)$-tuple of Boolean functions $g, w_i : GF(2)^m \to GF(2)$, $i = 1, \ldots, n$.

Let $\mathcal{G}$ be defined as in Definition 1 and let

$$\mathcal{W} = \text{ Set of all } n\text{-tuple functions } \{w(\cdot) = (w_1(\cdot), \ldots, w_n(\cdot)) | w_i \in \mathcal{G}\}$$
$$\text{such that } w(z) = (w_1(z), \ldots, w_n(z)) \neq (0, \ldots, 0) \text{ for all } z \in GF(2)^m.$$

The generalized nonlinearity is defined as:

$$GN_F = \min\{\min_{0 \neq u \in GF(2)^m} (wt(u \cdot F), 2^n - wt(u \cdot F)), nonlin_{gen}F\},$$

where

$$nonlin_{gen}F = 2^{n-1} - \frac{1}{2} \max_{g \in \mathcal{G}, w \in \mathcal{W}} \hat{F}(g(\cdot), w_1(\cdot), \ldots, w_n(\cdot)). \qquad (5)$$

Remark 2. We introduce the set $\mathcal{W}$ to give a meaningful definition to the generalized nonlinearity. This is because if there exists $z \in GF(2)^m$ such that $(w_1(z), \ldots, w_n(z)) = (0, \ldots, 0)$, then the equation (4) resulting from this value of z will not involve the input x and will therefore not be useful in the attack stage. Thus we let $w \in \mathcal{W}$ when computing $nonlin_{gen}F$. The other part $\min_{u \neq 0}(wt(u \cdot F), 2^n - wt(u \cdot F))$ ensures that $F(x)$ is close to balanced when GN_F is high.

From equation (5), we deduce that a high generalized nonlinearity is required for protection against generalized correlation attack.

In Proposition 1, we show that the generalized nonlinearity is lower than the other nonlinearity measures and thus provides linear approximations with better bias for correlation attack. The proof follows naturally from the definitions of the various nonlinearities.

Proposition 1. *Let $F : GF(2)^n \to GF(2)^m$. Then the nonlinearity, unrestricted nonlinearity and generalized nonlinearity are related by the following inequality:*

$$GN_F \leq UN_F \leq N_F. \qquad (6)$$

I.e., the generalized correlation attack is more effective than the Zhang-Chan's correlation attack, which itself is more effective than the usual correlation attack.

Remark 3. A vector function $F : GF(2)^n \to GF(2)^m$ is said to be balanced if $|F^{-1}(z)| = 2^{n-m}$ for all $z \in GF(2)^m$. It can be deduced that $wt(u \cdot F) = 2^{n-1}$ for all $u \in GF(2)^m - \{0\}$ if and only if F is balanced [3]. Thus

$$GN_F = \min\{\min_{0 \neq u \in GF(2)^m} (wt(u \cdot F), 2^n - wt(u \cdot F)), nonlin_{gen}F\}$$

$$= \min(2^{n-1}, nonlin_{gen}F) = nonlin_{gen}F,$$

because $GN_F \le N_F \le 2^{n-1} - 2^{(n-1)/2}$ [3]. Therefore $GN_F = nonlin_{gen}F$ if F is balanced. In a similar way, $UN_F = nonlin_{UN}F$ if F is balanced.

3 Efficient Computation of the Generalized Nonlinearity

To compute the generalized nonlinearity GN_F, we first compute $\min_{u \ne 0}(wt(u \cdot F), 2^n - wt(u \cdot F))$ with complexity 2^{m+n}. Then we need to compute $nonlin_{gen}F$ which requires computation of the generalized Hadamard transform over all input. But the complexity of computing $\hat{F}(g(\cdot), w_1(\cdot), \dots, w_n(\cdot))$ directly for all possible $(n+1)$-tuple of m-bit functions is $\approx 2^n \times 2^{2^m \times (n+1)}$. This is because for each fixed $(g(\cdot), w_1(\cdot), \dots, w_n(\cdot))$, we sum over 2^n elements x to compute $\hat{F}$ and there are approximately[1] $2^{2^m \times (n+1)}$ tuples of functions $g, w_i : GF(2)^m \to GF(2)$, $i = 1, \dots, n$. This computation quickly becomes unmanageable even for small values of n, m. Since the bulk of the computational time comes from $nonlin_{gen}F$, we need to make it more efficient to compute. Theorem 1 below gives an efficient way to compute $nonlin_{gen}F$.

Theorem 1. *Let $F : GF(2)^n \to GF(2)^m$ and let $w(\cdot)$ denote the n-tuple of m-bit Boolean functions $(w_1(\cdot), \dots, w_n(\cdot))$. Then*

$$nonlin_{gen}F = 2^{n-1} - 1/2 \sum_{z \in GF(2)^m} \max_{w(z) \in GF(2)^n - \{0\}} \left| \sum_{x \in F^{-1}(z)} (-1)^{w(z) \cdot x} \right|.$$

Because of editorial constraints, the proof of Theorem 1 can be found in the Appendix, Section 7.2.

Remark 4. The proof of Theorem 1 also provides the functions $g(\cdot), w_i(\cdot)$, $i = 1, \dots, n$, for the best generalized linear approximation. At each z, the optimal $g(z)$ is the one that makes the inner sum positive while the optimal tuple $(w_1(z), \dots, w_n(z))$ is the n-bit vector that maximizes the inner sum.

3.1 Reduction in Complexity

To compute $nonlin_{gen}F$ based on Theorem 1, we first perform a pre-computation to identify the sets $\{x : x \in F^{-1}(z)\}$ with complexity 2^n and store them with memory of size $n \times 2^n$. This is needed in computing the sum $\sum_{x \in F^{-1}(z)}(-1)^{w(z) \cdot x}$.

To compute $nonlin_{gen}F$, we consider the 2^m elements $z \in GF(2^m)$. For each z, we find $w(z) \in GF(2)^n$ which maximizes the sum $\left| \sum_{x \in F^{-1}(z)}(-1)^{w(z) \cdot x} \right|$. Thus the computational complexity is:

$$\text{Complexity} = \sum_{z \in GF(2)^m} 2^n \times |\{x : x \in F^{-1}(z)\}| = 2^n \sum_{z \in GF(2)^m} |\{x : x \in F^{-1}(z)\}|$$

$$= 2^n \times |Domain(F)| = 2^n \times 2^n = 2^{2n}.$$

---

[1] We say approximately $2^{2^m \times (n+1)}$ functions because we do not range over all tuples of functions $(w_1(\cdot), \dots, w_n(\cdot))$ but only over those in the set $\mathcal{W}$ of Defintion 2.

Together with a complexity of 2^{m+n} to compute $\min_{0 \neq u \in GF(2)^m} (wt(u \cdot F), 2^n - wt(u \cdot F))$, the total complexity for computing GN_F is:

Precomputation $= 2^n$, Memory $= n \times 2^n$, Time Complexity $= 2^{m+n} + 2^{2n}$.

This is much less than a time complexity of $2^{m+n} + 2^{n+2^m \times (n+1)}$ by the direct approach.

3.2 Experimental Results

Based on Theorem 1, we can compute the generalized nonlinearity of some highly nonlinear functions. We also computed the unrestricted nonlinearity of these functions for comparison. We shall apply Proposition 6 (in Section 7.3 of the Appendix) to help us compute $nonlin_{UN}F$ efficiently. First, let us look at bent functions, which have the highest nonlinearity.

Example 1. Consider the bent function $F : GF(2)^4 \rightarrow GF(2)^2$ (i.e. the function whose component functions $u \cdot F$, $u \neq 0$, are all bent) defined by $F(x_1, x_2, x_3, x_4) = (z_1, z_2) = (x_1 + x_1 x_4 + x_2 x_3, x_1 + x_1 x_3 + x_1 x_4 + x_2 x_4)$. The truth table of F (which lists the output $F(0000), F(0001), \ldots, F(1111)$ where every number represents its binary representation) is as follows.

$$0 \quad 0 \quad 0 \quad 0 \quad 0 \quad 1 \quad 2 \quad 3 \quad 3 \quad 0 \quad 2 \quad 1 \quad 3 \quad 1 \quad 0 \quad 2$$

The various nonlinearity and bias take the following values:

$$\text{Usual nonlinearity } N_F = 6 \Rightarrow Bias = 0.125$$
$$\text{unrestricted nonlinearity } UN_F = 5 \Rightarrow Bias = 0.1875$$
$$\text{Generalized nonlinearity } GN_F = 2 \Rightarrow Bias = 0.375.$$

From Remark 4, we deduce that the following approximation holds with bias 0.375.
$$Pr(z_1 + z_2 = (z_1 + 1)(z_2 + 1)x_2 + z_1 x_3 + z_2 x_4) = \frac{14}{16},$$

where $x = 0100, 1110$ are the only two points not satisfying the relation.

Next we look at the inverse S-box on $GF(2^8)$ with truncated output.

Example 2. Let $GF(2^8)$ be the finite field defined by the relation $\alpha^8 + \alpha^4 + \alpha^3 + \alpha^2 + 1 = 0$. Consider the S-box $Inv : GF(2)^8 \rightarrow GF(2)^8$ defined by $Inv(0) = 0$ and $Inv(x) = x^{-1}$ if $x \neq 0$. We use the correspondence:

$$(x_1, x_2, x_3, x_4, x_5, x_6, x_7, x_8) \leftrightarrow x_1 \alpha^7 + x_2 \alpha^6 + \cdots + x_7 \alpha + x_8$$

Consider $Inv(x)$ restricted to the least significant m bits. Then the nonlinearity, unrestricted nonlinearity and generalized nonlinearity are given by Table 1. We see that the generalized nonlinearity for the inverse function restricted to m output bits is lower than the usual and unrestricted nonlinearities. Therefore generalized correlation attack works better in this case. Moreover, for $m \geq 5$ output bits, the generalized nonlinearity is already 0 which means the system can be broken by linear algebra with very few keystream bits.

Table 1. Nonlinearities for x^{-1} on $GF(2^8)$ restricted to m least significant output bits

m	1	2	3	4	5	6	7
N_F	112	112	112	112	112	112	112
UN_F	112	108	100	94	84	70	56
GN_F	112	80	66	40	0	0	0

Example 3. Lastly in Table 2, we tabulate the average nonlinearity measures for 100 randomly generated balanced functions $F : GF(2)^n \to GF(2)^m$, $n = 2m$, for various n. Again, we see that the average generalized nonlinearity is much lower than the unrestricted and usual nonlinearities. Therefore generalized correlation attack can be more effective.

Table 2. Average nonlinearity for randomly generated balanced functions, $n = 2m$

n	6	8	10	12	14
N_F	18	100	443	1897	7856
UN_F	16	88	407	1768	7454
GN_F	6	36	213	1101	5224

4 Upper Bound on Generalized Nonlinearity

In this Section, we prove an upper bound for the generalized nonlinearity. This allows us to gauge theoretically the effectiveness of the generalized correlation attack.

Theorem 2. *Let* $F : GF(2)^n \to GF(2)^m$. *Then the following inequality holds.*

$$nonlin_{gen}F \le 2^{n-1} - \frac{1}{4} \sum_{z \in GF(2)^m} \sqrt{\frac{2^{n+2}|F^{-1}(z)| - 4|F^{-1}(z)|^2}{2^n - 1}}.$$

Furthermore if $F(x)$ *is balanced, then we have:*

$$GN_F \le 2^{n-1} - 2^{n-1}\sqrt{\frac{2^m - 1}{2^n - 1}}$$

Proof. According to Theorem 1, we have:

$$nonlin_{gen}F = 2^{n-1} - 1/2 \sum_{z \in GF(2)^m} \max_{a \in GF(2)^n - \{0\}} \left| \sum_{x \in F^{-1}(z)} (-1)^{a \cdot x} \right|.$$

Let $\phi_z(x)$ be the indicator function of $F^{-1}(z)$. I.e., $\phi_z(x) = 1$ if $F(x) = z$ else $\phi_z(x) = 0$.

$$\sum_{x \in F^{-1}(z)} (-1)^{a \cdot x} = \sum_{x \in GF(2)^n} \phi_z(x)(-1)^{a \cdot x} = \sum_{x \in GF(2)^n} \frac{1 - (-1)^{\phi_z(x)}}{2}(-1)^{a \cdot x}$$

$$= -\frac{1}{2} \sum_{x \in GF(2)^n} (-1)^{\phi_z(x) + a \cdot x} = -\frac{1}{2}\widehat{\phi_z}(a), \text{ when } a \neq 0.$$

Thus

$$nonlin_{gen} F = 2^{n-1} - 1/4 \sum_{z \in GF(2)^m} \max_{a \in GF(2)^n - \{0\}} \left|\widehat{\phi_z}(a)\right|.$$

In a similar way to the computation of $\sum_{x \in F^{-1}(z)} (-1)^{a \cdot x}$, we can prove that $|F^{-1}(z)| = \sum_{x \in F^{-1}(z)} (-1)^{0 \cdot x} = 2^{n-1} - \frac{1}{2}\widehat{\phi_z}(0)$. This implies $\widehat{\phi_z}(0) = 2^n - 2|F^{-1}(z)|$.

By Parseval's relation,

$$\sum_{a \in GF(2)^n - \{0\}} \widehat{\phi_z}(a)^2 = 2^{2n} - \widehat{\phi_z}(0)^2$$

$$= 2^{2n} - (2^n - 2|F^{-1}(z)|)^2 = 2^{n+2}|F^{-1}(z)| - 4|F^{-1}(z)|^2.$$

By the pigeon hole principle, we deduce that

$$\max_{a \in GF(2)^n - \{0\}} \widehat{\phi_z}(a)^2 \geq \frac{2^{n+2}|F^{-1}(z)| - 4|F^{-1}(z)|^2}{2^n - 1}.$$

and therefore

$$nonlin_{gen} F \leq 2^{n-1} - \frac{1}{4} \sum_{z \in GF(2)^m} \sqrt{\frac{2^{n+2}|F^{-1}(z)| - 4|F^{-1}(z)|^2}{2^n - 1}}.$$

When $F(x)$ is balanced, $nonlin_{gen} F = GN_F$, $|F^{-1}(z)| = 2^{n-m}$ for all $z \in GF(2)^m$ and we deduce:

$$GN_F \leq 2^{n-1} - 2^{n-1} \sqrt{\frac{2^m - 1}{2^n - 1}}$$

$\square$

This upper bound is much lower than the covering radius bound $2^{n-1} - 2^{n/2-1}$ and the upper bound for UN_F deduced in [4]:

$$UN_F \leq 2^{n-1} - \frac{1}{2}\left(\frac{2^{2m} - 2^m}{2^n - 1} + \sqrt{\frac{2^{2n} - 2^{2n-m}}{2^n - 1} + \left(\frac{2^{2m} - 2^m}{2^n - 1} - 1\right)^2 - 1}\right).$$

when $F : GF(2)^n \to GF(2)^m$ is balanced. Thus Theorem 2 provides further evidence that generalized correlation attack can be more effective than the usual and Zhang-Chan correlation attacks on vector Boolean functions.

5   Spectral Characterization and Generalized Correlation Immunity

In Theorem 3, we express the generalized correlation in terms of the Hadamard transform (also called the spectrum) of $F(x)$. This allows us to deduce general correlation properties based on the spectral distribution.

Theorem 3. *Let $F : GF(2)^n \rightarrow GF(2)^m$ and $w_i : GF(2)^m \rightarrow GF(2)$. Let $w(\cdot)$ denote the n-tuple of m-bit Boolean functions $(w_1(\cdot), \ldots, w_n(\cdot))$. Then the generalized Hadamard transform can be expressed as:*

$$\hat{F}(g(\cdot), w_1(\cdot), \ldots, w_n(\cdot)) = \frac{1}{2^m} \sum_{z \in GF(2)^m} (-1)^{g(z)} \sum_{v \in GF(2)^m} (-1)^{v \cdot z} \widehat{v \cdot F}(w(z)).$$

The proof of Theorem 3 is easy and can be found in the Appendix, Section 7.4.

Remark 5. Based on Theorem 3 and equation (5), we get

$$nonlin_{gen} F = 2^{n-1} - \frac{1}{2^{m+1}} \sum_{z \in GF(2)^m} \max_{\substack{0 \neq w(z) \in \\ GF(2)^n}} \left| \sum_{v \in GF(2)^m} (-1)^{v \cdot z} \widehat{v \cdot F}(w(z)) \right|. \quad (7)$$

If the Hadamard transform distribution of $F(x)$ is known, then we can have a more efficient computation of GN_F. By equation (7), we compute $nonlin_{gen} F$ by an outer sum over 2^m elements z, each of which finds the maximum inner sum (over 2^m elements v) for 2^n choices of $w(z)$. Thus the complexity of computing $nonlin_{gen} F$ is 2^{n+2m}. Together with a complexity of 2^{m+n} for determining the balanceness of $F(x)$, the complexity for computing GN_F is $2^{m+n} + 2^{n+2m}$. This is more efficient than the computation of Theorem 1 because usually, m is much smaller than n in applications. Furthermore, we do not need pre-computation and memory to store the sets $\{x : x \in F^{-1}(z)\}$ as in Theorem 1. Some examples of vector functions with known spectral distribution is the Maiorana-McFarland class which can be used to construct bent functions and highly nonlinear resilient functions, e.g. see [3,5].

5.1   Equivalence of Generalized Correlation Immunity and Usual Correlation Immunity

In this section, we extend the definition of correlation immunity (resiliency) for vectorial Boolean function to the generalized case with respect to the correlation attack based on equation (4). Then we show that the usual correlation immunity (resiliency) implies generalized correlation immunity (resiliency). First let us recall the definition of correlation immune vectorial Boolean functions. For a vector $w \in GF(2)^n$, denote by $wt(w)$ the number of ones in w.

Definition 3. *The vector function $F : GF(2)^n \rightarrow GF(2)^m$ is correlation immune of order t (denoted $CI(t)$) if*

$$u \cdot F(x) + w \cdot x \text{ is balanced, or equivalently } \widehat{u \cdot F}(w) = 0$$

whenever $1 \leq wt(w) \leq t$. Moreover if $F(x)$ is balanced, then $F(x)$ is t-resilient.

Resiliency is essential for protection against divide-and-conquer correlation attack on combinatorial generator (for more details, please see Siegenthaler [14]).

Next, let us describe a generalized divide-and-conquer attack against vector combinatorial stream ciphers. In a combinatorial generator involving n LFSR's, suppose there is a subset of outputs $z \in GF(2)^m$ for which the linear approximations $Pr(w_{i_1}(z)x_{i_1} + \cdots + w_{i_t}(z)x_{i_t} = g(z)) = p_z \neq 1/2$, involve the corresponding set of t linear feedback shift registers, $LFSR_{i_1}, \ldots, LFSR_{i_t}$. The attacker guesses the initial state of $LFSR_{i_1}, \ldots, LFSR_{i_t}$. If the guess is correct, then this relation should hold between the t LFSR's states and the relevant output[2] z with probability $p_z \neq 1/2$. If the guess is wrong, then the LFSR states and the output are uncorrelated. Thus the complexity of guessing the secret initial state is reduced because we only need to guess the content of t instead of n LFSR's. To protect against such an attack, we define the concept of generalized correlation immunity and resiliency as follows.

Definition 4. Let $F : GF(2)^n \to GF(2)^m$ and $g, w_i : GF(2)^m \to GF(2)$. We say $F(x)$ is generalized correlation immune of order t (generalized CI(t)) if

$$g(F(x)) + w_1(F(x))x_1 + \cdots + w_n(F(x))x_n \text{ is balanced,}$$

or equivalently,

$$\hat{F}(g(\cdot), w_1(\cdot), \ldots, w_n(\cdot)) = 0,$$

whenever $1 \leq wt(w_1(z), \ldots, w_n(z)) \leq t$ for all $z \in GF(2)^m$. Moreover if $F(x)$ is balanced, then we say $F(x)$ is generalized t-resilient.

Generalized t-resiliency ensures protection against generalized divide-and-conquer correlation attack on t or less LFSR's in a combinatorial stream cipher.

Theorem 4. Let $F : GF(2)^n \to GF(2)^m$. Then $F(x)$ is CI(t) (t-resilient) if and only if $F(x)$ is generalized CI(t) (generalized t-resilient).

Proof. If $F(x)$ is generalized $CI(t)$ (resp. generalized t-resilient), then it follows from Definitions 3 and 4 that $F(x)$ is $CI(t)$ (resp. t-resilient). Now assume $F(x)$ is $CI(t)$, we shall prove that $F(x)$ is generalized $CI(t)$. Suppose $1 \leq wt(w_1(z), \ldots, w_n(z)) \leq t$ for all $z \in GF(2)^m$. Then $\widehat{v \cdot F}(w(z)) = 0$ for all $v, z \in GF(2)^m$ because $F(x)$ is $CI(t)$. By Theorem 3, we see that

$$\hat{F}(g(\cdot), w_1(\cdot), \ldots, w_n(\cdot)) = \frac{1}{2^m} \sum_{z \in GF(2)^m} (-1)^{g(z)} \sum_{v \in GF(2)^m} (-1)^{v \cdot z} \widehat{v \cdot F}(w(z)) = 0.$$

This is because the inner summands is a sum of $\widehat{v \cdot F}(w(z))$ which are zeroes for all $v, z \in GF(2)^m$. Thus $F(x)$ is generalized $CI(t)$. The proof that t-resiliency implies generalized t-resiliency is identical to the $CI(t)$ case except that $F(x)$ is now balanced. □

Thus we see that usual resiliency is sufficient to ensure generalized resiliency.

---

[2] By relevant output, we mean those $z \in GF(2)^m$ for which there exist a linear approximation with positive bias involving the same set of input $x_{i_1}, \ldots, x_{i_t}$.

6 Generalized Nonlinearity of Secondary Constructions

Secondary constructions produce Boolean functions with high nonlinearity, resiliency and other good cryptographic properties from other Boolean functions as building blocks. With respect to the generalized correlation attack, it would be useful to check if these constructions yield functions with high generalized nonlinearity. Moreover by Theorem 4, vector functions that satisfy the usual correlation immunity are also generalized correlation immune. Thus we would also like to check that secondary construction for resilient functions have high generalized nonlinearity.

The first secondary construction we look at is output composition. One common candidate for output composition is the projection function, i.e. dropping output bits. For example, there are many known permutations with high nonlinearity [1] and by dropping output bits, we form vectorial Boolean functions with the same or higher nonlinearity.

Proposition 2. Let $F : GF(2)^n \to GF(2)^m$ and $G : GF(2)^m \to GF(2)^k$ be balanced functions. Then $GN_{G \circ F} \geq GN_F$. If $G(z)$ is a permutation, then $GN_{G \circ F} = GN_F$.

The proof of Proposition 2 can be found in the Appendix, Section 7.5. By Proposition 2, we see that output composition, e.g. dropping output bits, is good for enhancing security as it may increase the generalized nonlinearity.

Another common construction for vectorial resilient functions is concatenation. Let us look at the known results on this construction.

Proposition 3. ([16, Corollary 4]) Let $F_1 : GF(2)^{n_1} \to GF(2)^{m_1}$ be a t_1-resilient function and $F_2 : GF(2)^{n_2} \to GF(2)^{m_2}$ be a t_2-resilient function. Then $F_1 \| F_2 : GF(2)^{n_1+n_2} \to GF(2)^{m_1+m_2}$ defined by

$$F_1 \| F_2(x, y) = (F_1(x), F_2(y))$$

is a t-resilient function where $t = \min(t_1, t_2)$.

By Proposition 3, given two smaller vector Boolean functions which are t-resilient, we can form a bigger Boolean function which is t-resilient. With respect to generalized correlation attack, we would like to know its generalized nonlinearity.

Proposition 4. Let $F_1 : GF(2)^{n_1} \to GF(2)^{m_1}$ and $F_2 : GF(2)^{n_2} \to GF(2)^{m_2}$ be balanced functions. Then the generalized nonlinearity of their concatenation $F(x, y) = F_1(x) \| F_2(y)$ satisfies:

$$GN_F \leq 2^{n_1+n_2-1} - \frac{1}{2}(2^{n_1} - 2GN_{F_1})(2^{n_2} - 2GN_{F_2}).$$

The proof of Proposition 4 can be found in the Appendix, Section 7.6. By Proposition 4, we see that for a concatenated function to possess high generalized nonlinearity, both the component functions have to possess high generalized nonlinearity.

References

1. Canteaut, A., Charpin, P., Dobbertin, H.: Binary m-sequences with three-valued cross correlation: a proof of Welch's conjecture. IEEE Trans. Inform. Theory 46(1), 4–8 (2000)
2. Canteaut, A., Trabbia, M.: Improved fast correlation attacks using parity-check equations of weight 4 and 5. In: Preneel, B. (ed.) EUROCRYPT 2000. LNCS, vol. 1807, pp. 573–588. Springer, Heidelberg (2000)
3. Carlet, C.: Vectorial Boolean Functions for Cryptography. In: Crama, E.Y., Hammer, P. (eds.) Boolean Methods and Models, Cambridge University Press, Cambridge (to appear), http://www-rocq.inria.fr/codes/Claude.Carlet/chap-vectorial-fcts.pdf
4. Carlet, C., Prouff, E.: On a New Notion of Nonlinearity Relevant to Multi-Output Pseudo-Random Generators. In: Matsui, M., Zuccherato, R.J. (eds.) SAC 2003. LNCS, vol. 3006, pp. 291–305. Springer, Heidelberg (2004)
5. Carlet, C., Prouff, E.: Vectorial Functions and Covering Sequences. In: Mullen, G.L., Poli, A., Stichtenoth, H. (eds.) Finite Fields and Applications. LNCS, vol. 2948, pp. 215–248. Springer, Heidelberg (2004)
6. Chor, B., Goldreich, O., Hastad, J., Friedman, J., Rudich, S., Smolensky, R.: The Bit Extraction Problem or t-resilient Functions. In: IEEE Symposium on Foundations of Computer Science 26, pp. 396–407. IEEE Computer Society Press, Los Alamitos (1985)
7. Dillon, J.F.: Multiplicative Difference Sets via Additive Characters. Designs, Codes and Cryptography 17, 225–235 (1999)
8. Gold, R.: Maximal Recursive Sequences with 3-valued Cross Correlation Functions. IEEE Trans. Inform. Theory 14, 154–156 (1968)
9. Gupta, K.C., Sarkar, P.: Improved Construction of Nonlinear Resilient S-boxes. In: Zheng, Y. (ed.) ASIACRYPT 2002. LNCS, vol. 2501, pp. 466–483. Springer, Heidelberg (2002)
10. Nyberg, K.: On the Construction of Highly Nonlinear Permutations. In: Rueppel, R.A. (ed.) EUROCRYPT 1992. LNCS, vol. 658, pp. 92–98. Springer, Heidelberg (1993)
11. Pasalic, E., Maitra, S.: Linear Codes in Constructing Resilient Functions with High Nonlinearity. In: Vaudenay, S., Youssef, A.M. (eds.) SAC 2001. LNCS, vol. 2259, pp. 60–74. Springer, Heidelberg (2001)
12. Rueppel, R.: Analysis and Design of Stream Ciphers. Springer, Heidelberg (1986)
13. Sarkar, P.: The Filter-Combiner Model for Memoryless Synchronous Stream Ciphers. In: Yung, M. (ed.) CRYPTO 2002. LNCS, vol. 2442, pp. 533–548. Springer, Heidelberg (2002)
14. Siegenthaler, T.: Decrypting a Class of Stream Ciphers using Ciphertexts only. IEEE Transactions on Computers C34(1), 81–85 (1985)
15. Zhang, M., Chan, A.: Maximum Correlation Analysis of Nonlinear S-boxes in Stream Ciphers. In: Bellare, M. (ed.) CRYPTO 2000. LNCS, vol. 1880, pp. 501–514. Springer, Heidelberg (2000)
16. Zhang, X.M., Zheng, Y.: On Cryptographically Resilient Functions. EUROCRYPT 1995 43(5), 1740–1747 (1997) (Also presented at Eurocrypt'95, LNCS 921, pp. 274–288, Springer, Heidelberg (1995))

7 Appendix

7.1 The Single-Bit Output Case and Bilinear Cryptanalysis

It is easy to see that in the single output case ($m = 1$), the Zhang-Chan correlation attack is equivalent to the usual correlation attack, i.e. $UN_F = N_F$. However, it is not so obvious whether the generalized correlation attack is better than the usual correlation attack. The four functions from $GF(2)$ to $GF(2)$ are of the form $w(z) = az + b$, where $a, b \in GF(2)$. Hence, the expression used for the generalized correlation attack is a bilinear approximation:

$$Pr(a_0 z + b_0 + (a_1 z + b_1)x_1 + (a_2 z + b_2)x_2 + \cdots + (a_n z + b_n)x_n = 0), \quad a_i, b_i \in GF(2),$$

where for any $z \in GF(2)$, we have $(a_1 z + b_1, \ldots, a_n z + b_n) \neq (0, \ldots, 0)$. The above equation can also be written as:

$$Pr(za'(x) = a(x)) \text{ where } a(x), a'(x) \text{ are affine functions}, \tag{8}$$

such that $za'(x) + a(x)$ is a non-constant function for every $z \in GF(2)$. In Proposition 5, we show that generalized nonlinearity is equal to the usual nonlinearity in the single output case.

Proposition 5. Let $f : GF(2)^n \to GF(2)$. Then $GN_f = N_f$.

Proof. In this proof, $a(x)$, $a'(x)$ and $a''(x) = a(x) + a'(x) + 1$ are affine functions. We also require that $a(x)$ and $a''(x)$ be non-constant functions, so that the approximation in equation (8) is useful for correlation attack. From equation (5) and the discussion in Section 7.1, we deduce that:

$$nonlin_{gen}F = \min_{a(x), a'(x)} |\{x : f(x)a'(x) = a(x)\}|$$

$$= \min_{a(x), a'(x)} (|\{x : f(x) = a(x) = 0\}| + |\{x : f(x) = a(x) + a'(x) + 1 = 1\}|)$$

$$= \min_{a(x)} |\{x : f(x) = a(x) = 0\}| + \min_{a''(x)} |\{x : f(x) = a''(x) = 1\}|.$$

On the other hand, we see from equation (1) that:

$$N_f = \min_{a(x)} |\{x : f(x) = a(x)\}|$$

$$= \min_{a(x)} (|\{x : f(x) = a(x) = 0\}| + |\{x : f(x) = a(x) = 1\}|).$$

But

$$|\{x : f(x) = a(x) = 1\}| = |f^{-1}(1)| + |\{x : f(x) = a(x) = 0\}| - |a^{-1}(0)|$$

$$= |f^{-1}(1)| + |\{x : f(x) = a(x) = 0\}| - 2^{n-1}.$$

Thus

$$N_f = \min_{a(x)} (2 \times |\{x : f(x) = a(x) = 0\}| + c) \text{ where } c = |f^{-1}(1)| - 2^{n-1}.$$

From this, we deduce that:

$$\min_{a(x)} |\{x : f(x) = a(x) = 0\}| = \frac{N_f - c}{2}, \quad \min_{a''(x)} |\{x : f(x) = a''(x) = 1\}| = \frac{N_f + c}{2}.$$

By combining the above two expressions, we get:

$$N_f = \min_{a(x)} |\{x : f(x) = a(x) = 0\}| + \min_{a''(x)} |\{x : f(x) = a''(x) = 1\}| = nonlin_{gen}F.$$

Also $\min_{0 \neq u \in GF(2)^m} (wt(u \cdot F), 2^n - wt(u \cdot F)) \geq N_f$. Thus $GN_f = N_f$.   □

Although generalized correlation attack does not improve on the usual correlation attack when $m = 1$, we can see in Section 3.2 many examples where generalized correlation attack yields better results than the usual and Zhang-Chan correlation attack when the number of output bits is $m \geq 2$.

7.2 Proof of Theorem 1

Proof. We have:

$$\max_{g \in \mathcal{G}, w \in \mathcal{W}} \hat{F}(g(\cdot), w_1(\cdot), \ldots, w_n(\cdot))$$

$$= \max_{g \in \mathcal{G}, w \in \mathcal{W}} \sum_{z \in GF(2)^m} (-1)^{g(z)} \sum_{x \in F^{-1}(z)} (-1)^{w(z) \cdot x}$$

$$= \sum_{z \in GF(2)^m} \max_{g(z) \in GF(2), w(z) \in GF(2)^n - \{0\}} (-1)^{g(z)} \sum_{x \in F^{-1}(z)} (-1)^{w(z) \cdot x}.$$

To maximize this expression, we choose $g(z) = 0$ if $\sum_{x \in F^{-1}(z)} (-1)^{w(z) \cdot x} > 0$, else we choose $g(z) = 1$. Thus we can equivalently write the expression as:

$$\max_{g \in \mathcal{G}, w \in \mathcal{W}} \hat{F}(g(\cdot), w_1(\cdot), \ldots, w_n(\cdot)) = \sum_{z \in GF(2)^m} \max_{w(z) \in GF(2)^n - \{0\}} \left| \sum_{x \in F^{-1}(z)} (-1)^{w(z) \cdot x} \right|.$$

By substituting this expression in equation (5), we get $nonlin_{gen}F$.   □

7.3 Efficient Computation of Unrestricted Nonlinearity

The bulk of the work in computing UN_F comes from the computation of $nonlin_{UN}F$. Proposition 6 gives an efficient way to compute $nonlin_{UN}F$.

Proposition 6. *Let* $F : GF(2)^n \rightarrow GF(2)^m$. *Then* $nonlin_{UN}F$ *can be computed as:*

$$nonlin_{UN}F = 2^{n-1} - \frac{1}{2} \max_{w \neq 0} \sum_{z \in GF(2)^m} \left| \sum_{x \in F^{-1}(z)} (-1)^{w \cdot x} \right|.$$

Proof.

$$\max_{w \neq 0, g \in \mathcal{G}} \widehat{g \circ F}(w) = \max_{w \neq 0, g \in \mathcal{G}} \sum_{x \in GF(2)^n} (-1)^{g \circ F(x) + w \cdot x}$$

$$= \max_{w \neq 0, g \in \mathcal{G}} \sum_{z \in GF(2)^m} (-1)^{g(z)} \sum_{x \in F^{-1}(z)} (-1)^{w \cdot x}$$

$$= \max_{w \neq 0} \sum_{z \in GF(2)^m} \left| \sum_{x \in F^{-1}(z)} (-1)^{w \cdot x} \right|.$$

where we choose $g(z) = 0$ if the inner sum is positive and $g(z) = 1$ is the inner sum is negative. By substituting this expression in equation (3), Proposition 6 is proved.   □

7.4   Proof of Theorem 3

Proof. Let $\phi_z(x)$ be as defined in the proof of Theorem 2. For a fixed $z \in GF(2)^m$,

$$\sum_{x \in F^{-1}(z)} (-1)^{w(z) \cdot x} = \frac{1}{2^m} \sum_{x \in GF(2)^n} (-1)^{w(z) \cdot x} \times 2^m \phi_z(x)$$

$$= \frac{1}{2^m} \sum_{x \in GF(2)^n} (-1)^{w(z) \cdot x} \times \sum_{v \in GF(2)^m} (-1)^{v \cdot (F(x) + z)}$$

$$\left(\text{because} \sum_{v \in GF(2)^m} (-1)^{v \cdot a} = 2^m \text{ if and only if } a = 0 \right)$$

$$= \frac{1}{2^m} \sum_{v \in GF(2)^m} (-1)^{v \cdot z} \times \sum_{x \in GF(2)^n} (-1)^{w(z) \cdot x + v \cdot F(x)}$$

$$= \frac{1}{2^m} \sum_{v \in GF(2)^m} (-1)^{v \cdot z} \widehat{v \cdot F}(w(z)).$$

By substituting this expression in Lemma 1, the proof is complete.   □

7.5   Proof of Proposition 2

Proof. Let $\mathcal{G}, \mathcal{W}$ and $\mathcal{G}', \mathcal{W}'$ be the set of m-bit and k-bit Boolean functions in Definitions 1 and 2 respectively.

$$\max_{g' \in \mathcal{G}', w \in \mathcal{W}'} \widehat{G \circ F}(g', w_1', \dots, w_n') = \max_{g' \in \mathcal{G}', w' \in \mathcal{W}'} \hat{F}(g' \circ G, w_1' \circ G, \dots, w_n' \circ G)$$

$$\leq \max_{g \in \mathcal{G}, w \in \mathcal{W}} \hat{F}(g, w_1, \dots, w_n).$$

Therefore by equation (5), $nonlin_{gen} G \circ F \geq nonlin_{gen} F$. Note that $w' \in \mathcal{W}'$ implies $w' \circ G \in \mathcal{W}$ in the above inequality.

Since $F(x)$ is balanced, $nonlin_{gen}F = GN_F$ by remark 3. It is easy to deduce that $G \circ F$ is balanced if both F and G are balanced. Thus $nonlin_{gen}G \circ F = GN_{G \circ F}$ by remark 3 and we have $GN_{G \circ F} \geq GN_F$.

If $G(z)$ is a permutation, then $\{g \circ G | g \in \mathcal{G}\} = \mathcal{G}$ and $\{(w_1 \circ G, \ldots, w_n \circ G) | w \in \mathcal{W}\} = \mathcal{W}$. Thus we have $nonlin_{gen}G \circ F = nonlin_{gen}F$ which implies $GN_{G \circ F} = GN_F$. □

7.6   Proof of Proposition 4

Proof. Consider any $g_i : GF(2)^{m_i} \rightarrow GF(2)$, $i = 1, 2$ and any $w_{i,1}, \ldots, w_{i,n_i} :$ $GF(2)^{m_i} \rightarrow GF(2)$, $i = 1, 2$ where for all $z \in GF(2)^{m_i}$, $(w_{i,1}(z), \ldots, w_{i,n_i}(z)) \neq (0, \ldots, 0)$. We see that:

$$\widehat{F_1}(g_1(\cdot), w_{1,1}(\cdot), \ldots, w_{1,n_1}(\cdot)) \times \widehat{F_2}(g_2(\cdot), w_{2,1}(\cdot), \ldots, w_{2,n_2}(\cdot))$$

$$= \sum_x (-1)^{g_1(F_1(x)) + w_{1,1}(F_1(x))x_1 + \ldots + w_{1,n_1}(F_1(x))x_{n_1}}$$

$$\times \sum_y (-1)^{g_2(F_2(y)) + w_{2,1}(F_2(y))y_1 + \ldots + w_{2,n_2}(F_2(y))y_{n_2}}$$

$$= \sum_{x,y} (-1)^{g(F_1(x), F_2(y)) + w_1(F_1(x), F_2(y))x_1 + \ldots + w_{n_1+n_2}(F_1(x), F_2(y))y_{n_2}}$$

$$= \widehat{(F_1, F_2)}(g(\cdot), w_1(\cdot), \ldots, w_{n_1+n_2}(\cdot)).$$

where we let $g : GF(2)^{m_1+m_2} \rightarrow GF(2)$ be defined by $g(z_1, z_2) = g_1(z_1) + g_2(z_2)$. Let

$$w_1(z_1, z_2) = w_{1,1}(z_1), \ldots, w_{n_1}(z_1, z_2) = w_{1,n_1}(z_1),$$

$$w_{n_1+1}(z_1, z_2) = w_{2,1}(z_2), \ldots, w_{n_1+n_2}(z_1, z_2) = w_{2,n_2}(z_2).$$

For all $(z_1, z_2) \in GF(2)^{m_1+m_2}$, it is obvious that $(w_1(z_1, z_2), \ldots, w_{n_1+n_2}(z_1, z_2)) \neq (0, \ldots, 0)$.

Since on the left hand side of the above equations $g(\cdot)$ and $w_{i,j}(\cdot)$ can be any functions while the g, w_i defined on the right hand side are only functions on $(z_1, z_2) \in GF(2)^{m_1+m_2}$ of a special form, we have:

$$\max_{g_1, w_{1,i}} \widehat{F_1}(g_1(\cdot), w_{1,i}(\cdot)) \times \max_{g_2, w_{2,i}} \widehat{F_2}(g_2(\cdot), w_{2,i}(\cdot))$$

$$\leq \max_{g, w_i} \widehat{(F_1 || F_2)}(g(\cdot), w_1(\cdot), \ldots, w_{n_1+n_2}(\cdot)).$$

By substituting this inequality in equation (5), we get

$$nonlin_{gen}(F_1 || F_2) \leq 2^{n_1+n_2-1} - \frac{1}{2}(2^{n_1} - 2nonlin_{gen}F_1)(2^{n_2} - 2nonlin_{gen}F_2).$$
$$(9)$$

Since $F_1(x)$ and $F_2(y)$ are balanced functions, we have $nonlin_{gen}F_i = GN_{F_i}$ by remark 3. Furthermore, it is easy to see that $(F_1(x), F_2(y))$ is a balanced function. Thus $nonlin_{gen}(F_1 || F_2) = GN_{(F_1 || F_2)}$ by remark 3. Thus we can substitute all the $nonlin_{gen}F$ in equation (9) by GN_F and we are done. □

An Analytical Model
for Time-Driven Cache Attacks

Kris Tiri[1], Onur Acıiçmez[3,*], Michael Neve[1], and Flemming Andersen[2]

[1] Platform Validation Architecture
[2] Visual Computing Group
Intel Corporation
2111 NE 25th Avenue, Hillsboro Oregon 97124, USA
{kris.tiri,michael.neve.de.mevergnies,flemming.l.andersen}@intel.com
[3] Computer Science Lab
Samsung Information Systems America, USA
o.aciicmez@samsung.com

Abstract. Cache attacks exploit side-channel information that is leaked by a microprocessor's cache. There has been a significant amount of research effort on the subject to analyze and identify cache side-channel vulnerabilities since early 2002. Experimental results support the fact that the effectiveness of a cache attack depends on the particular implementation of the cryptosystem under attack and on the cache architecture of the device this implementation is running on. Yet, the precise effect of the mutual impact between the software implementation and the cache architecture is still an unknown. In this manuscript, we explain the effect and present an analytical model for time-driven cache attacks that accurately forecasts the strength of a symmetric key cryptosystem based on 3 simple parameters: (1) the number of lookup tables; (2) the size of the lookup tables; (3) and the length of the microprocessor's cache line. The accuracy of the model has been experimentally verified on 3 different platforms with different implementations of the AES algorithm attacked by adversaries with different capabilities.

1 Introduction

Exploiting cache behavior was presumed possible by early works [7,8]. Yet, Page was the first to study the cache-based side-channel attacks. In a theoretical work [14], he classified the cache attacks based on the method of leakage observation into two types of attacks: trace-driven and time-driven cache attacks.

In trace-driven attacks (e.g. [1,4,9]), the adversary observes the succession of cache hits and cache misses during a cryptographic cipher operation. Given (some of) the implementation details of the cipher, she is then able to derive key material from whether a particular key-dependent memory access results in a cache hit or a cache miss. In time-driven attacks (e.g. [2,3,5,16]), the adversary observes the total execution time of a cryptographic cipher operation.

---

[*] Work done while being with Intel Corporation for an internship.

A. Biryukov (Ed.): FSE 2007, LNCS 4593, pp. 399–413, 2007.

She is then able to derive key material as the execution time depends on the number of key-dependent memory accesses that result in cache misses. Since the cache behavior is only one of the many elements that affect the overall execution time of a cryptosystem, time-driven attacks require statistical analysis using a large number of samples to infer key material. However, they are more generic and easier to apply than trace-driven attacks, because simple and pure software methods are sufficient to carry out time-driven attacks. On the other hand, the current trace-driven attacks are based on power analysis and mandate physical access and alteration of the processing device, and thus their use is very restricted. To the best of our knowledge, trace-driven attacks have not been demonstrated in practice.

Recently, another type of cache attack has been demonstrated: the access-driven cache attacks [15, 13, 11]. In this attack type, the adversary observes the individual cache lines accessed by the cryptographic cipher operation. She derives key material from knowing which cache lines and thus also which table entries have been touched during the key-dependent memory accesses. A single observation caries a lot of information and access-driven attacks generally require a significantly lower number of measurements than other cache attacks.

Numerous mitigations have been proposed alongside and as a reaction to the cache-attacks (e.g. [14, 9, 3, 13, 6, 5]). Most of them are software based and alter the size or the deployment of the lookup tables. For instance, with compact lookup tables [6], where a cache line contains relatively more table entries, or with dynamic lookup table permutations [6], where a cache line contains different table entries over time, less information is disclosed through knowledge of a cache line access. Yet, it is never specified how strong these mitigations are. The mitigations appear to be ad hoc and we are unaware of any formal proofs attesting their effectiveness.

Limited experimental results, however, do exist to support the mitigations [6, 5]. Yet the results are not available for all microprocessors, are only valid for a single algorithm and are unclear with respect to the actual strength of the mitigation. In this manuscript, we will present an analytical model that allows to quickly evaluate any mitigated or non-mitigated implementation on any microprocessor of any symmetric key encryption algorithm using lookup tables that are being exploited in time-driven cache attacks. The model provides a strict lower bound on the required number of measurements for a successful time-driven cache attack. The analytical model allows further to modify the threat level: it can incorporate adversaries with a sampling resolution as small as a single encryption round or as large as several complete encryptions.

The remainder of this document is organized as follows. The next section briefly describes time-driven attacks and illustrates them with the last round correlation attack. Section 3 first derives the analytical model and then shows that it is universal and valid for any time-driven cache attack as the resulting metric is based on the signal-to-noise ratio present in the measurements. In section 4, the analytical model is experimentally verified on different platforms using the unmitigated OpenSSL and several mitigated implementations of the AES algorithm. Finally a conclusion will be formulated.

2 Time-Driven Cache Attacks

The cache is a processor component that stores recently used data in a fast memory block close to the microprocessor. Whenever the processor tries to retrieve data from the main memory, it can be delivered more quickly if this data is already stored in the cache (a.k.a. cache hit). On the other hand, if the data is not available in the cache (a.k.a. cache miss), it has to be fetched from the main memory (or a higher level of cache), which has a much larger latency compared to the cache. The difference between both cache access events, i.e. a cache hit and a cache miss, is measurable and provides the attacker with information on the state and the execution of the algorithm to extract secret key material.

Observing each single cache access event, however, is extremely hard when performing a local attack and even impossible for remote attacks. Hence in a time-driven cache attack, the adversary observes the aggregated effect of all the memory accesses in a cryptographic operation, i.e. the total number of cache misses and hits or at least its effect on the execution time of the operation. To infer key material, she then analyzes cache collisions in a lookup table of interest.

In this paper, we define a cache collision as the situation involving two different memory accesses attempting to access the same memory location or different but very close memory locations that are stored in the same cache line. A cache line, aka. cache block or entry, is the smallest unit of memory that can be transferred between the main memory and the cache. The effect is ideally that the first access ensures that the data is in the cache such that the second access results in a cache hit. The attack is based on the assumption that when there is a cache collision between two particular table lookup operations, the total number of cache misses tends to be lower. On the other hand, if there is no collision between these two table lookup operations, then the overall number of cache misses tends be higher.

Note that even when there is no cache collision between the two particular table lookup operations under investigation, the second cache access might still result in a cache hit because of a cache collision with another table lookup operation. However, if enough observations are taken into account, the distribution of the number of cache misses when the cache collisions are always correctly predicted will be different from the distribution obtained when they are not. Hence, the resistance against an attack is measured as the required number of samples for a successful attack or in other words the number of samples that must be analyzed to distinguish the correct key guess from the incorrect key guesses.

An adversary tries to estimate whether a cache collision occurs between two particular table lookup operations by examining the indices of the lookup operations. She computes these indices based on the known and observable data and also based on a guess on a fragment of the secret key. Note that the secret key fragment is relatively small and that the computational complexity of a cache attack is significantly reduced compared to a brute-force attack on the entire secret key. The complete secret key is revealed by finding all of the composing key fragments.

Several statistical techniques are available to decide on the correct key value. For instance, Bonneau et al. suggest the t-test to find statistical significant different averages between distributions [5]. We use the correlation coefficient between the measurements and the estimations. This method is common practice in side-channel analysis, and especially in power analysis. This method allows us to deduce our analytical model for time-driven cache attacks that can be used to compute the strength of a given implementation on a given platform instead of relying on cumbersome empirical assessments to estimate the required number of measurements for a successful attack.

With the correlation coefficient method, the secret key fragment K_{secret} is found by evaluating the following cost function:

$$K_{secret} = max_K(|corr(M, E_K)|) \tag{1}$$

The vector M consists of the measurement scores that represent the number of cache misses that occur during the cryptographic cipher operation on the messages in a sample set. To be more precise, a measurement score is a value that approximates the number of cache misses realized during the operation such as the execution time of the operation or the cache miss count obtained via the use of performance counters. The vector E_K consists of the corresponding estimations of the adversary on the actual number of cache misses. As mentioned above, she computes these estimations by using the known values of the ciphertext or the plaintext and her guess K on a portion of the secret key.

2.1   Last Round Correlation Attack

For a better understanding, we now describe the correlation attack on the last round of AES-128. We successfully mounted this attack on 10 different platforms (2 servers and 8 PCs) from 3 different processor manufacturers with 7 different operating systems.

Figure 1 shows a snippet of the last round of the OpenSSL AES implementation [12]. This implementation uses four lookup tables Te0, Te1, Te2, and Te3 for the first 9 rounds and a single table Te4 for the 10th (i.e. last) round. Each table contains 256 4-byte words. The input to the tables is a single byte output of the preceding key addition.

The attack will estimate a single cache miss of the last round accesses and compare it with the measurement score for the total number of cache misses of the complete AES encryption. The estimation assumes that if 2 inputs to the table of interest Te4 point to the same cache line, there is a cache hit; while if they point to different cache lines, there will be a cache miss. The table indexes –i.e. (t0>>24), (t1>>16)&0xff, (t2>>8)&0xff), etc. in figure 1– are a single byte of the input to the last encryption round. The model thus estimates whether $< p_i^{(10)} >$ equals $< p_j^{(10)} >$, where $p_i^{(10)}$ is the ith byte input to the 10th and

```
void AES_encrypt(const unsigned char *in,
                 unsigned char *out, const AES_KEY *key) {
...
// apply last round and
// map cipher state to byte array block:
s0 =
     (Te4[(t0 >> 24)       ] & 0xff000000) ^
     (Te4[(t1 >> 16) & 0xff] & 0x00ff0000) ^
     (Te4[(t2 >>  8) & 0xff] & 0x0000ff00) ^
     (Te4[(t3      ) & 0xff] & 0x000000ff) ^
     rk[0];
PUTU32(out     , s0);
s1 =
     (Te4[(t1 >> 24)       ] & 0xff000000) ^
     (Te4[(t2 >> 16) & 0xff] & 0x00ff0000) ^
     (Te4[(t3 >>  8) & 0xff] & 0x0000ff00) ^
     (Te4[(t0      ) & 0xff] & 0x000000ff) ^
     rk[1];
...
}
```

Fig. 1. Last round snippet of OpenSSL `AES_encrypt` function [10]

last encryption round and where the $<>$ operator selects the most significant bits to account for the fact that a cache line contains several table elements ordered in function of the table index. The values $< p_i^{(10)} >$ and $< p_j^{(10)} >$ can be calculated as $< sbox^{-1}(RK_i^{(10)} \oplus C_i) >$ and $< sbox^{-1}(RK_j^{(10)} \oplus C_j) >$ based on the ciphertext bytes C_i and C_j, which are known to the attacker, and a guess on the last round key bytes $RK_i^{(10)}$ and $RK_j^{(10)}$. The correct values of $RK_i^{(10)}$ and $RK_j^{(10)}$ are the ones that result in the highest correlation coefficient between the measurements and the estimations. The key search space equals 2^{16} to find the 2 initial key bytes and $14 \cdot 2^8$ to find the remaining 14 key bytes. In the remainder of this manuscript, we will refer to this attack as the cache line estimation (CLE) attack. The attack can be simplified if the estimation neglects the fact that a cache line contains several table elements and simply assumes that only when two inputs to the table Te4 are equal, there is a cache hit; while if they are different, there will be a cache miss. In that case, the model estimates if $p_i^{(10)}$ equals $p_j^{(10)}$ or not. The substitution box is a nonlinear bijection and can be removed from the equality estimation. The model thus estimates whether $RK_i^{(10)} \oplus C_i$ equals $RK_j^{(10)} \oplus C_j$, which can be simplified further to estimating whether C_i equals $RK_{ij}^{(10)} \oplus C_j$ with $RK_{ij}^{(10)} = RK_i^{(10)} \oplus RK_j$. The correct value of $RK_{ij}^{(10)}$ is the one that results in the highest correlation coefficient between the measurements and the estimations. The key search space now equals $15 \cdot 2^8$ to find the 15 offsets $RK_{ij}^{(10)}$ from $RK_i^{(10)}$ and 2^8 to brute-force $RK_i^{(10)}$. In the remainder of this manuscript, we will refer to this attack as the table index estimation (TIE) attack. Note that the idea behind the last round correlation attack is universal and can be applied to any implementation of a symmetric key encryption algorithm using lookup tables. In the remainder of this manuscript, we will refer to the lookup table of which the collisions are analyzed as the lookup table of interest.

3   Analytical Model

For a correlation attack, Mangard derived that the number of measurements N required for a successful attack can be computed as follows [10]:

$$N = 3 + 8 \cdot \left(\frac{Z_\alpha}{\ln\left(\frac{1+\rho}{1-\rho}\right)} \right)^2 \approx \frac{2 \cdot Z_\alpha^2}{\rho^2} \tag{2}$$

The parameter ρ is the correlation coefficient between the measurement vector M and the vector $E_{K_{secret}}$, which is the estimation vector computed from the correct secret key fragment K_{secret}. The parameter α is the probability to discover the secret key, while Z_α is the quantile of the standard normal distribution for a probability α. With a probability of 0.99 to discover the secret key fragment, N can be approximated as $11/\rho^2$.

Given equation 2, the attack resistance of an implementation can be determined by (1) modeling the measurements; and by (2) computing the correlation coefficient between the estimations and the modeled measurements:

$$\rho = \frac{\mathbb{E}\left(E_{K_{secret}} \cdot M\right) - \mathbb{E}\left(E_{K_{secret}}\right) \cdot \mathbb{E}\left(M\right)}{\sqrt{\mathbb{E}\left(E_{K_{secret}}^2\right) - \mathbb{E}\left(E_{K_{secret}}\right)^2} \sqrt{\mathbb{E}\left(M^2\right) - \mathbb{E}\left(M\right)^2}} \tag{3}$$

We will construct the analytical model of the measurements based on the following 5 assumptions:

1. *The cache does not contain any data related to the cryptographic cipher operation until the operation begins.* In other words, the cache is assumed to be "clean" before the operation starts. Note that this is a valid assumption as in order for the adversary to analyze the observed side-channel information she must hypothesize a known initial state of the cache.
2. *There are no collisions between different tables used in the cryptographic cipher operation.* The cache areas on which each table maps to are mutually disjoint. Note that this is a valid assumption as the cache size of contemporary platforms is large enough to store all the tables of nearly all common symmetric key algorithms.
3. *The cache accesses during the cryptographic cipher operation are random and independent from each other.* Note that this is a valid assumption because of the avalanche effect of cryptographic algorithms.
4. *The cryptographic cipher operation operates uninterrupted.* There is no outside effect on the operation and its cache access pattern. Note that this is a valid assumption as the model models a cryptographic cipher operation. The operation can be a single encryption for a normal adversary; multiple encryptions for a limited adversary; and a single round or even a more atomic operation for a powerful adversary.
5. *The execution time of the cryptographic cipher operation is proportional to the number of cache misses.* Note that this is the foundation for time-driven cache attacks. Figure 2 shows the linear relationship between the encryption

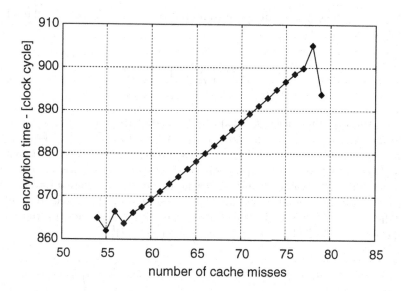

Fig. 2. Linear relationship between number of cache misses and execution time

time and the total number of cache misses for the AES algorithm on an arbitrary platform. The results were computed based on 100 million encryptions of random data and averaging the encryption times of those that yield the same number of cache misses. The non-linear effects of the outliers are due to an insufficient number of those events.

Since the execution time is proportional to the number of cache misses, the analytical model can draw on the number of cache misses instead of on the execution time. Furthermore, using the exact number of cache misses will assure a lower bound on N since any practical measurement score for the number of cache misses will be noisier than the actual number as can be observed from the outliers in figure 2 for which insufficient samples were available to average out the noise.

We will now derive the individual components of equation 3. Since the cache accesses are independent, the 1st and 2nd moment of the number of cache misses equal the sum of the moments of the T tables used in the cryptographic cipher operation, as shown in equation 4. M_t denotes the number of cache misses due to accesses to table t.

$$\mathbb{E}(M) = \sum_{t=1}^{T} \mathbb{E}(M_t) \quad ; \quad \mathbb{E}(M^2) = \sum_{t=1}^{T} \mathbb{E}(M_t^2) \tag{4}$$

As a result, we can concentrate on calculating the moments for a single lookup table and combine them afterwards. It can be shown that the probability $P_{k,l}(j)$ that of a table occupying l cache lines exactly j cache lines are accessed after k accesses to the table is expressed by equation 5, where C_n^r: is the binomial coefficient expressing the number of combinations of r items that can be selected from a set of n items.

$$P_{k,l}(j) = \frac{C_l^j \left(\sum_{i=1}^{j} (-1)^{j-i} C_j^i i^k \right)}{l^k} \tag{5}$$

Using these probabilities, the expected value of the number of cache misses $\mathbb{E}(M)$ and the variance on the number of cache misses $\mathbb{E}(M^2) - \mathbb{E}(M)^2$ for k accesses to a table occupying l cache lines can be calculated as in equations 6 and 7 respectively. A simple expression for the expected value can be derived by noting that the probability of a single cache line not being loaded into the cache after k accesses to l cache lines is $(1 - 1/l)^k$. As a result, the expected number of cache lines that are not loaded becomes $l - (1 - 1/l)^k$ and the expected number of lines that are loaded $\mu_M(k, l)$ equals $l - l \cdot (1 - 1/l)^k$.

$$\mu_M(k, l) = \sum_{j=1}^{l} j \cdot P_{(k,l)}(j) = l - l \cdot \left(\frac{l-1}{l} \right)^k \tag{6}$$

$$\sigma_M^2(k, l) = \sum_{j=1}^{l} j^2 \cdot P_{(k,l)}(j) - \mu_M^2(k, l) \tag{7}$$

For ease of calculation, we now adjust the estimation to output a 1 if a cache hit has been predicted and a 0 if a cache miss has been predicted. This only changes the sign of the correlation coefficient with respect to our earlier assumption that the estimation outputs a 0 for a cache hit and a 1 for a cache miss as $\rho(A - X, Y) = -\rho(X, Y)$ with X, Y random variables and A constant.

To find the 1st and 2nd moment of the estimations, we note that the estimation is either a 1 or a 0. Only the 1 value will contribute to the moments, which are thus equal to $1 \cdot P(E = 1)$ and $1^2 \cdot P(E = 1)$, with $P(E = 1)$ the probability of having an estimation equal to 1. To find $P(E = 1)$, we need to differentiate between the 2 estimation models. For the table index estimation, the probability that 2 independent accesses to a table use the same table index is equal to $1/r$, with r the number of elements in the table. For the cache line estimation, the probability that 2 independent accesses use the same cache line is equal to $1/l$, with l the number of cache lines occupied by the table. The expected value of the estimated number of cache misses $\mathbb{E}(E)$ and the variance on the estimated number of cache misses $\mathbb{E}(E^2) - \mathbb{E}(E)^2$ can be calculated as in equations 8 and 9 respectively.

$$\mu_E|_{TIE} = \frac{1}{r} \quad ; \quad \mu_E|_{CLE} = \frac{1}{l} \tag{8}$$

$$\sigma_E^2 = \mu_E - \mu_E^2 \tag{9}$$

Finally, to find $\mathbb{E}(E.M)$, we need to differentiate between the table of interest and the other tables that are part of the cryptographic cipher operation. For the tables that are not of interest, the estimation and the measurements are independent and $\mathbb{E}(E \cdot M)$ is simply $\mathbb{E}(E) \cdot \mathbb{E}(M)$. For the table of interest, only the estimation with value 1 will contribute to the moment and $\mathbb{E}(E \cdot M)$ is equal to $P(E = 1) \cdot 1 \cdot \mu_H(k, l)$ where $\mu_H(k, l)$ is the expected number of cache misses with k accesses to l cache lines when a cache hit is correctly estimated. Since a cache hit will occur we can ignore the first access and the expected number of cache misses is the expected number of cache misses for the remaining $k - 1$ accesses:

$$\mu_H(k, l) = \mu_M(k - 1, l) \tag{10}$$

Using the previous equations and derivations, the correlation coefficient between the estimations and the measurement score observing the cache misses of T tables, with table T the table of interest, is equal to:

$$\rho = \frac{\mu_E \cdot \left(\sum_{t=1}^{T-1} \mu_M(k_t, l_t) + \mu_H(k_T, l_T) \right) - \mu_E \cdot \left(\sum_{t=1}^{T-1} \mu_M(k_t, l_t) \right)}{\sigma_E \sqrt{\sum_{t=1}^{T} \sigma_M^2(k_t, l_t)}} \tag{11}$$

$$= \frac{\mu_E \cdot \mu_D(k_T, l_T)}{\sigma_E \sqrt{\sum_{t=1}^{T} \sigma_M^2(k_t, l_t)}} \quad \text{where } \mu_D(k_T, l_T) = \mu_H(k_T, l_T) - \mu_M(k_T, l_T)$$

$$= \left(\frac{l_T - 1}{l_T} \right)^{k_r - 1}$$

Combining equation 2 and equation 11 results in the analytical model for time-driven cache attacks:

$$N = \frac{2 \cdot Z_\alpha^2}{\frac{\mu_E^2}{\sigma_E^2} \cdot \frac{\mu_D^2(k_T, l_T)}{\sum_{t=1}^{T} \sigma_M^2(k_t, l_t)}} \tag{12}$$

As an example, for the last round correlation attack using the cache line estimation to attack the OpenSSL AES implementation running on a processor with cache lines of length 64 bytes, N can be found by noting that in addition to the 16 accesses to the table of interest Te4, which occupies 16 cache lines, there are 36 accesses to each of the other 4 tables Te0, Te1, Te2 and Te3, which each also occupy 16 cache lines. With a probability for success equal to 0.99, the resulting expected number of measurements for success N is forecasted to be 6592:

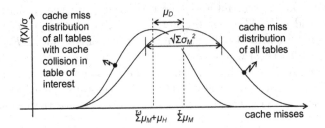

Fig. 3. Cache miss probability distribution (not drawn to scale)

$$N|_{\alpha=0.99} = \frac{11}{\frac{1/16^2}{1/16-1/16^2} \cdot \frac{\mu_D^2(16,16)}{4\cdot\sigma_M^2(36,16)+\sigma_M^2(16,16)}} = 6592 \qquad (13)$$

Equation 14 tells us that for the same implementation on the same platform, the cache line estimation attack is about r/l times more effective than the table index estimation attack. This means for instance that in order to attack a regular OpenSSL implementation on a platform with 64 byte cache lines 16 times less measurements are needed when the cache miss estimation is based on cache collisions instead of internal collisions.

$$\frac{N|_{TIE}}{N|_{CLE}} = \frac{\frac{\mu_E^2}{\sigma_E^2}\big|_{CLE}}{\frac{\mu_E^2}{\sigma_E^2}\big|_{TIE}} \approx \frac{r}{l} \qquad (14)$$

Equation 15 tells us that the analytical model is based on the signal-to-noise ratio that is present in the measurements and hence is independent of the actual method a time-driven attack with a given estimation model uses to select the secret key fragment. Indeed, for the same attack on two different implementations A and B on the same platform, the increase in resistance can be expressed as follows:

$$\frac{N_B}{N_A} = \frac{\frac{\mu_D^2(k_{T_A},l_{T_A})}{\sum_{t_A=1}^{T_A}\sigma_M^2(k_{t_A},l_{t_A})}}{\frac{\mu_D^2(k_{T_B},l_{T_B})}{\sum_{t_B=1}^{T_B}\sigma_M^2(k_{t_B},l_{t_B})}} = \frac{SNR_A}{SNR_B} \qquad (15)$$

As can be seen in figure 3, μ_D is the distance between the distributions of the expected number of cache misses and the expected number of cache misses when a cache hit is correctly predicted. This is the signal that the adversary tries to observe and μ_D^2 is the power that is present in this signal. The noise of the measurements is the variance of the expected number of cache misses of all tables which is equal to $\sum \sigma_M^2$. The increase in resistance N_B/N_A is thus

SNR_A/SNR_B, which is the ratio between the signal-to-noise ratios of the two implementations.

4 Experimental Results

The following 5 AES implementations or attack scenarios have been used to experimentally verify the correctness of the analytical model:

- OSSL (OpenSSL): The original OpenSSL implementation of the AES algorithm using 5 lookup tables Te0, Te1, Te2, Te3, and Te4, each of length 1024 bytes.
- CLR (Compact Last Round): OSSL but the last round employs a single compact S-box table of length 256 bytes instead of table Te4.
- NOT4 (NO table Te4): OSSL but the last round is implemented without the use of table Te4. Instead tables Te0, Te1, Te2, and Te3 are used.
- 2ENC (2 ENCryptions): OSSL observed by a limited adversary that is only able to observe the aggregated effect of 2 encryptions.
- OT4 (observe Only table Te4): OSSL observed by a powerful adversary who is able to observe a single round. Note that the same effect can be obtained by only cleaning out table Te4 but leaving tables Te0, Te1, Te2, and Te3 in the cache.

All experiments have been repeated on 3 different platforms from 2 different processor manufacturers with 2 different operating systems. Platform A and platform B each have L1 cache lines of length 32 bytes, while platform C has L1 cache lines of length 64 bytes. The length of the cache line is important as it specifies the number of cache lines that a lookup table occupies. If perfectly aligned in a processor with cache lines of length 32 bytes, a table of 1024 bytes occupies 32 lines, while a table of 256 bytes occupies 8 cache lines. With 64 byte cache lines, these tables occupy 16 and 4 cache lines respectively.

Since we are only interested in validating the correctness of the analytical model, the last round correlation attack of section 2.1 has been mounted in a single process setup and not with a spy and target process as would be the case for an actual attack. In the process, the cache is cleaned before the encryption of random plaintext data and the state of the performance counters, which have been programmed to count the L1 cache misses, is read just before and just after the encryption. A single measurement output consists of the ciphertext and the measurement score for the number of L1 cache misses.

Unlike the timestamp counter, the performance counters are only accessible in a high-privilege mode. Yet, a device driver and an application program interface can be installed to use the counters at any privilege level. Hence, the security evaluation of an implementation should be based on the measurement scores provided by the performance counters as this is the worst case scenario. If an adversary with access to the performance counters cannot succeed, then a more realistic adversary cannot succeed either.

Table 1 shows the required number of measurements for a successful table index estimation attack, in which the cache miss estimation is based on the table index of 2 table lookup operations. Table 2 shows the experimental results for the cache line estimation attack, in which the cache miss estimation is based on the cache line of 2 table lookup operations. The predicted results have been computed with a probability of 0.99 to discover the secret key. The measurement results are based on 100 assessments, except those that required a massive amount of measurement and processing time, which have been marked with a double dagger. For each assessment a new and random key has been generated. Note that each table entry contains 2 results: the minimum and the median required number of measurements over the 100 assessments to successfully extract the full 128-bit key. Figures 4 and 5 in the appendix are a graphic representation of the data in tables 1 and 2.

The results attest that the model accurately predicts a strict lower bound on the required number of measurements. Independent of the platform or the implementation or the adversary's observation capability, the prediction is lower than but very close to the minimum required number of measurements. Note that in order to obtain a prediction for the median required number of measurements, the confidence level, or in other words Z_α in equation 2 should be increased to account for the fact that in the majority of the experiments the correct key fragment would be extracted. Increasing the probability to discover the secret key fragment to 0.999 returns an approximate number for the median required number of measurements.

Table 1. Required number of measurements for a successful table index estimation attack

	32 byte cache line			64 byte cache line	
	predicted	measured (min/median)[†]		predicted	measured (min/median)[†]
		A	B		C
OSSL	105K	150K/230K	140K/220K	112K	160K/270K
CLR	2.06M	2.23M/3.85M	2.03M/4.00M	66.8M	109M/160M[‡]
NOT4	430K	420K/880K	440K/830K	1.52M	1.64M/3.30M
2ENC	234K	290K/520K	340K/510K	278K	450K/825K
OT4	12.4K	15.0K/26.0K	17.0K/26.0K	30.1K	38.0K/79.0K

[†]based on 100 experiments
[‡]based on 10 experiments

Given that the analytical model accurately predicts the required number of measurements, it is now possible to evaluate mitigated implementations for which experimental results can not be obtained as too many measurements would be required to extract the full 128-bit key. For instance the model forecasts that changing the OpenSSL implementation to use a single compact S-box table of length 256 bytes in the first and the last round increases the required number of measurements to $9 \cdot 10^9$ ($\approx 2^{33}$) for an adversary that is capable to observe a single encryption on contemporary platforms with 64 byte cache lines, while using an implementation that uses a single compact table in each round increases the required number of measurements to 10^{23} ($\approx 2^{76}$).

Table 2. Required number of measurements for a successful cache line estimation attack

	32 byte cache line			64 byte cache line	
	predicted	measured (min/median)†		predicted	measured (min/median)†
		A	B		C
OSSL	12.7K	21.1K/35.6K	20.0K/32.5K	6.59K	10.0K/17.5K
CLR	56.5K	55.0K/125K	82.0K/148K	787K	1.23M/2.18M‡
2ENC	28.5K	35.0K/75.6K	33.7K/75.0K	16.3K	28.7K/49.4K
OT4	1.50K	1.87K/3.75K	2.12K/3.75K	1.77K	4.12K/7.00K

†*based on 100 experiments*
‡*based on 25 experiments*

5 Conclusions

We have provided cryptographers and software developers with a tool to evaluate the strength of their encryption algorithm or software implementation against time-driven cache attacks. The analytical model accurately forecasts the strength of a symmetric key cryptosystem based on a few simple parameters that describe the adversary's observation capabilities, the software implementation, and the platform the algorithm is running on. The accuracy of the model has been confirmed with concrete measurement results for different implementations, attack scenarios and platforms.

References

1. Acıiçmez, O., Schindler, W., Koç, Ç.K.: Trace Driven Cache Attack on AES. e-print of the IACR (2006), Available online at http://eprint.iacr.org/2006/138.pdf
2. Acıiçmez, O., Schindler, W., Koç, Ç.K.: Cache based remote timing attack on the aes. In: Abe, M. (ed.) CT-RSA 2007. LNCS, vol. 4377, pp. 271–286. Springer, Heidelberg (2006)
3. Bernstein, D.J.: Cache-timing attacks on AES (2004), Available online at http://cr.yp.to/papers.html#cachetiming
4. Bertoni, G., Zaccaria, V., Breveglieri, L., Monchiero, M., Palermo, G.: AES Power Attack Based on Induced Cache Miss and Countermeasure. In: ITCC (1), pp. 586–591. IEEE Computer Society, Los Alamitos (2005)
5. Bonneau, J., Mironov, I.: Cache-collision timing attacks against aes. In: Goubin, L., Matsui, M. (eds.) CHES 2006. LNCS, vol. 4249, pp. 201–215. Springer, Heidelberg (2006)
6. Brickell, E., Graunke, G., Neve, M., Seifert, J.-P.: Software mitigations to hedge AES against cache-based software side channel vulnerabilities. Cryptology ePrint Archive, Report 2006/052 (2006), Available online at http://eprint.iacr.org/
7. Kelsey, J., Schneier, B., Wagner, D., Hall, C.: Side channel cryptanalysis of product ciphers. Journal of Computer Security 8(2/3) (2000)
8. Kocher, P.C.: Timing Attacks on Implementations of Diffie-Hellman, RSA, DSS, and Other Systems. In: Koblitz, N. (ed.) CRYPTO 1996. LNCS, vol. 1109, pp. 104–113. Springer, Heidelberg (1996)

9. Lauradoux, C.: Collision attacks on processors with cache and countermeasures. In: Wolf, C., Lucks, S., Yau, P.-W. (eds.) Proceedings of Western European Workshop on Research in Cryptplogy (WeWorc 2005). GI edn. Lecture Notes in Informatics (LNI), p. 74. Bonner Köllen Verlag (2005)

10. Mangard, S.: Hardware countermeasures against dpa? a statistical analysis of their effectiveness. In: CT-RSA, pp. 222–235 (2004)

11. Neve, M., Seifert, J.-P.: Advances on access-driven cache attacks on aes. Selected Areas of Cryptography – SAC 2006, LNCS, vol. 4356, Springer, Heidelberg (to appear, 2007)

12. OpenSSL. OpenSSL: the Open-source toolkit for SSL / TLS, Available online at http://www.openssl.org/

13. Osvik, D., Shamir, A., Tromer, E.: Cache Attacks and Countermeasures: The Case of AES. In: Pointcheval, D. (ed.) CT-RSA 2006. LNCS, vol. 3860, pp. 1–20. Springer, Heidelberg (2006)

14. Page, D.: Theoretical use of cache memory as a cryptanalytic side-channel. Technical Report CSTR-02-003, Department of Computer Science, University of Bristol (June 2002)

15. Percival, C.: Cache missing for fun and profit (2005), Available online at http://www.daemonology.net/hyperthreading-considered-harmful/

16. Tsunoo, Y., Saito, T., Suzaki, T., Shigeri, M., Miyauchi, H.: Cryptanalysis of DES Implemented on Computers with Cache. In: Walter, C.D., Koç, Ç.K., Paar, C. (eds.) CHES 2003. LNCS, vol. 2779, pp. 62–76. Springer, Heidelberg (2003)

Appendix

Figures 4 and 5 show the required number of measurements for a successful time-driven cache attack obtained through the analytical model and obtained through experimental results for the table index estimation and the cache line estimation attack respectively. Note that these figures are a graphic representation of the data in tables 1 and 2. For completeness, the description of the attack scenarios is repeated below:

- OSSL: OpenSSL implementation of the AES algorithm.
- CLR: OSSL with compact S-box table in last round.
- NOT4: OSSL without table Te4 in last round.
- 2ENC: OSSL attacked by a limited adversary observing 2 encryptions.
- OT4: OSSL attacked by a powerful adversary observing only last round.

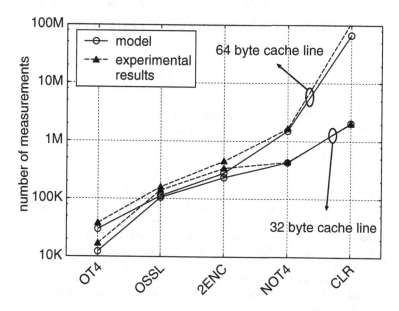

Fig. 4. Required number of measurements for a successful table index estimation attack

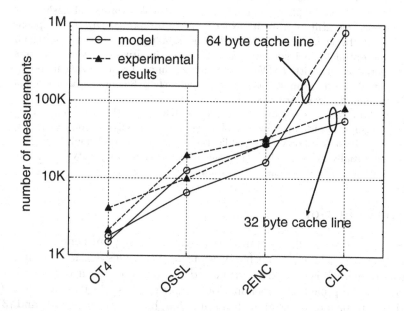

Fig. 5. Required number of measurements for a successful cache line estimation attack

Improving the Security of MACs Via Randomized Message Preprocessing

Yevgeniy Dodis[1] and Krzysztof Pietrzak[2]

[1] New York University
dodis@cs.nyu.edu
[2] CWI Amsterdam
k.z.pietrzak@cwi.nl

Abstract. "Hash then encrypt" is an approach to message authentication, where first the message is hashed down using an ε-universal hash function, and then the resulting k-bit value is encrypted, say with a block-cipher. The security of this scheme is proportional to εq^2, where q is the number of MACs the adversary can request. As ε is at least 2^{-k}, the best one can hope for is $O(q^2/2^k)$ security. Unfortunately, such small ε is not achieved by simple hash functions used in practice, such as the polynomial evaluation or the Merkle-Damgård construction, where ε grows with the message length L.

The main insight of this work comes from the fact that, by using *randomized message preprocessing* via a short random salt p (which must then be sent as part of the authentication tag), we can use the "hash then encrypt" paradigm with suboptimal "practical" ε-universal hash functions, and still improve its exact security to optimal $O(q^2/2^k)$. Specifically, by using at most an $O(\log L)$-bit salt p, one can *always* regain the optimal exact security $O(q^2/2^k)$, even in situations where ε grows polynomially with L. We also give very simple preprocessing maps for popular "suboptimal" hash functions, namely polynomial evaluation and the Merkle-Damgård construction.

Our results come from a general extension of the classical Carter-Wegman paradigm, which we believe is of independent interest. On a high level, it shows that public randomization allows one to use the potentially much smaller "average-case" collision probability in place of the "worst-case" collision probability ε.

1 Introduction

HASH THEN ENCRYPT. A popular paradigm to message authentication is "hash then encrypt", where the authentication tag for a message m is computed as $f(h(m))$ where h is a hash function and f a pseudorandom permutation (say AES). This approach is appealing for several reasons: (1) it is stateless, (2) h needs not to be a cryptographic hash function, but only ε-universal[1] and (3) the "slow" cryptographic f is then only applied on a short input (i.e. on the range of h).

---

[1] A family of hash functions $\mathcal{H}$ is ε-universal, if for any $x \neq x'$, $\Pr_{h \leftarrow \mathcal{H}}[h(x) = h(x')] \leq \varepsilon$.

A. Biryukov (Ed.): FSE 2007, LNCS 4593, pp. 414–433, 2007.

As f is indistinguishable from a uniformly random permutation, everything an attacker learns about $h(m_1), h(m_2), \ldots$ from the authentication tags $f(h(m_1))$, $f(h(m_2)), \ldots$ is whether there is a collision (as $f(h(m_i)) = f(h(m_j))$ iff $h(m_i) = h(m_j)$). Since h is not cryptographic, finding such a collision is usually enough for an adversary to come up with a forgery for a new message. If f is over $\{0,1\}^k$ (say, f is AES-128 where $k = 128$) then by the birthday bound the best security we can hope for is something in the order of $q^2/2^k$ where $q = q_{mac} + q_{forge}$ is the number of MAC queries and forgery attempts the adversary is allowed to make. More precisely, assuming that f is ideal (i.e. a uniform random permutation), the probability of a successful forgery can be upper bounded by[2]

$$\varepsilon \cdot q_{mac}^2 + \varepsilon \cdot q_{forge}$$

As $\varepsilon \geq 1/2^k$ the best one can hope for by the "hash then encrypt" approach is $O(q^2/2^k)$ security. In the sequel, we will call this *optimal* security.

USING SUBOPTIMAL HASH FUNCTIONS. Unfortunately, hash functions used in practice, such as polynomial evaluation and the cascade (aka Merkle-Damgård) construction[3], do not yield optimal security, since the value ε they achieve grows linearly with the length of the message. There are several ways to improve the exact security with such hash functions. Most obviously, one can increase the security parameter k. However, this does not regain the optimal security relative to this *larger* k (although the absolute exact security is improved). More critically, increasing k is typically not an option in practice, since k is tied to the block length of the block cipher which is usually fixed (say, to 128 bits for AES) and pretty inflexible to any changes.[4] Another option is to design a different ε-universal hash function achieving optimal $\varepsilon = O(1/2^k)$. For one thing, replacing existing and popular implementations is not so easy in practice, so this option is usually ruled out anyway. More importantly, and this is part of the reason practical hash functions are not "optimally universal", $O(1/2^k)$-universal hash functions tend to be much less efficient or convenient than their slightly suboptimal counter-parts. For example, achieving perfect $\varepsilon = 1/2^k$ requires a key of size L [23], where L is the length of the message, which is very impractical. In theory, one can achieve $\varepsilon = 2/2^k$ by composing a "practical" δ-universal hash function from L to $k + \ell$ bits, where ℓ is chosen large enough to bring δ below $1/2^k$ (typically $\delta = O(L)/2^{k+\ell}$ which gives $\ell = \log(L) + O(1)$), with the

---

[2] If f is not ideal but a pseudorandom permutation, then there should also be a term counting the insecurity of f. However, since this term is always the same and is independent of the hash component, we will omit it from all our bounds.

[3] This construction uses a fixed input length shrinking function iteratively in order to get a function for arbitrary long inputs. In this paper we always assume that the iterated function is ideal, i.e. a uniformly random function.

[4] Though, if one is willing to make two invocations to the block-cipher, one can use it in CBC-mode in order to get a $\{0,1\}^{2k} \rightarrow \{0,1\}^k$ PRF, which can then be used instead of the block-cipher. This mode of operation should not be called "hash then encrypt", as the application of the (shrinking) PRF is not invertible, i.e. not an encryption scheme.

perfectly universal hash function from $k+\ell$ to k bits, whose key is then only $k+\ell$ bits long. In practice, however, this composition is quite inconvenient to implement. Aside from the obvious inefficiency that the key size is at least doubled compared to using practical hash functions, the latter are usually optimized to operate on a specific value of k, and do not extend easily to larger outputs $k+\ell$, even if ℓ is very small. For example, the polynomial evaluation function with $k = 128$ corresponds to performing fast field operations over $GF(2^{128})$, which could be implemented in hardware. In contrast, evaluating field operations over $GF(2^{128+\ell})$ would be much slower even for $\ell = 1$, since a 129-bit element would not fit into a register. Similarly, the cascade construction typically uses a very specific compression function with fixed k, which is usually undefined for larger k.

To summarize, in practice the "hash then encrypt" paradigm does not achieve optimal exact security $O(q^2/2^k)$.

The question addressed in this paper is whether one can reclaim — with little extra cost — the optimal exact security of the "hash then encrypt" MAC when using such popular but sub-optimal ε-universal hash functions. Moreover, we would like a solution which does not change the value of k and does not modify the internals of the underlying hash function, so that our solution can be easily applied to existing implementations. Finally, the solution should remain stateless. Quite surprisingly, we answer this question in the affirmative for the popular polynomial evaluation and the cascade constructions, by using randomized message preprocessing before applying the actual MAC. We motivate our approach below.

USING RANDOMIZATION. Recall, a hash function is ε-universal if the collision probability of two messages is at most ε *for all* possible message pairs. But the actual collision probability could be much smaller for most pairs. To illustrate this on an example, let us consider the polynomial-based ε-universal hash function: here the message $m \in \{0,1\}^{Lk}$ is viewed as a degree $(L-1)$ polynomial f_m over $GF(2^k)$, the secret key s is an element of $GF(2^k)$, and $h_s(m)$ is the value of the polynomial f_m at s. Given two distinct messages m_1 and m_2, their probability of collision is $r/2^k$, where r is the number of roots of the non-zero polynomial $g = f_{m_1} - f_{m_2}$. Although g can have up to $(L-1)$ roots for some pairs (m_1, m_2), Hartman and Raz [11] showed that the fraction of polynomials (of any degree) with t roots decays proportional to $1/t!$, which means that a vast majority of polynomials have at most a constant number of roots. Thus, a vast majority of pairs (m_1, m_2) have collision probability $O(1/2^k)$ rather than the worst-case bound $(L-1)/2^k$.

This brings up the following idea to improve the security of a MAC based on the "hash then encrypt" paradigm: apply some randomized preprocessing function $m' = Pre(m, p)$ to the message m (where p is a fresh random salt), and return $(p, f(h_s(m')))$ as the randomized MAC of m. The hope is that preprocessing should thwart the attempts of the adversary to choose messages which have a large collision probability and thus increase the security of the MAC. To put it differently, by designing a clever randomized message preprocessing, we will try to ensure that two *processed* messages cannot collide with probability more than $O(1/2^k)$, even if the worst-case ε is much larger. Another advantage of preprocessing comes from the fact that the original "suboptimal" MAC is used in a "black-box" manner, while suddenly becoming more secure!

Of course, there is a price we need to pay for regaining optimal exact security: we need to tell the receiving party how we randomized the message. Thus, aside from showing that randomized message preprocessing *can always* yield optimal exact security, — which is a non-obvious statement we will prove later, — the secondary objective is to minimize the number of random bits needed to avoid "bad" message pairs in a given ε-universal family.

THE SALTED HASH FUNCTION PARADIGM. To capture the intuition just described, we introduce the concept of *salted* $(\varepsilon_{forge}, \varepsilon_{mac})$-universal hash functions. As the name suggests, the key of such a function has two parts, a "secret" part and a "public" salt. If the salt is empty, then $\varepsilon_{forge} = \varepsilon_{mac}$ and we have a usual ε-universal hash function with $\varepsilon = \varepsilon_{forge}$. Here ε_{forge} is an upper bound on the probability (over a choice of the secret key) that the hash values of two messages collide when the adversary can choose the messages and the public salt for both hashes, and ε_{mac} is the same probability but where at least one of the two public salts is chosen at random (clearly we always have $\varepsilon_{forge} \geq \varepsilon_{mac}$).

From such a salted hash function we can construct a MAC-scheme like from the usual ε-universal hash function (i.e. "hash then encrypt"), but where for every message to be authenticated the public salt is chosen at random and must be sent as a part of the authentication tag. As our first result, we generalize the standard "one-key" hash then encrypt MAC and show that the generalized MAC has security

$$\varepsilon_{mac} \cdot q_{mac}^2 + \varepsilon_{forge} \cdot q_{forge}$$

In particular, to get optimal $O(q^2/2^k)$ security here it is sufficient to have ε_{mac} in the order of $1/2^k$, while ε_{forge} can be considerably larger.

We also remark that the above salted hash paradigm is strictly more general than the "randomized preprocessing paradigm" advertised earlier. In particular, the latter corresponds to the salted hash functions of the form $h_s(Pre(m,p))$, where $Pre(m,p)$ is the preprocessing map. However, we find it more intuitive to state some of results for the general salted hash paradigm (which also makes them more general).

SALTED HASH FUNCTIONS WITH SHORT SALT. In this paper we propose randomized preprocessing mechanisms for several ε-universal hash functions and show that this turns them into $(\varepsilon_{forge}, \varepsilon_{mac})$-universal ones with $\varepsilon_{forge} \approx \varepsilon$ and $\varepsilon_{mac} = O(1/2^k)$. Moreover, in each case we use a very short public salt p.

Our first result is very general. In §4 we show (non-constructively) that for every balanced ε-universal hash function (where $\varepsilon = \varepsilon(L)$ can grow polynomially in the length of the messages L; i.e., $\varepsilon(L) = L^c/2^k$ for some constant c) there is a randomized preprocessing *using only* $O(\log L)$ *random bits* which gives an $(\varepsilon_{forge}, \varepsilon_{mac})$-universal hash function with $\varepsilon_{forge} \approx \varepsilon$ and $\varepsilon_{mac} = O(1/2^k)$. Although this result is non-constructive (i.e., the preprocessing map is inefficient and is only shown to exist), we believe the result is interesting given its generality.

Construction	ε		length hash key	length of public salt	domain
	ε_{forge}	ε_{mac}			

$(\varepsilon_{forge}, \varepsilon_{mac})$-universal hash functions from ε-universal ones

Generic: for each balanced $H(.)$ as below there is a permutation $g(.)$ such that

	$H(.)$	$L^c/2^k$		key for H	$-$	$\{0,1\}^L$
$\star$	$H(g(.))$	$(L+\ell)^c/2^k$	$2/2^k$	key for H	$\ell \approx (2c+1)\log(L)$	$\{0,1\}^L$

Construction based on polynomial evaluation over $GF(2^k)$

	poly	$(L-1)/2^k$		k	$-$	$(\{0,1\}^k)^{\le L}$
	poly$^{GF(2^k)}$	$L/2^k$	$1/2^k$	k	k	$(\{0,1\}^k)^{\le L}$
$\diamond$	poly$^{\mathcal{P}}$	$L/2^k$	$2/2^k$	k	$3\log(L)+\log(k)$	$(\{0,1\}^k)^{\le L}$

Cascade (Merkle-Damgård) based on function $func: \{0,1\}^{k+b} \to \{0,1\}^k$

	MD	$L/2^k$		key for $func$	$-$	$\{0,1\}^{bL}$
	MDR	$(L+1)/2^k$	$2/2^k$	key for $func$	$\log(L)$	$\{0,1\}^{bL}$

$(\varepsilon_{forge}, \varepsilon_{mac})$-$\Delta$ universal hash functions from ε-Δ universal ones

Generic: for each $H(.)$ as below there is a "small" set $\mathcal{P}$ such that

	$H(.)$	$L^c/2^k$		key for H	$-$	$\{0,1\}^L$
$\diamond$ generic	$H^{\mathcal{P}}(.)$	$(L+\ell)^c/2^k$	$2/2^k$	key for H	$\ell \approx (2c+1)\log(L)$	$\{0,1\}^L$

Fig. 1. Parameters for the hash functions considered in this paper. A leading $\star$ denotes a non-constructive, and a leading $\diamond$ a non-uniform result. The bounds for the cascade construction assume that $func$ is a uniform random function, also the higher order terms that appear in the bounds for this construction are omitted in this table.

Our next results involve simple and *efficient* implementations of the above generic result for popular ε-universal hash functions, such as the polynomial evaluation (denoted poly in the sequel) and the cascade (aka. Merkle-Damgård) construction. We describe these results in more details in §5 and §6, but mention that in each case we manage to design extremely simple preprocessing maps using an $O(\log L)$-bit public salt promised by the generic existence result: roughly, in the cascade construction we simply append a random string, and for poly the map consists of prepending a string chosen from a small set. Unfortunately the construction for poly is only non-uniform, i.e. we prove that maps with a succinct description (of length $O(L^3)$) exist, but do not give a specific map that works. Fortunately, we can show that almost all such succinct descriptions yield a good map, so one can just sample and "hardwire" a random string which will then define a good map almost certainly.

As our final result in §7, for hash functions which satisfy the slightly stronger property of being ε-Δ universal [16], we also show a *constructive* general result stating that we can always regain the optimal security by doing $O(\log L)$-bit *post*processing instead of preprocessing. By this we mean randomizing the hash value $h_s(m)$, rather than the message m, with an $O(\log L)$-bit public salt p. Moreover, we do not need the hash function to be balanced (which is necessary for the general result with preprocessing).

Figure 1 summarizes our results (see future sections for some of the notation). In Section §8 we give a numerical example to illustrate what security can be gained by using "salted" hash then encrypt. In Section §9 we review some related work and give some open problems.

2 Notation and Basic Definitions

For any integer $x \geq 0$ we denote by $\langle x \rangle_k$ its binary representation padded with leading 0's to length k, e.g. $\langle 7 \rangle_5 = 00111$. For two strings A, B we denote with $A \| B$ their concatenation. For $k \in \mathbb{N}$ we define $B_k \stackrel{\text{def}}{=} \{0,1\}^k$ and $B_k^{\leq i} \stackrel{\text{def}}{=} \cup_{j=1...i} B_k^j$ to be the set of all strings of length at most i k-bit blocks. For a set $\mathcal{X}$ we denote with $x \leftarrow \mathcal{X}$ that x is sampled uniformly at random from $\mathcal{X}$.

Definition 1 (ε-almost 2-universal hash function). *A hash function $H : \mathcal{S} \times \mathcal{X} \to \mathcal{Y}$ is ε-almost 2-universal if for all $x, y \in \mathcal{X}, x \neq y$ and $h_s(.) \stackrel{\text{def}}{=} H(s,.)$*

$$\Pr[s \leftarrow \mathcal{S}; h_s(x) = h_s(y)] \leq \varepsilon$$

To save on notation we only write "ε-universal" for "ε-almost 2-universal". We also often consider the case where $\varepsilon = \varepsilon(\mathcal{X})$ can be a function of the message space $\mathcal{X}$.

3 Salted Hashing and MACs

In this section we first review the "hash then encrypt" approach for message authentication. We then define *salted universal hash functions*, which are a randomized version of normal ε-universal hash functions. Based on such hash functions, we propose a randomized version of "hash then encrypt" and show that its security is mainly bounded by the "average-case" collision probability of the salted hash function, which can be much smaller than the "worst-case" probability which appears in the bound of the standard hash then encrypt approach. We now define what we mean by (the security of) a randomized MAC. Let us note that the definition given below becomes the standard definition for (deterministic) MACs when $\mathcal{R}$ is a singleton set.

Definition 2 (Randomized MAC). *A randomized message authentication scheme MAC is a function $\mathcal{S} \times \mathcal{X} \times \mathcal{P} \to \mathcal{Y}$. $\mathcal{P}$ is the randomness-space, $\mathcal{S}$ is the (secret) key-space, $\mathcal{X}$ the message domain and $\mathcal{Y}$ the tag-space.*

We denote with $\mathrm{FRG}_{\mathsf{MAC}}(q_{mac}, q_{forge})$ the advantage of any adversary A in finding an existential forgery for MAC where A is allowed to ask for at most q_{mac} MACs and make q_{forge} forgery attempts. More formally we consider the following experiment: first $s \leftarrow \mathcal{S}$ is sampled, then A may query the MACing oracle, which

$$\textit{on input } m \textit{ outputs } (\mathsf{MAC}(s, m, p), p) \quad \textit{where} \quad p \leftarrow \mathcal{P}$$

at most q_{mac} times and a verification oracle which

on input (m, a, p) outputs 1 if $\mathsf{MAC}(s, m, p) = a$ and 0 otherwise.

at most q_{forge} times. Now $\mathsf{FRG}_{\mathsf{MAC}}(q_{mac}, q_{forge})$ is an upper bound on the probability that any A succeeds in receiving 1 from the verification oracle on an input (m, a, p) where he did not already receive the output (a, p) on input m from the MACing oracle. Note that the adversary may choose the salt p when querying the verification oracle, but the MACing oracle chooses the p at random.

This definition is an information theoretic one as we did only bound the number of queries A is allowed to make but we make no other computational assumption. We can do this as we will consider only MACs which as a final step involve an application of a uniform random permutation. In reality one would have to replace this uniform random permutation (URP) with a pseudorandom permutation (PRP), as otherwise the construction is not practical, and to restrict the above definition to computationally bounded adversaries (as unbounded adversaries can distinguish a PRP from a URP). The security of such a computational MAC then can be upper bounded by $\mathsf{FRG}_{\mathsf{MAC}}(q_{mac}, q_{forge}) + \mathsf{Adv}_{\mathsf{PRP}}$ where $\mathsf{Adv}_{\mathsf{PRP}}$ is the distinguishing advantage for the pseudorandom permutation of the adversary considered. From now on, we will no longer mention this simple fact. The following proposition is well known.

Proposition 1 (Security of hash then encrypt). *Let $H : \mathcal{S} \times \mathcal{X} \to \mathcal{Y}$ be ε-universal and $f(.)$ a uniform random permutation over $\mathcal{Y}$.*

If $1/(|\mathcal{Y}| - q_{mac}) \le \varepsilon$, then the MAC scheme with secret key $s \leftarrow \mathcal{S}$ where the authentication tag for a message $m \in \mathcal{X}$ is computed as

$$\mathsf{MAC}(s, m) = f(h_s(m))$$

has security

$$\mathsf{FRG}_{\mathsf{MAC}}(q_{mac}, q_{forge}) \le \varepsilon \cdot q_{mac}^2 + \varepsilon \cdot q_{forge}$$

We do not prove this proposition, as it is just a special case of Theorem 1 below.

We now define the concept of salted hash functions described in the introduction.

Definition 3 (($\varepsilon_{forge}, \varepsilon_{mac}$)-almost 2-universal salted hash function).
A hash function $H : \mathcal{S} \times \mathcal{X} \times \mathcal{P} \to \mathcal{Y}$ is ($\varepsilon_{forge}, \varepsilon_{mac}$)-universal if (below $x_1, x_2 \in \mathcal{X}$, $p_1, p_2 \in \mathcal{P}$ and $h_s(.,.) \stackrel{\text{def}}{=} H(s, ., .)$)

$$\varepsilon_{forge} \ge \max_{(x_1, p_1) \ne (x_2, p_2)} \Pr[s \leftarrow \mathcal{S}; h_s(x_1, p_1) = h_s(x_2, p_2)]$$

$$\varepsilon_{mac} \ge \max_{x_1, x_2, p_1} \Pr[s \leftarrow \mathcal{S}; p_2 \leftarrow \mathcal{P}; h_s(x_1, p_1) = h_s(x_2, p_2) \wedge (x_1, p_1) \ne (x_2, p_2)]$$

Every ε-universal hash function is an (ε, ε)-universal salted hash function with $\mathcal{P} = \emptyset$. We can now generalize the hash then encrypt paradigm to salted hash functions.

Theorem 1 (Security of salted hash then encrypt). *Let $H : \mathcal{S} \times \mathcal{X} \times \mathcal{P} \to \mathcal{Y}$ be $(\varepsilon_{forge}, \varepsilon_{mac})$-universal and $f(.)$ a uniform random permutation over $\mathcal{Y}$.*

If $1/(|\mathcal{Y}| - q_{mac}) \leq \varepsilon_{forge}$,[5] then the MAC scheme with secret key $s \leftarrow \mathcal{S}$ where the authentication tag for a message $m \in \mathcal{X}$ is computed by first sampling $p \leftarrow \mathcal{P}$ and then setting

$$\mathsf{MAC}(s, m, p) = (f(h_s(m, p)), p)$$

has security

$$\mathsf{FRG}_{\mathsf{MAC}}(q_{mac}, q_{forge}) \leq \varepsilon_{mac} \cdot q_{mac}^2 + \varepsilon_{forge} \cdot q_{forge}$$

Proof. Instead of bounding $\mathsf{FRG}_{\mathsf{MAC}}(q_{mac}, q_{forge})$ we bound the (larger) probability P that any adversary A can forge a MAC or he finds a collision. By a collision we mean that two outputs $(a_1, p_1), (a_2, p_2)$ from the MACing oracle on two (not necessarily distinct) queries m_1 and m_2, satisfy $a_1 = a_2$ and $(m_1, p_1) \neq (m_2, p_2)$.

Let P_{col} denote the probability that a collision occurs before A found a forgery, and P_{frg} be the probability that A found a forgery before any collision occurred, so $P = P_{col} + P_{frg}$. Below we bound $P_{col} \leq \varepsilon_{mac} \cdot q_{mac}^2$ and $P_{frg} \leq \varepsilon_{forge} \cdot q_{forge}$ which then proves the theorem.

We can bound $P_{col} \leq \varepsilon_{mac} \cdot q_{mac}^2$ as follows: first, we can assume that A makes no forgery attempts (as trying to forge can only lower the probability of finding a collision *before* there was a successful forgery). Now we will show that for any $1 \leq i < j \leq q_{mac}$, the probability that the *first* collision is amongst the i'th and j'th query is at most ε_{mac}: as we are interested in the first collision, we can assume that $i = 1$ and $j = 2$, as making any intermediate queries can only lower the success probability (to see this, note that because of the application of the uniform random permutation, all the adversary learns about the outputs of the hash function is whether there were collisions or not). Next, for an adversary which makes only two queries, ε_{mac} is a trivial upper bound on the collision probability (even if we allow A to choose the salt for the first query). Applying the union bound we get that the probability that there are any $i, j, 1 \leq i < j \leq q_{mac}$ such that the i'th and j'th output collide, is at most $\varepsilon_{mac} \cdot q_{mac} \cdot (q_{mac} - 1)/2$, which thus is an upper bound for P_{col}.

We will now prove the bound $P_{frg} \leq \varepsilon_{forge} \cdot q_{forge}$. For any $j, 1 \leq j \leq q_{forge}$, let P_j denote the probability that the j'th forgery query is the first successful forgery and there was no collision before this forgery attempt. We will show $P_j \leq \varepsilon_{forge}$, this proves $P_{frg} \leq \varepsilon_{forge} \cdot q_{forge}$ as $P_{frg} = \sum_{j=1}^{q_{forge}} P_j$. To upper bound P_j we can assume that A skips the $j - 1$ first forgery attempts (this can only increase the success probability of the considered forgery attempt to be the *first* successful forgery). Moreover we allow A to chose all the salts for his

---

[5] This is satisfied if the hash function is input shrinking and ε_{forge} is at least slightly bigger than the optimal $1/|\mathcal{Y}|$, which is the setting that interests us. To see that the restriction is necessary, consider a hash function which is a permutation on $\mathcal{X} \times \mathcal{P} \equiv \mathcal{Y}$, then $\varepsilon_{forge} = \varepsilon_{mac} = 0$, but the forgery probability is not 0 as a random guess will always be successful with prob. $1/|\mathcal{Y}|$.

(up to q_{mac}) MACing queries he can ask before his forgery attempt. Again, this can only increase his success probability.

So we must upper bound (by ε_{forge}) the probability of any A winning the following game: A gets $\sigma_i = f(h_s(x_i))$ for $i = 1, \ldots, k$ (with $k \leq q_{mac}$) and x_i's of his choice. Then he must come up with a σ, x (where $x \neq x_i$ for all $1 \leq i \leq k$). He wins if $f(h_s(x)) = \sigma$ and $\sigma_i \neq \sigma_j$ for all $1 \leq i < j \leq k$.

If A chooses a σ where $\sigma \neq \sigma_i$ for all $i = 1, \ldots, k$, then the success probability (even conditioned on all σ_i being distinct), is at most $1/(|\mathcal{Y}| - k) \leq 1/(|\mathcal{Y}| - q_{mac}) \leq \varepsilon_{forge}$ (the last step by assumption), as then $\Pr[f(h_s(x)) = \sigma] \leq \Pr[f(h_s(x)) = \sigma | \forall i, 1 \leq i \leq k : h_s(x_i) \neq h_s(x)] = 1/(|\mathcal{Y}| - k)$, here the first step used that if $h_s(x_i) = h_s(x)$, then $\sigma = \sigma_i$, and the second used that f is a uniform random permutation and thus if $h_s(x)$ is new, then $f(h_s(x))$ is uniformly random over $\mathcal{Y} \setminus \{\sigma_i, \ldots, \sigma_k\}$.

Now consider the other case, i.e. when A chooses a σ where for some $i : \sigma = \sigma_i$. From this A we construct an adversary A' which has at least the same probability of winning as follows. A' runs A and answers each of A's queries $x_1, \ldots, x_k$ with uniformly random but distinct $\sigma_1', \ldots, \sigma_k'$. When now A outputs his forgery attempt ($\sigma = \sigma_i', x$), A' makes the single MACing query x_i, gets σ_i and outputs the forgery attempt ($\sigma = \sigma_i, x$). It's not hard to verify that A' success probability is at least the one of A (it can be larger as A' will never lose the game due to collisions amongst the $\sigma_1, \ldots, \sigma_k$, as he only asks one for each σ_i). Moreover A''s success probability is at most ε_{forge} as A' just chooses two values x, x_i before even using any oracle, and then wins if $h_s(x) = h_s(x_i)$.   □

4   A Generic Construction

In this section we show that for every balanced ε-universal hash function H, where ε can even grow polynomially in the message length L, there exists a preprocessing using only $O(\log L)$ random bits, which makes the hash function ($\varepsilon_{forge}, \varepsilon_{mac}$)-universal where $\varepsilon_{forge} \approx \varepsilon$ and ε_{mac} is of the smallest possible order.

Let $H : \mathcal{S} \times \{0,1\}^* \to \{0,1\}^k$ be $\varepsilon(.)$-almost 2-universal, by this we mean that for all $x_1, x_2 \in \{0,1\}^*$ where $x_1 \neq x_2$, $\ell = \max\{|x_1|, |x_2|\}$ and $h_s(.) \stackrel{\text{def}}{=} H(s, .)$

$$\Pr[s \leftarrow \mathcal{S}; h_s(x_1) = h_s(x_2)] \leq \varepsilon(\ell)$$

Definition 4. *A hash function H as above is balanced if for all $\ell \geq k$, $s \in \mathcal{S}$ and $y \in \{0,1\}^k$*

$$\Pr[x \leftarrow \{0,1\}^\ell; h_s(x) = y] = 2^{-k}$$

The following lemma states that from any such hash function H which is $\varepsilon(.)$-almost 2-universal and balanced we can get (non-constructively) a ($\varepsilon_{forge}, \varepsilon_{mac}$) salted hash function (with domain $\{0,1\}^L$ for any $L > k$) where $\varepsilon_{mac} = O(2/2^k)$ and $\varepsilon_{forge} = \varepsilon(L + r)$. Here r is the length of the public salt and if $\varepsilon(L) = O(L^c/2^k)$ we will get $r \approx (2c + 1)\log(L)$.

The construction is very simple, the salted hash function with key $s \in \mathcal{S}$, salt $p \in \{0,1\}^r$ and message x is computed as $h_s(g(p\|x))$ for some permutation g (we show that a random permutation is appropriate with high probability).

Lemma 1. *Let $H : \mathcal{S} \times \{0,1\}^* \to \{0,1\}^k$ be a balanced $\varepsilon(.)$-almost 2-universal hash function. Fix some integer $L \geq k$ and let r be the smallest integer satisfying*

$$2^r \geq \frac{2^{2k} \cdot \varepsilon(L+r)^2 \cdot (2L+r)}{\log(e)} \tag{1}$$

then there exists a permutation g over $\{0,1\}^{L+r}$ s.t. the salted hash function $H' : \mathcal{S} \times \{0,1\}^L \times \mathcal{P} \to \{0,1\}^k$ with $\mathcal{P} = \{0,1\}^r$ defined as

$$H'(s, m, p) \stackrel{def}{=} H(s, g(p\|m))$$

is $(\varepsilon_{forge}, \varepsilon_{mac}) - universal$ with $\varepsilon_{forge} = \varepsilon(L+r)$ $\varepsilon_{mac} = 2/2^k$.

Proof. Let $R \stackrel{def}{=} 2^r$ and $g_i(m) \stackrel{def}{=} g(i\|m)$. The bound on ε_{forge} is straightforward:

$$
\begin{aligned}
\varepsilon_{forge} &= \max_{x_1,x_2\in\{0,1\}^L, p_1,p_2\in\{0,1\}^r,(p_1,x_1)\neq(p_2,x_2)} \Pr[s \leftarrow \mathcal{S}; h_s(g_{p_1}(x_1)) = h_s(g_{p_2}(x_2))] \\
&= \max_{y_1,y_2\in\{0,1\}^{L+r}, y_1\neq y_2} \Pr[s \leftarrow \mathcal{S}; h_s(y_1) = h_s(y_2)] \\
&= \varepsilon(L+r)
\end{aligned}
$$

Above we used that $(p_1, x_1) \neq (p_2, x_2)$ implies $y_1 \neq y_2$ which holds as $y_1 = g(p_1\|x_1)$ and $y_2 = g(p_2\|x_2)$ and g is a permutation.

The proof for ε_{mac} is by the probabilistic method. We will show that a permutation g chosen at random has the desired property with probability > 0. For any $a, b \in \{0,1\}^L$ and $i, j \in \{0,1\}^r$ let $C_{i,j,a,b}$ denote the random variable (the probability is over g)

$$C_{i,j,a,b} = \Pr[s \leftarrow \mathcal{S}; h_s(g_i(a)) = h_s(g_j(b))]$$

As h_s is balanced we have for any $(i, a) \neq (j, b) : \mathsf{E}[C_{i,j,a,b}] \leq 1/2^k$, and as h_s is $\varepsilon(.)$-almost 2-universal and $|g_i(a)| = |g_j(b)| = L + r$

$$C_{i,j,a,b} \leq \varepsilon(L+r)$$

Let

$$C_{i,a,b} = \sum_{j\in\{0,1\}^r,(i,a)\neq(j,b)} C_{i,j,a,b} \tag{2}$$

As the sum ranges over R terms (resp. $R - 1$ if $a = b$) we have $\mathsf{E}[C_{i,a,b}] \leq R/2^k$. We can apply Hoeffding's inequality (see Appendix A), for the case $a \neq b$ (then the sum in eq.(2) has exactly R terms, if $a = b$ then we have only $R - 1$ terms and can use the same bound as proven below) we get

$$\Pr[C_{i,a,b} \geq 2 \cdot \mathsf{E}[C_{i,a,b}]] < \exp\left(-\frac{2 \cdot (R/2^k)^2}{R \cdot \varepsilon(L+r)^2}\right) \leq 2^{-2L-r}$$

Where in the last step we used (1) (recall that $R \stackrel{\text{def}}{=} 2^r$). So there is a g such that $C_{i,a,b} \leq 2 \cdot \mathsf{E}[C_{i,a,b}] \leq 2R/2^k$ for all $i \in \{0,1\}^r$ and $a,b \in \{0,1\}^L$, for this g

$$
\begin{aligned}
\varepsilon_{mac} &= \max_{\substack{i \in \{0,1\}^r \\ a,b \in \{0,1\}^L}} \Pr[j \leftarrow \{0,1\}^r; s \leftarrow \mathcal{S}; h_s(g_i(a)) = h_s(g_j(b)) \wedge ((i,a) \neq (j,b))] \\
&= \max_{i \in \{0,1\}^r, a,b \in \{0,1\}^L} R^{-1} \sum_{j \in \{0,1\}^r, (i,a) \neq (j,b)} C_{i,j,a,b} \\
&= R^{-1} \max_{i \in \{0,1\}^r, a,b \in \{0,1\}^L} C_{i,a,b} \\
&\leq 2^{-k+1} \qquad\qquad\qquad\qquad\qquad\qquad\qquad\qquad \square
\end{aligned}
$$

5   poly: Hashing by Polynomial Evaluation

A popular way of ε-almost universal hashing is to parse the message into coefficients of a polynomial over some field (we will use $GF(2^k)$) and evaluate it on a random point. We propose a simple randomized preprocessing for this hash function: just set the constant coefficient at random. If this coefficient is set uniformly at random, this gives a salted hash function with an optimal $\varepsilon_{mac} = 1/2^k$. We then show that one can also sample the coefficient from a small set, thus using fewer randomness, and still achieve an almost optimal $\varepsilon_{mac} \leq 2/2^k$. We will come back to this construction later in Section 7, where we prove some generic results which imply Lemma 3 and Lemma 4 from this section (though with somewhat worse parameters).

Definition 5. *For $M = (M_1, \ldots, M_m)$ (each $M_i \in GF(2^k) \cong B_k$) we denote with $f_M(.)$ the polynomial of degree $m-1$ over $GF(2^k)$ given by*

$$f_M(x) = \sum_{i=1}^{m} M_i \cdot x^{i-1}$$

Definition 6. *With poly we denote the hash function which on input $M \in GF(2^k)^* \cong B_k^*$ with key $s \leftarrow GF(2^k)$ is computed as $\mathsf{poly}_s(M) = f_M(s)$.*

Lemma 2 (see [22]). *poly with domain $B_k^{\leq L}$ is ε-almost universal with $\varepsilon = (L-1)/2^k$.*

Definition 7. *$\mathsf{poly}^{GF(2^k)}$ is the salted hash function with secret key part $s \leftarrow GF(2^k)$ and public salt $p \leftarrow GF(2^k)$ which on input $M \in B_k^*$ is computed as*

$$\mathsf{poly}_{s,p}^{GF(2^k)}(M) = f_{(p,M)}(s)$$

Lemma 3. *$\mathsf{poly}^{GF(2^k)}$ with domain $B_k^{\leq L}$ is $(\varepsilon_{forge}, \varepsilon_{mac})$-universal where*

$$\varepsilon_{forge} = L/2^k \qquad\qquad\qquad (3)$$
$$\varepsilon_{mac} = 1/2^k \qquad\qquad\qquad (4)$$

The bound on ε_{forge} follows from Lemma 2, and the bound on ε_{mac} is obvious. We now consider another salted version of the poly hash function which is similar to $\mathsf{poly}^{GF(2^k)}$ but where the public salt is not chosen from the whole of $GF(2^k)$ but only from a subset $\mathcal{P} \subset GF(2^k)$.

Definition 8. *For any $\mathcal{P} \subset GF(2^k)$ we denote with $\mathsf{poly}^{\mathcal{P}}$ the salted hash function with secret key part $s \leftarrow GF(2^k)$ and public salt $p \leftarrow \mathcal{P}$ which on input $M \in GF(2^k)^* \cong B_k^*$ is computed as*

$$\mathsf{poly}^{\mathcal{P}}_{s,p}(M) = f_{(p,M)}(s)$$

We will show (constructively, but "non-uniformly") that there is a "small" $\mathcal{P}$ such that the construction is $(\varepsilon_{mac}, \varepsilon_{forge})$-universal with $\varepsilon_{mac} = 2/2^k$. Namely, a random "small" $\mathcal{P}$ works with all but negligible probability (in particular, once such $\mathcal{P}$ is chosen *once*, it can be fixed *forever* and "hardwired" into the implementation).

Lemma 4. *For any $L \in \mathbb{N}$ and a random subset $\mathcal{P} \subset GF(2^k)$ of size $|\mathcal{P}| = \frac{k(L+2)L^2}{\log(e)}$, with probability $1 - 2^{-k}$ (over the choice of $\mathcal{P}$) the hash function $\mathsf{poly}^{\mathcal{P}}$ with domain $B_k^{\leq L}$ is $(\varepsilon_{forge}, \varepsilon_{mac})$-universal with*

$$\varepsilon_{forge} = L/2^k \tag{5}$$
$$\varepsilon_{mac} = 2/2^k \tag{6}$$

Proof. The bound (5) on ε_{forge} follows from Lemma 2 (and holds for any $\mathcal{P}$).

To prove the bound (6) on ε_{mac} we must show that a $\mathcal{P}$ chosen at random has the claimed property with probability $1 - 2^{-k}$. For a polynomial f over $GF(2^k)$ we denote with $z(f) = |\{x \in GF(2^k) : f(x) = 0\}|$ the number of zeros of f. Let f be any polynomial over $GF(2^k)$ of degree at most L and (for some m to be defined) let $\mathcal{P} = (p_1, \ldots, p_m)$ denote a subset of $GF(2^k)$ sampled uniformly at random (with repetition).

X_i denotes the random variable $z(f_i)$ where f_i is $f + p_i$ (i.e. f with p_i added to the constant coefficient). As p_i is random we have $\Pr[f_i(x) = 0] = 1/2^k$ for any $x \in GF(2^k)$ and thus

$$\mathsf{E}[X_i] = \sum_{x \in GF(2^k)} \Pr[f_i(x) = 0] = 1$$

and as any polynomial of degree L has at most L roots

$$0 \leq X_i \leq L$$

Let $S = X_1 + X_2 + \ldots + X_m$, we have $\mathsf{E}[S] = m$ and by the Hoeffding bound [12]

$$\Pr[S - \mathsf{E}[S] \geq m] = \Pr[S \geq 2m] \leq \exp\left(-\frac{m^2}{m \cdot L^2}\right) \tag{7}$$

which for $m = \frac{k(L+2)L^2}{\log(e)}$ is less than $2^{-k(L+2)}$. Taking the union bound over all $2^{k(L+1)}$ polynomials of degree $\leq L$, we get the probability that (7) is not satisfied for at least one of them is at most $2^{-k(L+2)} \cdot 2^{k(L+1)} = 2^{-k}$.

To conclude the proof we must still show that any $\mathcal{P}$ which satisfies (7) for all polynomials of degree $\leq L$ also satisfies (6). ε_{mac} is the maximum over $M \in GF(2^k)^{L+1}$ (with the first element from $\mathcal{P}$, but we will not use that) and $M' \in GF(2^k)^L$ of

$$\varepsilon_{mac} \geq \Pr_{s \leftarrow GF(2^k), p \leftarrow \mathcal{P}}[f_M(s) = f_{(p,M')}(s)]$$

Which for $f = f_{(0^k, M')} - f_M$ and $f_i = f + p_i$ we can write as

$$\Pr_{s \leftarrow GF(2^k), p \leftarrow \mathcal{P}}[f(s) + p = 0] = \sum_{i=1}^{m} \Pr[p = p_i] z(f_i)/2^k \leq 2/2^k$$

In the last step we used that we chose our $\mathcal{P}$ such that $\sum_{i=1}^{m} z(f_i) \leq 2m$ for all f and $\Pr[p = p_i] = 1/m$.   □

6   Cascade Construction

In his section we consider the Merkle-Damgård construction. Here a preprocessing which simply appends a few random bits to the message gives a salted hash function with good parameters.

For a function $\xi : \{0,1\}^k \times \{0,1\}^b \rightarrow \{0,1\}^k$ we denote with $\mathsf{MD}_\xi : B_b^* \rightarrow B_k$ the cascade (aka. Merkle-Damgård) construction based on ξ which on input $M = M_1 \| \dots \| M_m$, each $M_i \in B_b$, outputs X_m which is recursively defined as $X_0 = 0^k$, and $X_i = \xi(X_{i-1}, M_i)$.

Definition 9 (MDR). *For $k, b, L \in \mathbb{N}$ where $b \geq \log(L)$ we denote with $\mathsf{MDR}^L : B_b^{L-1} \rightarrow B_k$ the salted hash function whose secret key part is a uniformly random function $\xi : \{0,1\}^k \times \{0,1\}^b \rightarrow \{0,1\}^k$ and the public salt is $r \leftarrow \{0,1\}^{\lceil \log(L) \rceil}$. The randomized hash value on input $M \in B_b^*$ is computed as*

$$\mathsf{MDR}^L(M) = \mathsf{MD}_\xi(M \| \langle r \rangle_b)$$

Lemma 5. $\mathsf{MDR}^L : B_b^{L-1} \rightarrow B_k$ *is $(\varepsilon_{forge}, \varepsilon_{mac})$-universal with*

$$\varepsilon_{mac} = \frac{2}{2^k} + O(L^3/2^{2k}) \tag{8}$$

$$\varepsilon_{forge} = \frac{L}{2^k} + O(L^3/2^{2k}) \tag{9}$$

Proof. The proof follows almost directly form Propositions 1 and 2 from [8]. Proposition 2 from [8] states that for a random ξ and any $M \neq M' \in B_b^L$

$$\Pr_\xi[\mathsf{MD}_\xi(M) = \mathsf{MD}_\xi(M')] \leq L/2^k + O(L^3/2^{2k})$$

which directly gives the bound (9) on ε_{forge}. Proposition 1 from [8] states that if M and M' differ in the last b-bit block, then an even better bound

$$\Pr_\xi[\mathsf{MD}_\xi(M) = \mathsf{MD}_\xi(M')] \le 1/2^k + O(L^2/2^{2k})$$

applies. To bound ε_{mac} we can use that MDR adds a random last block $r \leftarrow \{0,1\}^{\lceil \log(L) \rceil}$ to the message, which then will be equivalent to the last block of the other message with prob. at most $\phi \le 1/L$ and we get

$$\varepsilon_{mac} \le (1 - \phi) \cdot \frac{1}{2^k} + \phi \cdot \frac{L}{2^k} + O(L^3/2^{2k}) \le \frac{2}{2^k} + O(L^3/2^{2k}). \qquad \square$$

7   A Generic Construction from ϵ-Δ Universal Hash Functions

In this section we consider hash functions which are not only ε-universal, but satisfy the stronger notion of ε-Δ universality.

Definition 10 (ε-Δ universal hash function [16]). *A hash function* $H : \mathcal{S} \times \mathcal{X} \to \mathcal{Y}$, *where* $\mathcal{Y}$ *is an additive Abelian group, is* ε-Δ *universal if for all* $x, y \in \mathcal{X}, x \ne y$ *and* $c \in \mathcal{Y}$

$$\Pr[s \leftarrow \mathcal{S}; h_s(x) - h_s(y) = c] \le \varepsilon$$

It is easy to see (and stated as Proposition 2 below) that adding a value chosen uniformly at random to the output of a ε-Δ universal hash function gives a $(\varepsilon_{forge}, \varepsilon_{mac})$-universal hash function with $\varepsilon_{forge} = \varepsilon$ and an optimal $\varepsilon_{mac} = 1/|\mathcal{Y}|$.

The main result of this section is a theorem which states that for every ε-Δ universal hash function, there always exists a randomized *post*processing, which only uses a logarithmic number of random bits and makes the hash function $(\varepsilon_{forge}, \varepsilon_{mac})$-universal where $\varepsilon_{forge} = \varepsilon$ and ε_{mac} is close to optimal. By postprocessing we mean that only the hash value of the message, but not the message itself, must be randomized.

Let us remark that the polynomial construction from Section 5 is $L/2^k$-Δ universal if the constant coefficient (i.e. the first message block) is fixed, say 0^k. With this observation Lemma 3 follows directly from Proposition 2, and Lemma 4 (with somewhat worse parameters) follows from Theorem 2 we prove below.

Definition 11 ($(\varepsilon_{forge}, \varepsilon_{mac})$-$\Delta$ universal hash function). *For an* ε-Δ *universal hash function* $H : \mathcal{S} \times \mathcal{X} \to \mathcal{Y}$ *and a set* $\mathcal{P} \subseteq \mathcal{Y}$ *we say that* $H^{\mathcal{P}}$ *is* $(\varepsilon_{forge}, \varepsilon_{mac})$-$\Delta$ *universal if for any* $p_1, p_2 \in \mathcal{P}, x_1, x_2 \in \mathcal{X}$ *where* $(p_1, x_1) \ne (p_2, x_2)$

$$\varepsilon_{forge} \ge \Pr[s \leftarrow \mathcal{S}; h_s(x_1) + p_1 = h_s(x_2) + p_2]$$

and

$$\varepsilon_{mac} \ge \Pr[s \leftarrow \mathcal{S}; p \leftarrow \mathcal{P}; h_s(x_1) + p_1 = h_s(x_2) + p \wedge (p_1, x_1) \ne (p, x_2)]$$

Proposition 2. *If $H : S \times X \to Y$ is ε-Δ universal, then H^Y is $(\varepsilon, 1/|Y|)$-Δ universal.*

Theorem 2. *If $H : S \times X \to Y$ is ε-Δ universal, then there exists a $\mathcal{P} \subset Y$ of size $m = |\mathcal{P}|$ such that*

$$m \leq \ln(|X|^2 \cdot m) \cdot |Y|^2 \cdot \varepsilon^2 \tag{10}$$

and $H^{\mathcal{P}}$ is $(\varepsilon, 2/|Y|)$-Δ universal.

Proof. The proof is by the probabilistic method. We show that a random subset $\mathcal{P} = \{p_1, \ldots, p_m\}$ of Y of size m which satisfies (10) has the claimed property with probability > 0 and thus exists.

For $i, j : 1 \leq i, j \leq m$ and $a, b \in X$ where $(i, a) \neq (j, b)$ let let $C_{i,j,a,b}$ denote the random variable (the probability is over the choice of $\mathcal{P}$)

$$C_{i,j,a,b} = \Pr[s \leftarrow S; h(a) + p_i = h(b) + p_j]$$

and for $(i, a) = (j, b)$ we set $C_{i,j,a,b} = 0$. Clearly for $(i, a) \neq (j, b)$

$$\mathsf{E}[C_{i,j,a,b}] = 1/|Y|$$

Now consider the random variable

$$C_{i,j,b} = \sum_{j \in X} C_{i,j,a,b}$$

We have $\mathsf{E}[C_{i,a,b}] \leq m/|Y|$ and for $a \neq b$ we get by the Hoeffding bound (see Appendix A) an upper bound for the probability that $C_{i,a,b}$ is more that twice its expected value (for $a = b$ the bound is even slightly better)

$$\Pr[C_{i,a,b} \geq 2 \cdot \mathsf{E}[C_{i,a,b}]] < \exp\left(-\frac{2 \cdot (m/|Y|)^2}{m \cdot \varepsilon^2}\right)$$

$$\leq \exp\left(-\frac{m}{|Y|^2 \cdot \varepsilon^2}\right)$$

$$\leq \frac{1}{|X|^2 \cdot m}$$

So there is a $\mathcal{P}$ such that $C_{i,a,b} \leq 2 \cdot \mathsf{E}[C_{i,a,b}] \leq 2 \cdot m/|Y|$ is satisfied for all $i \in [m]$ and $a, b \in X$. For this $\mathcal{P}$ we get

$$\varepsilon_{mac} = \max_{i \in [m], a, b \in X} \Pr[j \leftarrow [m]; s \leftarrow S; h_s(a) + p_i = h(b) + p_j \wedge ((p_i, a) \neq (p_j, b))]$$

$$= m^{-1} \max_{i \in [m], a, b \in X} \sum_{j \in [m]} C_{i,j,a,b}$$

$$= m^{-1} \max_{i \in [m], a, b \in X} C_{i,a,b}$$

$$\leq 2/|Y| \qquad \qquad \square$$

To get an intuition what eq. (10) means, assume we start with a hash function which maps L-bit strings to k-bit strings and which is $L^c/2^k$-Δ universal for some $c > 0$, so $|\mathcal{X}| = 2^L$ and $|\mathcal{Y}| = 2^k$. Now (10) means

$$m \leq \frac{(2 \cdot L + \log m) \cdot L^{2c}}{\log e}$$

or assuming $\log(m) \leq L$

$$\log(m) \leq \log 3 - \log e + (2c+1)\log L < 2 + (2c+1)\log L \qquad (11)$$

The assumption $\log(m) \leq L$ holds for all $L \geq 2 + (2c+1)\log L$, thus for such L also (11) is satisfied. So to sample from $\mathcal{P}$ we need $O(1) + (2c+1)\log L$ random bits.

8   Numerical Example

In this section we give a numerical example to illustrate how the classical upper bound $\varepsilon \cdot q_{mac}^2 + \varepsilon \cdot q_{forge}$ for the forging probability from Proposition 1 compares to the $\varepsilon_{mac} \cdot q_{mac}^2 + \varepsilon_{forge} \cdot q_{forge}$ upper bound (for salted hash then encrypt) given in Theorem 1. We will use polynomial hashing poly and its salted version $\mathsf{poly}^{\mathcal{P}}$ as considered in Section §5. We take a (standard) tag-length of $k := 128$ bits and the messages to be signed are of size 128MB (so each message has $L := 2^{23}$ blocks of size 128 bits). Now consider an adversary which can request $q_{mac} := 2^{40}$ MACs and make up to $q_{forge} := 2^{40}$ forgery attempts. For these parameters, poly is ε-universal with $\epsilon = (L-1)/2^{128} \approx 2^{-105}$, and the bound from Proposition 1 gives an upper bound on the forging probability for poly of

$$\varepsilon \cdot q_{mac}^2 + \varepsilon \cdot q_{forge} \approx \frac{2^{80}}{2^{105}} + \frac{2^{40}}{2^{105}} \approx 2^{-25}. \qquad (12)$$

For the salted version $\mathsf{poly}^{\mathcal{P}}$, which as shown in Lemma 4 has $\varepsilon_{forge} = L/2^{128} = 2^{-105}$ and $\varepsilon_{mac} = 2^{-127}$, we get with Theorem 1 an upper bound of

$$\varepsilon_{mac} \cdot q_{mac}^2 + \varepsilon_{forge} \cdot q_{forge} = \frac{2^{80}}{2^{127}} + \frac{2^{40}}{2^{105}} \approx 2^{-47}. \qquad (13)$$

So, compared to the classical bound (12), the security guarantee is better by a factor of $\approx 2^{22} \approx 4 \cdot 10^6$. Note that this gap is basically $L/2$, this will always be the case for (the most interesting) range of parameters where $\varepsilon_{mac} \cdot q_{mac}^2 \gg \varepsilon_{forge} \cdot q_{forge}$. The length of the random salt used in this construction (with these parameters) is $3\log(L) + \log(k) = 3 \cdot 23 + 7 = 76$ bits.

For the cascade construction MD and its salted version MDR as considered in Section §6 (with $b = k = 128$), we get the same bounds (12) and (13) respectively,[6] but the length of the random salt needed for this construction is considerably shorter, $\log(L) = 23$ bits compared to the 76 bits needed for $\mathsf{poly}^{\mathcal{P}}$.

---

[6] The reason is that the bounds on $\varepsilon_{forge}, \varepsilon_{mac}$ for MDR as given in Lemma 5 are identical to the bounds for $\mathsf{poly}^{\mathcal{P}}$ from Lemma 4 up to a $O(L^3/2^k)$ term, which for the parameters considered here will be irrelevant.

9   Related Work and Open Problems

Following the initial papers of Carter and Wegman [7,26], foundations of universal hash function-based authentication were laid by [23,16,19,24].

The analysis of the folklore polynomial construction is well known (see [5] for some history). The Merkle-Damgård functions was analyzed as an ε-universal hash function by [2,8], the latter proving a bound of $\varepsilon \approx L/2^k$ for hashing an Lb-block input using the compression function $\xi : \{0,1\}^k \times \{0,1\}^b \to \{0,1\}^k$ (modeled as a truly random function; here $L < 2^{k/2}$).

It is also interesting to compare the "hash then encrypt" approach we study here with a variant of this approach studied by [26,16], which actually led to the introduction of ε-Δ universal hash functions. Here one replaces a block cipher by a "fresh one-time pad". In modern terminology, the MAC has a tag of the form $(r, h_s(m) \oplus f_t(r))$, where f_t is a pseudorandom function with a new key t, and r is a fresh "nonce" which is not supposed to repeat. In practice, this means that r is either a counter, or a fresh random value. In the first case, we get *perfect* security (aside from the insecurity of f).[7] However, maintaining a counter introduces state, and stateful MACs are extremely inconvenient in many situations (see [21,15] for several good reasons). Correspondingly, to make a fair comparison we should only consider the case when r is a fresh random salt (in which case the MAC is indeed stateless, but r has to be part of the tag). To make such a comparison, let us fix the output of the hash function h_s to $\{0,1\}^k$, and replace the PRF by a truly random function. Then, we get that the length of the tag in the "XOR-scheme" is $|r| + k$, while its exact security is $\varepsilon + O(q^2/2^{|r|})$ (where the last term comes from the birthday bound measuring collisions on r). Thus, to get to the desired overall security of $O(q^2/2^k)$, this randomized MAC must use $|r| = \Omega(k)$ random bits and increase the tag length by this amount as well. On the other hand, we demonstrated that our randomized MACs can achieve the same level of security using only $|p| = O(\log L)$ bits of randomness, which is asymptotically smaller than k. In other words, although using the one-time pad has other advantages over using the block cipher,[8] it is *provably inferior* to our method for achieving $O(q^2/2^k)$ security (via *stateless* MACs).

We also mention that message preprocessing has been used previously in other contexts. For example, we already mentioned the work of Jaulmes et al. [15] on RMAC. Semanko [21] investigated the security of iterated MACs (like the cascade or CBC construction), which are randomized by prepending a random string. Here finding collisions does not necessarily imply a new forgery, but Semanko showed some non-obvious forgery attacks. In particular, even when prepending up to $k/2$ random bits one can find a forgery after $2^{k/2}$ queries (and not just a collision, which by the birthday bound is trivial). Let us stress that in

---

[7] Bernstein [4] investigates how much security one loses when f is a *permutation* (like AES).

[8] E.g., it can go below the $O(q^2/2^k)$ barrier, even when the hash function output is fixed to k.

our case, the length of random salt is small enough so that the above distinction between collisions and forgeries does not play any significant role.

Bellare et al. [2] also used randomness to improve the security of the cascade construction for a (stronger than ours) task of domain extension of a pseudorandom function (in which case there is no need to add the encryption step at the end). In particular, the message preprocessing used there is different from ours (prepend instead of append), and the exact security is weaker as well. The same paper also considers changing the cascade construction by appending a fixed but random secret string t to each message (instead of choosing a fresh public t per each message). However, this is done for a different purpose of achieving "prefix-freeness" of messages.

Recently, Halevi and Krawczyk [9] also proposed to use randomized message preprocessing in the design of signature schemes. Their goal is to lower the *computational* assumptions on the hash function used in signature schemes. In particular, they show that randomized versions of some signature schemes based on hash functions only require second-preimage resistance while the original scheme needed (the stronger) collision resistance. This works continues and optimizes the more general direction of replacing a fixed collision-resistant hash function in the "hash then sign" paradigm by a universal one-way hash function [13] chosen at random for each new message signed (and appended to the message before signing). Notice, while the main goal of [9] is to reduce the needed computational assumption on the hash function (and not the length of the salt or the actual exact security), our setting is information-theoretic with the primary goal of improving the exact security while simultaneously minimizing the salt length.

AMORTIZED COLLISION PROBABILITY. Another possibility to get better bounds on the exact security of the "hash then encrypt" paradigm is to consider "amortized" collision probability. For any q messages $x_1, \ldots, x_q$ and a ε-universal hash-function, the probability that $h(x_i) = h(x_j)$ for any $i \neq j$ is in $O(\varepsilon q^2)$. This $O(\varepsilon q^2)$ bound is proven by applying the union bound to all $\binom{q}{2}$ pairs of authenticated messages, which introduces some slackness: even if there are some pairs of messages with collision probability ε, it is not clear whether there exist $q \gg 2$ messages where each (or most) pairs collide with probability $\Omega(\varepsilon)$, and, moreover, the collision probabilities for distinct pairs are sufficiently independent.[9] Thus, it is possible that the actual collision probability is much less than $O(\varepsilon q^2)$. We know of one example where this has been proven to be the case, namely the CBC-function (based on uniformly random permutations). This function is ε-universal with $\varepsilon = \Theta(L^{1/\ln \ln L})$ [3], but using the amortized approach described above (assuming $q \geq L^2$ and $L \leq 2^{k/8}$) one can prove [18] an optimal $O(q^2/2^k)$ security bound, despite $\varepsilon = \omega(1/2^k)$.

Of course, this raises the question if the other constructions we consider also already achieve $O(q^2/2^k)$ security (for a non-trivial range of parameters) *without* randomization. As to the cascade construction, this is easily seen not to be

---

[9] To see why those probabilities must be independent, consider an h where either all or none of the messages collide, then the collision probability for all q messages is only $O(\varepsilon)$ and not $O(\varepsilon q^2)$.

the case, as there are q messages of length L where the collision probability is $\Theta(Lq^2/2^k)$. As to the polynomial construction, we do not know the answer, but conjecture that one cannot achieve the optimal security $O(q^2/2^k)$ (as we do with randomization).

References

1. Bellare, M., Canetti, R., Krawczyk, H.: Keying hash functions for message authentication. In: Koblitz, N. (ed.) CRYPTO 1996. LNCS, vol. 1109, pp. 1–15. Springer, Heidelberg (1996)
2. Bellare, M., Canetti, R., Krawczyk, H.: Pseudorandom functions revisited: The cascade construction and its concrete security. In: FOCS, pp. 514–523 (1996)
3. Bellare, M., Pietrzak, K., Rogaway, P.: Improved Security Analyses for CBC MACs. In: Shoup, V. (ed.) CRYPTO 2005. LNCS, vol. 3621, Springer, Heidelberg (2005)
4. Bernstein, D.J.: Stronger Security Bounds for Wegman-Carter-Shoup Authenticators. In: Cramer, R.J.F. (ed.) EUROCRYPT 2005. LNCS, vol. 3494, pp. 164–180. Springer, Heidelberg (2005)
5. Bernstein, D.J.: The Poly1305-AES Message-Authentication Code. In: Gilbert, H., Handschuh, H. (eds.) FSE 2005. LNCS, vol. 3557, pp. 32–49. Springer, Heidelberg (2005)
6. Integrity Primitives for Secure Information Systems, Final Report of RACE Integrity Primitives Evaluation RIPE-RACE 1040. In: Bosselaers, A., Preneel, B. (eds.) Integrity Primitives for Secure Information Systems. LNCS, vol. 1007, Springer, Heidelberg (1995)
7. Carter, L., Wegman, M.: Universal classes of hash functions. Journal of Computer and System Sciences (JCSS) 18, 143–154 (1979)
8. Dodis, Y., Gennaro, R., Håstad, J., Krawczyk, H., Rabin, T.: Randomness Extraction and Key Derivation Using the CBC, Cascade and HMAC Modes. In: Franklin, M. (ed.) CRYPTO 2004. LNCS, vol. 3152, pp. 494–510. Springer, Heidelberg (2004)
9. Halevi, S., Krawczyk, H.: Strengthening digital signatures via randomized hashing. In: Dwork, C. (ed.) CRYPTO 2006. LNCS, vol. 4117, Springer, Heidelberg (2006)
10. Hardy, G., Wright, E.: An Introduction to the Theory of Numbers. Oxford University Press, Oxford, UK (1980)
11. Hartman, T., Raz, R.: On the distribution of the number of roots of polynomials and explicit weak designs. Random Struct. Algorithms 23(3), 235–263 (2003)
12. Hoeffding, W.: Probability inequalities for sums of bounded random variables. J. Amer. Statist. Assoc. 58, 13–30 (1963)
13. Naor, M., Yung, M.: Universal one-way hash functions and their cryptographic applications. In: STOC, pp. 33–43 (1989)
14. Janson, S., Luczak, T., Rucinski, A.: Random Graphs. Wiley, Chichester (2000)
15. Jaulmes, É., Joux, A., Valette, F.: On the Security of Randomized CBC-MAC Beyond the Birthday Paradox Limit: A New Construction. In: FSE, pp. 237–251 (2002)
16. Krawczyk, H.: LFSR-Based Hashing and Authentication. In: Desmedt, Y.G. (ed.) CRYPTO 1994. LNCS, vol. 839, pp. 129–139. Springer, Heidelberg (1994)
17. Petrank, E., Rackoff, C.: CBC MAC for real-time data sources. Journal of Cryptology, 315–338 (2000)
18. Pietrzak, K.: A tight bound for EMAC. In: Bugliesi, M., Preneel, B., Sassone, V., Wegener, I. (eds.) ICALP 2006. LNCS, vol. 4052, pp. 168–179. Springer, Heidelberg (2006)

19. Rogaway, P.: Bucket Hashing and Its Application to Fast Message Authentication. In: Coppersmith, D. (ed.) CRYPTO 1995. LNCS, vol. 963, pp. 29–42. Springer, Heidelberg (1995)
20. Schwartz, J.T.: Fast probabilistic algorithms for verification of polynomial identities. J. ACM 27(4), 701–717 (1980)
21. Semanko, M.: L-collision Attacks against Randomized MACs. In: Bellare, M. (ed.) CRYPTO 2000. LNCS, vol. 1880, pp. 216–228. Springer, Heidelberg (2000)
22. Shoup, V.: A Computational Introduction to Number Theory and Algebra. Cambridge University Press, Cambridge (2005)
23. Stinson, D.R.: Universal Hashing and Authentication Codes. Designs, Codes and Cryptography 4, 369–380 (1994)
24. Stinson, D.R.: On the connections between universal hashing, combinatorial designs and error-correcting codes. Congressus Numerantium 114, 7–27 (1996)
25. Stinson, D.R.: Personal Communication (2005)
26. Wegman, M.N., Carter, L.: New hash functions and their use in authentication and set equality. Journal of Computer and System Sciences (JCSS) 22(3), 265–279 (1981)

A Hoeffding Bound

Let $X_1, \ldots, X_n$ be independent random variables. Further assume the X_i are bounded, i.e. for each $i = 1, \ldots, n$ there are a_i, b_i such that

$$\Pr[a_i \leq X_i \leq b_i] = 1$$

then for the sum $S = X_1 + \ldots X_n$ we have for any $t \geq 0$

$$\Pr[S - \mathsf{E}[S] \geq nt] \leq \exp\left(-\frac{2n^2 t^2}{\sum_{i=1}^n (b_i - a_i)^2}\right)$$

We will only use the case where the X_i are identically distributed such that $0 \leq X_i \leq \varepsilon$ and are only interested in the probability that $S \geq 2\mathsf{E}[S]$, so if γ is an upper bound on $\mathsf{E}[S]$ we use

$$\Pr[S \geq 2 \cdot \mathsf{E}[S]] \; \leq \; \Pr[S - \mathsf{E}[S] \geq \gamma] \; \leq \; \exp\left(-\frac{2 \cdot \gamma}{n \cdot \varepsilon^2}\right) \tag{14}$$

In our applications the X_i's are not completely independent, but chosen at random from some finite set *without repetition*. Fortunately the Hoeffding bound also applies in this case as proven in section 6 of the original paper [12] (see also Theorem 2.10 of [14]).

New Bounds for PMAC, TMAC, and XCBC

Kazuhiko Minematsu[1,2] and Toshiyasu Matsushima[2]

[1] NEC Corporation, 1753 Shimonumabe, Nakahara-Ku, Kawasaki, Japan
k-minematsu@ah.jp.nec.com
[2] Waseda University, 3-4-1 Okubo Shinjuku-ku Tokyo, Japan

Abstract. We provide new security proofs for PMAC, TMAC, and XCBC message authentication modes. The previous security bounds for these modes were $\sigma^2/2^n$, where n is the block size in bits and σ is the total number of queried message blocks. Our new bounds are $\ell q^2/2^n$ for PMAC and $\ell q^2/2^n + \ell^4 q^2/2^{2n}$ for TMAC and XCBC, where q is the number of queries and ℓ is the maximum message length in n-bit blocks. This improves the previous results under most practical cases, e.g., when no message is exceptionally long compared to other messages.

1 Introduction

Message authentication code (MAC) is a symmetric-keyed function used for ensuring the authenticity of messages. Many studies have been done on MACs built using blockciphers (i.e., MAC modes of operation) including the CBC-MAC and its variants. The theoretical security of stateless (i.e., no counter or nonce is used) MAC mode $F[E_K]$ using blockcipher E_K can be measured using the maximum advantage of an adversary trying to distinguish $F[E_K]$ from the random oracle, which provides an independent and uniform output for any distinct input, using a chosen-plaintext attack (CPA). Typically, the key task in proving the maximum advantage is to prove the maximum information-theoretic (IT) advantage for the target MAC, where the adversary has infinite computational power and the MAC is built using the uniform random permutation (URP), which is the ideal functionality of a blockcipher. Improving the maximum IT-advantage is important, because it will contribute to better understanding of the target function and to expanding the scope of application.

Bellare, Pietrzak, and Rogaway [5] analyzed the IT-advantage for the CBC-MAC and the encrypted CBC-MAC called EMAC [2]. Neglecting constants, the previous EMAC bound using the n-bit URP was $\ell^2 q^2/2^n$ [7] for any (q, ℓ)-CPA, which uses q chosen messages with lengths less than $n\ell$ bits. Bellare et al. investigated whether this could be improved, particularly with respect to ℓ. They proved the improved bound $\mathsf{d}(\ell)q^2/2^n + \ell^4 q^2/2^{2n}$, where $\mathsf{d}(\ell)$ is a function that grows very slowly with ℓ (see Sect. 4). A similar result was obtained for CBC-MAC for prefix-free messages. Recently, Pietrzak [18] proved EMAC bound $q^2/2^n$ for a range of (q, ℓ) (in fact, the result was $q^2/2^n + \ell^8 q^2/2^{2n}$ for any $q \geq \ell^2$).

Given these findings, it is quite natural to ask if similar improvements can be obtained for modes other than EMAC, especially more sophisticated ones.

A. Biryukov (Ed.): FSE 2007, LNCS 4593, pp. 434–451, 2007.

EMAC uses two blockcipher keys, and only messages with a length multiple of n are supported. In this paper, we describe several MAC modes and provide new security bounds for them. Our first target is PMAC, which was proposed by Black and Rogaway [7] and Rogaway [19]. It is a one-key MAC; i.e., the MAC function uses one blockcipher key, and messages of any lengths are supported, and is fully parallelizable. The original security bound was $\sigma^2/2^n$ [7,19], where σ is the total number of queried message blocks, which immediately implies $\ell^2 q^2/2^n$ for any (q, ℓ)-CPA, as $\sigma \leq \ell q$ holds. Here, we demonstrate a new bound $\ell q^2/2^n$ by taking an approach different from that of the previous proof.

We also provide new bounds for two successors of EMAC called TMAC [13] and XCBC [7]. Like EMAC, they are based on CBC-MAC. However, they do not use two blockcipher keys, and can efficiently handle messages of arbitrary length. Our bounds are obtained by combining our PMAC proof technique and the CBC-MAC collision analysis provided by Bellare et al.[5]. For TMAC and XCBC, the previous bounds are $\sigma^2/2^n$ shown by Iwata and Kurosawa [10], and our bound is $\ell q^2/2^n + \ell^4 q^2/2^{2n}$. We also investigated OMAC [11] (i.e., CMAC [1]), which is an optimized version of TMAC and XCBC. Although some part of TMAC proof can also be applied to OMAC, we could not obtain a new bound at this moment. The analysis of OMAC is briefly described in Sect. 5.

We have to emphasize that our results are not always better than the previous ones. Since all of our targets have $\sigma^2/2^n$ bounds, ours are worse if message length distribution is heavily biased to the left, e.g., one ℓ-block message and $(q - 1)$ one-block messages. For other cases, ours are better. A detailed comparison is given in Sect. 5.

2 Preliminaries

NOTATION. $\{0,1\}$ and $\{0,1\}^n$ are denoted by Σ and Σ^n. The set of i-bit sequences for all $i = 1, \ldots, n$ is denoted by $\Sigma^{\leq n} \overset{\text{def}}{=} \bigcup_{i=1,\ldots n} \Sigma^i$. $(\Sigma^n)^{\leq m}$ is the set of binary sequences with lengths that are a multiple of n and at most nm. Σ^* is the set of all finite-length bit sequences. The bit length of x is denoted by $|x|$.

The n-bit uniform random permutation (URP), denoted by P_n, is a random permutation with a uniform distribution over all permutations on Σ^n. The random oracle (RO), which has n-bit output and is denoted by O_n, is a random function that accepts any $x \in \Sigma^*$ and outputs independent and uniformly random n-bit values for any distinct inputs. For any two colliding inputs, RO outputs the same value.

FIELD WITH 2^n POINTS. We consider the elements of field $\mathrm{GF}(2^n)$ as n-bit coefficient vectors of the polynomials in the field. We represent n-bit coefficient vectors by integers $0, 1, \ldots, 2^n - 1$, e.g., 2 corresponds to coefficient vector $(00\ldots010)$, which corresponds to x in the polynomial representation, and 3 denotes $(00\ldots011)$, which corresponds to $\mathsf{x}+1$. For any $x, y \in \Sigma^n$, xy denotes the field multiplication of two elements represented by x and y. For simplicity, we assume $n = 128$ throughout the paper.

SECURITY NOTIONS. We used the standard security notion for symmetric cryptography [3,4,9].

Definition 1. *Let F and G be two random (here, random means it is probabilistic) functions. The oracle has implemented H, which is identical to one of F or G. An adversary, A, guesses if H is F or G using a θ-chosen-plaintext attack (θ-CPA), where θ is a list of parameters, such as the number of queries. The maximum advantage in distinguishing F from G is defined as*

$$\text{Adv}_{F,G}^{\text{cpa}}(\theta) \overset{\text{def}}{=} \max_{A:\theta\text{-CPA}} \left| \Pr[A^F = 1] - \Pr[A^G = 1] \right|,$$

where $A^F = 1$ denotes that A's guess is 1, which indicates one of F or G. The probabilities are determined by the randomness of F or G and A.

THE GOAL OF OUR ANALYSIS. In this paper, we consider only the information-theoretic security, where the adversary has infinite computational power (thus θ contains no computational restrictions), and the target is realized by the ideal n-bit blockcipher, i.e., P_n. In many cases, including ours, once the information-theoretic security is proved, the computational counterpart, where the adversary is computationally restricted and a real blockcipher is used, is quite easy.

Our target modes are stateless and variable-input-length (VIL) functions with n-bit output (VIL means that the domain is Σ^*). Therefore, for mode $F[E_K]$, where E_K is a blockcipher, we evaluate $\text{Adv}_{F[\mathsf{P}_n]}^{\text{vilqrf}}(\theta) \overset{\text{def}}{=} \text{Adv}_{F[\mathsf{P}_n],O_n}^{\text{cpa}}(\theta)$. vilqrf denotes a VIL quasi-random function [15] that cannot be information-theoretically distinguished from RO without a negligibly small success probability. If $\text{Adv}_{F[\mathsf{P}_n]}^{\text{vilqrf}}(\theta)$ is small, the maximum success probability of a MAC forgery for all θ-CPAs against $F[\mathsf{P}_n]$ is also small (e.g., see Proposition 2.7 of [3]). In this paper, θ contains one of two additional parameters in addition to the number of queries, q: the total number of n-bit blocks for all q queries, σ, and the maximum length of a query (in n-bit blocks), ℓ. We focus on the $\theta = (q, \ell)$ case.

3   PMAC

3.1   Description and Previous Security Proof

PMAC has two versions; we focus on the later version [12,19]. We call it simply "PMAC". The main idea of PMAC is as follows.

Lemma 1. *(Proposition 5 of [19]) Assume that the representation of $\text{GF}(2^n)$ ($n = 128$) is based on the lexicographically first primitive polynomial (see [19] for details). Let $\mathbb{I} = \{1, .., 2^{n/2}\}$ and $\mathbb{J} = \{0, 1, 2\}$ be the set of integers used as indices for distinct elements ("bases") of $\text{GF}(2^n)$. Then, for any $(\alpha, \beta), (\alpha', \beta') \in \mathbb{I} \times \mathbb{J}$, $2^\alpha 3^\beta \neq 2^{\alpha'} 3^{\beta'}$ holds if $(\alpha, \beta) \neq (\alpha', \beta')$ holds[1].*

---

[1] Actually, Proposition 5 of [19] proved this for a wider range of indices. The original PMAC uses $\mathbb{J} = \{2, 3, 4\}$ instead of $\{0, 1, 2\}$, so our PMAC definition is slightly different from the original. However, the security proofs are essentially the same.

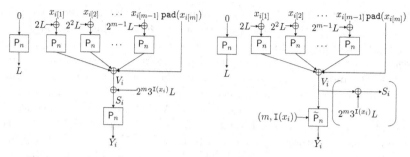

Fig. 1. PMAC[P_n] (left) and the modified PMAC (MPMAC) (right)

Multiplication of a constant and a variable is generally much simpler than multiplication of variables. For example, multiplication with 2 (i.e., a doubling operation) requires a bit shift followed by a conditional xor. Therefore, computation of $2^\alpha 3^\beta$ from $2^{\alpha-1}3^\beta$ or $2^\alpha 3^{\beta-1}$ is significantly faster than one blockcipher invocation. The idea of PMAC [19], called the "powering-up construction", is to use $2^\alpha 3^\beta$ as a masking value for every blockcipher input, incrementing α or β.

For blockcipher E_K, PMAC is defined as follows. We say "partition $x \in \Sigma^*$ into $(x_{[1]}, \ldots, x_{[m]})$" to set $m = \|x\|_n \stackrel{\text{def}}{=} \max\{\lceil |x|/n \rceil, 1\}$ and $x = (x_{[1]}, \ldots, x_{[m]})$ with $x_{[i]} \in \Sigma^n$ for $i = 1, \ldots, m-1$ and $x_{[m]} \in \Sigma^{\leq n}$. First, compute $L = E_K(0)$, where 0 corresponds to the all-zero n-bit sequence, in the preprocessing. Then, for input $x \in \Sigma^*$, partition it into $(x_{[1]}, \ldots, x_{[m]})$. The tag for x is $Y = E_K(\mathsf{Psum} \oplus \mathsf{pad}(x_{[m]}) \oplus 2^m 3^{\mathtt{I}(x)})$, where $\mathsf{Psum} = \bigoplus_{\alpha=1}^{m-1} E_K(x_{[\alpha]} \oplus 2^\alpha L)$ if $m > 1$, and if $m = 1$, $\mathsf{Psum} = 0$. Here, $\mathtt{I}(x) = 1$ if $|x|$ is a multiple of n and $\mathtt{I}(x) = 2$ otherwise, and $\mathsf{pad}(x_{[m]}) = x_{[m]}$ if $|x_{[m]}| = n$ and $\mathsf{pad}(x_{[m]}) = x_{[m]}\|10^*$ otherwise, where $x_{[m]}\|10^*$ is a concatenation of $x_{[m]}$ and the $(n - |x_{[m]}|)$-bit sequence $(100\ldots 0)$.

Rogaway [19] proved the security of PMAC, which is as follows.

Theorem 1. *(Corollary 17 of [19]) Let* PMAC[P_n] *be the PMAC using* P_n *(see the left of Fig. 1). We then have* $\mathrm{Adv}^{\mathtt{vilqrf}}_{\mathrm{PMAC}[P_n]}(q, \sigma) \leq 5.5\sigma^2/2^n$ *and* $\mathrm{Adv}^{\mathtt{vilqrf}}_{\mathrm{PMAC}[P_n]}(q, \ell) \leq 5.5\ell^2 q^2/2^n$, *where* q, σ, *and* ℓ *are as defined in Sect. 2.*

Corollary 17 of [19] only proved the first claim. The second follows from the first and $\sigma \leq \ell q$.

3.2 New Security Bound for PMAC

Our security bound of PMAC is the following. The proof will be provided later.

Theorem 2. *Let* PMAC[P_n] *be the PMAC using* P_n. *We then have*

$$\mathrm{Adv}^{\mathtt{vilqrf}}_{\mathrm{PMAC}[P_n]}(q, \ell) \leq \frac{5\ell q^2}{2^n - 2\ell}.$$

From this theorem, we have $\mathrm{Adv}^{\mathtt{vilqrf}}_{\mathrm{PMAC}[P_n]}(q, \ell) \leq 10\ell q^2/2^n$ if $\ell \leq 2^{n-2}$.

Notation for Proof. Since we use Maurer's methodology[2] (e.g., see [15]) to make our proofs intuitive and simple, we briefly describe his notation. For completeness, part of his results that we used for our proof is cited in Appendix A. Consider event a_i defined for i input/output pairs, and possibly some internal variables, of random function F. Let $\overline{a_i}$ be the negation of a_i. We assume a_i is monotone; i.e., a_i never occurs if $\overline{a_{i-1}}$ occurs. For instance, a_i is monotone if it indicates that all i outputs are distinct. An infinite sequence of monotone events, $\mathcal{A} = a_0 a_1 \ldots$, is called a *monotone event sequence* (MES) [15]. Here, a_0 denotes some tautological event. Note that $\mathcal{A} \wedge \mathcal{B} = (a_0 \wedge b_0)(a_1 \wedge b_1) \ldots$ is an MES if $\mathcal{A} = a_0 a_1 \ldots$ and $\mathcal{B} = b_0 b_1 \ldots$ are both MESs. For any sequence of variables, $X_1, X_2, \ldots$, let X^i denote $(X_1, \ldots, X_i)$. We use $\mathrm{dist}(X^i)$ (or, equivalently, $\mathrm{dist}(\mathbf{X}^{(i)})$, where $\mathbf{X}^{(i)}$ is set $\{X_j\}_{j=1,\ldots,i}$) to denote an event where $X_1, X_2, \ldots, X_i$ are distinct.

Let MESs $\mathcal{A}$ and $\mathcal{B}$ be defined for two random functions, F and G, respectively. Let X_i and Y_i be the i-th input and output. Let P^F be the probability space defined by F. For example, $P^F_{Y_i|X^i Y^{i-1}}(y^i, x^i)$ means $\Pr[Y_i = y_i | X^i = x^i, Y^{i-1} = y^{i-1}]$, where $Y_j = F(X_j)$ for $j \geq 1$.

Definition 2. *Let θ contain q. For MES $\mathcal{A}$ defined for F, $\nu_\theta(F, \overline{a_q})$ denotes the maximum probability of $\overline{a_q}$ for any θ-CPA that interacts with F. Similarly, $\mu_\theta(F, \overline{a_q})$ denotes the maximum probability of $\overline{a_q}$ for any non-adaptive θ-CPA. For $\theta = (q, \ell)$, they are abbreviated to $\nu_\ell(F, \overline{a_q})$ and $\mu_\ell(F, \overline{a_q})$. If $\theta = q$, the subscript is omitted, e.g., we write $\nu(F, \overline{a_q})$.*

Here, $\mu_\theta(F, \overline{a_q})$ can be rewritten as $\max_{x^q} P^F_{\overline{a_q}|X^q}(x^q)$, where the maximum is taken for all (non-adaptively chosen) x^q satisfying θ (e.g., if $\theta = (q, \ell)$, $|x_i| \leq n\ell$ for all $i \leq q$), hereafter abbreviated to $\max_{X^q} P^F_{\overline{a_q}|X^q}$.

Analysis of PHASH. Proving Theorem 2 requires an analysis of the message-hashing part of PMAC$[\mathsf{P}_n]$, which we call PHASH. For $x = (x_{[1]}, \ldots, x_{[m]}) \in (\Sigma^n)^m$, it is defined as:

$$\mathrm{PHASH}(x) \overset{\text{def}}{=} \bigoplus_{i=1,\ldots,m} \mathsf{P}_n(x_{[i]} \oplus 2^i L), \text{ where } L = \mathsf{P}_n(0) .$$

Lemma 2. *For any $x = (x_{[1]}, \ldots, x_{[m]}) \in (\Sigma^n)^m$ and $x' = (x'_{[1]}, \ldots, x'_{[m']}) \in (\Sigma^n)^{m'}$, $x \neq x'$, and for any $f : \Sigma^n \to \Sigma^n$, we have*

$$\Pr[\mathrm{PHASH}(x) \oplus \mathrm{PHASH}(x') = f(L)] \leq \frac{m + m'}{2^n} + \frac{1}{2^n - (m + m')}, \text{ and} \quad (1)$$

$$\Pr[\mathrm{PHASH}(x) = f(L)] \leq \frac{m}{2^n} + \frac{1}{2^n - m}, \text{ where } L = \mathsf{P}_n(0) . \quad (2)$$

---

[2] It is known that some information-theoretic results obtained by Maurer's methodology can not be converted into computational ones (for instance, see [16,17]). However, we do not encounter such difficulties in this paper.

Proof. We only prove Eq. (1) as Eq. (2) can be similarly proved. Fix x and x'. Let $U_i = x_{[i]} \oplus 2^i L$ for $i = 1, \ldots, m$, and $U_i = x'_{[i-m]} \oplus 2^{i-m} L$ for $i = m+1, \ldots, m+m'$. Then, $\text{PHASH}(x) \oplus \text{PHASH}(x')$ equals $\text{Sum} \stackrel{\text{def}}{=} P_n(U_1) \oplus \ldots \oplus P_n(U_{m+m'})$. Let $\mathbf{U} = \{U_1, \ldots, U_{m+m'}\} \setminus \mathbf{U}_{\text{coll}}$, where $\mathbf{U}_{\text{coll}}$ is the set of all *trivial collisions*, e.g., U_1 and U_{1+m} when $x_{[1]} = x'_{[1]}$. Note that $\mathbf{U}$ can not be the empty set as $x \neq x'$. For simplicity, we assume no trivial collision (thus $\mathbf{U} = \{U_1, \ldots, U_{m+m'}\}$), however the following analysis works even if some trivial collisions exist. For index subset $\{i_1, \ldots, i_k\} \subseteq \{1, \ldots, q\}$, $\mathbf{U}_{\text{sub}} = \{U_{i_j}\}_{j=1,\ldots,k}$ is an *equivalent set* if $U_{i_1} = U_{i_2} = \cdots = U_{i_k}$ and $U_{i_1} \neq U_h$ for all $h \notin \{i_1, \ldots, i_k\}$. Here, the sum of all equivalent sets is a decomposition of $\mathbf{U}$. Whether $\mathbf{U}_{\text{sub}}$ is an equivalent set or not depends on the value of L. If k is odd (even), we say $\mathbf{U}_{\text{sub}}$ is an odd (even) equivalent set. Let odd_k be the event such that there are k odd equivalent sets having non-zero values (the value of an equivalent set is the value of its members). We have

$$
\begin{aligned}
&\Pr[\text{Sum} = f(L) | \text{odd}_k] \\
&\leq \max_{c \text{ satisfies } \text{odd}_k} \Pr[P_n(u_1(c)) \oplus \ldots \oplus P_n(u_{m+m'}(c)) = f(c) | \text{odd}_k, L = c], \quad (3)
\end{aligned}
$$

where $u_i(c)$ is $x_{[i]} \oplus 2^i c$ for $i = 1, \ldots, m$ and $x'_{[i-m]} \oplus 2^{i-m} c$ for $i = m+1, \ldots, m+m'$. In Eq. (3), note that $P_n(u_i(c))$ is canceled out if $u_i(c)$ is in an even equivalent set. Therefore, given $L = c$ and odd_k for some $k > 0$, $P_n(u_1(c)) \oplus \ldots \oplus P_n(u_{m+m'}(c))$ is either the sum of k URP outputs for k non-zero distinct inputs or the sum of c and k URP outputs for k non-zero distinct inputs (note that odd_k does not exclude an odd equivalent set with value 0). Then, the property of P_n shows that, for any non-zero distinct k inputs, $z_1, \ldots, z_k$,

$$
\begin{aligned}
&\Pr[P_n(z_1) \oplus P_n(z_2) \oplus \ldots \oplus P_n(z_k) = f(c) | P_n(0) = c] \\
&= \sum_{c_1,\ldots,c_k, \text{dist}(\{c_1,\ldots,c_k,c\}), c_1 \oplus \ldots \oplus c_k = f(c)} \Pr[P_n(z_1) = c_1, \ldots, P_n(z_k) = c_k | P_n(0) = c] \\
&= \frac{|\{(c_1, \ldots, c_k) : \text{dist}(\{c_1, \ldots, c_k, c\}), c_1 \oplus \ldots \oplus c_k = f(c)\}|}{(2^n - 1) \cdot \ldots \cdot (2^n - k)} \leq \frac{1}{2^n - k} \quad (4)
\end{aligned}
$$

holds, where the inequality holds since c_k is uniquely determined (or does not exist) if $c_1, \ldots, c_{k-1}$ are fixed. From Eqs. (3) and (4), we obtain

$$
\Pr[\text{Sum} = f(L) | \text{odd}_k] \leq \frac{1}{2^n - k} \text{ for any } 0 < k \leq m + m' \text{ and for any } f. \quad (5)
$$

Next, we analyze $\Pr[\text{odd}_0]$. We have

$$
\begin{aligned}
\Pr[\text{odd}_0] &= \Pr[\text{odd}_0, U_1 \notin \{U_2, \ldots, U_{m+m'}\}] \\
&\quad + \Pr[\text{odd}_0, U_1 = U_j \text{ for some } j = 2, \ldots, m + m'] \\
&\leq \Pr[U_1 = 0] + \sum_{j=2,\ldots,m+m'} \Pr[U_1 = U_j] \leq (m + m') \frac{1}{2^n}, \quad (6)
\end{aligned}
$$

where the first inequality holds since if U_1 is unique (i.e., U_1 is in an odd equivalent set) and odd_0 holds, U_1 must be 0. The second holds since both U_1 and $U_1 \oplus U_j$ for any $j \neq 1$ are permutations of L from Lemma 1. From Eqs. (5) and (6), we obtain

$$\Pr[\mathsf{Sum} = f(L)] = \sum_{k=0,\ldots,m+m'} \Pr[\mathsf{Sum} = f(L)|\mathsf{odd}_k] \cdot \Pr[\mathsf{odd}_k]$$

$$\leq \Pr[\mathsf{odd}_0] + \sum_{k=1}^{m+m'-1} \frac{\Pr[\mathsf{odd}_k]}{2^n - k} + \frac{1 - \sum_{k=0}^{m+m'-1} \Pr[\mathsf{odd}_k]}{2^n - (m+m')}$$

$$\leq \Pr[\mathsf{odd}_0] + \frac{1}{2^n - (m+m')} \leq \frac{m+m'}{2^n} + \frac{1}{2^n - (m+m')}.$$

This concludes the proof of Lemma 2.

Proof of Theorem 2. First, we introduce the tweakable[14] n-bit URP, $\widetilde{\mathsf{P}}_n$. It has tweak space $\mathcal{T} = \mathbb{I} \times \mathbb{J}'$, where $\mathbb{I} = \{1, \ldots, 2^{n/2}\}$ and $\mathbb{J}' = \{1, 2\}$. It consists of $|\mathcal{T}|$ independent n-bit URPs; $\widetilde{\mathsf{P}}_n(t, x)$ is the output of an n-bit URP indexed by $t \in \mathcal{T}$ and having input $x \in \Sigma^n$. Using $\widetilde{\mathsf{P}}_n$ and P_n, independent of $\widetilde{\mathsf{P}}_n$, we define the modified PMAC (MPMAC) as follows. First, compute $L = \mathsf{P}_n(0)$. For input $x \in \Sigma^* = (x_{[1]}, \ldots, x_{[m]})$, compute Psum using PHASH (i.e., $\mathsf{Psum} = \mathsf{PHASH}(\widehat{x})$, where $\widehat{x} = (x_{[1]}, \ldots, x_{[m-1]})$, if $m > 1$, and $\mathsf{Psum} = 0$ otherwise). The tag is $Y = \widetilde{\mathsf{P}}_n((m, \mathtt{I}(x)), \mathsf{Psum} \oplus \mathsf{pad}(x_{[m]}))$. Here, $(m, \mathtt{I}(x))$ is the tweak. Note that a tweak is a function of x.

PROOF IDEA. Since the advantage is the absolute difference between two probabilities, we can use a triangle inequality, $\mathsf{Adv}^{\mathsf{cpa}}_{\mathsf{PMAC}[\mathsf{P}_n], \mathsf{O}_n}(\theta)$ is not larger than $\mathsf{Adv}^{\mathsf{cpa}}_{\mathsf{PMAC}[\mathsf{P}_n], H}(\theta) + \mathsf{Adv}^{\mathsf{cpa}}_{H, \mathsf{O}_n}(\theta)$ for any VIL function H, and for any θ. Here, H is an intermediate function. Theorem 1 was derived using "PMAC with an ideal tweakable blockcipher", which invokes an independent URP for each message block in the message-hashing part as well as in the finalization, as the intermediate function. Here, our proof uses MPMAC as the intermediate function.

We start by proving the advantage between $\mathsf{PMAC}[\mathsf{P}_n]$ and MPMAC, which requires defining some random variables. Let $X_i \in \Sigma^*$ be the i-th query of the adversary. If $m = \|X_i\|_n$, we write $X_i = (X_{i[1]}, \ldots, X_{i[m]})$. Note that X_i is a random variable, and its distribution is determined by the adversary and the target MAC. Fixed queries (and other random variables) are written in lower case, e.g., $x_i = (x_{i[1]}, \ldots, x_{i[m]})$. For $\mathsf{PMAC}[\mathsf{P}_n]$, let $\mathbf{M}^{(q)}$ ($\mathbf{C}^{(q)}$) be the set of inputs (outputs) to P_n generated in the PHASH for all q queries. We do not include the result of preprocessing, i.e., $L = \mathsf{P}_n(0)$, in $\mathbf{M}^{(q)}$ and $\mathbf{C}^{(q)}$. We also define $Y_i \in \Sigma^n$ as the i-th tag, and $\mathbf{Y}^{(q)} \stackrel{\text{def}}{=} \{Y_i\}_{i=1,\ldots,q}$. If $m = \|X_i\|_n > 1$, we define V_i as the XOR of the i-th PHASH output and $\mathsf{pad}(X_{i[m]})$. If $m = 1$, $V_i = \mathsf{pad}(X_i)$. Moreover, $S_i \stackrel{\text{def}}{=} V_i \oplus 2^m 3^{\mathtt{I}(X_{i[m]})} L$, and $\mathbf{S}^{(q)} \stackrel{\text{def}}{=} \{S_i\}_{i=1,\ldots,q}$. Thus, in $\mathsf{PMAC}[\mathsf{P}_n]$, $Y_i = \mathsf{P}_n(S_i)$. These variables are similarly defined for MPMAC except Y_i; in MPMAC, Y_i is $\widetilde{\mathsf{P}}_n((m, \mathtt{I}(X_{i[m]})), V_i)$ when the i-th query has m blocks. Also, S_i is defined as a dummy variable in MPMAC. See Fig. 1 for reference.

Lemma 3. *Let event* $a_q \overset{\text{def}}{=} [\mathbf{M}^{(q)} \cap \mathbf{S}^{(q)} = \emptyset] \wedge [\text{dist}(\mathbf{S}^{(q)})] \wedge [0 \notin \mathbf{M}^{(q)} \cup \mathbf{S}^{(q)}].$
Moreover, $b_q \overset{\text{def}}{=} [\text{dist}(\mathbf{Y}^{(q)})]$, $d_q \overset{\text{def}}{=} [\mathbf{C}^{(q)} \cap \mathbf{Y}^{(q)} = \emptyset]$, *and* $e_q \overset{\text{def}}{=} [L \notin \mathbf{Y}^{(q)}]$, *where*
$L = \mathsf{P}_n(0)$. *We then have*

$$\text{Adv}^{\text{cpa}}_{\text{PMAC}[\mathsf{P}_n],\text{MPMAC}}(q,\ell) \le \nu_\ell(\text{MPMAC}, \overline{a_q \wedge b_q \wedge d_q \wedge e_q})$$

$$\le \nu_\ell(\text{MPMAC}, \overline{a_q \wedge b_q}) + \nu_\ell(\text{MPMAC}, \overline{d_q \wedge e_q})$$

$$\le \frac{1}{2^n - 2\ell} \left((4\ell - 2.5)q^2 + 1.5q \right). \tag{7}$$

Proof. (of Lemma 3) The first and second inequalities are derived from Maurer's methodology. See Appendix B for the proof. In the following, we prove the third. ANALYSIS FOR $\nu_\ell(\text{MPMAC}, \overline{a_q \wedge b_q})$. We use the following lemma. The proof is in Appendix C.

Lemma 4

$$\nu_\ell(\text{MPMAC}, \overline{a_q \wedge b_q}) = \mu_\ell(\text{MPMAC}, \overline{a_q \wedge b_q}) = \max_{X^q} P^{\text{MPMAC}}_{\overline{a_q}|X^q} + \max_{X^q} P^{\text{MPMAC}}_{\overline{b_q}|a_q X^q},$$

where the maximums are taken for all $X^q = x^q$ *with* $|x_i| \le n\ell$ *for all* i.

Note that $\max_{X^q} P^{\text{MPMAC}}_{\overline{a_q}|X^q}$ denotes $\max_{x^q} P^{\text{MPMAC}}_{\overline{a_q}|X^q}(x^q)$. Let $\mathbf{M}_i$ denote the input set to P_n that occur in the i-th PHASH call, except the all-zero input used to obtain L. Note that $\mathbf{M}^{(q)} = \mathbf{M}_1 \cup \cdots \cup \mathbf{M}_q$. For $i = 1, \ldots, q$, we have

$$\chi_1 \overset{\text{def}}{=} \text{dist}(\mathbf{S}^{(q)}), \quad \chi_{2,i} \overset{\text{def}}{=} [S_i \notin \mathbf{M}^{(q)}], \quad \chi_{3,i} \overset{\text{def}}{=} [S_i \ne 0], \quad \text{and} \quad \chi_{4,i} \overset{\text{def}}{=} [0 \notin \mathbf{M}_i].$$

Note that $\overline{a_q} \equiv \overline{\chi_1 \wedge \chi_2 \wedge \chi_3 \wedge \chi_4}$ where $\chi_i \overset{\text{def}}{=} \chi_{i,1} \wedge \cdots \wedge \chi_{i,q}$ for $i = 2, 3, 4$. Using the union bound and its variant, we have

$$\max_{X^q} P^{\text{MPMAC}}_{\overline{a_q}|X^q} \le \max_{X^q} P^{\text{MPMAC}}_{\overline{\chi_1}|X^q}$$

$$+ \sum_{i=1,\ldots,q} \left(\max_{X^q} P^{\text{MPMAC}}_{\overline{\chi_{2,i}}|\chi_{4,i},X^q} + \max_{X^q} P^{\text{MPMAC}}_{\overline{\chi_{3,i}}|\chi_{4,i},X^q} + \max_{X^q} P^{\text{MPMAC}}_{\overline{\chi_{4,i}}|X^q} \right). \tag{8}$$

Now we analyze each term in Eq. (8). For this analysis, for some $i \ne j$, we fix the i-th and j-th queries to $x_i = (x_{i[1]}, \ldots, x_{i[m]})$ and $x_j = (x_{j[1]}, \ldots, x_{j[m']})$ with $x_i \ne x_j$. We start with the first term. Collision $S_i = S_j$ is equivalent to

$$V_i \oplus V_j \oplus 2^m 3^{\text{I}(x_i)} L \oplus 2^{m'} 3^{\text{I}(x_j)} L = 0. \tag{9}$$

To prove the maximum probability of Eq. (9), we need to use a case analysis.

Case 1: $m = m' = 1$. In this case, $V_i \oplus V_j = \text{pad}(x_i) \oplus \text{pad}(x_j)$. If $\text{I}(x_i) \ne \text{I}(x_j)$, the L.H.S. of Eq. (9) is a permutation of L from Lemma 1. Thus, the probability of Eq. (9) is $1/2^n$. If $\text{I}(x_i) = \text{I}(x_j)$, the probability is zero as $\text{pad}(x_i) \ne \text{pad}(x_j)$.

Case 2: $m > 1, m' = 1$. In this case, the probability of Eq. (9) is obviously at most $(m-1)/2^n + 1/(2^n - (m-1)) \le (\ell-1)/2^n + 1/(2^n - (\ell-1))$ from the second claim of Lemma 2.

Case 3: $m = m' > 1$. If the first $m-1$ blocks of x_i and x_j are the same, the probability is at most $1/2^n$, which is the same as in Case 1. Otherwise, Eq. (9)

occurs with a probability of at most $(2\ell - 2)/2^n + 1/(2^n - (2\ell - 2))$ from the first claim of Lemma 2.

Case 4: $m > 1, m' > 1, m \neq m'$. The bound of Case 3 also holds true.

Thus, we have

$$\max_{X^q} P^{\mathrm{MPMAC}}_{\overline{\chi_1}|X^q} \leq \sum_{i<j} \max_{X^q} P^{\mathrm{MPMAC}}_{[S_i=S_j]|X^q} \leq \binom{q}{2}\left(\frac{2\ell - 2}{2^n} + \frac{1}{2^n - (2\ell - 2)}\right). \quad (10)$$

For the second term, observe that $\overline{\chi_{2,i}}$ is the logical sum of events such that $S_i = 2^h L \oplus x_{i'[h]}$ for some i' including i, and $1 \leq h \leq \ell-1$. As x_i has m blocks, this is equivalent to $V_i = x_{i'[h]} \oplus 2^m 3^{\mathrm{I}(x_i)} L \oplus 2^h L$. From Lemma 1, we have $2^m 3^{\mathrm{I}(x_i)} \neq 2^h$. Thus, it is enough to evaluate the maximum of $\Pr[V_i = \mathsf{u}_1 L \oplus \mathsf{u}_2 | \chi_{4,i}]$ for all $\mathsf{u}_1 \in \Sigma^n \setminus \{0\}, \mathsf{u}_2 \in \Sigma^n$. We fix $\mathsf{u}_1 \neq 0$ and u_2, and let $f(z) = \mathsf{u}_1 z \oplus \mathsf{u}_2$ and $\mathsf{u}_3 \overset{\mathrm{def}}{=} (\mathsf{u}_2 \oplus \mathsf{pad}(x_{i[m]}))/\mathsf{u}_1$, where $/$ denotes field division. We assume that $m > 1$ and $L = \mathsf{u}_3$ satisfies $\chi_{4,i}$ with x_i. Note that $\Pr[V_i = f(L)|\chi_{4,i}]$ equals:

$$\sum_c \Pr[V_i = f(L)|L = c]\Pr[L = c|\chi_{4,i}] + \Pr[V_i = 0|L = \mathsf{u}_3]\Pr[L = \mathsf{u}_3|\chi_{4,i}],$$

$$\leq \max_c \Pr[V_i = f(c)|L = c] + \Pr[L = \mathsf{u}_3|\chi_{4,i}], \quad (11)$$

$$= \max_c \Pr[\mathsf{Sum}(c) = f(c) \oplus \mathsf{pad}(x_{i[m]})|L = c] + \Pr[L = \mathsf{u}_3|\chi_{4,i}], \quad (12)$$

where the sum and maximums are taken for all $c \neq \mathsf{u}_3$ that satisfies $\chi_{4,i}$, and $\mathsf{Sum}(c) = \mathsf{P}_n(x_{i[1]} \oplus 2c) \oplus \ldots \oplus \mathsf{P}_n(x_{i[m-1]} \oplus 2^{m-1}c)$. If every element in $\{(x_{i[1]} \oplus 2c), \ldots, (x_{i[m-1]} \oplus 2^{m-1}c)\}$ is in an even equivalent set, $\mathsf{Sum}(c)$ is 0 while $f(c) \oplus \mathsf{pad}(x_{i[m]}) \neq 0$ from $c \neq \mathsf{u}_3$. If there exists any element which is in an odd equivalent set, $\mathsf{Sum}(c)$ is the sum of k URP outputs for distinct inputs, for some $1 \leq k \leq m - 1$. These inputs are not 0 as c satisfies $\chi_{4,i}$. Therefore, the first term of the R.H.S. of Eq. (12) is at most $1/(2^n - (\ell - 1))$ from Eq. (4). Also, the second term of the R.H.S. of Eq. (12) is at most $1/(2^n - (\ell - 1))$. From these observations, $\max_{\mathsf{u}_1 \neq 0, \mathsf{u}_2} \Pr[V_i = \mathsf{u}_1 L \oplus \mathsf{u}_2 | \chi_{4,i}]$ is at most $2/(2^n - (\ell - 1))$ if $m > 1$ and $L = \mathsf{u}_3$ satisfies $\chi_{4,i}$. For other cases (i.e., when $m = 1$ or $m > 1$ and $L = \mathsf{u}_3$ does not satisfy $\chi_{4,i}$), this bound also holds true. Therefore,

$$\max_{X^q} P^{\mathrm{MPMAC}}_{\overline{\chi_{2,i}}|\chi_{4,i},X^q} \leq (\ell-1)q \cdot \max_{\mathsf{u}_1 \neq 0, \mathsf{u}_2} \Pr[V_i = \mathsf{u}_1 L \oplus \mathsf{u}_2 | \chi_{4,i}] = \frac{2(\ell-1)q}{2^n - (\ell - 1)} \quad (13)$$

holds for any $1 \leq i \leq q$, where the inequality holds since $\mathbf{M}^{(q)}$ contains at most $(\ell - 1)q$ distinct elements. For the third and fourth terms of Eq. (8), we have

$$\max_{X^q} P^{\mathrm{MPMAC}}_{\overline{\chi_{3,i}}|\chi_{4,i},X^q} \leq \frac{2}{2^n - (\ell - 1)}, \text{ and } \max_{X^q} P^{\mathrm{MPMAC}}_{\overline{\chi_{4,i}}|X^q} \leq \frac{(\ell - 1)}{2^n}, \quad (14)$$

where the first inequality follows from the same analysis as for the second term, and the second inequality holds since $\overline{\chi_{4,i}}$ occurs if L takes one of (at most) $\ell-1$

values defined by x_i. Combining Eqs. (10),(13), and (14), we get

$$\max_{X^q} P^{\mathrm{MPMAC}}_{\overline{a_q}|X^q} \leq \binom{q}{2}\left(\frac{2\ell-2}{2^n} + \frac{1}{2^n-(2\ell-2)}\right) + \frac{2(\ell-1)q^2+2q}{2^n-(\ell-1)} + \frac{(\ell-1)q}{2^n}$$

$$\leq \frac{1}{2^n-(2\ell-2)}((3\ell-2.5)q^2+1.5q). \tag{15}$$

Note that a_q implies $V_i \neq V_j$ if i-th and j-th tweaks are the same, for all $1 \leq i < j \leq q$. Therefore, if a_q is given, the collision probability between Y_i and Y_j is at most $1/2^n$ for all fixed q queries. Thus we have

$$\max_{X^q} P^{\mathrm{MPMAC}}_{\overline{b_q}|a_qX^q} \leq \sum_{i<j} \max_{X^q} P^{\mathrm{MPMAC}}_{[Y_i=Y_j]|a_qX^q} \leq \binom{q}{2}\frac{1}{2^n}. \tag{16}$$

ANALYSIS FOR $\nu_\ell(\mathrm{MPMAC}, \overline{d_q \wedge e_q})$. We consider a tweakable function, G, having n-bit input and output and tweak space $\mathcal{T} = \mathbb{I} \times \mathbb{J}$, where $\mathbb{I} = \{1, .., 2^{n/2}\}$ and $\mathbb{J} = \{0, 1, 2\}$. For any input $x \in \Sigma^n$ and tweak $t = (t_{[1]}, t_{[2]}) \in \mathcal{T}$, it is defined as $G(t,x) \overset{\text{def}}{=} \mathsf{P}_n(x)$ if $t_{[2]} = 0$, otherwise $G(t,x) \overset{\text{def}}{=} \widetilde{\mathsf{P}}_n((t_{[1]}, t_{[2]}), x)$, where P_n and $\widetilde{\mathsf{P}}_n$ are independent. If we allow an adversary against G to make $(\ell-1)q$ queries for P_n and q queries for $\widetilde{\mathsf{P}}_n$ (the order of query is arbitrary), he can *simulate* any (q, ℓ)-CPA against MPMAC. Here, we assume that $L = \mathsf{P}_n(0)$ is publicly available, so that ℓq queries are enough to simulate an attack. Moreover, if a G-based simulation generates distinct ℓq outputs of G, this implies the occurrence of d_q in MPMAC[3]. From these observations, $\nu_\ell(\mathrm{MPMAC}, \overline{d_q})$ is at most

$$\nu_{\widetilde{\ell q}}(G, \overline{\mathrm{dist}(\mathbf{Y}^{(\ell q)})}) = \mu_{\widetilde{\ell q}}(G, \overline{\mathrm{dist}(\mathbf{Y}^{(\ell q)})}) \leq \frac{(\ell-1)q^2}{2^n} + \binom{q}{2}\frac{1}{2^n},$$

where $\mathbf{Y}^{(\ell q)}$ is the set of ℓq outputs, and $\widetilde{\ell q}$ means that the adversary can make $(\ell-1)q$ queries for P_n and q queries for $\widetilde{\mathsf{P}}_n$, and the equality follows from an analysis similar to the one used for the proof of Lemma 4. The last inequality is trivial. Similarly, we can prove $\nu_\ell(\mathrm{MPMAC}, \overline{e_q}) \leq q/2^n$ using G. Thus we have

$$\nu_\ell(\mathrm{MPMAC}, \overline{d_q \wedge e_q}) \leq \frac{(\ell-1)q^2}{2^n} + \binom{q}{2}\frac{1}{2^n} + \frac{q}{2^n} \tag{17}$$

using Lemma 9. Combining Eqs. (15),(16), and (17) and Lemma 4, Lemma 3 is proved.

PROVING THEOREM 2. Deriving an upper bound of $\mathrm{Adv}^{\mathrm{vilqrf}}_{\mathrm{MPMAC}}(q, \ell)$ is easy since MPMAC can be seen as an instance of the Carter-Wegman MAC [20](CW-MAC). Since the following lemma is almost the same as previous CW-MAC lemmas (e.g., Lemma 4 of [5]), we omit the proof here.

---

[3] We assume that the adversary never makes colliding queries and a pair of queries such as $((t_{[1]}, t_{[2]}), x)$ and $((t'_{[1]}, t'_{[2]}), x')$ with $t_{[2]} = t'_{[2]} = 0$, $x = x'$, and $t_{[1]} \neq t'_{[1]}$. These queries are obviously useless for simulation.

Lemma 5. $\mathrm{Adv}_{\mathrm{MPMAC}}^{\mathrm{vilqrf}}(q, \ell) \leq \binom{q}{2}\mathrm{dp}(\ell - 1) + \binom{q}{2}/2^n$, *where* $\mathrm{dp}(m)$ *denotes* $\max_{x, x' \in (\Sigma^n)^{\leq m}, x \neq x', u \in \Sigma^n} \Pr[\mathrm{PHASH}(x) \oplus \mathrm{PHASH}(x') = u]$.

Finally, combining Lemmas 2, 3, and 5, we obtain

$$\mathrm{Adv}_{\mathrm{PMAC}[\mathsf{P}_n]}^{\mathrm{vilqrf}}(q, \ell) \leq \mathrm{Adv}_{\mathrm{PMAC}[\mathsf{P}_n], \mathrm{MPMAC}}^{\mathrm{cpa}}(q, \ell) + \mathrm{Adv}_{\mathrm{MPMAC}}^{\mathrm{vilqrf}}(q, \ell), \text{ and} \qquad (18)$$

$$\leq \frac{(4\ell - 2.5)q^2 + 1.5q}{2^n - 2\ell} + \binom{q}{2}\left(\frac{2\ell - 2}{2^n} + \frac{1}{2^n - (2\ell - 2)} + \frac{1}{2^n}\right)$$

$$\leq \frac{(5\ell - 2.5)q^2 + (1.5 - \ell)q}{2^n - 2\ell} \leq \frac{5\ell q^2}{2^n - 2\ell}, \qquad (19)$$

where the last inequality holds since $q, \ell \geq 1$. This concludes the proof of Theorem 2.

4   TMAC and XCBC

4.1   New Security Bounds for TMAC and XCBC

Since CBC-MAC provides no security if two messages with the same prefix are processed, a number of modifications have been proposed to make CBC-MAC secure for any message. EMAC, an early attempt, uses two blockcipher keys; TMAC [13] and XCBC [7] were later proposed as better solutions: they use one blockcipher key and some additional keys, and thus avoid two blockcipher key schedulings. TMAC and XCBC are defined as follows. Let CBC be the CBC-MAC function using P_n; that is, for input $x = (x_{[1]}, \ldots, x_{[m]}) \in (\Sigma^n)^m$, $\mathrm{CBC}(x) = C_m$, where $C_i = \mathsf{P}_n(x_{[i]} \oplus C_{i-1})$ and $C_0 = 0$. Let $\mathrm{TMAC}[\mathsf{P}_n]$ denote the TMAC using P_n. For input $x \in \Sigma^*$, $\mathrm{TMAC}[\mathsf{P}_n]$ works as follows. First, we partition x into $x = (x_{[1]}, \ldots, x_{[m]})$, where $m = \|x\|_n$. If $m > 1$, the tag for x is $Y = \mathsf{P}_n(\mathrm{CBC}(\widehat{x}) \oplus \mathrm{pad}(x_{[m]}) \oplus 2^{\mathrm{I}(x)-1}L)$, where $\widehat{x} = (x_{[1]}, \ldots, x_{[m-1]})$ and L is independent and uniform over Σ^n. If $m = 1$, Y is $\mathsf{P}_n(\mathrm{pad}(x) \oplus 2^{\mathrm{I}(x)-1}L)$. Note that the P_n used in CBC and the one used in the finalization are identical. Therefore, in practice, TMAC has one blockcipher key and an additional n-bit key L. XCBC is similar to TMAC, but uses two n-bit keys, L_1 and L_2, as masking values instead of L and $2L$. The previous bound of $\mathrm{TMAC}[\mathsf{P}_n]$ is $(3\ell^2 + 1)q^2/2^n$ [13] against (q, ℓ)-CPA, and $3\sigma^2/2^n$ against (q, σ)-CPA [10]. Almost the same results are obtained for XCBC [10,13]. However, using our proof approach in Sect. 3 and Bellare et al.'s analysis of the CBC function [5], we obtain the following.

Theorem 3. *Let* $\mathrm{TMAC}[\mathsf{P}_n]$ *be the TMAC using* P_n. *We then have*

$$\mathrm{Adv}_{\mathrm{TMAC}[\mathsf{P}_n]}^{\mathrm{vilqrf}}(q, \ell) \leq \frac{4\ell q^2}{2^n} + \frac{64\ell^4 q^2}{2^{2n}}.$$

The proof of Theorem 3 is in the next section. The bound of Theorem 3 is also applicable to XCBC. The proof for XCBC is the same as the proof of Theorem 3, thus we omit it here.

4.2   Proof of Theorem 3

Since the proof structure is the same as that of Theorem 2, we give only a sketch of the proof. We define a modified TMAC, denoted by MTMAC, that uses an independent tweakable URP for its finalization. In MTMAC, we partition message x into $(x_{[1]}, \ldots, x_{[m]})$, where $m = \|x\|_n$, and when $m > 1$, the tag is $Y = \widetilde{\mathsf{P}}_n(\mathtt{I}(x), \mathrm{CBC}(\widehat{x}) \oplus \mathtt{pad}(x_{[m]}))$, where $\mathtt{I}(x) \in \{1, 2\}$ is a tweak. When $m = 1$, $Y = \widetilde{\mathsf{P}}_n(\mathtt{I}(x), \mathtt{pad}(x))$. For both $\mathrm{TMAC}[\mathsf{P}_n]$ and MTMAC, let $X_i \in \Sigma^*$ be the i-th query and $\mathbf{M}^{(q)}$ ($\mathbf{C}^{(q)}$) be the set of inputs (outputs) to P_n generated in the CBC function for all q queries. We also define Y_i as the i-th tag, and $\mathbf{Y}^{(q)} \stackrel{\text{def}}{=} \{Y_i\}_{i=1,\ldots,q}$. When $\|X_i\|_n = m > 1$, we define V_i as the XOR of the i-th CBC output and $\mathtt{pad}(X_{i[m]})$, and when $m = 1$, $V_i = \mathtt{pad}(X_i)$. Moreover, $S_i \stackrel{\text{def}}{=} V_i \oplus 2^{\mathtt{I}(X_i)-1}L$, and $\mathbf{S}^{(q)} \stackrel{\text{def}}{=} \{S_i\}_{i=1,\ldots,q}$. In MTMAC, S_i is a dummy variable. Note that $Y_i = \mathsf{P}_n(S_i)$ in $\mathrm{TMAC}[\mathsf{P}_n]$ and that $Y_i = \widetilde{\mathsf{P}}_n(\mathtt{I}(X_i), V_i)$ in MTMAC, where $m = \|X_i\|_n$. We define $a_q \stackrel{\text{def}}{=} \mathrm{dist}(\mathbf{S}^{(q)}) \wedge [\mathbf{M}^{(q)} \cap \mathbf{S}^{(q)} = \emptyset]$ and $b_q \stackrel{\text{def}}{=} \mathrm{dist}(\mathbf{Y}^{(q)})$, $d_q \stackrel{\text{def}}{=} [\mathbf{C}^{(q)} \cap \mathbf{Y}^{(q)} = \emptyset]$. We then obtain

$$\mathbf{Adv}^{\mathrm{cpa}}_{\mathrm{TMAC}[\mathsf{P}_n], \mathrm{MTMAC}}(q, \ell) \leq \nu_\ell(\mathrm{MTMAC}, \overline{a_q \wedge b_q}) + \nu_\ell(\mathrm{MTMAC}, \overline{d_q}) \qquad (20)$$

for any (q, ℓ) using an argument similar to that used for Lemma 3. Note that a_q does not contain $[0 \notin \mathbf{M}^{(q)} \cup \mathbf{S}^{(q)}]$, as we do not have to care about 0 being an input to P_n. Since Lemma 4 does not depend on the structure of message-hashing part, it also applies to MTMAC and we have

$$\nu_\ell(\mathrm{MTMAC}, \overline{a_q \wedge b_q}) = \mu_\ell(\mathrm{MTMAC}, \overline{a_q \wedge b_q}) \leq \max_{X^q} P^{\mathrm{MTMAC}}_{\overline{a_q}|X^q} + \max_{X^q} P^{\mathrm{MTMAC}}_{\overline{b_q}|a_q X^q}.$$
$$(21)$$

To obtain bounds of last two terms of Eq. (21), we need the following lemma[4]. It generalizes a lemma of Bellare et al.[5].

Lemma 6.

$$\max_{x \in (\Sigma^n)^m, x' \in (\Sigma^n)^{m'}, x \neq x', u \in \Sigma^n} \Pr[\mathrm{CBC}(x) \oplus \mathrm{CBC}(x') = u] \leq \frac{2d(m^*)}{2^n} + \frac{64(m^*)^4}{2^{2n}},$$

$$\max_{x \in (\Sigma^n)^m, u \in \Sigma^n} \Pr[\mathrm{CBC}(x) = u] \leq \frac{2d(m+1)}{2^n} + \frac{64(m+1)^4}{2^{2n}},$$

where $\mathsf{d}(m)$ *is the maximum number of positive integers that divide* h, *for all* $h \leq m$, *and* $m^* = \max\{m, m'\} + 1$.

Proof. (of Lemma 6) For any $z \in \Sigma^n$, $\mathrm{CBC}(x) \oplus \mathrm{CBC}(x') = u$ is equivalent to $\mathsf{P}_n(\mathrm{CBC}(x) \oplus z) = \mathsf{P}_n(\mathrm{CBC}(x') \oplus z \oplus u)$, which is equivalent to $\mathrm{CBC}(x\|z) = \mathrm{CBC}(x'\|(z \oplus u))$. From Lemma 5 of [5], we see that the collision probability of

---

[4] Pietrzak [18] proved that the collision probability of CBC among q messages could be smaller than the union bound applied to Lemma 6 for some (q, ℓ). Since our analysis is based on the union bound, we do not know if the result of [18] can be combined into our proof to obtain other proofs.

$CBC(x\|z)$ and $CBC(x'\|(z \oplus u))$ is at most $2d(m^*)/2^n + 64(m^*)^4/2^{2n}$ for any z (note that $x\|z \neq x'\|(z \oplus u)$ holds for any z and u as we assumed $x \neq x'$). Therefore, the first claim is proved. The second can be similarly proved.

We analyze $\max_{X^q} P^{MTMAC}_{\overline{a_q}|X^q}$. If the i-th and j-th queries are fixed to x_i and x_j with $x_i \neq x_j$, collision $S_i = S_j$ is equivalent to $V_i \oplus 2^{I(x_i)}L = V_j \oplus 2^{I(x_j)}L$. If $I(x_i) \neq I(x_j)$, the collision occurs with probability $1/2^n$ since L is independent of V_i and V_j and $2^{I(x_i)}L \oplus 2^{I(x_j)}L$ is a permutation of L from Lemma 1. If $I(x_i) = I(x_j)$, $S_i = S_j$ implies $V_i = V_j$, which has a probability of at most $2d(\ell)/2^n + 64\ell^4/2^{2n}$ from Lemma 6 and a case analysis similar to the one used to derive Eq. (10). Therefore, the probability of $\overline{dist(S^{(q)})}$ is at most $\binom{q}{2}(2d(\ell)/2^n + 64\ell^4/2^{2n})$. Note that any collision event consisting of $[\mathbf{M}^{(q)} \cap \mathbf{S}^{(q)} = \emptyset]$ has probability $1/2^n$ since L is independent of all members of $\mathbf{M}^{(q)}$. From these observations, we have

$$\max_{X^q} P^{MTMAC}_{\overline{a_q}|X^q} \leq \max_{X^q} P^{MTMAC}_{\overline{dist(S^{(q)})}|X^q} + \max_{X^q} P^{MTMAC}_{[\mathbf{M}^{(q)} \cap \mathbf{S}^{(q)} = \emptyset]|X^q}$$
$$\leq \binom{q}{2}\left(\frac{2d(\ell)}{2^n} + \frac{64\ell^4}{2^{2n}}\right) + \frac{(\ell-1)q^2}{2^n}. \tag{22}$$

The analyses of $\max_{X^q} P^{MTMAC}_{\overline{b_q}|a_q X^q}$ and $\nu_\ell(MTMAC, \overline{d_q})$ are the same as those used for the proof of Lemma 3. We obtain

$$\max_{X^q} P^{MTMAC}_{\overline{b_q}|a_q X^q} \leq \binom{q}{2}\frac{1}{2^n}, \text{ and } \nu_\ell(MTMAC, \overline{d_q}) \leq \frac{(\ell-1)q^2}{2^n} + \binom{q}{2}\frac{1}{2^n}. \tag{23}$$

As with MPMAC, MTMAC is an instance of CW-MAC. Thus, we have

$$Adv^{vilqrf}_{MTMAC}(q, \ell) \leq \binom{q}{2}\left(\frac{2d(\ell)}{2^n} + \frac{64\ell^4}{2^{2n}} + \frac{1}{2^n}\right). \tag{24}$$

Combining the bound of $Adv^{cpa}_{TMAC[P_n],MTMAC}(q, \ell)$, which can be derived from Eqs. (20),(21),(22), and (23), and the bound of $Adv^{vilqrf}_{MTMAC}(q, \ell)$ by Eq. (24), $Adv^{vilqrf}_{TMAC[P_n]}(q, \ell)$ is at most $((2d(\ell) + 2\ell)q^2)/2^n + 64\ell^4q^2/2^{2n}$. Since $d(\ell) \leq \ell$, this concludes the proof of Theorem 3.

5   Conclusion and Future Work

In this paper, we have provided new security bounds for PMAC, TMAC, and XCBC. Our result demonstrates that the security degradation with respect to the maximum length of a message is linear for PMAC and almost linear (unless message is impractically long) for TMAC and XCBC, while previous analyses of these modes proved quadratic security degradation.

A COMPARISON OF BOUNDS. As we mentioned, our new bounds improve the old ones under most (but not all) cases. Here, we give a detailed comparison between new and old bounds. For simplicity, we ignore the constants. Thus, the new PMAC bound is $\ell q^2/2^n$, the new TMAC (and XCBC) bound is $\ell q^2/2^n + \ell^4 q^2/2^{2n}$, and the old bounds are $\sigma^2/2^n$ for all. For PMAC, the new bound is better if and only if $\sqrt{\ell}q < \sigma$, i.e., the mean message block length (σ/q) is larger than $\sqrt{\ell}$. Similarly, for TMAC and XCBC, the new bound is better if and only if the mean message block length is larger than $\sqrt{\ell(1+c)}$, where $c = \ell^3/2^n$, which can be small in practice. Thus, the criterion for choosing a bound is the distance between the mean block length and the square root of the maximum block length.

As a concrete example, let $n = 128$, $q = 2^{40}$, and $\ell = 2^{16}$. Then the new PMAC bound is 2^{-32} (the new TMAC and XCBC bounds are almost 2^{-32}), while the old bound ranges from 2^{-48} to 2^{-16}. The old bound is better if 99.9% of the messages are one-block, as $\sigma^2/2^n \leq (1 \cdot 0.999q + \ell \cdot 0.001q)^2/2^n < 2^{-35}$. In this case, the mean block length is smaller than 2^6, which is smaller than $\sqrt{\ell} = 2^8$. In contrast, if 1% of the messages are ℓ-block, the new bound is better since $\sigma^2/2^n \geq (\ell \cdot 0.01q + 1 \cdot 0.99q)^2/2^n > 2^{-30}$ and the mean block length is at least 2^9. Generally, the new bounds are better when only a tiny fraction of the message length distribution is concentrated on the right.

ON THE SECURITY OF OMAC. OMAC [11], i.e., CMAC [1], is similar to TMAC, but uses a different finalization. In OMAC using P_n, denoted by $OMAC[P_n]$, L is $P_n(0)$, and, instead of using $2^{I(x)-1}L$, it uses $2^{I(x)}L$ as the masking value. Thus OMAC has only one blockcipher key. The known security bound of $OMAC[P_n]$ is $(5\ell^2+1)q^2/2^n$[11] against (q, ℓ)-CPA, and $4\sigma^2/2^n$ against (q, σ)-CPA [10]. Unfortunately, we have not yet succeeded in showing new bounds. In a manner similar to that for TMAC, we define a modified[5] OMAC (MOMAC), using P_n and $\widetilde{P}_n$, and define sets of variables, ($\mathbf{M}^{(q)}$, $\mathbf{C}^{(q)}$, $\mathbf{S}^{(q)}$, and $\mathbf{Y}^{(q)}$), for both $OMAC[P_n]$ and MOMAC. By defining events $a_q \stackrel{\text{def}}{=} \text{dist}(\mathbf{S}^{(q)}) \wedge [\mathbf{M}^{(q)} \cap \mathbf{S}^{(q)} = \emptyset] \wedge [0 \notin \mathbf{S}^{(q)}]$, $b_q \stackrel{\text{def}}{=} \text{dist}(\mathbf{Y}^{(q)})$, $d_q \stackrel{\text{def}}{=} [\mathbf{C}^{(q)} \cap \mathbf{Y}^{(q)} = \emptyset]$, and $e_q \stackrel{\text{def}}{=} [L \notin \mathbf{Y}^{(q)}]$, we can prove that $\text{Adv}^{\text{cpa}}_{OMAC[P_n],MOMAC}(q, \ell)$ is at most $\nu_\ell(MOMAC, \overline{a_q \wedge b_q \wedge d_q \wedge e_q})$. However, to obtain a bound of $\nu_\ell(MOMAC, \overline{a_q \wedge b_q})$, we need the maximum probability of $[CBC(x) \oplus CBC(x') = u_1 L \oplus u_2]$ for $u_1 = (2 \oplus 2^2)$, which corresponds to the sum of two distinct masking values, and for all u_2, i.e., we need a generalization of Lemma 6. We think that this is an interesting open problem.

Acknowledgments

We would like to thank Tetsu Iwata and the anonymous referees for very useful comments and suggestions.

References

1. Recommendation for Block Cipher Modes of Operation: The CMAC Mode for Authentication. NIST Special Publication 800-38B, available from http://csrc.nist.gov/CryptoToolkit/modes/

---

[5] The "modified OMAC" was also described in [11]; however, our definition is different.

2. den Boer, B., Boly, J.P., Bosselaers, A., Brandt, J., Chaum, D., Damgård, I., Dichtl, M., Fumy, W., van der Ham, M., Jansen, C.J.A., Landrock, P., Preneel, B., Roelofsen, G., de Rooij, P., Vandewalle, J.: RIPE Integrity Primitives, final report of RACE Integrity Primitives Evaluation (1995)
3. Bellare, M., Kilian, J., Rogaway, P.: The Security of the Cipher Block Chaining Message Authentication Code. Journal of Computer and System Science 61(3) (2000)
4. Bellare, M., Desai, A., Jokipii, E., Rogaway, P.: A Concrete Security Treatment of Symmetric Encryptiont. In: FOCS '97. Proceedings of the 38th Annual Symposium on Foundations of Computer Science, pp. 394–403 (1997)
5. Bellare, M., Pietrzak, K., Rogaway, P.: Improved Security Analyses for CBC MACs. In: Shoup, V. (ed.) CRYPTO 2005. LNCS, vol. 3621, pp. 527–541. Springer, Heidelberg (2005)
6. Bernstein, D.J.: Stronger Security Bounds for Wegman-Carter-Shoup Authenticators. In: Cramer, R.J.F. (ed.) EUROCRYPT 2005. LNCS, vol. 3494, pp. 164–180. Springer, Heidelberg (2005)
7. Black, J., Rogaway, P.: CBC MACs for Arbitrary-Length Messages: The Three-Key Constructions. In: Bellare, M. (ed.) CRYPTO 2000. LNCS, vol. 1880, pp. 197–215. Springer, Heidelberg (2000)
8. Carter, L., Wegman, M.: Universal Classes of Hash Functions. Journal of Computer and System Science 18, 143–154 (1979)
9. Goldreich, O.: Modern Cryptography, Probabilistic Proofs and Pseudorandomness. In: Algorithms and Combinatorics, vol. 17, Springer, Heidelberg (1998)
10. Iwata, T., Kurosawa, K.: Stronger Security Bounds for OMAC, TMAC, and XCBC. In: Johansson, T., Maitra, S. (eds.) INDOCRYPT 2003. LNCS, vol. 2904, pp. 402–415. Springer, Heidelberg (2003)
11. Iwata, T., Kurosawa, K.: OMAC: One-Key CBC MAC. In: Johansson, T. (ed.) FSE 2003. LNCS, vol. 2887, pp. 129–153. Springer, Heidelberg (2003)
12. Krovetz, T., Rogaway, P.: The OCB Authenticated-Encryption Algorithm. Internet Draft (2005)
13. Kurosawa, K., Iwata, T.: TMAC: Two-Key CBC MAC. In: Joye, M. (ed.) CT-RSA 2003. LNCS, vol. 2612, pp. 33–49. Springer, Heidelberg (2003)
14. Liskov, M., Rivest, R.L., Wagner, D.: Tweakable Block Ciphers. In: Yung, M. (ed.) CRYPTO 2002. LNCS, vol. 2442, pp. 31–46. Springer, Heidelberg (2002)
15. Maurer, U.: Indistinguishability of Random Systems. In: Knudsen, L.R. (ed.) EUROCRYPT 2002. LNCS, vol. 2332, pp. 110–132. Springer, Heidelberg (2002)
16. Maurer, U., Pietrzak, K.: Composition of Random Systems: When Two Weak Make One Strong. In: Naor, M. (ed.) TCC 2004. LNCS, vol. 2951, pp. 410–427. Springer, Heidelberg (2004)
17. Pietrzak, K.: Composition Does Not Imply Adaptive Security. In: Shoup, V. (ed.) CRYPTO 2005. LNCS, vol. 3621, pp. 55–65. Springer, Heidelberg (2005)
18. Pietrzak, K.: A Tight Bound for EMAC. In: Bugliesi, M., Preneel, B., Sassone, V., Wegener, I. (eds.) ICALP 2006. LNCS, vol. 4052, pp. 168–179. Springer, Heidelberg (2006)
19. Rogaway, P.: Efficient Instantiations of Tweakable Blockciphers and Refinements to Modes OCB and PMAC. In: Lee, P.J. (ed.) ASIACRYPT 2004. LNCS, vol. 3329, pp. 16–31. Springer, Heidelberg (2004), (September 24, 2006) http://www.cs.ucdavis.edu/~rogaway/papers/offsets.pdf
20. Wegman, M., Carter, L.: New Hash Functions and Their Use in Authentication and Set Equality. Journal of Computer and System Sciences 22, 265–279 (1981)

A   Lemmas from Maurer's Methodology

We describe some lemmas developed by Maurer (e.g., [15]) that we used. We assume that F and G are two random functions with the same input/output size; we define MESs $\mathcal{A} = a_0 a_1 \ldots$ and $\mathcal{B} = b_0 b_1 \ldots$ for F and G. The i-th input and output are denoted by X_i and Y_i for F (or G), respectively. Equality of (possibly conditional) probability distributions means equality as functions, i.e., equality holds for all possible arguments. For example, we write $P^F_{Y^i | X^i a_i} = P^G_{Y^i | X^i b_i}$ to mean $P^F_{Y^i | X^i a_i}(y^i, x^i) = P^G_{Y^i | X^i b_i}(y^i, x^i)$ for all (x^i, y^i), where $P^F_{a_i | X^i}(x^i)$ and $P^G_{b_i | X^i}(x^i)$ are positive. Inequalities, such as $P^F_{Y^i | X^i a_i} \le P^G_{Y^i | X^i b_i}$, are similarly defined.

Lemma 7. *(A corollary from Theorem 1 (i), Lemma 1 (iv), and Lemma 4 (ii) of [15]) Let $\mathbb{F}$ be the function of F or G (i.e., $\mathbb{F}[F]$ is a function that internally invokes F, possibly several times, to process its inputs). Here, $\mathbb{F}$ can be probabilistic, and, if so, $\mathbb{F}$ is independent of F or G. Suppose that $P^F_{Y_i | X^i Y^{i-1} a_i} = P^G_{Y_i | X^i Y^{i-1} b_i}$ and $P^F_{a_i | X^i Y^{i-1} a_{i-1}} \le P^G_{b_i | X^i Y^{i-1} b_{i-1}}$ holds for $i \ge 1$. We then have*

$$\mathsf{Adv}^{\mathrm{cpa}}_{F,G}(q) \le \nu(F, \overline{a_q}), \quad \text{and} \quad \mathsf{Adv}^{\mathrm{cpa}}_{\mathbb{F}[F], \mathbb{F}[G]}(q) \le \nu(\mathbb{F}[F], \overline{a_q^*}).$$

Here, MES $\mathcal{A}^* = a_0^* a_1^* \ldots$ is defined such that a_i^* denotes $\mathcal{A}$-event is satisfied for time period i. For example, if $\mathbb{F}[F]$ always invokes F k times for any input, then $a_i^* \equiv a_{ki}$.

Lemma 8. *(Theorem 2 of [15]) If $P^F_{a_i | X^i Y^{i-1} a_{i-1}} = P^F_{a_i | X^i a_{i-1}}$ holds for $i \ge 1$, the maximum probabilities of $\overline{a_q}$ for all adaptive and non-adaptive attacks are the same, i.e., $\nu(F, \overline{a_q}) = \mu(F, \overline{a_q})$.*

Lemma 9. *(Lemma 6 (iii) of [15]) If MESs $\mathcal{A} = a_0 a_1 \ldots$ and $\mathcal{B} = b_0 b_1 \ldots$ are defined for F, we have $\nu(F, \overline{a_q \wedge b_q}) \le \nu(F, \overline{a_q}) + \nu(F, \overline{b_q})$.*

These lemmas are easily extended even if the adversary's parameter θ contains ℓ (or σ) in addition to q.

B   Proof of the First and Second Inequalities of Lemma 3

We define two tweakable functions having n-bit input/output and tweak space $\mathcal{T} = \{1, .., 2^{n/2}\} \times \{0, 1, 2\}$. For any input $x \in \Sigma^n$ and tweak $t = (t_{[1]}, t_{[2]}) \in \mathcal{T}$,

$$\mathrm{XE}(t, x) \stackrel{\text{def}}{=} \mathsf{P}_n(x \oplus 2^{t_{[1]}} 3^{t_{[2]}} L), \text{ where } L = \mathsf{P}_n(0), \text{ and}$$

$$\widetilde{\mathrm{XE}}(t, x) \stackrel{\text{def}}{=} \begin{cases} \mathsf{P}_n(x \oplus 2^{t_{[1]}} 3^{t_{[2]}} L), & \text{if } t_{[2]} = 0, \text{ where } L = \mathsf{P}_n(0); \\ \widetilde{\mathsf{P}}_n((t_{[1]}, t_{[2]}), x), & \text{otherwise.} \end{cases}$$

In the definition of $\widetilde{\mathrm{XE}}$, P_n and $\widetilde{\mathsf{P}}_n$ are assumed to be independent. It is obvious that $\mathrm{PMAC}[\mathsf{P}_n]$ and MPMAC can be realized by using XE and $\widetilde{\mathrm{XE}}$ in a black-box manner. We consider a game in which an adversary tries to distinguish XE from $\widetilde{\text{XE}}$ using q queries. Note that a query is in $\mathcal{T} \times \Sigma^n$. Let $(T_i, X_i) \in \mathcal{T} \times \Sigma^n$ be the i-th query, and $Y_i \in \Sigma^n$ be the i-th output. In addition, let S_i be $X_i \oplus 2^{T_{i[1]}} 3^{T_{i[2]}} L$, where $T_i = (T_{i[1]}, T_{i[2]})$. For $\widetilde{\text{XE}}$, S_i is defined as a dummy variable when $T_{i[2]} \neq 0$. We define the following two events:

$$a_i^* \overset{\text{def}}{=} [S_j \neq S_k \text{ for all } (j,k) \in \xi(i)] \wedge [S_j \neq 0 \text{ for all } j = 1, \ldots, i].$$

$$b_i^* \overset{\text{def}}{=} [Y_j \neq Y_k \text{ for all } (j,k) \in \xi(i)] \wedge [Y_j \neq L \text{ for all } j = 1, \ldots, i],$$

$$\text{where } \psi(i) \overset{\text{def}}{=} \{j : 1 \leq j \leq i, T_{j[2]} \in \{1,2\}\}, \text{ and}$$

$$\xi(i) \overset{\text{def}}{=} \{(j,k) \in \{1, \ldots, i\}^2 : j \neq k, \text{ at least one of } j \text{ or } k \text{ is in } \psi(i)\}.$$

Note that $\psi(i)$ and $\xi(i)$ depend on $\{T_{j[2]}\}_{j=1,\ldots,i}$. Clearly, $\mathcal{A}^* = a_0^* a_1^* \ldots$ and $\mathcal{B}^* = b_0^* b_1^* \ldots$ are MESs and they are equivalent in XE (but not in $\widetilde{\text{XE}}$). Also, note that $\mathcal{A}^*$ and $\mathcal{B}^*$ defined for XE ($\widetilde{\text{XE}}$) are compatible with MESs defined for PMAC[P_n] (MPMAC). For example, if one uses XE to simulate an attack against PMAC[P_n] and observes a_i^* in time period i, $a_{i'}$ is occurring for the i'-th query to PMAC[P_n], for some $i' \leq i$. We then have

$$P_{Y_i|X^i T^i Y^{i-1} a_i^*}^{\text{XE}} = \sum P_{Y_i|L X^i T^i Y^{i-1}}^{\text{XE}} \cdot P_{L|X^i T^i Y^{i-1} a_i^*}^{\text{XE}}, \tag{25}$$

where the summation is taken for all $L \in \Gamma(x^i, t^i, y^{i-1})$, which is the set of $L = c$ such that the rightmost term is non-zero. The equality of Eq. (25) holds since S^i is completely determined if X^i and L are fixed.

We focus on the rightmost two terms of Eq. (25) for some fixed $X^i = x^i$, $T^i = t^i$, $Y^{i-1} = y^{i-1}$ satisfying b_{i-1}^*, and $L = c \in \Gamma(x^i, t^i, y^{i-1})$ (thus $S^i = s^i$ is also fixed). It is clear that $P_{L|X^i T^i Y^{i-1} a_i^*}^{\text{XE}}(c, x^i, t^i, y^{i-1})$ is the uniform distribution over $\Gamma(x^i, t^i, y^{i-1})$. The conditional probability $P_{Y_i|L X^i T^i Y^{i-1}}^{\text{XE}}(y_i, c, x^i, t^i, y^{i-1})$ is 1 if $y_i = y_j$, and $i \notin \psi(i)$ and $\exists j \notin \psi(i)$ such that $s_i = s_j$. If $i \in \psi(i)$ or $i \notin \psi(i)$ but $s_i \neq s_j$ for $j \leq i-1$, Y_i is uniform over $\Sigma^n \setminus \{y_1, \ldots, y_{i-1}, c\}$.

Similarly, for $\widetilde{\text{XE}}$, we have

$$P_{Y_i|X^i T^i Y^{i-1} a_i^* b_i^*}^{\widetilde{\text{XE}}} = \sum P_{Y_i|L X^i T^i Y^{i-1} b_i^*}^{\widetilde{\text{XE}}} \cdot P_{L|X^i T^i Y^{i-1} a_i^* b_{i-1}^*}^{\widetilde{\text{XE}}}, \tag{26}$$

where the summation is taken for all $L \in \Gamma(x^i, t^i, y^{i-1})$. Then, a simple case analysis shows that

$$P_{Y_i|X^i T^i Y^{i-1} a_i^*}^{\text{XE}} = P_{Y_i|X^i T^i Y^{i-1} a_i^* b_i^*}^{\widetilde{\text{XE}}}. \tag{27}$$

Moreover, we have

$$P_{a_i^*|X^i T^i Y^{i-1} a_{i-1}^*}^{\text{XE}} = \sum P_{a_i^*|L X^i T^i Y^{i-1} a_{i-1}^*}^{\text{XE}} \cdot P_{L|X^i T^i Y^{i-1} a_{i-1}^*}^{\text{XE}}, \text{ and} \tag{28}$$

$$P_{a_i^* b_i^*|X^i T^i Y^{i-1} a_{i-1}^* b_{i-1}^*}^{\widetilde{\text{XE}}} = \sum P_{a_i^* b_i^*|L X^i T^i Y^{i-1} a_{i-1}^* b_{i-1}^*}^{\widetilde{\text{XE}}} \cdot P_{L|X^i T^i Y^{i-1} a_{i-1}^* b_{i-1}^*}^{\widetilde{\text{XE}}}, \tag{29}$$

where the summations are taken for all $L \in \Gamma'(x^{i-1}, t^{i-1}, y^{i-1})$, which is the set of $L = c$ such that the last term of Eq. (28) (or Eq. (29)) is non-zero. It is easy to find that the last terms of Eqs. (28) and (29) are the same conditional distributions. However, we have $P^{\widetilde{XE}}_{a_i^* b_i^* | L X^i T^i Y^{i-1} a_{i-1}^* b_{i-1}^*} \leq P^{XE}_{a_i^* | L X^i T^i Y^{i-1} a_{i-1}^*}$ since both sides are 0 if $L \notin \Gamma(x^i, t^i, y^{i-1})$, and otherwise the R.H.S. is 1. Thus we have

$$P^{\widetilde{XE}}_{a_i^* b_i^* | X^i T^i Y^{i-1} a_{i-1}^* b_{i-1}^*} \leq P^{XE}_{a_i^* | X^i T^i Y^{i-1} a_{i-1}^*}. \tag{30}$$

From Eqs. (27) and (30) and the second claim of Lemma 7, the first inequality of Lemma 3 is proved. The second follows from the first and Lemma 9.

C   Proof of Lemma 4

Note that $\mathbf{M}^{(i)}$ ($\mathbf{C}^{(i)}$) denotes the set of P_n inputs (outputs) generated in PHASH up to the i-th query. Let $\mathbf{Z}^{(i)}$ be the set of random variables $(L, \mathbf{C}^{(i)})$. If $\mathbf{Z}^{(i)}$ and X^i are fixed, $\mathbf{M}^{(i)}$, V^i, and S^i are uniquely determined. We have

$$P^{\text{MPMAC}}_{a_i b_i | X^i Y^{i-1} a_{i-1} b_{i-1}} = \sum_{\mathbf{Z}^{(i)}} P^{\text{MPMAC}}_{b_i | \mathbf{Z}^{(i)} X^i Y^{i-1} a_i b_{i-1}}$$
$$\cdot P^{\text{MPMAC}}_{a_i | \mathbf{Z}^{(i)} X^i Y^{i-1} a_{i-1} b_{i-1}} \cdot P^{\text{MPMAC}}_{\mathbf{Z}^{(i)} | X^i Y^{i-1} a_{i-1} b_{i-1}}, \tag{31}$$

where the summations are taken for all $\mathbf{Z}^{(i)} = \mathbf{z}^{(i)}$ such that $(\mathbf{z}^{(i)}, x^i)$ satisfies a_{i-1}. Note that a_i implies that, if the j-th and j'-th tweaks (recall that the i-th tweak is a function of X_i) are the same, $V_j \neq V_{j'}$ holds for all $j, j' \leq i$, $j \neq j'$. From this, $P^{\text{MPMAC}}_{b_i | \mathbf{Z}^{(i)} X^i Y^{i-1} a_i b_{i-1}}(\mathbf{z}^{(i)}, x^i, y^{i-1})$ does not depend on y^{i-1}, and it is $(2^n - (i-1))/(2^n - \pi(x^i))$, where $\pi(x^i)$ is the number of indices $j \in \{1, \ldots, i-1\}$ such that the j-th and i-th tweaks are the same. Moreover, $P^{\text{MPMAC}}_{a_i | \mathbf{Z}^{(i)} X^i Y^{i-1} a_{i-1} b_{i-1}}(\mathbf{z}^{(i)}, x^i, y^{i-1})$ does not depend on y^{i-1} as it is 1 if $(\mathbf{z}^{(i)}, x^i)$ satisfies a_i, and otherwise 0. Finally, it is easy to see that $P^{\text{MPMAC}}_{\mathbf{Z}^{(i)} | X^i Y^{i-1} a_{i-1} b_{i-1}}$ equals $P^{\text{MPMAC}}_{\mathbf{Z}^{(i)} | X^i a_{i-1}}$ (here, b_{i-1} implies $V_j \neq V_{j'}$ whenever j-th and j'-th tweaks are the same, however, this is already implied by a_{i-1}). Thus, $P^{\text{MPMAC}}_{a_i b_i | X^i Y^{i-1} a_{i-1} b_{i-1}}$ does not depend on y^{i-1}, and, for any x^i and $\widehat{y}^{i-1}$ satisfying b_{i-1}, we have

$$P^{\text{MPMAC}}_{a_i b_i | X^i a_{i-1} b_{i-1}}(x^i) = \sum P^{\text{MPMAC}}_{a_i b_i | X^i Y^{i-1} a_{i-1} b_{i-1}}(x^i, y^{i-1}) \cdot P^{\text{MPMAC}}_{Y^{i-1} | X^i a_{i-1} b_{i-1}}(y^{i-1}, x^i),$$
$$= P^{\text{MPMAC}}_{a_i b_i | X^i Y^{i-1} a_{i-1} b_{i-1}}(x^i, \widehat{y}^{i-1}) \sum P^{\text{MPMAC}}_{Y^{i-1} | X^i a_{i-1} b_{i-1}}(y^{i-1}, x^i),$$
$$= P^{\text{MPMAC}}_{a_i b_i | X^i Y^{i-1} a_{i-1} b_{i-1}}(x^i, \widehat{y}^{i-1}), \tag{32}$$

where the summations are taken for all y^{i-1} satisfying b_{i-1}. From this and Lemma 8, we prove the first claim of Lemma 4. The second follows from the first and the union bound.

Perfect Block Ciphers with Small Blocks*

Louis Granboulan[1] and Thomas Pornin[2]

[1] École Normale Supérieure; EADS
louis.granboulan@eads.net
[2] Cryptolog International, Paris, France
thomas.pornin@cryptolog.com

Abstract. Existing symmetric encryption algorithms target messages consisting of elementary binary blocks of at least 64 bits. Some applications need a block cipher which operates over smaller and possibly non-binary blocks, which can be viewed as a pseudo-random permutation of n elements. We present an algorithm for selecting such a random permutation of n elements and evaluating efficiently the permutation and its inverse over arbitrary inputs. We use an underlying deterministic RNG (random number generator). Each evaluation of the permutation uses $O(\log n)$ space and $O((\log n)^3)$ RNG invocations. The selection process is "perfect": the permutation is uniformly selected among the $n!$ possibilities.

1 Introduction

Block ciphers such as AES[1] or DES[2] typically operate on large input data blocks, each consisting of 64 or more bits (128 or 256 bits are now preferred). Using smaller blocks leads to important security issues when encrypting large messages or using the block cipher for a MAC over such a large message. However, some applications need smaller blocks, and possibly non-binary blocks (i.e., a block space size which is not a power of two).

An example of such an application is the generation of unique unpredictable decimal numbers, destined to be typed in by users. Such numbers must be short; this precludes the use of a simple PRNG, which would imply collisions with a non-negligeable probability. A solution is to use a block cipher operating over the set of decimal numbers of the required length: simply encrypt successive values of a counter.

Building a secure block cipher is known to be a tricky task. Small blocks and non-binary alphabets are even more challenging in several ways:

- There are theoretical results on the construction of secure block ciphers, e.g. [3]; more recent examples include [4] and [5]. These constructions implicitly rely on huge block sizes to hide some biases which our security model (which we detail in section 5) cannot tolerate.

---

* This work has been supported in part by the French government through X-Crypt, in part by the European Commission through ECRYPT.

A. Biryukov (Ed.): FSE 2007, LNCS 4593, pp. 452–465, 2007.

- Usual security analysis techniques try to model the attacker as having access to an encryption or decryption oracle, with some limitations on the number of queries. With a small block size, the complete code book is too small for this model to be adequate.
- Some classes of attacks (differential and linear cryptanalysis) are much more difficult to express on non-binary alphabets. For instance, several distinct ring structures can be applied on the set of decimal digits. Moreover, there is no field of size 10, only rings with some non-invertible elements, which makes things even more complex[6].

For these reasons, there have been only few attempts at designing such block ciphers, e.g. [7] which proposes to build ciphers of arbitrary domains by iterating a larger block cipher or by Feistel-like structures.

In this paper, we investigate another entirely different way of generating permutations, based of the following SHUFFLE algorithm. Let's consider a very small space of input data, for instance the two-digit decimal numbers. The input space has size 100, and there are 100! possible permutations. Each of them may be represented as an array of 100 numbers from 0 to 99, with no two array elements being equal: if the array element of index x contains y, then the permutation maps x to y (we number array elements from 0). If an appropriate source of randomness is available, then the following algorithm uniformly selects a random permutation of 100 elements:

SHUFFLE:
1. fill the array with element w containing the number w
2. $i \leftarrow 0$
3. select a random number j between i and 99 (inclusive)
4. if $i \neq j$, swap array elements of index i and j
5. increment i
6. if $i < 99$, return to step 3
7. the array contents represent the permutation

This algorithm was first published by Fisher and Yates[8], then published again by Durstenfeld[9], and popularized by Knuth[10], to the point that this algorithm is usually known as the "Knuth shuffle". If the random number selection at step 3 is uniform, then it can be shown that all possible permutations have an equal probability of being generated. Thus, for very small blocks, we have a practical candidate for a block cipher: use the key in a cryptographically secure seeded PRNG, then generate the complete permutation using the Knuth shuffle, and apply it. A simple linear pass can be used to compute the inverse permutation, which is used for decryption. The uniform selection of the random number at step 3 is a bit tricky if the PRNG outputs bits. Here is an algorithm which outputs uniformly random numbers between 0 and $d - 1$ using a PRNG which outputs bits:

RANDOMNUMBER:
1. chose an integer r such that $2^r \geq d$
2. obtain r random bits and interpret them as a number x such that $0 \leq x < 2^r$
3. compute $t = \lfloor \frac{x}{d} \rfloor$
4. if $d + td > 2^r$, return to step 2
5. output $x - td$

Depending on r, the probability that the loop occurs can be made arbitrarily low, to the point of being impossible to achieve in practice. A practical implementation may decide to use $r = 256$ and ignore the possibility that the test in step 4 is verified. Ultimately, this would imply a slight selection bias which is sufficiently small to be neglected.

The permutation generated by SHUFFLE can be viewed as a block cipher, where the key is the PRNG output (or, equivalently, the seed used in the PRNG). Since a truly random generator implies a truly random permutation, any successful attack on that selection system would be a successful distinguisher on the PRNG. Cryptographically speaking, the "block" cipher thus obtained is as perfect as the PRNG used. Unfortunately, this procedure has a cost $O(n)$ (for a permutation over a set of size n), both in CPU and in space, which makes it impractical for values of n beyond, for instance, 10^4. Generating a full permutation means computing each of the n output values; but we do not need the full permutation. Our goal is to evaluate the block cipher (i.e. the permutation) over one given input block. Hence, what we need here is a deterministic procedure which, from the output of a PRNG, may generate a permutation ϕ over n elements such that:

- assuming a perfect random source instead of the PRNG, all possible permutations of n elements have an equal chance of being selected for ϕ;
- for any given x, $\phi(x)$ can be computed with an average computational cost much lower than $O(n)$.

The purpose of this article is to describe such a procedure. It computes $\phi(x)$ for any x in the input data set in time $O((\log n)^3)$, and may also compute $\phi^{-1}(y)$ for any y in the output data set in time $O((\log n)^3)$. The notion of time we use here is the number of requests to the PRNG; we assume a *seekable* PRNG output[1]. Our algorithms are described as recursive for simplicity, but they may easily be transformed into tail recursions; the real space requirements are for a fixed number of integer values lower than n, hence implying a $O(\log n)$ space complexity. We describe our notations more formally in the next section; then we proceed with the algorithm description. Finally, we specify our security model.

---

[1] Note that the overall number of requests to the PRNG that are needed to define the whole permutation is $O(n(\log n)^3)$ which is bigger than $\log(n!)$.

2   Notations

We consider the problem of selecting a random permutation ϕ which operates over a set of n elements. n is an integer greater than 1; the targeted values of n are between 10^3 and 10^{20}, although our construction may be used with smaller and greater blocks as well.

We need a deterministic PRNG, which constitutes the key for our cipher. We model the PRNG as a function:

$$\mathcal{R} : \mathbb{N}^r \longrightarrow \{0,1\}$$
$$(n_1, n_2, \dots n_r) \longmapsto \mathcal{R}(n_1, n_2, \dots n_r)$$

The PRNG is assumed to be an unbiased random oracle, in that the output for a given set of input integers cannot be computationally predicted from the knowledge of the output of $\mathcal{R}$ for all other input value sets. From such a PRNG $\mathcal{R}$ with r input parameters, one can easily build a PRNG $\mathcal{R}'$ with $r+1$ input parameters, by definining:

$$\mathcal{R}'(n_1, \dots n_r, n_{r+1}) = \mathcal{R}(n_1, \dots n_{r-1}, (n_r + n_{r+1} + 1)(n_r + n_{r+1} + 2)/2 - n_r - 1)$$

In practice, we will need a keyed PRNG with three input parameters, for which we will provide workable bounds.

Our PRNG must be *seekable*, which means that computing the output of $\mathcal{R}$ for some given input parameters must be efficient, regardless of which other parameter values were previously input into $\mathcal{R}$. Unfortunately, this characteristic precludes the use of some of the "provably secure" PRNG such as BBS[11] or QUAD[12]: these PRNG output bits sequentially, and are not seekable.

We will use "log" for the base-2 logarithm, such that $\log 2^x = x$ for all x.

The algorithm description uses binomial terms; in order to simplify the equations, we use extended binomial coefficients, which are defined thus:

$$\binom{n}{p} = \frac{n!}{p!(n-p)!} \quad \text{if } 0 \le p \le n, \quad \text{and } 0 \text{ otherwise.}$$

3   Random Permutation Selection

In this section, we describe our algorithm for computing $\phi(x)$ on input x, as well as the inverse algorithm which computes $\phi^{-1}(y)$ from y. The first algorithm is an extension of a random permutation selection algorithm described in [13] in the context of a massively parallel architecture; we add the ability to compute only $\phi(x)$ and not the whole description of ϕ, as well as an algorithm for ϕ^{-1}.

Our two main algorithms are called PERMUTATOR and INVPERMUTATOR; they compute, respectively, ϕ and ϕ^{-1} on a given input. PERMUTATOR implements ϕ by using a binary tree of "splits", where each split segregates the inputs into two groups. The SPLITTER algorithm implements the split; it itself uses a binary tree of repartitions that REPARTITOR computes. REPARTITOR is the algorithm which uses the PRNG output. INVPERMUTATOR uses INVSPLITTER, which itself relies on REPARTITOR. We describe all those algorithms in the following sections.

3.1 Repartitor

The REPARTITOR has the following inputs:

- integers n and p such that $0 \le p \le n$ and $n \ne 0$;
- a PRNG index i.

We define $a = \lfloor n/2 \rfloor$.

Using only $O(\log n)$ PRNG requests $\mathcal{R}(i, \ldots)$, all with the value i as first index, REPARTITOR outputs an integer value u following this distribution:

$$P(u = k) = \frac{\binom{a}{k}\binom{n-a}{p-k}}{\binom{n}{p}}$$

This distribution is known as the hypergeometric distribution. It can easily be shown that values u which have a non-zero probability of being returned must be such that:

$$0 \le u \le a$$
$$(a + p) - n \le u \le p$$

Note that in the degenerate cases where $p = 0$ or $p = n$, then only one return value is possible ($u = 0$ or $u = a$, respectively), allowing for a fast return.

When $p > a$, one may implement REPARTITOR by using:

$$\text{REPARTITOR}(n, p, i) = a - \text{REPARTITOR}(n, n - p, i)$$

which means that one may always assume that $p \le a$ when implementing REPARTITOR.

REPARTITOR models the following experiment: from a set of n elements, split into two subsets of size a and $n - a$, p distinct elements are randomly selected. REPARTITOR computes and returns how many of those p elements come from the first subset (of size a). Implementing REPARTITOR efficiently, with a sufficiently unbiased output is tricky; we give a slow but precise algorithm in section 4.

3.2 Splitter

The SPLITTER algorithm has the following inputs:

- integers n and p such that $0 \le p \le n$ and $n \ne 0$;
- an input index x ($0 \le x < n$);
- a PRNG index i.

Using only requests to REPARTITOR$(*, *, j)$ for indexes j such that $i \le j < i + n - 1$, SPLITTER returns an integer $y = \text{SPLITTER}(n, p, x)$ such that:

$$0 \le \text{SPLITTER}(n, p, x) < n$$
$$x_1 \ne x_2 \Rightarrow \text{SPLITTER}(n, p, x_1) \ne \text{SPLITTER}(n, p, x_2)$$
$$\#\{x | \text{SPLITTER}(n, p, x) < p\} = p$$
$$\text{SPLITTER}(n, p, x_1) < \text{SPLITTER}(n, p, x_2) \le p \Rightarrow x_1 < x_2$$
$$p \le \text{SPLITTER}(n, p, x_1) < \text{SPLITTER}(n, p, x_2) \Rightarrow x_1 < x_2$$

These equations mean that SPLITTER extracts p elements from the set of integer values from 0 to $n-1$; these p elements are given the output indexes 0 to $p-1$, but their order is preserved. The $n-p$ unextracted elements are given the indexes p to $n-1$, and their order is also preserved. There are $\binom{n}{p}$ such "splits" and we require that SPLITTER selects one of those with uniform probability.

SPLITTER is implemented with a "distribution tree". This is a binary tree whose leaves are the n elements; each leaf is thus "extracted" (part of the p elements) or "not extracted" (part of the $n-p$ other elements). Each node in the tree records how many elements among those descending from this node are "extracted". Hence, the root is the ancestor for all n elements, and records the value p. The tree is built from the root: each node manages k elements, and gets f "extractions" to distribute evenly among those k elements; it breaks the k elements into two halves and distributes the f extractions between the two halves (using REPARTITOR to follow the right probability distribution); the same process is invoked recursively on both halves. Computing the whole tree costs $O(n)$ calls to REPARTITOR, but knowing the status (extracted or not) of a given input element x, and computing its final index, can be done by exploring only the tree path from the root to the leaf corresponding to x. If the tree is balanced this costs $O(\log n)$ calls to REPARTITOR, hence $O((\log n)^2)$ PRNG invocations.

Here is an implementation of SPLITTER:

```
SPLITTER:
input: n, p, x and i
output: y
 1. if n = 1, then return x
 2. compute a ← ⌊n/2⌋
 3. compute u ← REPARTITOR(n, p, i)
 4. if x ≥ a, go to step 7
 5. compute t ← SPLITTER(a, u, x, i + 1)
 6. if t < u, return t, else return p + (t − u)
 7. compute t ← SPLITTER(n − a, p − u, x − a, i + a)
 8. if t < p − u, return t + u, else return a + t
```

Since a is computed such that the tree is well balanced, then the maximum recursion depth is $\lceil \log n \rceil$. It is easily shown that SPLITTER uses the PRNG only for indexes j such that $i \le j \le i+n-2$: this is true for $n = 1$ (no PRNG invocation at all), and is extended by recursion (the first recursive SPLITTER invocation may use only indexes from $i+1$ to $i+a-1$, and the second invocation may use only indexes from $i+a$ to $i+n-2$). Moreover, for a given x value, the same PRNG index is never used twice.

The hypergeometric distribution implemented by REPARTITOR ensures that SPLITTER selects uniformly the split among the possible splits of n elements into sets of length p and $n-p$.

3.3 InvSplitter

The INVSPLITTER implements a reversal of SPLITTER: INVSPLITTER(n, p, y, i) computes x such that SPLITTER$(n, p, x, i) = y$. Since SPLITTER implements a permutation of n elements, a unique solution always exists for all y from 0 to $n - 1$.

INVSPLITTER first checks whether the targeted y is part of the p extracted elements, or not. Then it invokes REPARTITOR to know how the elements, at that tree level, are split between the left and right parts, and it follows the correct node, depending on whether y was extracted, and at which rank. When the leaf x is reached, then x is the value looked for. Here is the INVSPLITTER algorithm:

INVSPLITTER:
input: n, p, y and i
output: x
1. if $n = 1$, then return y
2. compute $a \leftarrow \lfloor n/2 \rfloor$
3. if $y \geq p$, then go to step 5
4. if $y < u$, return INVSPLITTER$(a, u, y, i + 1)$;
   otherwise, return $a + $ INVSPLITTER$(n - a, p - u, y - u, i + a)$
5. if $y < a + p - u$, return INVSPLITTER$(a, u, y - (p - u), i + 1)$;
   otherwise, return $a + $ INVSPLITTER$(n - a, p - u, y - a, i + a)$

Note that if $y < p$ (respectively $y \geq p$) then all nested invocations of INVS-PLITTER will also have $y < p$ (respectively $y \geq p$): this is because "$y < p$" is equivalent to "y is from the set of p extracted elements". INVSPLITTER has cost $O((\log n)^2)$ ($\lceil \log n \rceil$ calls to REPARTITOR).

3.4 Permutator

The PERMUTATOR algorithm selects ϕ and computes $\phi(x)$. ϕ is defined as a binary tree of splits, as defined by SPLITTER: each tree node splits its input into two halves, and uses SPLITTER to decide in which half each input element goes. It then invokes itself recursively on the half where the actual input x has gone.

The inputs are n (the input space size), x (the actual input element, between 0 and $n-1$) and i (the base index for PRNG access). For the complete permutation, these parameters are, respectively, n (the total input space size), x and 0.

PERMUTATOR:
input: n, x and i
output: $\phi(x)$
1. if $n = 1$, then return x
2. compute $a \leftarrow \lfloor n/2 \rfloor$
3. compute $t \leftarrow $ SPLITTER(n, a, x, i)
4. if $t < a$, return PERMUTATOR$(a, t, i + n - 1)$
5. otherwise, return $a + $ PERMUTATOR$(n - a, t - a, i + n - 1 + \gamma(a))$

In this description, γ is a function which computes how many PRNG input indexes are used by PERMUTATOR: PERMUTATOR may use indexes from i to $i + \gamma(n) - 1$ (inclusive). γ must be such that:

$$\gamma(1) \geq 0$$
$$\gamma(n) \geq n - 1 + \gamma(a) + \gamma(n - a) \text{ for } n > 1 \text{ and } a = \lfloor n/2 \rfloor$$

The "optimal" γ (where these inequalities are equalities) is achieved with the following formula for $n > 1$:

$$\gamma(n) = nf(n) - 2^{f(n)} + 1$$

where:

$$f(n) = 1 + \lfloor \log(n - 1) \rfloor$$

($f(n)$ is the bit length of the binary representation of $n - 1$.)

The depth of the binary tree of splits that PERMUTATOR implements is $\lceil \log n \rceil$, because a is computed to be as close as possible to $n/2$. The number of invocations of REPARTITOR is $O((\log n)^2)$, hence the cost of $O((\log n)^3)$ PRNG invocations.

3.5   InvPermutator

The INVPERMUTATOR algorithm must select the same ϕ than PERMUTATOR with the same PRNG, and then compute $\phi^{-1}(y)$ for a given value y. INVPERMUTATOR explores the same binary tree of splits than PERMUTATOR, but upwards, from the leaves to the root, instead of downwards. At each level, PERMUTATOR uses SPLITTER to know where its current input value goes. For INVPERMUTATOR, we want to know from where a given output comes; we thus use INVSPLITTER.

INVPERMUTATOR:
input: n, y and i
output: $\phi^{-1}(y)$
1. if $n = 1$, then return y
2. compute $a \leftarrow \lfloor n/2 \rfloor$
3. if $y < a$, compute $t \leftarrow$ INVPERMUTATOR$(a, y, i + n - 1)$; otherwise, compute $t \leftarrow a +$ INVPERMUTATOR$(n - a, y - a, i + n - 1 + \gamma(a))$
4. return INVSPLITTER(n, a, t, i)

INVPERMUTATOR first goes recursively to the relevant leaf, and then works upwards, back to the root, where $\phi^{-1}(y)$ will be known.

4   Sampling Following the Hypergeometric Distribution

The hypergeometric distribution models the selection of p distinct elements among n, and returns how many of these p elements are taken from the first $a = \lfloor n/2 \rfloor$ elements. For small values of p, this exact procedure can be simulated. When p

increases, this becomes impractical, and we need a more efficient sampling method. As outlined in section 3.1, we may assume that $p \leq a$.

4.1 Acceptance-Rejection Method Aka. Rejection Sampling

This classical method, due to von Neumann, generates sampling values from an arbitrary probability distribution function $f(k)$ by using an instrumental distribution $g(k)$, under the only restriction that $f(k) < Mg(k)$ where $M > 1$ is an appropriate bound on $f(k)/g(k)$.

The algorithm runs as follows: k is sampled following the distribution g. Then $f(k)$ is computed and a random integer $y \in [0,1]$ is generated. If $y \leq \frac{f(k)}{Mg(k)}$, then k is output, else nothing is output and a new k must be sampled.

If g can be perfectly sampled, and if $\frac{f(k)}{Mg(k)}$ and y can be computed with arbitrary precision, the output of the algorithm follows exactly the distribution f. In practice, since one only needs to decide whether $y \leq \frac{f(k)}{Mg(k)}$ or not, the average precision needed is only a few bits.

We want to select according to the hypergeometric distribution $HG_{n,\alpha,p}$ where $\alpha = \lfloor n/2 \rfloor / n$, which we bound with the binomial distribution $B_{\alpha,p}$, itself being bounded by the Cauchy-Lorentz distribution $CL_{\mu,\nu}$ where $\mu = \alpha p$ and $\nu = 2\alpha(1-\alpha)p$:

$$HG_{n,\alpha,p}(k) = \frac{\binom{\alpha n}{k}\binom{(1-\alpha)n}{p-k}}{\binom{n}{p}}$$

$$B_{\alpha,p}(k) = \binom{p}{k}\alpha^k(1-\alpha)p - k$$

$$CL_{\mu,\nu}(x) = \frac{1}{\pi}\left(\frac{\sqrt{\nu}}{(x-\mu)^2 + \nu}\right)$$

HG and B are discrete, whereas CL is continuous.

4.2 Upper-Bounding Distributions

Theorem 1. If $\alpha \leq 1/2$, then:

$$\frac{HG_{n,\alpha,p}(k)}{B_{\alpha,p}(k)} < \frac{2^{(1-2\alpha)n}}{(2\alpha)^p}\sqrt{1 + \frac{p}{n-p}}.$$

Theorem 2. If $k \in \mathbb{N}$ and $\alpha = 1/2$, then:

$$\frac{B_{\alpha,p}(k)}{CL_{\alpha p, 2\alpha(1-\alpha)p}(k)} \leq \sqrt{\pi}.$$

Proofs for these theorems are provided in appendix A and B, respectively.

The values of interest to us are $\alpha = \lfloor n/2 \rfloor / n$ for large n and sufficiently large p. For these values, and for any x, there exists a constant c such that:

$$\frac{B_{\alpha,p}(\lfloor x + 1/2 \rfloor)}{CL_{\alpha p, 2\alpha(1-\alpha)p}(x)} \leq c\sqrt{\pi}.$$

We decide that if $p \leq 10$ or $n \leq 10$, then we will use another method for REPARTITOR, namely performing p random selections. Under those conditions, $c = 1.2$ is an appropriate constant.

4.3   Cauchy-Lorentz Distribution by Inverse Method

If $u \in [0, 1[$ is selected uniformly (and $u \neq 1/2$), then the value $\mu + \sqrt{\nu}\tan(\pi u)$ follows the $CL_{\mu,\nu}$ distribution. If $x = \sqrt{\nu}\tan(\pi u)$ then we need, for REPARTI-TOR, the integer closest to $x + \mu$, and we have $CL_{\mu,\nu} = \frac{\sqrt{\nu}}{\pi(x^2 + \nu)}$.

4.4   Description of Repartitor

The full REPARTITOR handles the situation where $p > a$, and where $p \leq 10$ (if $n \leq 10$ then $p \leq 10$):

```
REPARTITOR:
input: n, p and i
output: k
  1. a ← ⌊n/2⌋
  2. if p > a then return a − REPARTITOR(n, n − p, i)
  3. if p > 10 then return REPARTITORREJECT(n, p, i)
  4. n₁ ← a, n₂ ← n − a
  5. if p = 0, then return a − n₁
  6. select randomly r between 0 and n₁ + n₂ − 1 (inclusive)
  7. if r < n₁, then n₁ ← n₁ − 1, else n₂ ← n₂ − 1
  8. p ← p − 1
  9. go to step 5
```

The random selection of r is done using an algorithm similar to RANDOM-NUMBER, working with the stream of random bits output by $\mathcal{R}(i, j, 0)$ for values of j beginning with 0. The average number of invocations of $\mathcal{R}$ for this process will be less than $p \log n$; it is extremely improbable that more than 10 times this value is actually needed.

REPARTITORREJECT implements the general selection with the rejection method. We have shown that, for the parameters used with REPARTITOR:

$$\frac{HG_{n,\alpha,p}(\lfloor x + 1/2 \rfloor)}{CL_{\alpha p, 2\alpha(1-\alpha)p}(x)} < 1.2\sqrt{\pi}\frac{2^{(1-2\alpha)n}}{(2\alpha)^p}\sqrt{1 + \frac{p}{n - p}}$$

For positive integers i and j, we define the value $u_{i,j}$: $u_{i,j} = \sum_{k \geq 0} \mathcal{R}(i, j, k)2^{-k-1}$ (the conceptually infinite stream $\mathcal{R}(i, j, *)$ defines the binary digits of $u_{i,j}$).

REPARTITORREJECT:
input: n, p and i
output: k
1. define $a = \lfloor n/2 \rfloor$, $\mu = ap/n$ and $\nu = 2(a/n)(1 - a/n)p$
2. define $M = 1.2\sqrt{\frac{\nu}{\pi}}\frac{2^{(1-2\alpha)n}}{(2\alpha)^p}\sqrt{1 + \frac{p}{n-p}}$
3. $l \leftarrow 0$
4. $l \leftarrow l + 1$
5. $x \leftarrow \sqrt{\nu}\tan(\pi u_{i,2l-1})$
6. $k \leftarrow \lfloor x + \mu + 1/2 \rfloor$
7. if $k < 0$ or $k > p$, then go to step 4
8. $H \leftarrow \binom{a}{k}\binom{n-a}{p-k}/\binom{n}{p}$
9. if $u_{i,2l} \leq (x^2 + \nu)H/M$, then return k
10. go to step 4

The value of k needs to be determined exactly, which means that x will have to be computed with enough precision. Required precision increases linearly with the derivative of tan over the range of accepted values. Here, the maximum tangent value is on the order of $\sqrt{\mu}$, with a derivative of μ, whose size is that of n. Therefore, we need about $2\log n$ random bits (from $u_{i,2l-1}$) to compute k.

The other computations need be precise enough only for the final test against $u_{i,2l}$ to be meaningful; precision must be raised only if the test is not clear cut, which happens with a probability on the order of 2^{-r} if r bits of precision are used. On the average, only $O(1)$ random bits are used for $u_{i,2l}$.

5   Security Model

We use the following security model: the attacker is given access to a black box implementing PERMUTATOR for an internal secret key, with a message space of size n. The attacker may perform up to $n-2$ encryption or decryption (adaptive) queries. The attacker must then predict the output of PERMUTATOR for the encryption of an hitherto unseen input. His probability of success is at most 0.5 if the black box implements a truly random permutation.

Using the terminology of [3], our attacker is an $(n-2)$-*limited adaptive distinguisher* in the model of *super-pseudorandomness* (the oracle accepts both encrypt and decrypt queries). However, this terminology is here quite stretched. As an illustration, the Luby-Rackoff construction offers *no* security against our attacker: since a Feistel network can only implement an even permutation, knowing the outputs for $n-2$ inputs implies only a single possible permutation, and our attacker has a probability 1 of success.

We consider that we achieve a security level of r bits if the attacker cannot succeed with probability greater than $0.5 + 2^{-r}$, except by demonstrating that the PRNG output is not truly random. With the REPARTITOR implementation described in section 4, we achieve an "infinite" level of security, i.e. PERMUTATOR is as robust as the PRNG used: any successful attack is mechanically a proof that the PRNG is not a truly random source, i.e. a distinguisher attack on the PRNG.

However, PERMUTATOR may tolerate a slightly biased REPARTITOR. PERMUTATOR is, basically, an acyclic graph of REPARTITOR invocations, containing about $n \log n$ nodes. For a given permutation ϕ, there is a single set of output values for these nodes which yields the ϕ permutation. Each REPARTITOR should follow the hypergeometric distribution $H_{m,p}(u)$ to ensure uniform permutations selection ($H_{m,p}(u)$ is the probability that the unbiased REPARTITOR outputs u on inputs m and p). We now consider a biased REPARTITOR with the distribution $H'_{m,p}$ such that:

$$\sum_{H'_{m,p}(u)=0} H_{m,p}(u) \leq \epsilon$$
$$H'_{m,p}(u) \neq 0 \Rightarrow \left| \frac{H_{m,p}(u)}{H'_{m,p}(u)} - 1 \right| \leq \epsilon$$

for a value $\epsilon \ll 1/n$. Then, for a given permutation ϕ:

- either one of the $H'_{m,p}(u)$ is 0 for ϕ, which means that ϕ cannot be selected; this happens with probability at most $\epsilon n \log n$;
- or ϕ is selected with probability $P = \prod H'_{m,p}(u)$, which is close to the theoretical unbiased probability: $|P/n! - 1| \leq \epsilon n \log n$.

To achieve a security level of r bits, it is sufficient that these two probabilities be lower than 2^{-r}, i.e. $\log \epsilon \leq -(r + \log n + \log \log n)$. Practically, for 128 bits of security and $n = 2^{32}$, ϵ is 2^{-165}. Any implementation of REPARTITOR which achieves that level of unbiasness will fulfill our security requirements.

6   Conclusion

We have presented a new algorithm for constructing a pseudo-random permutation over a space of size n from a seekable PRNG. The PRNG must take three numerical indexes as parameters, the first never exceeding $\gamma(n) \approx n \log n$, while the other two parameters remain below $100 \log n$ with an overwhelming probability. This is a "perfect" block cipher, up to the computational indistinguishability of the PRNG.

We implemented PERMUTATOR in C, using the MPFR[14] library for arbitrary precision computations. Values of n up to $2^{32} - 1$ are supported. The underlying PRNG uses the AES[1], by encrypting the concatenation of the three PRNG indexes to produce 128 bits of random. Performance is quite inadequate: for $n = 10^9$, a 2 GHz PC requires about 0.48 seconds to compute the image of a single value by PERMUTATOR. Profiling shows that more than 98% of the time is spent in the floating-point computations.

We believe that a much simpler implementation of REPARTITOR can be designed, for much improved performance. As outlined in section 5, a slightly biased REPARTITOR can be tolerated, which opens the way to all kinds of approximations.

References

1. Advanced Encryption Standard, National Institute of Standards and Technology (NIST), FIPS 197 (2001)
2. Data Encryption Standard, National Institute of Standards and Technology (NIST), FIPS 46(3) (1999)
3. How to construct pseudo-random permutations from pseudo-random functions. In: Luby, M., Rackoff, C.(eds.) Lecture Notes in Computer Science, Proceedings of Crypto'85 (1985)
4. Baignères, T., Finiasz, M.: Dial C for Cipher. Proceedings of SAC 2006, LNCS, vol. 4356, Springer, Heidelberg (to appear, 2007)
5. Baignères, T., Finiasz, M.: KFC - the Krazy Feistel Cipher. In: Lai, X., Chen, K. (eds.) ASIACRYPT 2006. LNCS, vol. 4284, pp. 380–395. Springer, Heidelberg (2006)
6. Pseudo random Permutation Families over Abelian Groups. In: Robshaw, M. (ed.) FSE 2006. LNCS, vol. 4047, pp. 15–17. Springer, Heidelberg (2006)
7. Ciphers with Arbitrary Finite Domains. In: Preneel, B. (ed.) CT-RSA 2002. LNCS, vol. 2271, pp. 114–130. Springer, Heidelberg (2002)
8. Statistical Tables, Fisher, R.A., Yates, F. London, example 12 (1938)
9. CACM. Durstenfeld, R.: 7, p.420 (1964)
10. The Art of Computer Programming, Knuth, D.: vol. 2, 3rd edn. p. 145 (1997)
11. Blum, L., Blum, M., Shub, M.: A simple unpredictable pseudorandom number generator. SIAM Journal on Computing 15, 364–383 (1986)
12. Berbain, C., Gilbert, H., Patarin, J.: QUAD: A Practical Stream Cipher with Provable Security. In: Vaudenay, S. (ed.) EUROCRYPT 2006. LNCS, vol. 4004, pp. 109–128. Springer, Heidelberg (2006)
13. Czumaj, A., Kanarek, P., Kutylowski, M., Lorys, K.: Fast Generation of Random Permutations via Networks Simulation. In: Díaz, J. (ed.) ESA 1996. LNCS, vol. 1136, pp. 246–260. Springer, Heidelberg (1996)
14. The MPFR Library, http://www.mpfr.org/

A   Proof of Theorem 1

We now prove theorem 1. Let's compute this:

$$\frac{HG_{n,\alpha,p}(k)}{B_{\alpha,p}(k)} = \frac{\binom{n-p}{\alpha n-k}}{\binom{n}{\alpha n}} \frac{1}{\alpha^k(1-\alpha)^{p-k}}.$$

If $\alpha \leq 1/2$, then $\alpha \leq 1-\alpha$, therefore $\frac{1}{\alpha^k(1-\alpha)^{p-k}}$ reaches its maximum for $k = p$. Hence $\frac{1}{\alpha^k(1-\alpha)^{p-k}} \leq \alpha^{-p}$.

Define $m = \alpha n$ and $\beta = n - 2m = (1 - 2\alpha)n$. We have three possibilities:

- $p \geq \beta$ and $p - \beta$ is even;
- $p \geq \beta$ and $p - \beta$ is odd;
- $p < \beta$.

If $p = 2q + \beta$, the maximum of $\binom{n-p}{\alpha n - k} = \binom{2m-2q}{m-k}$ is achieved for $k = q$; hence:

$$\frac{HG_{n,\alpha,p}(k)}{B_{\alpha,p}(k)} \leq \alpha^{-p} \frac{\binom{2m-2q}{m-q}}{\binom{2m}{m}} = \frac{2^{-2q}}{\alpha^p} \frac{F(m-q)\sqrt{m}}{F(m)\sqrt{m-q}} < \frac{2^\beta}{(2\alpha)^p} \sqrt{1 + \frac{p}{n-p}}.$$

If $p = 2q + \beta + 1$, the maximum of $\binom{n-p}{\alpha n - k} = \binom{2m-2q-1}{m-k}$ is achieved for $k = q$ or $k = q + 1$ (same value); hence:

$$\frac{HG_{n,\alpha,p}(k)}{B_{\alpha,p}(k)} \leq \alpha^{-p} \frac{\binom{2m-2q-1}{m-q}}{\binom{2m}{m}} = \frac{2^{-2q-1}}{\alpha^p} \frac{F(m-q)\sqrt{m}}{F(m)\sqrt{m-q}} < \frac{2^\beta}{(2\alpha)^p} \sqrt{1 + \frac{p}{n-p}}.$$

If $p < \beta$, the maximum of $\binom{n-p}{\alpha n - k} = \binom{2m+\beta-p}{m-k}$ is achieved for $k = 0$; hence:

$$\frac{HG_{n,\alpha,p}(k)}{B_{\alpha,p}(k)} \leq \alpha^{-p} \frac{\binom{2m+(\beta-p)}{m}}{\binom{2m}{m}} = \alpha^{-p} 2^{\beta-p} \prod_{i=1}^{\beta-p} \frac{1 + i/(2m)}{1 + i/m} < \frac{2^\beta}{(2\alpha)^p}.$$

B Proof of Theorem 2

We need the following lemma:

Lemma 1. *The function* $F(k) = \binom{2k}{k} 2^{-2k} \sqrt{k}$ *is strictly increasing, with limit* $\pi^{-1/2} = 0.56418\ldots$

Proof. The limit of $F(k)$ when $k \to \infty$ is easily deduced from Stirling's formula:

$$F(k) = \frac{(2k)!}{(k!)^2} 2^{-2k} \sqrt{k} \approx \frac{(2k)^{2k} e^{-2k} \sqrt{4\pi k} \sqrt{k}}{k^{2k} e^{-2k} 2\pi k 2^{2k}} = \frac{1}{\sqrt{\pi}}.$$

To prove that F is strictly increasing, it suffices to remark that:

$$\frac{F(k+1)^2}{F(k)^2} = 1 + \frac{1}{4k(1+k)} > 1.$$

We may now see that:

$$\frac{B_{1/2,p}(k)}{CL_{p/2,p/2}(k)} = \pi F(p/2) \frac{\binom{p}{k}}{\binom{p}{p/2}} \left(\left(\frac{k}{p/2} - 1 \right)^2 + 1 \right)$$

which reaches its maximum for $k = p/2$. This maximum is $\pi F(p/2) < \sqrt{\pi}$.

Author Index

Lecture Notes in Computer Science

For information about Vols. 1–4570

please contact your bookseller or Springer

Vol. 4616: A. Dress, Y. Xu, B. Zhu (Eds.), Combinatorial Optimization and Applications. XI, 390 pages. 2007.

Vol. 4615: R. de Lemos, C. Gacek, A. Romanovsky (Eds.), Architecting Dependable Systems IV. XIV, 435 pages. 2007.

Vol. 4613: F.P. Preparata, Q. Fang (Eds.), Frontiers in Algorithmics. XI, 348 pages. 2007.

Vol. 4612: I. Miguel, W. Ruml (Eds.), Abstraction, Reformulation, and Approximation. XI, 418 pages. 2007. (Sublibrary LNAI).

Vol. 4611: J. Indulska, J. Ma, L.T. Yang, T. Ungerer, J. Cao (Eds.), Ubiquitous Intelligence and Computing. XXIII, 1257 pages. 2007.

Vol. 4610: B. Xiao, L.T. Yang, J. Ma, C. Muller-Schloer, Y. Hua (Eds.), Autonomic and Trusted Computing. XVIII, 571 pages. 2007.

Vol. 4609: E. Ernst (Ed.), ECOOP 2007 – Object-Oriented Programming. XIII, 625 pages. 2007.

Vol. 4608: H.W. Schmidt, I. Crnkovic, G.T. Heineman, J.A. Stafford (Eds.), Component-Based Software Engineering. XII, 283 pages. 2007.

Vol. 4607: L. Baresi, P. Fraternali, G.-J. Houben (Eds.), Web Engineering. XVI, 576 pages. 2007.

Vol. 4606: A. Pras, M. van Sinderen (Eds.), Dependable and Adaptable Networks and Services. XIV, 149 pages. 2007.

Vol. 4605: D. Papadias, D. Zhang, G. Kollios (Eds.), Advances in Spatial and Temporal Databases. X, 479 pages. 2007.

Vol. 4604: U. Priss, S. Polovina, R. Hill (Eds.), Conceptual Structures: Knowledge Architectures for Smart Applications. XII, 514 pages. 2007. (Sublibrary LNAI).

Vol. 4603: F. Pfenning (Ed.), Automated Deduction – CADE-21. XII, 522 pages. 2007. (Sublibrary LNAI).

Vol. 4602: S. Barker, G.-J. Ahn (Eds.), Data and Applications Security XXI. X, 291 pages. 2007.

Vol. 4600: H. Comon-Lundh, C. Kirchner, H. Kirchner (Eds.), Rewriting, Computation and Proof. XVI, 273 pages. 2007.

Vol. 4599: S. Vassiliadis, M. Berekovic, T.D. Hämäläinen (Eds.), Embedded Computer Systems: Architectures, Modeling, and Simulation. XVIII, 466 pages. 2007.

Vol. 4598: G. Lin (Ed.), Computing and Combinatorics. XII, 570 pages. 2007.

Vol. 4597: P. Perner (Ed.), Advances in Data Mining. XI, 353 pages. 2007. (Sublibrary LNAI).

Vol. 4596: L. Arge, C. Cachin, T. Jurdziński, A. Tarlecki (Eds.), Automata, Languages and Programming. XVII, 953 pages. 2007.

Vol. 4595: D. Bošnački, S. Edelkamp (Eds.), Model Checking Software. X, 285 pages. 2007.

Vol. 4594: R. Bellazzi, A. Abu-Hanna, J. Hunter (Eds.), Artificial Intelligence in Medicine. XVI, 509 pages. 2007. (Sublibrary LNAI).

Vol. 4593: A. Biryukov (Ed.), Fast Software Encryption. XI, 467 pages. 2007.

Vol. 4592: Z. Kedad, N. Lammari, E. Métais, F. Meziane, Y. Rezgui (Eds.), Natural Language Processing and Information Systems. XIV, 442 pages. 2007.

Vol. 4591: J. Davies, J. Gibbons (Eds.), Integrated Formal Methods. IX, 660 pages. 2007.

Vol. 4590: W. Damm, H. Hermanns (Eds.), Computer Aided Verification. XV, 562 pages. 2007.

Vol. 4589: J. Münch, P. Abrahamsson (Eds.), Product-Focused Software Process Improvement. XII, 414 pages. 2007.

Vol. 4588: T. Harju, J. Karhumäki, A. Lepistö (Eds.), Developments in Language Theory. XI, 423 pages. 2007.

Vol. 4587: R. Cooper, J. Kennedy (Eds.), Data Management. XIII, 259 pages. 2007.

Vol. 4586: J. Pieprzyk, H. Ghodosi, E. Dawson (Eds.), Information Security and Privacy. XIV, 476 pages. 2007.

Vol. 4585: M. Kryszkiewicz, J.F. Peters, H. Rybinski, A. Skowron (Eds.), Rough Sets and Intelligent Systems Paradigms. XIX, 836 pages. 2007. (Sublibrary LNAI).

Vol. 4584: N. Karssemeijer, B. Lelieveldt (Eds.), Information Processing in Medical Imaging. XX, 777 pages. 2007.

Vol. 4583: S.R. Della Rocca (Ed.), Typed Lambda Calculi and Applications. X, 397 pages. 2007.

Vol. 4582: J. Lopez, P. Samarati, J.L. Ferrer (Eds.), Public Key Infrastructure. XI, 375 pages. 2007.

Vol. 4581: A. Petrenko, M. Veanes, J. Tretmans, W. Grieskamp (Eds.), Testing of Software and Communicating Systems. XII, 379 pages. 2007.

Vol. 4580: B. Ma, K. Zhang (Eds.), Combinatorial Pattern Matching. XII, 366 pages. 2007.

Vol. 4579: B. M. Hämmerli, R. Sommer (Eds.), Detection of Intrusions and Malware, and Vulnerability Assessment. X, 251 pages. 2007.

Vol. 4578: F. Masulli, S. Mitra, G. Pasi (Eds.), Applications of Fuzzy Sets Theory. XVIII, 693 pages. 2007. (Sublibrary LNAI).

Vol. 4577: N. Sebe, Y. Liu, Y.-t. Zhuang, T.S. Huang (Eds.), Multimedia Content Analysis and Mining. XIII, 513 pages. 2007.

Vol. 4576: D. Leivant, R. de Queiroz (Eds.), Logic, Language, Information and Computation. X, 363 pages. 2007.

Vol. 4575: T. Takagi, T. Okamoto, E. Okamoto, T. Okamoto (Eds.), Pairing-Based Cryptography – Pairing 2007. XI, 408 pages. 2007.

Vol. 4574: J. Derrick, J. Vain (Eds.), Formal Techniques for Networked and Distributed Systems – FORTE 2007. XI, 375 pages. 2007.

Vol. 4573: M. Kauers, M. Kerber, R. Miner, W. Windsteiger (Eds.), Towards Mechanized Mathematical Assistants. XIII, 407 pages. 2007. (Sublibrary LNAI).

Vol. 4572: F. Stajano, C. Meadows, S. Capkun, T. Moore (Eds.), Security and Privacy in Ad-hoc and Sensor Networks. X, 247 pages. 2007.

Vol. 4571: P. Perner (Ed.), Machine Learning and Data Mining in Pattern Recognition. XIV, 913 pages. 2007. (Sublibrary LNAI).